A First Course in Causal Inference

The past decade has witnessed an explosion of interest in research and education in causal inference, due to its wide applications in biomedical research, social sciences, artificial intelligence etc. This textbook, based on the author's course on causal inference at UC Berkeley taught over the past seven years, only requires basic knowledge of probability theory, statistical inference, and linear and logistic regressions. It assumes minimal knowledge of causal inference, and reviews basic probability and statistics in the appendix. It covers causal inference from a statistical perspective and includes examples and applications from biostatistics and econometrics.

Key Features:

- All R code and data sets available at Harvard Dataverse.
- Solutions manual available for instructors.
- Includes over 100 exercises.

This book is suitable for an advanced undergraduate or graduate-level course on causal inference, or postgraduate and PhD-level course in statistics and biostatistics departments.

Peng Ding is an Associate Professor in the Department of Statistics at UC Berkeley. His research focuses on causal inference and its applications.

Recently Published Titles

Time Series for Data Science: Analysis and Forecasting
Wayne A. Woodward, Bivin Philip Sadler and Stephen Robertson

Statistical Theory
A Concise Introduction, Second Edition
Felix Abramovich and Ya'acov Ritov

Applied Linear Regression for Longitudinal Data
With an Emphasis on Missing Observations
Frans E.S. Tan and Shahab Jolani

Fundamentals of Mathematical Statistics
Steffen Lauritzen

Modelling Survival Data in Medical Research, Fourth Edition
David Collett

Applied Categorical and Count Data Analysis, Second Edition
Wan Tang, Hua He and Xin M. Tu

Geographic Data Science with Python
Sergio Rey, Dani Arribas-Bel and Levi John Wolf

Models for Multi-State Survival Data
Rates, Risks, and Pseudo-Values
Per Kragh Andersen and Henrik Ravn

Spatio–Temporal Methods in Environmental Epidemiology with R, Second Edition
Gavin Shaddick, James V. Zidek, and Alex Schmidt

A Course in the Large Sample Theory of Statistical Inference
W. Jackson Hall and David Oakes

Statistical Inference, Second Edition
George Casella and Roger Berger

Nonparametric Statistical Methods Using R, Second Edition
John Kloke and Joesph McKean

Generalized Linear Mixed Models
Modern Concepts, Methods and Applications, Second Edition
Walter W. Stroup, Marina Ptukhina and Julie Garai

For more information about this series, please visit: https://www.routledge.com/Chapman--HallCRC-Texts-in-Statistical-Science/book-series/CHTEXSTASCI

A First Course in Causal Inference

Peng Ding

CRC Press is an imprint of the
Taylor & Francis Group, an **informa** business
A CHAPMAN & HALL BOOK

Designed cover image: © Ding Peng

First edition published 2024
by CRC Press
2385 NW Executive Center Drive, Suite 320, Boca Raton FL 33431

and by CRC Press
4 Park Square, Milton Park, Abingdon, Oxon, OX14 4RN

CRC Press is an imprint of Taylor & Francis Group, LLC

Library of Congress Cataloging-in-Publication Data
Names: Ding, Peng (Statistician), author.
Title: A first course in causal inference / Peng Ding.
Description: First edition. | Boca Raton, FL : CRC Press, 2024. | Series:
Chapman & Hall/CRC texts in statistical science | Includes
bibliographical references and index.
Identifiers: LCCN 2023056503 (print) | LCCN 2023056504 (ebook) | ISBN
9781032758626 (hardback) | ISBN 9781032776316 (paperback) | ISBN
9781003484080 (ebook)
Subjects: LCSH: Mathematical statistics--Textbooks. |
Probabilities--Textbooks. | Causation--Mathematical models. |
Inference--Mathematical models.
Classification: LCC QA276 .D536 2024 (print) | LCC QA276 (ebook) | DDC
519.5/4--dc23/eng/20240316
LC record available at https://lccn.loc.gov/2023056503
LC ebook record available at https://lccn.loc.gov/2023056504

ISBN: 978-1-032-75862-6 (hbk)
ISBN: 978-1-032-77631-6 (pbk)
ISBN: 978-1-003-48408-0 (ebk)

DOI: 10.1201/9781003484080

Typeset in CMR10
by KnowledgeWorks Global Ltd.

Publisher's note: This book has been prepared from camera-ready copy provided by the authors.

To students and readers
who are interested in causal inference

Contents

Preface

Causal inference research and education in the past decade

The past decade has witnessed an explosion of interest in research and education in causal inference, due to its wide applications in biomedical research, social sciences, tech companies, etc. It was quite different even ten years ago when I was a Ph.D. student in statistics. At that time, causal inference was not a mainstream research topic in statistics and very few undergraduate and graduate programs offered courses in causal inference. In the academic world of statistics, many people were still very skeptical about the foundation of causal inference. Many leading statisticians were reluctant to accept causal inference because of the fundamental conceptual difficulties, which differ from the traditional training of mathematical statistics.

The applications of causal inference in empirical research have changed the field of statistics in both research and education. In the end, statistics is not only about abstract theory but also about solving real-world problems. Many talented researchers have joined the effort to advance our knowledge of causal inference. Many students are eager to learn state-of-the-art theory and methods in causal inference so that they are better equipped to solve problems from various fields.

Due to the needs of the students, my colleagues encouraged me to develop a course in causal inference. Initially, I taught a graduate-level course cross-listed under Political Science and Statistics, which was taught by my former colleague Jas Sekhon for many years at UC Berkeley. Later, I developed this course for both undergraduate and graduate students. At UC Berkeley, the course numbers for "Causal Inference" are Stat 156 and Stat 256, with undergraduate students in Stat 156 and graduate students in Stat 256. Students in both sessions used the same lecture notes and attended the same lectures given by me and my teaching assistants, although they needed to finish different homework problems, reading assignments, and final projects.

Given the mixed levels of technical preparations of my students, the most challenging part of my teaching was to balance the interests of both undergraduate and graduate students. On the one hand, I wanted to present the materials in an intuitive way and only required the undergraduate students to have the basic knowledge of probability, statistics, linear regression, and logistic regression. On the other hand, I also wanted to introduce recent research topics and results to the graduate students. This book is a product of my efforts in the past seven years.

Recommendations for instructors

This book contains 29 chapters in the main text and three chapters in the appendix. UC Berkeley is on the semester system and each semester has 14 weeks of lectures. I could not finish all 32 chapters in one semester. Here are some recommendations based on my own teaching experience:

Appendices

I started with the chapters in the main text but asked my teaching assistants to review the basics in Appendices A and B. To encourage the students to review Appendices A–C before reading the main text, I also assigned several homework problems from Appendices A–C at the beginning of the semester.

Part I

The key topic in Chapter 1 is the Yule–Simpson Paradox. Chapter 2 introduces the notion of potential outcomes, which is the foundation for the whole book.

Part II

Different researchers and instructors may have quite different views on the materials in Part II on randomized experiments. I have talked to many friends about the pros and cons of the current presentation in Part II. Causal inference in randomized experiments is relatively straightforward because randomization eliminates unmeasured confounding. So some friends feel that Chapters 3–9 are too long at least for beginners of causal inference. This also disappointed some of my students when I spent a month on randomized experiments. On the other hand, I was trained from the book of Imbens and Rubin (2015) and believed that to understand observational studies, it is better to understand randomized experiments first. Moreover, I am a big fan of the canonical research of Neyman (1923) and Fisher (1935). Therefore, Part II deeply reflects my own intellectual history and personal taste in statistics. Other instructors may not want to spend a month on randomized experiments and can cover Chapters 5, 7, 8, and 9 quickly.

Part III

Part III covers the key ideas in observational studies without unmeasured confounding. Four pillars of observational studies are outcome regression, inverse propensity score weighting, doubly robust, and matching estimators, which are covered in Chapters 10, 11, 12, and 15, respectively. Chapters 13 and 14 are optional in teaching. But the results in Chapters 13 and 14 are not uninteresting, so I sometimes covered one or two results there, asked the teaching assistants to cover more in the lab sessions, and encouraged students to read them by assigning some homework problems from those chapters.

Part IV

Part IV is a novel treatment of the fundamental difficulties of observational studies including unmeasured confounding and overlap. However, this part is far from perfect due to the complexities and subtleties of the issues. Chapters 17, 18, and 20 are central, whereas Chapters 16 and 19 are optional.

Part V

Part V discusses the idea of the instrumental variable. Chapters 21, 23, and 24 are key, whereas Chapters 22 and 25 are optional.

Part VI

Part VI discusses some special topics. They are all optional in some sense. Probably it is worth teaching Chapter 27 given the popularity of the Baron–Kenny method in mediation analysis.

Help from teaching assistants

My teaching assistants offered invaluable help for my courses at UC Berkeley. Since I could not cover everything in this book, I consistently relied on them to cover some technical details or `R` program issues in their labs.

Solution to some homework problems

I have also prepared the solutions to most theory problems. If you are an instructor for a causal inference course, please contact me for the solutions with detailed information about your course.

Additional recommendations for readers and students

Readers and students can first read my recommendations for instructors above. In addition, I have three other recommendations.

More simulation studies

Statistical theories are important but they often hold under strong assumptions. Simulation studies are crucial for understanding the properties of the proposed methods in this book. On the one hand, they can provide supporting evidence for the theories under the ideal assumptions. On the other hand, they can provide insights into the cases in which the assumptions for theories break down. I provide some basic simulation studies in this book and encourage the readers to do more in some homework problems. Nevertheless, those simulation studies are still quite limited for understanding the full picture of the proposed estimators. Therefore, I encourage the readers not only to replicate the simulation studies provided in this book but also to conduct more simulation studies under other data-generating processes. Applied researchers should calibrate their simulation studies based on their applied problems at hand to better choose the methods to analyze their data.

Homework problems

Each chapter of this book contains homework problems. To deepen the understanding, it is important to try some homework problems. Moreover, some homework problems contain useful theoretical results. Even if you do not have time to figure out the details for those problems, it is helpful to at least read the statements of the problems.

Recommended reading

Each chapter of this book contains recommended reading. If you want to do research in causal inference, those recommended papers can be useful background knowledge of the literature. When I taught the graduate-level causal inference course at UC Berkeley, I assigned the following papers to the students as weekly reading from week one to the end of the semester:

- Bickel et al. (1975);
- Holland (1986);
- Miratrix et al. (2013);
- Lin (2013);
- Li et al. (2018b);

- Rosenbaum and Rubin (1983b);
- Lunceford and Davidian (2004);
- Ding and VanderWeele (2016a);
- Pearl (1995);
- Angrist et al. (1996);
- Imbens (2014);
- Frangakis and Rubin (2002).

Many students gave me positive feedback about their experience of reading the papers above. I recommend reading the above papers even if you do not read this book.

Omitted topics

Econometrics for causal inference

This book covers some econometric methods for causal inference. The instrumental variable method is the key topic in Part V. Regression discontinuity appears in Chapters 20 and 24. Although it is very unusual, I introduce regression discontinuity together with the issue of overlap in observational studies in Chapter 20. However, this book does not cover many popular econometric methods including the difference in differences, panel data, and synthetic controls. Instructors can use Angrist and Pischke (2008) as a reference for those topics.

Variance estimation

In randomized experiments, I discuss variance estimation very carefully and recommend simple procedures that can be easily implemented via regressions. For observational studies and more complicated settings, I often omit the detailed discussion of variance estimation and recommend using the bootstrap to approximate the asymptotic variances because most estimators in this book can be written as solutions to estimating equations. Advanced readers can apply the calculus for estimating equations to obtain the close-form formulas for the asymptotic variances (Newey and McFadden, 1994; Stefanski and Boos, 2002).

Machine learning for causal inference

This book assumes minimal preparation for the background knowledge in probability and statistics. I expect that the readers know the basics of linear and logistic regressions, reviewed in Appendix B. When I talk about statistical models, I usually use the linear and logistic models to illustrate the ideas. More advanced readers can extend my discussion to more flexible statistical models, including the modern machine learning tools for estimating the conditional mean functions. Using machine learning for causal inference has been a fruitful research area in recent years, starting from the pioneer work of Hill (2011), Chernozhukov et al. (2018), and Wager and Athey (2018). I encourage the readers to extend my discussion and `R` code to allow for using machine learning methods.

Bayesian causal inference

Again, this book assumes minimal preparation for the background knowledge in probability and statistics. Since most introductory statistics courses use the frequentists' view that assumes the unknown parameters are fixed, I adopt this view in this book and omit the Bayesian view for causal inference. In fact, many fundamental ideas of causal inference are from the Bayesian view, starting from Rubin (1978). If readers and students are interested in Bayesian causal inference, please read the review paper by Li et al. (2023).

Features of the book

There are already many excellent causal inference books published in the last decade. Some of them have profound influences on me. When I was in college, I read some draft chapters of Imbens and Rubin (2015) from the internet. They completely challenged my way of thinking about statistics and helped to build my research interest in causal inference. I read Angrist and Pischke (2008) many times and have gained new insights each time I reread it. Rosenbaum (2002b), Morgan and Winship (2015), and Hernán and Robins (2020) are three other excellent books from leading researchers in causal inference. When I was preparing for the book, Cunningham (2021), Huntington-Klein (2022), Brumback (2022), and Huber (2023) appeared as four recent excellent books on causal inference.

Thanks to my teaching experience at UC Berkeley, this book has the following features that instructors, students, and readers may find attractive.

- This book assumes minimal preparation for causal inference and reviews the basic probability and statistics knowledge in the appendix.
- This book covers causal inference from the statistics, biostatistics, and econometrics perspectives, and draws applications from various fields.
- This book uses `R` code and data analysis to illustrate the ideas of causal inference. All the `R` code and datasets are publicly available at Harvard Dataverse:

  `https://doi.org/10.7910/DVN/ZX3VEV`.

  Dr. Apoorva Lal kindly provides the corresponding `Python` code at

  `https://github.com/apoorvalal/ding_causalInference_python`.

- This book contains homework problems and can be used as a textbook for both undergraduate and graduate students. Instructors can also ask me for solutions to some homework problems.

Acknowledgments

Professor Zhi Geng at Peking University introduced me to the area of causal inference when I was studying in college. Professors Luke Miratrix, Tirthankar Dasgupta, and Don Rubin served on my Ph.D. thesis committee at the Harvard Statistics Department. Professor Tyler VanderWeele supervised me as a postdoctoral researcher in Epidemiology at the Harvard T.H. Chan School of Public Health.

My colleagues at the Berkeley Statistics Department have created a critical and productive research environment. Bin Yu and Jas Sekhon have been very supportive since I was a junior faculty. My department chairs, Deb Nolan, Sandrine Dudoit, and Haiyan Huang, encouraged me to develop the "Causal Inference" course. It has been a rewarding experience for me.

I have been lucky to work with many collaborators, in particular, Avi Feller, Laura Forastiere, Zhichao Jiang, Fang Han, Fan Li, Xinran Li, Alessandra Mattei, Fabrizia Mealli,

Shu Yang, Fan Yang, and Anqi Zhao. I will report in this book what I have learned from them.

Many students at UC Berkeley made critical and constructive comments on early versions of my lecture notes. As teaching assistants for my "Causal Inference" course, Emily Flanagan and Sizhu Lu read early versions of my book carefully and helped me to improve the book a lot.

Professor Joe Blitzstein read an early version of the book carefully and made very detailed comments. Addressing his comments leads to significant improvement in the book. Professors Hongyuan Cao and Zhichao Jiang taught "Causal Inference" courses based on an early version of the book. They made very valuable suggestions.

I am also very grateful for the suggestions from Nianqiao Ju, Young Woong Min, Fangzhou Su, Chaoran Yu, Andy Shen, Yiqing Xu, and Lo-Hua Yuan.

The U.S. National Science Foundation partially supported my research over the years (grant numbers # 1712714, # 1745640, and # 1945136).

Contacting me

Please feel free to email me at

pengdingpku@berkeley.edu

if you identify any errors in the book, or if you use the book to teach causal inference and want the solutions to the homework problems.

Acronyms

To simplify the writing, I will use many acronyms in this book. The following table gives the acronyms, their full names, and the first chapters in which they appear.

acronym	full name	first chapter
ACE	average causal effect	2
AI	Abadie and Imbens (for matching estimators)	15
ANCOVA	analysis of covariance	6
BMI	body mass index	2
BRE	Bernoulli randomized experiment	3
CACE	complier average causal effect	21
CATE	conditional average causal effect	10
CDE	controlled direct effect	28
CLT	central limit theorem	3 and A
CRE	completely randomized experiment	3
EHW	Eicker–Huber–White (robust standard error)	4 and B
FAR	Fieller–Anderson–Rubin (confidence set)	21
FRT	Fisher randomization test	3
FWL	Frisch–Waugh–Lovell (theorem)	B
HT	Horvitz–Thompson (estimator)	11
IID	independent and identically distributed	3 and A
ILS	indirect least squares	23
IPW	inverse propensity score weighting	11
ITT	intention-to-treat (analysis)	21
IV	instrumental variable	21
LASSO	least absolute shrinkage and selection operator	6
LATE	local average treatment effect	21
MLE	maximum likelihood estimate	B
MPE	matched-pairs experiment	7
MR	Mendelian randomization	25
MSM	marginal structural model	28
NDE	natural direct effect	27
NHANES	National Health and Nutrition Examination Survey	10
NIE	natural indirect effect	27
OLS	ordinary least squares	4 and B
OR	odds ratio	1
RCT	randomized controlled trial	1
RD	risk difference	1
ReM	rerandomization using the Mahalanobis distance	6
RR	risk ratio or relative risk	1
SNP	single nucleotide polymorphism	25
SRE	stratified randomized experiment	5
SUTVA	stable unit treatment value assumption	2
TSLS	two-stage least squares	23
WLS	weighted least squares	14 and B

Notation

I use the following conventional notation in this book.

Math

$\binom{n}{m}$	"n choose m" which equals $\frac{n!}{m!(n-m)!}$
$\sum$	summation, e.g., $\sum_{i=1}^{n} a_i = a_1 + \cdots + a_n$
$I(\cdot)$	indicator function, i.e., $I(A) = 1$ if A happens and 0 otherwise
$\#$	counting the number of units in a set
$\approx$	approximately equal
$\propto$	proportional to (by dropping some unimportant constant)
logit	$\text{logit}(x) = \log \frac{x}{1-x}$
expit	$\text{expit}(x) = \frac{e^x}{1+e^x} = (1 + e^{-x})^{-1}$
$\mathbb{R}$	the set of all real numbers
$\mathbb{R}^K$	the set of K-dimensional vectors with real-valued elements

Basic probability and statistics

$\text{pr}(\cdot)$	probability
$E(\cdot)$	expectation of a random variable
$\text{var}(\cdot)$	variance of a random variable
$\text{cov}(\cdot)$	covariance between random variables
$\perp\!\!\!\perp$	independence or conditional independence between random variables
$\stackrel{\cdot}{\sim}$	"$A \stackrel{\cdot}{\sim} B$" means that A and B have the same asymptotic distribution
ρ_{YX}	Pearson correlation coefficient between Y and X
$\rho_{YX\mid Z}$	partial correlation coefficient between Y and X given Z
R^2_{YX}	squared multiple correlation coefficient between Y and X

Random variables

$\text{Bernoulli}(p)$	Bernoulli distribution with probability p
$\text{Binomial}(n, p)$	Binomial distribution with n trials and probability p
$\text{N}(\mu, \sigma^2)$	Normal distribution with mean μ and variance σ^2
t_ν	t distribution with degrees of freedom ν
χ^2_ν	chi-squared distribution with degrees of freedom ν
$z_{1-\alpha/2}$	the $1 - \alpha/2$ upper quantile of $\text{N}(0, 1)$, e.g., $z_{0.975} = 1.96$

Causal inference

X_i	pretreatment covariates
Z_i	treatment indicator
$Y_i(1), Y_i(0)$	potential outcomes of unit i under treatment and control
τ_i	individual causal effect $\tau_i = Y_i(1) - Y_i(0)$
τ	finite-population average causal effect $\tau = n^{-1}\sum_{i=1}^n \{Y_i(1) - Y_i(0)\}$
τ	super-population average causal effect $\tau = E\{Y(1) - Y(0)\}$
τ_T	average causal effect for the treated units $\tau = E\{Y(1) - Y(0) \mid Z = 1\}$
τ_C	average causal effect for the control units $\tau = E\{Y(1) - Y(0) \mid Z = 0\}$
$\mu_1(X)$	outcome model $\mu_1(X) = E(Y \mid Z = 1, X)$
$\mu_0(X)$	outcome model $\mu_0(X) = E(Y \mid Z = 0, X)$
$e(X)$	propensity score $e(X) = \mathrm{pr}(Z = 1 \mid X)$
τ_c	complier average causal effect $\tau_\mathrm{c} = E\{Y(1) - Y(0) \mid U = \mathrm{c}\}$
U	unmeasured confounder
U	latent compliance status $U = (D(1), D(0))$
$Y(z, M_{z'})$	nested potential outcome (for mediation analysis)

Part I

Introduction

1

Correlation, Association, and the Yule–Simpson Paradox

Causality is central to human knowledge. Two famous quotes from ancient Greeks are below.

> "I would rather discover one causal law than be King of Persia."
> — Democritus

> "We do not have knowledge of a thing until we grasped its cause."
> — Aristotle

However, the major part of classic statistics is about association rather than causation. This chapter will review some basic association measures and point out their fundamental limitations.

1.1 Traditional view of statistics

A traditional view of statistics is to infer correlation or association among variables. Based on this view, there is no role for causal inference in statistics. Two famous aphorisms associated with this view are as follows:

- "Correlation does not imply causation."
- "You can not prove causality with statistics."

This book has a very different view:

statistics is crucial for understanding causality.

The main focus of this book is to introduce the formal language for causal inference and develop statistical methods to estimate causal effects in randomized experiments and observational studies.

1.2 Some commonly-used measures of association

1.2.1 Correlation and regression

The Pearson correlation coefficient between two random variables Z and Y is

$$\rho_{ZY} = \frac{\text{cov}(Z, Y)}{\sqrt{\text{var}(Z)\text{var}(Y)}},$$

which measures the linear dependence of Z and Y.

The linear regression of Y on Z is the model

$$Y = \alpha + \beta Z + \varepsilon, \tag{1.1}$$

where $E(\varepsilon) = 0$ and $E(\varepsilon Z) = 0$. We can show that the regression coefficient β equals

$$\begin{aligned} \beta &= \frac{\text{cov}(Z, Y)}{\text{var}(Z)} \\ &= \rho_{ZY} \sqrt{\frac{\text{var}(Y)}{\text{var}(Z)}}. \end{aligned}$$

So β and ρ_{ZY} always have the same sign.

We can also define multiple regression of Y on Z and X:

$$Y = \alpha + \beta Z + \gamma X + \varepsilon, \tag{1.2}$$

where $E(\varepsilon) = 0, E(\varepsilon Z) = 0$, and $E(\varepsilon X) = 0$. We usually interpret β as the "effect" of Z on Y, *holding X constant* or *conditioning on X* or *controlling for X*. Appendix B reviews the basics of linear regression.

More interestingly, the β's in the above two regressions (1.1) and (1.2) can be different; they can even have different signs. The following `R` code reanalyzed the LaLonde observational data used by Hainmueller (2012). The main question of interest is the "causal effect" of a job training program on earnings. The regression controlling for all covariates gives a coefficient `1067.5461` for `treat`, whereas the regression not controlling for any covariates gives a coefficient `-8506.4954` for `treat`.

```
> dat <- read.table("cps1re74.csv", header = TRUE)
> dat$u74 <- as.numeric(dat$re74==0)
> dat$u75 <- as.numeric(dat$re75==0)
>
> ## linear regression on the outcome
> ## . means regression on all other variable in dat
> lmoutcome = lm(re78 ~ ., data = dat)
> round(summary(lmoutcome)$coef[2, ], 3)
  Estimate Std. Error    t value   Pr(>|t|)
  1067.546    554.060      1.927      0.054
>
> lmoutcome = lm(re78 ~ treat, data = dat)
> round(summary(lmoutcome)$coef[2, ], 3)
  Estimate Std. Error    t value   Pr(>|t|)
 -8506.495    712.766    -11.934      0.000
```

1.2.2 Contingency tables

We can represent the joint distribution of two binary variables Z and Y by a two-by-two contingency table. With $p_{zy} = \text{pr}(Z = z, Y = y)$, we can summarize the joint distribution in the following table:

	$Y = 1$	$Y = 0$
$Z = 1$	p_{11}	p_{10}
$Z = 0$	p_{01}	p_{00}

Viewing Z as the treatment or exposure and Y as the outcome, we can define the risk difference as

$$\begin{aligned} \text{RD} &= \text{pr}(Y=1 \mid Z=1) - \text{pr}(Y=1 \mid Z=0) \\ &= \frac{p_{11}}{p_{11}+p_{10}} - \frac{p_{01}}{p_{01}+p_{00}}, \end{aligned}$$

the risk ratio as

$$\begin{aligned} \text{RR} &= \frac{\text{pr}(Y=1 \mid Z=1)}{\text{pr}(Y=1 \mid Z=0)} \\ &= \frac{p_{11}}{p_{11}+p_{10}} \Big/ \frac{p_{01}}{p_{01}+p_{00}}, \end{aligned}$$

and the odds ratio[1] as

$$\begin{aligned} \text{OR} &= \frac{\text{pr}(Y=1 \mid Z=1)/\text{pr}(Y=0 \mid Z=1)}{\text{pr}(Y=1 \mid Z=0)/\text{pr}(Y=0 \mid Z=0)} \\ &= \frac{\frac{p_{11}}{p_{11}+p_{10}} \Big/ \frac{p_{10}}{p_{11}+p_{10}}}{\frac{p_{01}}{p_{01}+p_{00}} \Big/ \frac{p_{00}}{p_{01}+p_{00}}} \\ &= \frac{p_{11}p_{00}}{p_{10}p_{01}}. \end{aligned}$$

The terminology "risk difference", "risk ratio", and "odds ratio" comes from epidemiology. Because the outcomes in epidemiology are often diseases, it is natural to use the name "risk" for the probability of having diseases.

We have the following simple facts about these measures.

Proposition 1.1 *(1) The following statements are all equivalent*[2]*:* $Z \perp\!\!\!\perp Y$, $\text{RD} = 0$, $\text{RR} = 1$, *and* $\text{OR} = 1$. *(2) If* p_{zy}*'s are all positive, then* $\text{RD} > 0$ *is equivalent to* $\text{RR} > 1$ *and is also equivalent to* $\text{OR} > 1$. *(3)* $\text{OR} \approx \text{RR}$ *if* $\text{pr}(Y=1 \mid Z=1)$ *and* $\text{pr}(Y=1 \mid Z=0)$ *are small.*

I leave the proofs of statements (1) and (2) as Problem 1.1. Statement (3) is informal. The approximation holds because the $p/(1-p)$ is close to the probability p for rare diseases with $p \approx 0$: by Taylor expansion $p/(1-p) = p + p^2 + \cdots \approx p$. In epidemiology, if the outcome represents the occurrence of a rare disease, then it is reasonable to assume that $\text{pr}(Y=1 \mid X=1)$ and $\text{pr}(Y=1 \mid X=0)$ are small.

We can also define conditional versions of the RD, RR, and OR if the probabilities are replaced by the conditional probabilities given another variable X, i.e., $\text{pr}(Y=1 \mid Z=1, X=x)$ and $\text{pr}(Y=1 \mid Z=0, X=x)$.

With counts $n_{zy} = \#\{i : Z_i = z, Y_i = y\}$, we can summarize the observed data in the following two-by-two table:

	$Y=1$	$Y=0$
$Z=1$	n_{11}	n_{10}
$Z=0$	n_{01}	n_{00}

We can estimate RD, RR, and OR by replacing the true probabilities by the sample proportions $\hat{p}_{zy} = n_{zy}/n$, where n is the total sample size (see Section A.3.2). In R, the function `fisher.test` performs an exact test and the function `chisq.test` performs an asymptotic test for $Z \perp\!\!\!\perp Y$ based on a two-by-two table of observed data.

[1] In probability theory, the odds of an event are defined as the ratio of the probability that the event happens over the probability that the event does not happen.

[2] This book uses the notation $\perp\!\!\!\perp$ to denote independence or conditional independence of random variables. The notation is due to Dawid (1979).

Example 1.1 *Bertrand and Mullainathan (2004) conducted a randomized experiment on resumes to study the effect of perceived race on callbacks for interviews. They randomly assigned Black-sounding or White-sounding names on fictitious resumes to help-wanted ads in Boston and Chicago newspapers. The following two-by-two table summarizes perceived race and callback:*

```
> resume = read.csv("resume.csv")
> Alltable = table(resume$race, resume$call)
> Alltable

          0    1
  black 2278  157
  white 2200  235
```

The two rows have the same total count, so it is apparent that White-sounding names received more callbacks. Fisher's exact test below shows that this difference is statistically significant.

```
> fisher.test(Alltable)

        Fisher's Exact Test for Count Data

data:   Alltable
p-value = 4.759e-05
alternative hypothesis: true odds ratio is not equal to 1
95 percent confidence interval:
 1.249828 1.925573
sample estimates:
odds ratio
  1.549732
```

1.3 An example of the Yule–Simpson Paradox

1.3.1 Data

The classic kidney stone example is from Charig et al. (1986), where Z is the treatment with 1 for an open surgical procedure and 0 for a small puncture procedure, and Y is the outcome with 1 for success and 0 for failure. The treatment and outcome data can be summarized in the following two-by-two table:

	$Y=1$	$Y=0$
$Z=1$	273	77
$Z=0$	289	61

The estimated RD is

$$\begin{aligned}\widehat{\text{RD}} &= \frac{273}{273+77} - \frac{289}{289+61} \\ &= 78\% - 83\% \\ &= -5\% < 0.\end{aligned}$$

Treatment 0 seems better, that is, the small puncture leads to a higher success rate compared with the open surgical procedure.

However, the data were not from a randomized controlled trial (RCT).[3] Patients receiving treatment 1 can be very different from patients receiving treatment 0. A "lurking variable" in this study is the severity of the case: some patients have smaller stones and some patients have larger stones. We can split the data according to the size of the stones.

For patients with smaller stones, the treatment and outcome data can be summarized in the following two-by-two table:

	$Y=1$	$Y=0$
$Z=1$	81	6
$Z=0$	234	36

For patients with larger stones, the treatment and outcome data can be summarized in the following two-by-two table:

	$Y=1$	$Y=0$
$Z=1$	192	71
$Z=0$	55	25

The latter two tables must add up to the first table:

$$81+192=273, \quad 6+71=77, \quad 234+55=289, \quad 36+25=61.$$

From the table for patients with smaller stones, the estimated RD is

$$\begin{aligned}\widehat{\text{RD}}_{\text{smaller}} &= \frac{81}{81+6} - \frac{234}{234+36} \\ &= 93\% - 87\% \\ &= 6\% > 0,\end{aligned}$$

suggesting that treatment 1 is better. From the table for patients with larger stones, the estimated RD is

$$\begin{aligned}\widehat{\text{RD}}_{\text{larger}} &= \frac{192}{192+71} - \frac{55}{55+25} \\ &= 73\% - 69\% \\ &= 4\% > 0,\end{aligned}$$

also suggesting that treatment 1 is better.

The above data analysis leads to

$$\widehat{\text{RD}}_{\text{smaller}} > 0, \quad \widehat{\text{RD}}_{\text{larger}} > 0,$$

but

$$\widehat{\text{RD}} < 0.$$

Informally, treatment 1 is better for both patients with smaller and larger stones, but treatment 1 is worse for the whole population. This interpretation is quite confusing if the goal is to infer the treatment effect. In statistics, this is called the Yule–Simpson Paradox or Simpson's Paradox in which the marginal association has the opposite sign to the conditional associations at all levels.

[3] In an RCT, patients are randomly assigned to the treatment arms. Part II of this book will focus on RCTs.

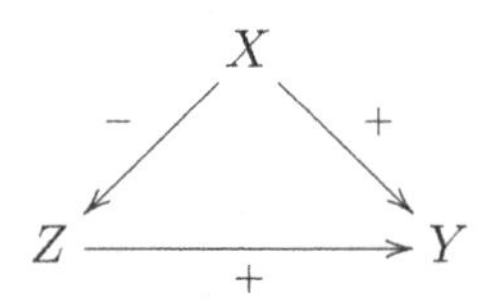

FIGURE 1.1: A diagram for the kidney stone example. The signs indicate the associations of two variables, conditioning on other variables pointing to the downstream variable.

1.3.2 Explanation

Let X be the binary indicator with $X = 1$ for smaller stones and $X = 0$ for larger stones. Let us first take a look at the X–Z relationship by comparing the probabilities of receiving treatment 1 among patients with smaller and larger stones:

$$\begin{aligned} & \hat{\text{pr}}(Z = 1 \mid X = 1) - \hat{\text{pr}}(Z = 1 \mid X = 0) \\ = & \frac{81 + 6}{81 + 6 + 234 + 36} - \frac{192 + 71}{192 + 71 + 55 + 25} \\ = & 24\% - 77\% \\ = & -53\% < 0. \end{aligned}$$

So patients with larger stones tend to take treatment 1 more frequently. Statistically, X and Z have a negative association.

Let us then take a look at the X–Y relationship by comparing the probabilities of success among patients with smaller and larger stones: under treatment 1,

$$\begin{aligned} & \hat{\text{pr}}(Y = 1 \mid Z = 1, X = 1) - \hat{\text{pr}}(Y = 1 \mid Z = 1, X = 0) \\ = & \frac{81}{81 + 6} - \frac{192}{192 + 71} \\ = & 93\% - 73\% \\ = & 20\% > 0; \end{aligned}$$

under treatment 0,

$$\begin{aligned} & \hat{\text{pr}}(Y = 1 \mid Z = 0, X = 1) - \hat{\text{pr}}(Y = 1 \mid Z = 0, X = 0) \\ = & \frac{234}{234 + 36} - \frac{55}{55 + 25} \\ = & 87\% - 69\% \\ = & 18\% > 0. \end{aligned}$$

So under both treatment levels, patients with smaller stones have higher success probabilities. Statistically, X and Y have a positive association conditional on both treatment levels.

We can summarize the qualitative associations in the diagram in Figure 1.1. In technical terms, the treatment has a positive path ($Z \rightarrow Y$) and a more negative path ($Z \leftarrow X \rightarrow Y$) to the outcome, so the overall association is negative between the treatment and outcome. In plain English, when the less effective treatment 0 is applied more frequently to the less severe cases, it can appear to be a more effective treatment.

In general, the association between Z and Y can differ qualitatively from the conditional association between Z and Y given X due to the association between X and Z and the association between X and Y. In Figure 1.1, we say that X is a *confounding variable* or *confounder* for the relationship between Z and Y, or Z and Y are *confounded* by X. See Chapters 10 and 17 for more in-depth discussion.

1.3.3 Geometry of the Yule–Simpson Paradox

Assume that the two-by-two table based on the aggregated data has counts

whole population	$Y = 1$	$Y = 0$
$Z = 1$	n_{11}	n_{10}
$Z = 0$	n_{01}	n_{00}

The two two-by-two tables based on subgroups have counts

subpopulation $X = 1$	$Y = 1$	$Y = 0$
$Z = 1$	$n_{11\|1}$	$n_{10\|1}$
$Z = 0$	$n_{01\|1}$	$n_{00\|1}$

for the subgroup with $X = 1$ and

subpopulation $X = 0$	$Y = 1$	$Y = 0$
$Z = 1$	$n_{11\|0}$	$n_{10\|0}$
$Z = 0$	$n_{01\|0}$	$n_{00\|0}$

for the subgroup with $X = 0$.

Figure 1.2 shows the geometry of the Yule–Simpson Paradox. The y-axis shows the count of successes with $Y = 1$ and the x-axis shows the count of failures with $Y = 0$. The two parallelograms OA_0AA_1 and OB_0BB_1 correspond to aggregating the counts of successes and failures under two treatment levels. The slope of OA_1 is larger than that of OB_1, and the slope of OA_0 is larger than that of OB_0. So the treatment seems beneficial to the outcome within both levels of X. However, the slope of OA is smaller than that of OB. So the treatment seems harmful to the outcome for the whole population. The Yule–Simpson Paradox arises.

1.4 The Berkeley graduate school admission data

Bickel et al. (1975) investigated the admission rates of male and female students into the graduate school of Berkeley. The `R` package `datasets` contains the original data `UCBAdmissions`. The raw data stratified by the six largest departments are shown below:

```
> library(datasets)
> UCBAdmissions = aperm(UCBAdmissions, c(2, 1, 3))
> UCBAdmissions
, , Dept = A

        Admit
Gender   Admitted Rejected
  Male        512      313
  Female       89       19

, , Dept = B

        Admit
```

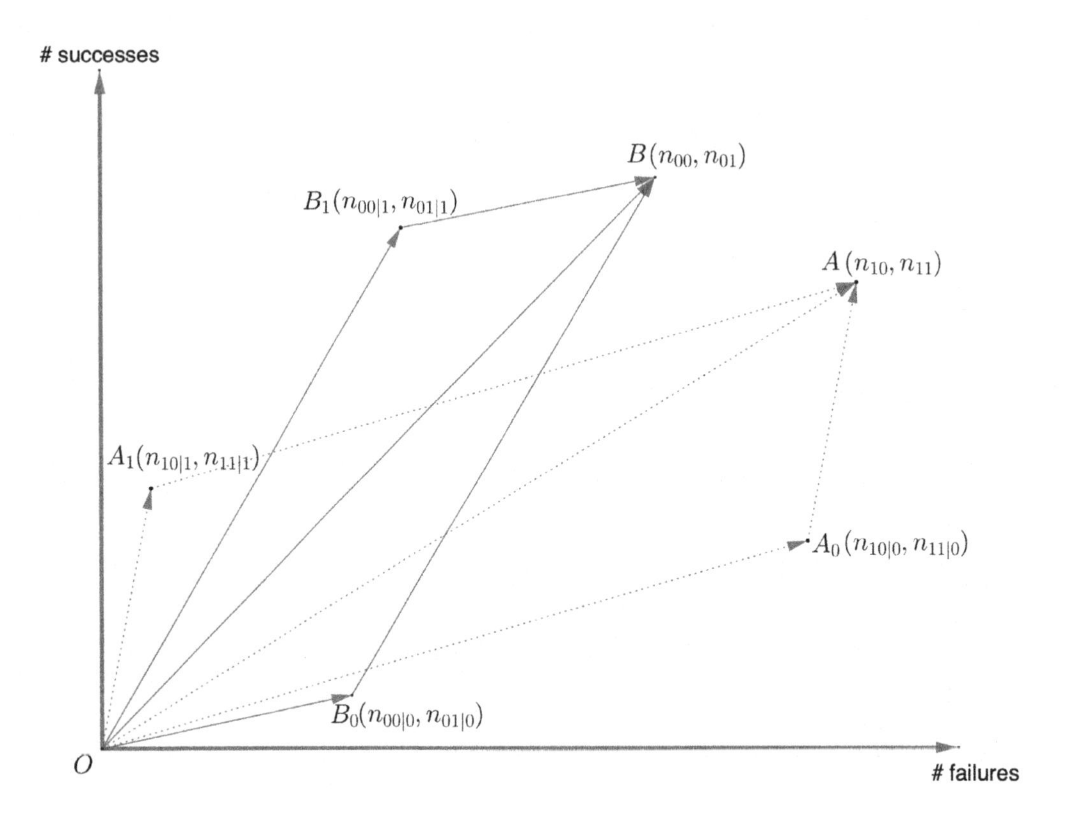

FIGURE 1.2: Geometry of the Yule–Simpson Paradox

```
Gender   Admitted Rejected
  Male        353      207
  Female       17        8

, , Dept = C

        Admit
Gender   Admitted Rejected
  Male        120      205
  Female      202      391

, , Dept = D

        Admit
Gender   Admitted Rejected
  Male        138      279
  Female      131      244

, , Dept = E

        Admit
Gender   Admitted Rejected
```

```
  Male           53        138
  Female         94        299

, , Dept = F

        Admit
Gender   Admitted Rejected
  Male         22      351
  Female       24      317
```

Aggregating the data over departments, we have a simple two-by-two table:

```
> UCBAdmissions.sum = apply(UCBAdmissions, c(1, 2), sum)
> UCBAdmissions.sum
        Admit
Gender   Admitted Rejected
  Male       1198     1493
  Female      557     1278
```

The following function, building upon `chisq.test`, has a two-by-two table as the input and the estimated RD and p-value as output:

```
> risk.difference = function(tb2)
+ {
+   p1      = tb2[1, 1]/(tb2[1, 1] + tb2[1, 2])
+   p2      = tb2[2, 1]/(tb2[2, 1] + tb2[2, 2])
+   testp   = chisq.test(tb2)
+
+   return(list(p.diff = p1 - p2,
+               pv = testp$p.value))
+ }
```

With this function, we find a large, significant difference between the admission rates of male and female students:

```
> risk.difference(UCBAdmissions.sum)
$p.diff
[1] 0.1416454

$pv
[1] 1.055797e-21
```

Stratifying on the departments, we find smaller and insignificant differences between the admission rates of male and female students. In department A, the difference is significant but negative.

```
> P.diff = rep(0, 6)
> PV     = rep(0, 6)
> for(dd in 1:6)
+ {
+        department = risk.difference(UCBAdmissions[, , dd])
+        P.diff[dd] = department$p.diff
+        PV[dd]     = department$pv
+ }
>
> round(P.diff, 2)
[1] -0.20 -0.05  0.03 -0.02  0.04 -0.01
> round(PV, 2)
[1] 0.00 0.77 0.43 0.64 0.37 0.64
```

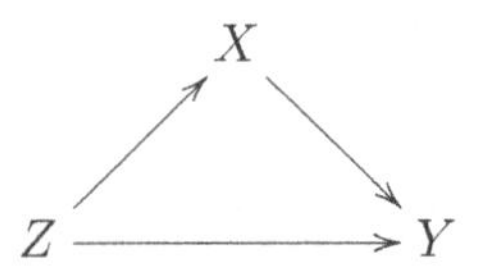

FIGURE 1.3: A diagram for the Berkeley graduate school admission data.

Let Z denote the indicator of gender, Y denote the indicator of admission, and X denote the indicator of departments. The Yule–Simpson Paradox in the Berkeley graduate school admission data has a different interpretation compared with that in the kidney stone example in Section 1.3. Figure 1.3 better illustrates the relationship among Z, Y, and X. In particular, X is a variable between Z and Y, which is called a *mediator*, not a *confounder*. The above analysis suggests that Z mainly affects Y indirectly through X whereas Z has a small effect directly on Y. Chapter 27 will discuss the notions of "mediator", "direct effect," and "indirect effect" in depth. Pearl and Mackenzie (2018, Chapter 9) also revisited this example.

1.5 Homework problems

1.1 Independence in two-by-two tables

Prove (1) and (2) in Proposition 1.1.

1.2 More examples of the Yule–Simpson Paradox

Give a numeric example of a two-by-two-by-two table in which the Yule–Simpson Paradox arises.

Find a real-life example in which the Yule–Simpson Paradox arises.

1.3 Correlation and partial correlation

Consider a three-dimensional Normal random vector:

$$\begin{pmatrix} X \\ Y \\ Z \end{pmatrix} \sim \mathrm{N}\left(\begin{pmatrix} 0 \\ 0 \\ 0 \end{pmatrix}, \begin{pmatrix} 1 & \rho_{XY} & \rho_{XZ} \\ \rho_{XY} & 1 & \rho_{YZ} \\ \rho_{XZ} & \rho_{YZ} & 1 \end{pmatrix} \right).$$

The correlation coefficient between Y and Z is ρ_{YZ}. There are many equivalent definitions of the partial correlation coefficient. For a multivariate Normal vector, let $\rho_{YZ|X}$ denote the partial correlation coefficient between Y and Z given X, which is defined as their correlation coefficient in the conditional distribution $(Y, Z) \mid X$. Show that

$$\rho_{YZ|X} = \frac{\rho_{YZ} - \rho_{YX}\rho_{ZX}}{\sqrt{1 - \rho_{YX}^2}\sqrt{1 - \rho_{ZX}^2}}.$$

Give a numerical example with $\rho_{YZ} > 0$ and $\rho_{YZ|X} < 0$.

Remark: This is the Yule–Simpson Paradox for a Normal random vector. You can use the results in Section A.1.2 to prove the formula for the partial correlation coefficient.

1.4 Specification searches

Section 1.2.1 reanalyzes the data used by Hainmueller (2012). We used the outcome named `re78` and the treatment named `treat` in the analysis. Moreover, the data also contain ten covariates and therefore $2^{10} = 1024$ possible subsets of covariates in the linear regression. Run 1024 linear regressions with all possible subsets of covariates, and report the regression coefficients of the treatment. How many coefficients of the treatment are positively significant, how many are negatively significant, and how many are not significant? You can also report other interesting findings from these regressions.

1.5 More on racial discrimination

Section 1.2.2 reanalyzes the data collected by Bertrand and Mullainathan (2004). Conduct analyses separately for males and females. What do you find from these subgroup analyses?

1.6 Recommended reading

Bickel et al. (1975) is the original paper for the paradox reported in Section 1.4. Pearl and Mackenzie (2018) recently revisited the study from the causal inference perspective.

2

Potential Outcomes

2.1 Experimentalists' view of causal inference

Rubin (1975) and Holland (1986) made up the aphorism:

> "no causation without manipulation."

Not everybody agrees with this point of view. However, it is quite helpful to clarify the ambiguity in thinking about causal relationships. This book follows this view and defines causal effects using the potential outcomes framework (Neyman, 1923; Rubin, 1974). In this framework, an experiment, or at least a thought experiment, has a treatment, and we are interested in its effect on an outcome or multiple outcomes. Sometimes, the *treatment* is also called an *intervention* or a *manipulation*.

Example 2.1 *If we are interested in the effect of taking aspirin or not on the relief of headaches, the intervention is taking aspirin or not.*

Example 2.2 *If we are interested in the effect of participating in a job training program or not on employment and wage, the intervention is participating in the job training program or not.*

Example 2.3 *If we are interested in the effect of studying in a small classroom or a large classroom on standardized test scores, the intervention is studying in a small classroom or not.*

Example 2.4 *Gerber et al. (2008) were interested in the effect of different get-out-to-vote messages on the voting behavior. The intervention is the different get-out-to-vote messages.*

Example 2.5 *Pearl (2018) claimed that we could infer the effect of obesity on life span. A popular measure of obesity is the body mass index (BMI), defined as the body mass divided by the square of the body height in units of* kg/m^2. *So the intervention can be different levels of BMI.*

However, there are different levels of ambiguity in the interventions above. The meanings of interventions in Examples 2.1–2.4 are relatively clear, but the meaning of intervention on BMI in Example 2.5 is less clear. In particular, we can imagine different versions of BMI reduction: healthier diet, more physical exercise, bariatric surgery, etc. These different versions of the intervention can have quite different effects on the outcome. In this book, we will view the intervention in Example 2.5 as ill-defined without further clarifications.

Another ill-defined intervention is race. Racial discrimination is an important issue in the labor market, but it is not easy to imagine an experiment to change the race of any experimental unit. Bertrand and Mullainathan (2004) give an interesting experiment that partially answers the question.

Example 2.6 *Recall Example 1.1. Bertrand and Mullainathan (2004) randomly change the names on the resumes, and compare the callback rates of resumes with African-American- or White-sounding names. For each resume, the intervention is the binary indicator of a Black-sounding or White-sounding name, and the outcome is the binary indicator of callback. I analyzed the following two-by-two table in Section 1.2.2:*

	callback	*no callback*
Black	*157*	*2278*
White	*235*	*2200*

From the above, we can compare the probabilities of being called back among Black-sounding and White-sounding names:

$$\begin{aligned} \frac{157}{2278+157} - \frac{235}{2200+235} &= 6.45\% - 9.65\% \\ &= -3.20\% < 0 \end{aligned}$$

with p-value from the Fisher exact test being much smaller than 0.001.

In the experiment of Bertrand and Mullainathan (2004), the treatment is the *perceived race* which can be manipulated by experimenters. They design an experiment to answer a well-defined causal question. However, critics may raise the concern that the causal effect of the *perceived race* may differ from the causal effect of the *actual race*. Moreover, other characteristics in the fake resumes may be systematically different from those in the real resumes. These issues may challenge the interpretation of the results in Bertrand and Mullainathan (2004).

2.2 Formal notation of potential outcomes

Consider a study with n experimental units indexed by $i = 1, \dots, n$. As a starting point, we focus on a treatment with two levels: 1 for the treatment and 0 for the control. For each unit i, the outcome of interest Y has two versions:

$$Y_i(1) \text{ and } Y_i(0),$$

which are potential outcomes under the hypothetical interventions 1 and 0. Neyman (1923) first used this notation. It seems intuitive but has some hidden assumptions. Rubin (1980) made the following clarifications on the hidden assumptions.

Assumption 2.1 (no interference) *Unit i's potential outcomes do not depend on other units' treatments. This is sometimes called the no-interference assumption.*

Assumption 2.2 (consistency) *There are no other versions of the treatment. Equivalently, we require that the treatment levels be well-defined, or have no ambiguity at least for the outcome of interest. This is sometimes called the consistency assumption.*[1]

Assumption 2.1 can be violated in infectious diseases or network experiments. For instance, if some of my friends receive flu shots, my chance of getting the flu decreases even

[1]This notion of consistency is totally different from the one in Definition A.4 in Appendix A.

if I do not receive the flu shot; if my friends see an advertisement on Facebook, my chance of buying that product increases even if I do not see the advertisement directly. It is an active research area to study situations with interfering units in modern causal inference literature (e.g., Hudgens and Halloran, 2008).

Assumption 2.2 can be violated for treatments with complex components. For instance, when studying the effect of cigarette smoking on lung cancer, the type of cigarettes may matter; when studying the effect of college education on income, the type and major of college education may matter.

Rubin (1980) called the Assumptions 2.1 and 2.2 above together the Stable Unit Treatment Value Assumption (SUTVA).

Assumption 2.3 (SUTVA) *Both Assumptions 2.1 and 2.2 hold.*

Under SUTVA, Rubin (2005) called the $n \times 2$ matrix of potential outcomes the Science Table:

i	$Y_i(1)$	$Y_i(0)$
1	$Y_1(1)$	$Y_1(0)$
2	$Y_2(1)$	$Y_2(0)$
$\vdots$	$\vdots$	$\vdots$
n	$Y_n(1)$	$Y_n(0)$

Due to the fundamental contributions of Neyman and Rubin to statistical causal inference, the potential outcomes framework is sometimes referred to as the Neyman Model, the Neyman–Rubin Model, or the Rubin Causal Model. I prefer using "the potential outcomes framework" in this book and my other scientific writings.[2]

Causal effects are functions of the Science Table. Inferring individual causal effects

$$\tau_i = Y_i(1) - Y_i(0), \quad (i = 1, \ldots, n)$$

is fundamentally challenging because we can only observe either $Y_i(1)$ or $Y_i(0)$ for each unit i, that is, we can observe only half of the Science Table. As a starting point, most parts of the book focus on the average causal effect (ACE):

$$\begin{aligned} \tau &= n^{-1}\sum_{i=1}^{n}\{Y_i(1) - Y_i(0)\} \\ &= n^{-1}\sum_{i=1}^{n} Y_i(1) - n^{-1}\sum_{i=1}^{n} Y_i(0). \end{aligned}$$

But we can extend our discussion to many other parameters (also called *estimands*). Problem 2.2 gives some examples.

2.2.1 Causal effects, subgroups, and the non-existence of Yule–Simpson Paradox

If we have two subgroups defined by a binary variable X_i, we can define the subgroup causal effects as

$$\tau_x = \frac{\sum_{i=1}^{n} I(X_i = x)\{Y_i(1) - Y_i(0)\}}{\sum_{i=1}^{n} I(X_i = x)}, \quad (x = 0, 1)$$

[2]Freedman (2009) used the name "response-schedule model" which is not popular in the literature.

where $I(\cdot)$ is the indicator function. A simple identity is that

$$\tau = \pi_1 \tau_1 + \pi_0 \tau_0,$$

where $\pi_x = \sum_{i=1}^n I(X_i = x)/n$ is the proportion of units with $X_i = x$ ($x = 0, 1$). Therefore, if $\tau_1 > 0$ and $\tau_0 > 0$, we must have $\tau > 0$. It is impossible to have $\tau_1 > 0$ and $\tau_0 > 0$ but $\tau < 0$, that is, the Yule–Simpson-like Paradox cannot happen to causal effects.

2.2.2 Subtlety of the definition of the experimental unit

I now discuss a subtlety related to the definition of the experimental unit. Simply speaking, the experimental unit can be different from the physical unit. For example, if I did not take aspirin before and my headache did not go away, but I take aspirin now and my headache goes away, then you might think that we can observe my potential outcomes under both the control and treatment. Let i index myself, and let $Y = 1$ denote the indicator of no headache. Then, the above heuristic suggests that $Y_i(0) = 0$ and $Y_i(1) = 1$, so it seems that aspirin kills my headache. But this logic is very wrong because of the misunderstanding of the definition of the experimental unit. At different time points, I, the same physical person, become two distinct experimental units, indexed by "i, before" and "i, after". Therefore, we have four potential outcomes

$$Y_{i,\text{before}}(0) = 0, \quad Y_{i,\text{before}}(1) = ?, \quad Y_{i,\text{after}}(0) = ?, \quad Y_{i,\text{after}}(1) = 1,$$

with two of them observed and two of them missing. The individual causal effects

$$Y_{i,\text{before}}(1) - Y_{i,\text{before}}(0) = ? - 0 \text{ and } Y_{i,\text{after}}(1) - Y_{i,\text{after}}(0) = 1 - ?$$

are unknown. It is possible that my headache goes away even if I do not take aspirin:

$$Y_{i,\text{after}}(0) = 1, \quad Y_{i,\text{after}}(1) = 1$$

which implies zero effect; it is also possible that my headache does not go away if I do not take aspirin:

$$Y_{i,\text{after}}(0) = 0, \quad Y_{i,\text{after}}(1) = 1$$

which implies a positive effect of aspirin.

The wrong heuristic argument might get the right answer if the control potential outcomes are stable at the before and after periods:

$$Y_{i,\text{before}}(0) = Y_{i,\text{after}}(0) = 0.$$

But this assumption is rather strong and fundamentally untestable. Rubin (2001) offered a related discussion in the context of self-experimentation for causal effects.

2.3 Treatment assignment mechanism

Let Z_i be the binary treatment indicator for unit i, vectorized as $\boldsymbol{Z} = (Z_1, \ldots, Z_n)$. The observed outcome of unit i is a function of the potential outcomes and the treatment indicator:

$$
\begin{aligned}
Y_i &= \begin{cases} Y_i(1), & \text{if } Z_i = 1 \\ Y_i(0), & \text{if } Z_i = 0 \end{cases} && (2.1)\\
&= Z_i Y_i(1) + (1 - Z_i) Y_i(0) && (2.2)\\
&= Y_i(0) + Z_i\{Y_i(1) - Y_i(0)\} && (2.3)\\
&= Y_i(0) + Z_i \tau_i. && (2.4)
\end{aligned}
$$

Equation (2.1) is the definition[3] of the observed outcome. Equation (2.2) is equivalent to (2.1). It is a trivial fact, but Pearl (2010b) viewed it as the fundamental bridge between the potential outcomes and the observed outcome. Equations (2.3) and (2.4) highlight the fact that the individual causal effect $\tau_i = Y_i(1) - Y_i(0)$ can be heterogeneous across units.

The experiment reveals only one of unit i's potential outcomes with the other one missing:

$$
\begin{aligned}
Y_i^{\text{mis}} &= \begin{cases} Y_i(0), & \text{if } Z_i = 1 \\ Y_i(1), & \text{if } Z_i = 0 \end{cases}\\
&= Z_i Y_i(0) + (1 - Z_i) Y_i(1).
\end{aligned}
$$

The missing potential outcome corresponds to the opposite treatment level of unit i. For this reason, the potential outcomes framework is also called the *counterfactual* framework.[4] Some people find the terminology "counterfactual framework" inaccurate because before the experiment, both the potential outcomes can be observed, and only after the experiment, one potential outcome is observed whereas the other potential outcome is counterfactual.

The treatment assignment mechanism, i.e., the probability distribution of $\boldsymbol{Z}$, plays an important role in inferring causal effects. The following simple numerical examples illustrate this point. We first generate potential outcomes from Normal distributions with the average causal effect close to -0.5.

```
> n   = 500
> Y0  = rnorm(n)
> tau = - 0.5 + Y0
> Y1  = Y0 + tau
```

A perfect doctor assigns the treatment to the patient if s/he knows that the individual causal effect is non-negative. This results in a positive difference in means of the observed outcomes:

```
> Z = (tau >= 0)
> Y = Z*Y1 + (1 - Z)*Y0
> mean(Y[Z==1]) - mean(Y[Z==0])
[1] 2.166509
```

A clueless doctor does not know any information about the individual causal effects and assigns the treatment to patients by flipping a fair coin. This results in a difference in means of the observed outcomes close to the true average causal effect:

```
> Z = rbinom(n, 1, 0.5)
> Y = Z*Y1 + (1 - Z)*Y0
> mean(Y[Z==1]) - mean(Y[Z==0])
[1] -0.552064
```

[3]It can be a philosophical issue whether $Y_i = Y_i(Z_i)$ is a definition or an assumption. For simplicity, we can define the observed outcome Y_i as $Y_i(Z_i)$. A similar issue will appear in Assumption 27.1 of Chapter 27.

[4]The terminology "counterfactual" is more popular in philosophy, probably due to Lewis (1973).

The above examples are hypothetical since no doctors perfectly know the individual causal effects. However, the examples do demonstrate the crucial role of the treatment assignment mechanism. This book will organize the topics based on the treatment assignment mechanism.

2.4 Homework problems

2.1 A perfect doctor

Following the first perfect doctor example in Section 2.3, assume the potential outcomes are random variables generated from

$$\begin{aligned} Y(0) &\sim \mathrm{N}(0,1), \\ \tau &= -0.5 + Y(0), \\ Y(1) &= Y(0) + \tau. \end{aligned}$$

The binary treatment is determined by the treatment effect as $Z = 1(\tau \geq 0)$, and the observed outcome is determined by the potential outcomes and the treatment by $Y = ZY(1) + (1 - Z)Y(0)$. Calculate the difference in means

$$E(Y \mid Z = 1) - E(Y \mid Z = 0).$$

Remark: You may find Lemma 2.1 below useful.

Lemma 2.1 *The mean of a truncated Normal random variable equals*

$$E(X \mid a < X < b) = \mu - \sigma \frac{\phi\left(\frac{b-\mu}{\sigma}\right) - \phi\left(\frac{a-\mu}{\sigma}\right)}{\Phi\left(\frac{b-\mu}{\sigma}\right) - \Phi\left(\frac{a-\mu}{\sigma}\right)},$$

where $X \sim \mathrm{N}(\mu, \sigma^2)$, and $\phi(\cdot)$ and $\Phi(\cdot)$ are the probability density and cumulative distribution functions of a standard Normal random variable $\mathrm{N}(0,1)$.

2.2 Nonlinear causal estimands

With potential outcomes $\{(Y_i(1), Y_i(0)\}_{i=1}^n$ for n units under the treatment and control, the difference in means equals the mean of the individual treatment effects:

$$\bar{Y}(1) - \bar{Y}(0) = n^{-1} \sum_{i=1}^{n} \{Y_i(1) - Y_i(0)\}.$$

Therefore, the average treatment effect is a *linear* causal estimand in the sense that the difference in average potential outcomes equals the average of the differences in individual potential outcomes.

Other estimands may not be linear. For instance, we can define the median treatment effect as

$$\delta_1 = \mathrm{median}\{(Y_i(1)\}_{i=1}^n - \mathrm{median}\{(Y_i(0)\}_{i=1}^n,$$

which is, in general, different from the median of the individual treatment effect

$$\delta_2 = \mathrm{median}\{(Y_i(1) - Y_i(0)\}_{i=1}^n.$$

1. Give numerical examples that have $\delta_1 = \delta_2$, $\delta_1 > \delta_2$, and $\delta_1 < \delta_2$, respectively.
2. Which estimand makes more sense, δ_1 or δ_2? Why? Use examples to justify your conclusion. If you feel that both δ_1 and δ_2 can make sense in different applications, you can also give examples to justify both estimands.

2.3 *Average and individual effects*

Give a numerical example in which $\tau = n^{-1}\sum_{i=1}^{n}\{Y_i(1) - Y_i(0)\} > 0$ but the proportion of units with $Y_i(1) > Y_i(0)$ is smaller than 0.5. That is, the average causal effect is positive, but the treatment benefits less than half of the units.

2.4 *Recommended reading*

Holland (1986) is a classic review article on statistical causal inference. It popularized the name "Rubin Causal Model" for the potential outcomes framework. At UC Berkeley, many people call it the "Neyman Model" because Neyman was the founder of the Department of Statistics.

Part II

Randomized experiments

3

The Completely Randomized Experiment and the Fisher Randomization Test

The potential outcomes framework has intrinsic connections with randomized experiments. Understanding causal inference in various randomized experiments is fundamental and helpful for understanding causal inference in more complicated non-experimental studies.

Part II of this book focuses on randomized experiments. This chapter focuses on the simplest experiment, the completely randomized experiment (CRE).

3.1 CRE

Consider an experiment with n units, with n_1 receiving the treatment and n_0 receiving the control. We can define the CRE based on its treatment assignment mechanism.[1]

Definition 3.1 (CRE) *Fix n_1 and n_0 with $n = n_1 + n_0$. A CRE has the treatment assignment mechanism:*

$$\text{pr}(\boldsymbol{Z} = \boldsymbol{z}) = 1 \Big/ \binom{n}{n_1},$$

where $\boldsymbol{z} = (z_1, \ldots, z_n)$ satisfies $\sum_{i=1}^{n} z_i = n_1$ and $\sum_{i=1}^{n}(1 - z_i) = n_0$.

In Definition 3.1, we assume that the potential outcome vector under treatment $\boldsymbol{Y}(1) = (Y_1(1), \ldots, Y_n(1))$ and the potential outcome vector under control $\boldsymbol{Y}(0) = (Y_1(0), \ldots, Y_n(0))$ are both fixed. Even if we view them as random, we can condition on them and the treatment assignment mechanism becomes

$$\text{pr}\{\boldsymbol{Z} = \boldsymbol{z} \mid \boldsymbol{Y}(1), \boldsymbol{Y}(0)\} = 1 \Big/ \binom{n}{n_1}$$

because $\boldsymbol{Z} \perp\!\!\!\perp \{\boldsymbol{Y}(1), \boldsymbol{Y}(0)\}$ in a CRE. In a CRE, the treatment vector $\boldsymbol{Z}$ is from a random permutation of n_1 1's and n_0 0's.

In his seminal book *Design of Experiments*, Fisher (1935) pointed out the following advantages of randomization:

1. It creates comparable treatment and control groups on average.
2. It serves as a "reasoned basis" for statistical inference.

[1] Readers may think that a CRE has Z_i's as independent and identically distributed (IID) Bernoulli random variables with probability π, in which n_1 is a Binomial(n, π) random variable. This is called the Bernoulli randomized experiment (BRE), which reduces to the CRE if we condition on (n_1, n_0). I will give more details for the BRE in Problem 4.8 in Chapter 4.

Point 1 is intuitive because the random treatment assignment does not bias toward the treatment or the control. Most people understand point 1 well. Point 2 is more subtle. What Fisher meant is that randomization justifies a statistical test, which is now called the Fisher Randomization Test (FRT). This chapter illustrates the basic idea of the FRT under a CRE.

3.2 FRT

Fisher (1935) was interested in testing the following null hypothesis:[2]

$$H_{0F} : Y_i(1) = Y_i(0) \text{ for all units } i = 1, \dots, n.$$

Rubin (1980) called it the *sharp null hypothesis* in the sense that it can determine all the potential outcomes based on the observed data: $\boldsymbol{Y}(1) = \boldsymbol{Y}(0) = \boldsymbol{Y} = (Y_1, \dots, Y_n)$, the vector of the observed outcomes. It is also called the *strong null hypothesis* (e.g., Wu and Ding, 2021).

Conceptually, under H_{0F}, the FRT works for any test statistic

$$T = T(\boldsymbol{Z}, \boldsymbol{Y}), \tag{3.1}$$

which is a function of the observed data. The observed outcome vector $\boldsymbol{Y}$ is fixed under H_{0F}, so the only random component in the test statistic T is the treatment vector $\boldsymbol{Z}$. The experimenter determines the distribution of $\boldsymbol{Z}$, which in turn determines the distribution of T under H_{0F}. This is the basis for calculating the p-value. I will give more details below.

In a CRE, $\boldsymbol{Z}$ is uniform over the set

$$\{\boldsymbol{z}^1, \dots, \boldsymbol{z}^M\}$$

where $M = \binom{n}{n_1}$, and the $\boldsymbol{z}^m$'s are all possible vectors with n_1 1's and n_0 0's. For instance, with $n = 5$ and $n_1 = 3$, we can enumerate $M = \binom{5}{3} = 10$ vectors as follows:

```
> permutation10 = function(n, n1){
+   M   = choose(n, n1)
+   treat.index = combn(n, n1)
+   Z = matrix(0, n, M)
+   for(m in 1:M){
+     treat   = treat.index[, m]
+     Z[treat, m] = 1
+   }
+   Z
+ }
>
> permutation10(5, 3)
     [,1] [,2] [,3] [,4] [,5] [,6] [,7] [,8] [,9] [,10]
[1,]    1    1    1    1    1    1    0    0    0     0
[2,]    1    1    1    0    0    0    1    1    1     0
[3,]    1    0    0    1    1    0    1    1    0     1
[4,]    0    1    0    1    0    1    1    0    1     1
[5,]    0    0    1    0    1    1    0    1    1     1
```

[2]Actually, Fisher (1935) did not use this form of H_{0F} since he did not use the notation of potential outcomes. This form was due to Rubin (1980).

As a consequence, T is uniform over the set (with possible duplications)

$$\{T(\boldsymbol{z}^1, \boldsymbol{Y}), \dots, T(\boldsymbol{z}^M, \boldsymbol{Y})\}.$$

That is, the distribution of T is known due to the design of the CRE. We will call this distribution of T the *randomization distribution.*

If larger values are more extreme for T, we can use the following tail probability to measure the extremeness of the test statistic with respect to its randomization distribution:[3]

$$p_{\text{FRT}} = M^{-1} \sum_{m=1}^{M} I\{T(\boldsymbol{z}^m, \boldsymbol{Y}) \geq T(\boldsymbol{Z}, \boldsymbol{Y})\}, \tag{3.2}$$

which is called the p-value by Fisher. Figure 3.1 illustrates the computational process of p_{FRT}.

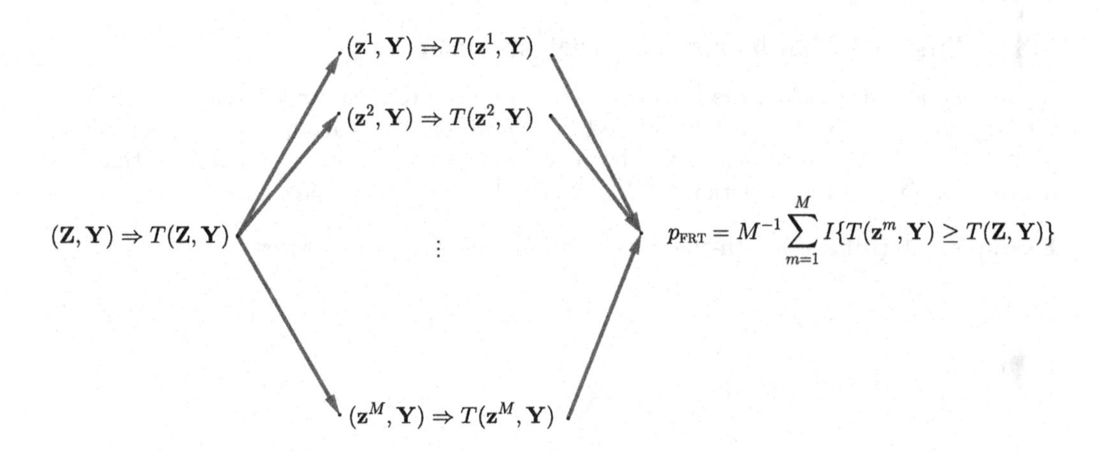

FIGURE 3.1: Illustration of the FRT

The p-value, p_{FRT}, in (3.2) works for any choice of test statistic and any outcome-generating process. It also extends naturally to any experiment, which will be a topic repeatedly discussed in the following chapters. Importantly, it is finite-sample exact in the sense[4] that under $H_{0\text{F}}$,

$$\text{pr}(p_{\text{FRT}} \leq u) \leq u \quad \text{for all} \quad 0 \leq u \leq 1. \tag{3.3}$$

In practice, M is often too large (e.g., with $n = 100, n_1 = 50$, we have $M > 10^{29}$), and it is computationally infeasible to enumerate all possible values of the treatment vector. We often approximate p_{FRT} by Monte Carlo (see Section A.5 for a review of the basic Monte

[3] Because of this, the p_{FRT} in (3.2) is one-sided. Sometimes, we may want to define two-sided p-values. One possibility is to use the absolute value of T if it is distributed around 0.

[4] This is the standard definition of the p-value in mathematical statistics. The inequality is often due to the discreteness of the test statistic, and when the equality holds, the p-value is Uniform$(0, 1)$ under the null hypothesis. Let $F(\cdot)$ be the distribution function of $T(\boldsymbol{Z}, \boldsymbol{Y})$. Even though it is a step function, we assume that it is continuous and strictly increasing as if it is the distribution function of a continuous random variable taking values on the whole real line. So $p_{\text{FRT}} = 1 - F(T)$, and

$$\text{pr}(p_{\text{FRT}} \leq u) = \text{pr}\{1 - F(T) \leq u\} = \text{pr}\{T \geq F^{-1}(1-u)\} = 1 - F(F^{-1}(1-u)) = u.$$

The discreteness of T does cause some technical issues in the proof, yielding an inequality instead of an equality. I leave the technical details to Problem 3.1.

Carlo method). To be more specific, we take independent random draws from all possible values of the treatment vector, or, equivalently, we randomly permute $\boldsymbol{Z}$, and approximate p_{FRT} by

$$\hat{p}_{\text{FRT}} = R^{-1} \sum_{r=1}^{R} I\{T(\boldsymbol{z}^r, \boldsymbol{Y}) \geq T(\boldsymbol{Z}, \boldsymbol{Y})\}, \tag{3.4}$$

where the $\boldsymbol{z}^r$'s are the R random permutations of $\boldsymbol{Z}$. The p-value in (3.4) has Monte Carlo error decreasing fast with an increasing R; see Problem 3.2. Because the calculation of the p-value in (3.4) involves permutations of $\boldsymbol{Z}$, the FRT is sometimes called the *permutation test* in the context of the CRE. However, the idea of FRT is more general than the permutation test in more complex experiments.

3.3 Canonical choices of the test statistic

From the above discussion, the FRT generates finite-sample exact p-value for any choice of the test statistic. This is a feature of the FRT. However, this feature should not encourage arbitrary choice of the test statistic. Intuitively, we must choose test statistics that give information for possible violations of $H_{0\text{F}}$. Below I will review some canonical choices.

Example 3.1 (difference-in-means) *The difference-in-means statistic is*

$$\hat{\tau} = \hat{\bar{Y}}(1) - \hat{\bar{Y}}(0)$$

where

$$\hat{\bar{Y}}(1) = n_1^{-1} \sum_{Z_i=1} Y_i = n_1^{-1} \sum_{i=1}^{n} Z_i Y_i$$

is the sample mean of the outcomes under the treatment and

$$\hat{\bar{Y}}(0) = n_0^{-1} \sum_{Z_i=0} Y_i = n_0^{-1} \sum_{i=1}^{n} (1 - Z_i) Y_i$$

is the sample mean of the outcomes under the control, respectively. Under $H_{0\text{F}}$, it has mean

$$\begin{aligned} E(\hat{\tau}) &= n_1^{-1} \sum_{i=1}^{n} E(Z_i) Y_i - n_0^{-1} \sum_{i=1}^{n} E(1 - Z_i) Y_i \\ &= 0 \end{aligned}$$

and variance

$$\begin{aligned} \text{var}(\hat{\tau}) &= \text{var}\left\{ n_1^{-1} \sum_{i=1}^{n} Z_i Y_i - n_0^{-1} \sum_{i=1}^{n} (1 - Z_i) Y_i \right\} \\ &= \text{var}\left(\frac{n}{n_0} \frac{1}{n_1} \sum_{i=1}^{n} Z_i Y_i \right) \\ &= \frac{n^2}{n_0^2} \left(1 - \frac{n_1}{n}\right) \frac{s^2}{n_1} \\ &= \frac{n}{n_1 n_0} s^2, \end{aligned}$$

where the second equation holds by dropping the terms that do not involve the random Z_i's, and the third equation follows from Lemma C.2 for simple random sampling with

$$\bar{Y} = n^{-1}\sum_{i=1}^{n} Y_i, \quad s^2 = (n-1)^{-1}\sum_{i=1}^{n}(Y_i - \bar{Y})^2.$$

Furthermore, the randomization distribution of $\hat{\tau}$ is approximately Normal due to the finite population central limit theorem (CLT) in Lemma C.4:

$$\frac{\hat{\tau}}{\sqrt{\frac{n}{n_1 n_0} s^2}} \to \mathrm{N}(0,1) \tag{3.5}$$

in distribution. Since s^2 is fixed under $H_{0\mathrm{F}}$, it is equivalent to use

$$\frac{\hat{\tau}}{\sqrt{\frac{n}{n_1 n_0} s^2}}$$

as the test statistic in the FRT, which is asymptotically $\mathrm{N}(0,1)$ *as shown in* (3.5)*. Then we can calculate an approximate p-value.*

The observed data are $\{Y_i : Z_i = 1\}$ and $\{Y_i : Z_i = 0\}$, so the problem is essentially a two-sample problem. Under the assumption of independent and identically distributed (IID) Normal outcomes (see Section A.4.1), the classic two-sample t-test assuming equal variance is based on

$$\frac{\hat{\tau}}{\sqrt{\frac{n}{n_1 n_0 (n-2)}\left[\sum_{Z_i=1}\{Y_i - \hat{\bar{Y}}(1)\}^2 + \sum_{Z_i=0}\{Y_i - \hat{\bar{Y}}(0)\}^2\right]}} \sim t_{n-2}. \tag{3.6}$$

Based on some algebra (see Problem 3.8), we have the expansion

$$(n-1)s^2 = \sum_{Z_i=1}\{Y_i - \hat{\bar{Y}}(1)\}^2 + \sum_{Z_i=0}\{Y_i - \hat{\bar{Y}}(0)\}^2 + \frac{n_1 n_0}{n}\hat{\tau}^2. \tag{3.7}$$

With a large sample size n, we can ignore the difference between $\mathrm{N}(0,1)$ and t_{n-2} and the difference between $n-1$ and $n-2$. Moreover, under $H_{0\mathrm{F}}$, $\hat{\tau}$ converges to zero in probability, so $n_1 n_0 / n\hat{\tau}^2$ can be ignored asymptotically. Therefore, under $H_{0\mathrm{F}}$, the approximate p-value in Example 3.1 is close to the p-value from the classic two-sample t-test assuming equal variance, which can be calculated by `t.test` with `var.equal = TRUE`. Under alternative hypotheses with nonzero τ, the additional term $\frac{n_1 n_0}{n}\hat{\tau}^2$ in the expansion (3.7) can make the FRT less powerful than the usual t-test; Ding (2016) made this point.

Based on the above discussion, the FRT with $\hat{\tau}$ effectively uses a pooled variance ignoring the heteroskedasticity between these two groups. In classical statistics, the two-sample problem with heteroskedastic Normal outcomes is called the Behrens–Fisher problem (see Section A.4.1). In the Behrens–Fisher problem, a standard choice of the test statistic is the studentized statistic below.

Example 3.2 (studentized statistic) *The studentized statistic*[5] *is*

$$t = \frac{\hat{\bar{Y}}(1) - \hat{\bar{Y}}(0)}{\sqrt{\frac{\hat{S}^2(1)}{n_1} + \frac{\hat{S}^2(0)}{n_0}}},$$

[5]The t notation is intentional because it is related to the t distribution. But it should not be confused with the notation t_ν, which denotes the t distribution with degrees of freedom ν.

where

$$\hat{S}^2(1) = (n_1 - 1)^{-1} \sum_{Z_i=1} \{Y_i - \hat{\bar{Y}}(1)\}^2,$$
$$\hat{S}^2(0) = (n_0 - 1)^{-1} \sum_{Z_i=0} \{Y_i - \hat{\bar{Y}}(0)\}^2$$

are the sample variances of the observed outcomes under the treatment and control, respectively. Under $H_{0\mathrm{F}}$, the finite population CLT again implies that t is asymptotically Normal:

$$t \to \mathrm{N}(0, 1)$$

in distribution. Then we can calculate an approximate p-value which is close to the p-value from `t.test` with `var.equal = FALSE`.

An extremely important point is that the FRT justifies the traditional t-tests using `t.test` with either `var.equal = TRUE` or `var.equal = FALSE`, even if the underlying distributions are not Normal. Standard statistics textbooks motivate the t-tests based on the Normality assumption, but the assumption can be too strong in practice. Fortunately, the t-test procedures can still be used as long as the finite population CLTs hold. Even if we do not believe the CLTs, we can still use $\hat{\tau}$ and t as test statistics in the FRT to obtain finite-sample exact p-values.

We will motivate this studentized statistic from another perspective in Chapter 8. The theory there shows that using t in FRT is more robust to heteroskedasticity across the two groups.

The following test statistic is robust to outliers resulting from heavy-tailed outcome data.

Example 3.3 (Wilcoxon rank sum) *The difference-in-means statistic $\hat{\tau}$ uses the sample means of the original outcomes, and its sampling distribution depends on the variances of the outcomes. The studentized statistic t uses the sample means and variances of the original outcomes, and its sampling distribution depends on higher moments of the outcomes. This makes $\hat{\tau}$ and t sensitive to outliers.*

Another popular test statistic is based on the ranks of the pooled observed outcomes. Let R_i denote the rank of Y_i in the pooled samples $(Y_1, \ldots, Y_n)$:

$$R_i = \#\{j : Y_j \le Y_i\}.$$

The Wilcoxon rank sum statistic is the sum of the ranks under treatment:

$$W = \sum_{i=1}^{n} Z_i R_i.$$

For algebraic simplicity, we assume that there are no ties in the outcomes.[6] *Because the sum of the ranks of the pooled samples is fixed at $1 + 2 + \cdots + n = n(n+1)/2$, the Wilcoxon statistic is equivalent to the difference in the means of the ranks under the treatment and*

[6]The FRT can be applied regardless of the existence of ties. With ties, we can either add small random noises to the original outcomes or use the average ranks for the tied outcomes. For the case with ties, see Lehmann (1975, Chapter 1 Section 4).

control. Under H_{0F}, the R_i's are fixed, so W has mean

$$\begin{aligned} E(W) &= \sum_{i=1}^{n} E(Z_i)R_i \\ &= \frac{n_1}{n}\sum_{i=1}^{n} i \\ &= \frac{n_1}{n} \times \frac{n(n+1)}{2} \\ &= \frac{n_1(n+1)}{2} \end{aligned}$$

and variance

$$\begin{aligned} \text{var}(W) &= \text{var}\left(n_1 \frac{1}{n_1}\sum_{i=1}^{n} Z_i R_i\right) \\ &=_* n_1^2\left(1-\frac{n_1}{n}\right)\frac{1}{n_1}\frac{1}{n-1}\sum_{i=1}^{n}\left(R_i - \frac{n+1}{2}\right)^2 \\ &= \frac{n_1 n_0}{n(n-1)}\sum_{i=1}^{n}\left(i-\frac{n+1}{2}\right)^2 \\ &= \frac{n_1 n_0}{n(n-1)}\left\{\sum_{i=1}^{n} i^2 - n\left(\frac{n+1}{2}\right)^2\right\} \\ &= \frac{n_1 n_0}{n(n-1)}\left\{\frac{n(n+1)(2n+1)}{6} - n\left(\frac{n+1}{2}\right)^2\right\} \\ &= \frac{n_1 n_0 (n+1)}{12}, \end{aligned}$$

where $=_*$ follows from Lemma C.2. Furthermore, under H_{0F}, the finite population CLT ensures that the randomization distribution of $\hat{\tau}$ is approximately Normal:

$$\frac{\sum_{i=1}^{n} Z_i R_i - \frac{n_1(n+1)}{2}}{\sqrt{\frac{n_1 n_0 (n+1)}{12}}} \rightarrow \text{N}(0,1) \tag{3.8}$$

in distribution. Based on (3.8), we can conduct an asymptotic test. In `R`*, the function* `wilcox.test` *can compute both exact and asymptotic p-values based on the statistic $W - n_1(n_1+1)/2$. Based on some asymptotic analyses, Lehmann (1975) showed that the FRT using W has reasonable power over a wide range of data-generating processes.*

Example 3.4 (Kolmogorov–Smirnov statistic) *The treatment may affect the outcome in different ways. It seems natural to summarize the treatment outcomes and control outcomes based on the empirical distributions:*

$$\begin{aligned} \hat{F}_1(y) &= n_1^{-1}\sum_{i=1}^{n} Z_i I(Y_i \le y), \\ \hat{F}_0(y) &= n_0^{-1}\sum_{i=1}^{n} (1-Z_i) I(Y_i \le y). \end{aligned}$$

Comparing these two empirical distributions yields the famous Kolmogorov–Smirnov statistic

$$D = \max_y \left| \hat{F}_1(y) - \hat{F}_0(y) \right|.$$

It is a challenging mathematics problem to derive the distribution of D. With large sample sizes $n_1 \to \infty$ and $n_0 \to \infty$, its distribution function converges to

$$\text{pr}\left(\frac{n_1 n_0}{n} D \le x\right) \to \frac{\sqrt{2\pi}}{x} \sum_{j=1}^{\infty} e^{-(2j-1)^2\pi^2/(8x^2)},$$

based on which we calculate an asymptotic p-value (Van der Vaart, 2000). In R, *the function* `ks.test` *can compute both exact and asymptotic p-values.*

3.4 A case study of the LaLonde experimental data

I use the experimental data of LaLonde (1986) to illustrate the FRT. The data are available in the `Matching` package (Sekhon, 2011):

```
> library(Matching)
> data(lalonde)
> z = lalonde$treat
> y = lalonde$re78
```

In the above, `z` denotes the binary treatment vector indicating whether a unit was randomly assigned to the job training program or not, and `y` is the outcome vector representing a unit's real earnings in 1978. Figure 3.2 shows the histograms of the outcomes under the treatment and control.

The following code computes the observed values of the test statistics using existing functions:

```
> tauhat = t.test(y[z == 1], y[z == 0],
+                 var.equal = TRUE)$statistic
> tauhat
       t
2.835321
> student = t.test(y[z == 1], y[z == 0],
+                 var.equal = FALSE)$statistic
> student
       t
2.674146
> W = wilcox.test(y[z == 1], y[z == 0])$statistic
> W
      W
27402.5
> D = ks.test(y[z == 1], y[z == 0])$statistic
> D
        D
0.1321206
```

By randomly permuting the treatment vector, we can obtain the Monte Carlo approximation of the randomization distributions of the test statistics, stored in four vectors: `Tauhat`, `Student`, `Wilcox`, and `Ks`.

```
> MC = 10^4
> Tauhat    = rep(0, MC)
> Student   = rep(0, MC)
> Wilcox    = rep(0, MC)
> Ks        = rep(0, MC)
> for(mc in 1:MC)
+ {
+   zperm        = sample(z)
+   Tauhat[mc]   = t.test(y[zperm == 1], y[zperm == 0],
+                         var.equal = TRUE)$statistic
+   Student[mc]  = t.test(y[zperm == 1], y[zperm == 0],
+                         var.equal = FALSE)$statistic
+   Wilcox[mc]   = wilcox.test(y[zperm == 1],
+                              y[zperm == 0])$statistic
+   Ks[mc]       = ks.test(y[zperm == 1],
+                          y[zperm == 0])$statistic
+ }
```

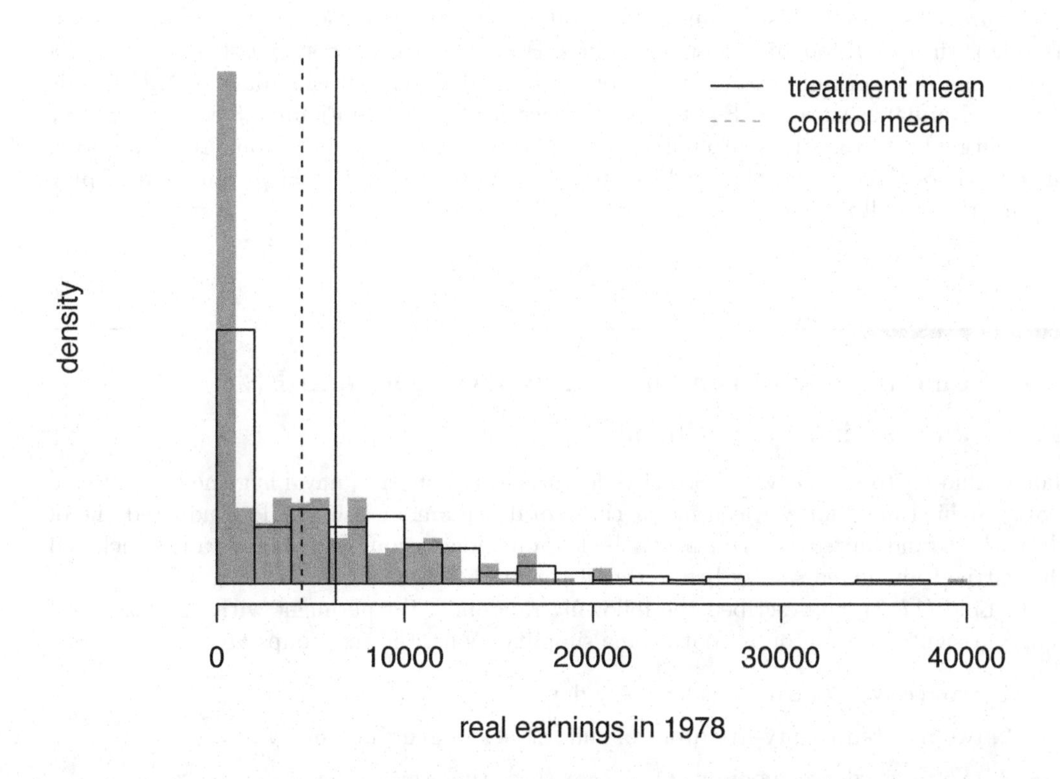

FIGURE 3.2: Histograms of the outcomes in the LaLonde experimental data: the treatment in white and the control in grey. The vertical lines are the sample means of the outcomes: the solid line for the treatment and the dashed line for the control.

The one-sided p-values based on the FRT are all smaller than 0.05:

```
> exact.pv = c(mean(Tauhat >= tauhat),
+              mean(Student >= student),
+              mean(Wilcox >= W),
+              mean(Ks >= D))
> round(exact.pv, 3)
[1] 0.002 0.002 0.006 0.040
```

Without using Monte Carlo, we can also compute the asymptotic p-values which are all smaller than 0.05:

```
> asym.pv = c(t.test(y[z == 1], y[z == 0],
+                    var.equal = TRUE)$p.value,
+             t.test(y[z == 1], y[z == 0],
+                    var.equal = FALSE)$p.value,
+             wilcox.test(y[z == 1], y[z == 0])$p.value,
+             ks.test(y[z == 1], y[z == 0])$p.value)
> round(asym.pv, 3)
[1] 0.005 0.008 0.011 0.046
```

The differences between the p-values are due to the asymptotic approximations as well as the fact that the default choices for `t.test` and `wilcox.test` are two-sided tests. To make fair comparisons, we need to multiply the first three p_{FRT}'s by a factor of 2.

Figure 3.3 shows the histograms of the randomization distributions of four test statistics, as well as their corresponding observed values. For the first three test statistics, the Normal approximations work quite well even though the underlying outcome data distribution is far from Normal as shown in Figure 3.2. In general, a figure like Figure 3.3 can give clearer information for testing the sharp null hypothesis. Recently, Bind and Rubin (2020) propose, in the title of their paper, that "when possible, report a Fisher-exact p-value and display its underlying null randomization distribution." I agree.

3.5 Some history of randomized experiments and FRT

3.5.1 James Lind's experiment

James Lind (1716—1794) was a Scottish doctor and a pioneer of naval hygiene in the Royal Navy. At his time, scurvy was a major cause of death among sailors. He conducted one of the earliest randomized experiments with clear documentation of the details and concluded that citrus fruits cured scurvy before the discovery of Vitamin C.

In Lind (1753), he described the following randomized experiment with 12 patients of scurvy assigned to six groups. With some simplifications, the six groups are:

1. two received a quart of cider every day;
2. two received twenty-five drops of sulfuric acid three times every day;
3. two received two spoonfuls of vinegar three times every day;
4. two received half a pint of seawater every day;
5. two received two oranges and one lemon every day;
6. two received a spicy paste plus a drink of barley water every day.

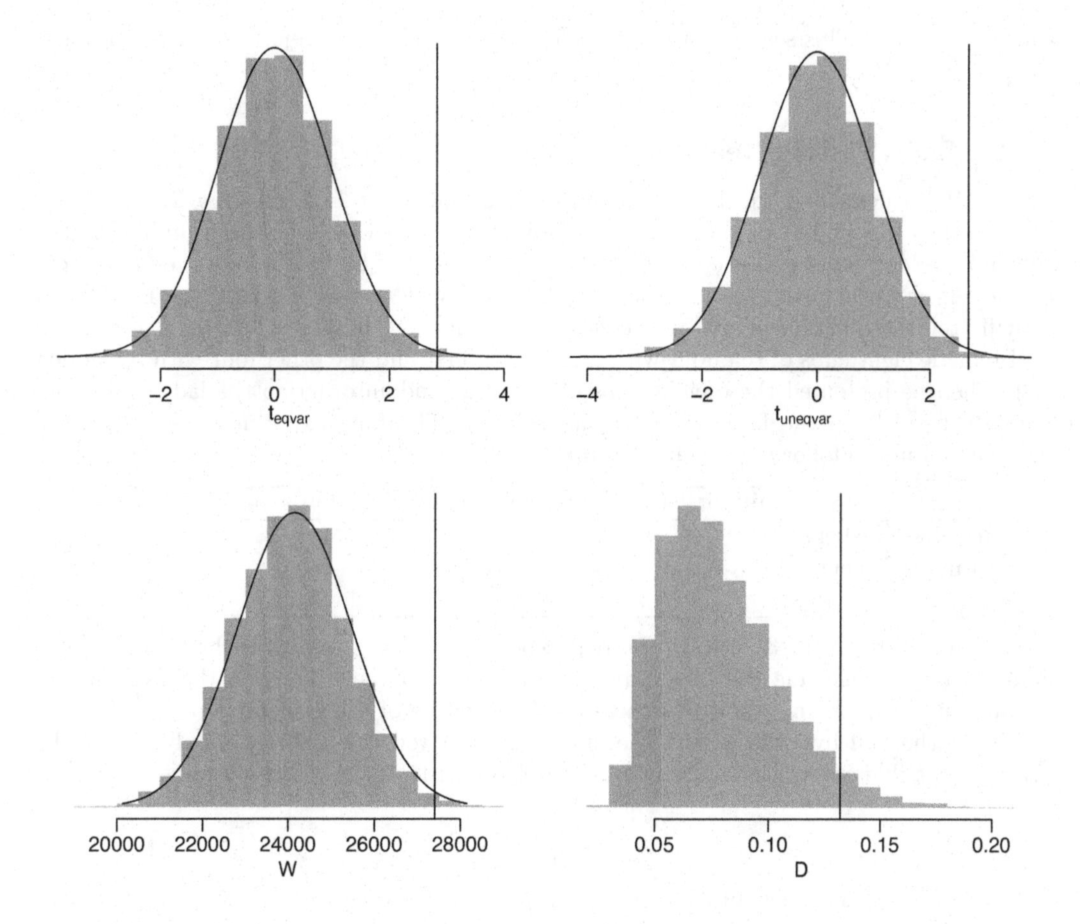

FIGURE 3.3: The randomization distributions of four test statistics based on the LaLonde experimental data

After six days, patients in the fifth group recovered, but patients in other groups did not. If we simplify the treatment as

$$Z_i = 1(\text{unit } i \text{ received citrus fruits})$$

and the outcome as

$$Y_i = 1(\text{unit } i \text{ recovered after six days}),$$

then we have a two-by-two table

	$Y=1$	$Y=0$
$Z=1$	2	0
$Z=0$	0	10

This is the most extreme two-by-two table we can observe under this experiment, and the data contain strong evidence for the positive effect of citrus fruits for curing scurvy. Statistically, how do we measure the strength of the evidence?

Following the logic of the FRT, if the treatment has no effect at all (under $H_{0\text{F}}$), this extreme two-by-two table will occur with probability

$$\frac{1}{\binom{12}{2}} = \frac{1}{66} = 0.015$$

which is the p_{FRT}. This seems a surprise under $H_{0\text{F}}$: we can easily reject $H_{0\text{F}}$ at the level of 0.05.

3.5.2 Lady Tasting Tea

Fisher (1935) described the following famous experiment of *Lady Tasting Tea.*[7] A lady claimed that she could tell the difference between the two ways of making milk tea: one with milk added first, and the other with tea added first. This might sound odd to most people. As a statistician, Fisher designed an experiment to test whether the lady could actually tell the difference between the two ways of making milk tea.

He made eight cups of tea, four with milk added first and the other four with tea added first. Then he presented these eight cups of tea in a random order to the lady and asked the lady to pick up the four with milk added first. The final experiment result can be summarized in the following two-by-two table:

	milk first (lady)	tea first (lady)	column sum
milk first (Fisher)	X	$4-X$	4
tea first (Fisher)	$4-X$	X	4
row sum	4	4	8

The X can be $0,1,2,3$, or 4. In the real experiment, $X=4$, which is the most extreme data, strongly suggesting that the lady could tell the difference between the two ways of making milk tea. Again, how do we measure the strength of the evidence?

Under the null hypothesis that the lady could not tell the difference, only one of the $\binom{8}{4}=70$ possible orders yields the two-by-two table with $X=4$. So the p-value is

$$p_{\text{FRT}}=\frac{1}{70}=0.014.$$

Given the significance level of 0.05, we reject the null hypothesis.

3.5.3 Two Fisherian principles for experiments

In the above two examples in Sections 3.5.1 and 3.5.2, the p_{FRT}'s are justified by the randomization of the experiments. This highlights the first Fisherian principle of experiments: *randomization.*

Moreover, the above two experiments are in some sense the smallest possible experiments that can yield statistically meaningful results. For instance, if Lind only assigned one patient to each of the six groups, then the smallest p-value would be

$$\frac{1}{\binom{6}{1}}=\frac{1}{6}=0.167;$$

if Fisher only made six cups of tea, three with milk added first and the other three with tea added first, then the smallest p-value would be

$$\frac{1}{\binom{6}{3}}=\frac{1}{20}=0.05.$$

We can never reject the null hypotheses at the level of 0.05. This highlights the second Fisherian principle of experiments: *replications.* This is, to ensure the FRT to have power, we must have enough units in experiments.

Chapter 5 will discuss the third Fisherian principle of experiments: *blocking.*

[7] This later became the title of a book on the modern history of statistics by Salsburg (2001).

3.6 Discussion

3.6.1 Other sharp null hypotheses and confidence intervals

I focus on the sharp null hypothesis H_{0F} above. In fact, the logic of the FRT also works for other sharp null hypotheses. For instance, we can test

$$H_0(\boldsymbol{\tau}) : Y_i(1) - Y_i(0) = \tau_i \text{ for all } i = 1, \ldots, n$$

for a known vector $\boldsymbol{\tau} = (\tau_1, \ldots, \tau_n)$. Because the individual causal effects are all known under $H_0(\boldsymbol{\tau})$, we can impute all missing potential outcomes based on the observed data. With known potential outcomes, the distribution of any test statistic $T = T(\boldsymbol{Z}, \boldsymbol{Y}(1), \boldsymbol{Y}(0))$ is completely determined by the treatment assignment mechanism, and therefore, we can compute the corresponding p_{FRT} as a function of $\boldsymbol{\tau}$, denoted by $p_{\text{FRT}}(\boldsymbol{\tau})$. If we can specify all possible $\boldsymbol{\tau}$'s, then we can compute a series of $p_{\text{FRT}}(\boldsymbol{\tau})$'s. By duality of hypothesis testing and confidence set (see Section A.2.5), we can obtain a $(1 - \alpha)$-level confidence set for the average causal effect:

$$\left\{ \tau = n^{-1} \sum_{i=1}^{n} \tau_i : p_{\text{FRT}}(\boldsymbol{\tau}) \geq \alpha \right\}.$$

Although this strategy is conceptually straightforward, it has practical complexities due to the large number of all possible $\boldsymbol{\tau}$'s. In the special case of a binary outcome, Rigdon and Hudgens (2015) and Li and Ding (2016) proposed some computationally feasible approaches to constructing confidence intervals for τ based on the FRT. For general unbounded outcomes, this strategy is often computationally infeasible.

A canonical simplification (Rosenbaum, 2002b, 2010) is to consider a subclass of the sharp null hypotheses with constant individual causal effects:

$$H_0(c) : Y_i(1) - Y_i(0) = c \text{ for all } i = 1, \ldots, n$$

for a known constant c. Given c, we can compute $p_{\text{FRT}}(c)$. By duality, we can obtain a $(1 - \alpha)$-level confidence set for the average causal effect:

$$\{c : p_{\text{FRT}}(c) \geq \alpha\}.$$

Because this procedure only involves a one-dimensional search, it is computationally feasible. However, the constant individual causal effect assumption is too strong. In particular, it does not hold for binary outcomes unless all units have effects 0, -1, or 1. In general, the constant individual causal effect assumption has testable implications that can be rejected by the observed data; Ding et al. (2016) proposed a formal statistical test.

3.6.2 Other test statistics

The FRT is a general strategy because it is applicable in any randomized experiment with any test statistic. I have given several examples of test statistics in Section 3.3. In fact, the definition of a test statistic can be much more general. For instance, with pretreatment covariate matrix $\boldsymbol{X}$ with the ith row being X_i for unit i $(i = 1, \ldots, n)$,[8] we can allow the

[8] In causal inference, we call X_i a covariate if it is not affected by the treatment. That is, if the covariate has two potential outcomes $X_i(1)$ and $X_i(0)$, then they must satisfy $X_i(1) = X_i(0)$. Standard statistics books often do not distinguish the treatment and covariates because they often appear on the right-hand side of a regression model for the outcome. They are both called covariates in those statistical models. This book distinguishes the treatment and covariates because they play different roles in causal inference.

test statistic $T(\boldsymbol{Z}, \boldsymbol{Y}, \boldsymbol{X})$ to be a function of the treatment vector, outcome vector, and the covariate matrix. Problem 3.6 gives an example.

3.6.3 Final remarks

For a general experiment, the probability distribution of $\boldsymbol{Z}$ is not uniform over all possible permutations of n_1 1's and n_0 0's. However, its distribution is completely known by the experimenter. Therefore, we can always simulate its distribution which in turn implies the distribution of any test statistic under the sharp null hypothesis. A finite-sample exact p-value equals

$$p_{\text{FRT}} = \text{pr}'\{T(\boldsymbol{Z}', \boldsymbol{Y}) \geq T(\boldsymbol{Z}, \boldsymbol{Y})\}$$

where pr$'$ is average over the distribution of $\boldsymbol{Z}'$ conditional on data. I will discuss other experiments in the subsequent chapters and I want to emphasize that the FRT works beyond the specific experiments discussed in this book.

The FRT works with any test statistic. However, this does not answer the practical question of how to choose a test statistic in the data analysis. If the goal is to find surprises with respect to the sharp null hypothesis, it is desirable to choose a test statistic that yields high power under alternative hypotheses. In general, no test statistic can dominate others in terms of power because power depends on the alternative hypothesis. The four test statistics in Section 3.3 are motivated by different alternative hypotheses. For instance, $\hat{\tau}$ and t are motivated by an alternative hypothesis with a nonzero average treatment effect; W is motivated by an alternative hypothesis with a constant causal effect with heavy-tailed outcomes. Specifying a working alternative hypothesis is often helpful for constructing a test statistic although it does not have to be precise to guarantee the validity of the FRT. Problems 3.6 and 3.7 illustrate the idea of using a working alternative hypothesis or statistical model to construct test statistics.

3.7 Homework problems

3.1 Exactness of p_{FRT}

Prove (3.3).

3.2 Monte Carlo error of $\hat{p}_{\text{FRT}}$

Given data, p_{FRT} is a fixed number while its Monte Carlo estimator $\hat{p}_{\text{FRT}}$ in (3.4) is random. Show that

$$E_{\text{mc}}(\hat{p}_{\text{FRT}}) = p_{\text{FRT}}$$

and

$$\text{var}_{\text{mc}}(\hat{p}_{\text{FRT}}) \leq \frac{1}{4R},$$

where the subscript "mc" signifies the randomness due to Monte Carlo, that is, $\hat{p}_{\text{FRT}}$ is random because $\boldsymbol{z}^r$'s are R independent random draws from all possible values of $\boldsymbol{Z}$.

Remark: p_{FRT} is random because $\boldsymbol{Z}$ is random. But in this problem, we condition on data, so p_{FRT} becomes a fixed number. $\hat{p}_{\text{FRT}}$ is random because the $\boldsymbol{z}^r$'s are random permutations of $\boldsymbol{Z}$. Problem 3.2 shows that $\hat{p}_{\text{FRT}}$ is unbiased for p_{FRT} over the Monte Carlo randomness and gives an upper bound on the variance of $\hat{p}_{\text{FRT}}$. Luo et al. (2021, Theorem 2) gives a more delicate bound on the Monte Carlo error.

3.3 *A finite-sample valid Monte Carlo approximation of* p_{FRT}

Although $\hat{p}_{\text{FRT}}$ is unbiased for p_{FRT} by the result in Problem 3.2, it may not be a valid p-value in the sense that $\text{pr}(\hat{p}_{\text{FRT}} \le u) \le u$ for all $u \in (0,1)$ due to Monte Carlo error with a finite R. The following modified Monte Carlo approximation is always a finite-sample valid p-value. Phipson and Smyth (2010) pointed out this trick in the permutation test.

Define

$$\tilde{p}_{\text{FRT}} = \frac{1 + \sum_{r=1}^{R} I\{T(\boldsymbol{z}^r, \boldsymbol{Y}) \ge T(\boldsymbol{Z}, \boldsymbol{Y})\}}{1+R}$$

where the $\boldsymbol{z}^r$'s are the R random permutations of $\boldsymbol{Z}$. Show that with an arbitrary R, the Monte Carlo approximation $\tilde{p}_{\text{FRT}}$ is always a finite-sample valid p-value in the sense that $\text{pr}(\tilde{p}_{\text{FRT}} \le u) \le u$ for all $u \in (0,1)$.

Remark: You can use the following two basic probability results to prove the claim in Problem 3.3.

Lemma 3.1 *For two Binomial random variables* $X_1 \sim$ *Binomial*(R, p_1) *and* $X_2 \sim$ *Binomial*(R, p_2) *with* $p_1 \ge p_2$*, we have* $\text{pr}(X_1 \le x) \le \text{pr}(X_2 \le x)$ *for all* x*.*

Lemma 3.2 *If* $p \sim$ *Uniform*$(0,1)$ *and* $X \mid p \sim$ *Binomial*(R, p)*, then, marginally,* X *is a uniform random variable over* $\{0, 1, \ldots, R\}$*.*

3.4 *Fisher's exact test*

Consider a CRE with a binary outcome, with data summarized in the following two-by-two table:

	$Y=1$	$Y=0$	total
$Z=1$	n_{11}	n_{10}	n_1
$Z=0$	n_{01}	n_{00}	n_0

Under $H_{0\text{F}}$, show that any test statistic $T(n_{11}, n_{10}, n_{01}, n_{00})$ is a function of n_{11} and other non-random fixed constants, and the exact distribution of n_{11} is Hypergeometric. Specify the parameters for the Hypergeometric distribution.

Remark: Barnard (1947) and Ding and Dasgupta (2016) pointed out the equivalence of Fisher's exact test (reviewed in Section A.3.1) and the FRT under a CRE with a binary outcome.

3.5 *More details for Lady Tasting Tea*

Recall the example in Section 3.5.2. Calculate $\text{pr}(X = k)$ for $k = 0, 1, 2, 3, 4$.

3.6 *Covariate-adjusted FRT*

This problem gives more details for Section 3.6.2.

Section 3.4 reanalyzed the LaLonde experimental data using the FRT with four test statistics. With additional covariates, the FRT can be more general with at least the following two additional strategies. Assume all potential outcomes and covariates are fixed numbers.

First, we can use test statistics based on residuals from the linear regression. Run a linear regression of the outcomes on the covariates, and obtain the residuals (i.e., view the residuals as the "pseudo outcomes"). Then define the four test statistics based on the residuals. Conduct the FRT using these four new test statistics. Report the corresponding p-values.

Second, we can define the test statistic as the coefficient in the linear regression of the outcomes on the treatment and covariates. Conduct the FRT using this test statistic. Report the corresponding p-value.

Why are the five p-values from the above two strategies finite-sample exact? Justify them.

3.7 FRT with a generalized linear model

Use the same dataset as Problem 3.6 but change the outcome to a binary indicator for whether `re78` is positive or not. Run logistic regression of the outcome on the treatment and covariates. Is the coefficient of the treatment significant and what is the p-value? Calculate the p-value from the FRT using the coefficient of the treatment as the test statistic.

3.8 An algebraic detail

Verify (3.7).

3.9 Recommended reading

Bind and Rubin (2020) is a recent paper advocating the use of p-values as well as the display of the corresponding randomization distributions in analyzing complex experiments.

4

Neymanian Repeated Sampling Inference in Completely Randomized Experiments

In his seminal paper, Neyman (1923) not only proposed the notation of potential outcomes but also derived rigorous mathematical results for making inference for the average causal effect under a CRE. In contrast to Fisher's idea of calculating the p-value under the sharp null hypothesis, Neyman (1923) proposed an unbiased point estimator and a conservative confidence interval based on the sampling distribution of the point estimator. This chapter will introduce Neyman's (1923) fundamental results, which are very important for understanding later chapters in Part II of this book.

4.1 Finite population quantities

Consider a CRE with n units, where n_1 of them receive the treatment and n_0 of them receive the control. For unit $i = 1, \dots, n$, we have potential outcomes $Y_i(1)$ and $Y_i(0)$, and individual effect $\tau_i = Y_i(1) - Y_i(0)$. The potential outcomes have finite population means

$$\bar{Y}(1) = n^{-1}\sum_{i=1}^{n} Y_i(1),$$
$$\bar{Y}(0) = n^{-1}\sum_{i=1}^{n} Y_i(0),$$

variances[1]

$$S^2(1) = (n-1)^{-1}\sum_{i=1}^{n}\{Y_i(1) - \bar{Y}(1)\}^2,$$
$$S^2(0) = (n-1)^{-1}\sum_{i=1}^{n}\{Y_i(0) - \bar{Y}(0)\}^2,$$

and covariance

$$S(1,0) = (n-1)^{-1}\sum_{i=1}^{n}\{Y_i(1) - \bar{Y}(1)\}\{Y_i(0) - \bar{Y}(0)\}.$$

The individual effects have mean

$$\tau = n^{-1}\sum_{i=1}^{n}\tau_i = \bar{Y}(1) - \bar{Y}(0)$$

[1]Here the divisor $n-1$ makes the main theorems in this chapter more elegant. Changing the divisor to n complicates the formulas but does not change the results fundamentally. With large n, the difference is minor.

and variance

$$S^2(\tau) = (n-1)^{-1} \sum_{i=1}^{n} (\tau_i - \tau)^2.$$

We have the following relationship between the variances and covariance.

Lemma 4.1 $2S(1,0) = S^2(1) + S^2(0) - S^2(\tau)$.

Lemma 4.1 is a basic result and will be useful for later discussion. The proof of Lemma 4.1 follows from elementary algebra. I leave it as Problem 4.1.

These fixed quantities are functions of the Science Table $\{Y_i(1), Y_i(0)\}_{i=1}^n$. We are interested in estimating the average causal effect τ based on the data $(Z_i, Y_i)_{i=1}^n$ from a CRE.

4.2 Neyman's (1923) theorem

Based on the observed outcomes, we can calculate the sample means

$$\begin{aligned} \hat{\bar{Y}}(1) &= n_1^{-1} \sum_{i=1}^{n} Z_i Y_i, \\ \hat{\bar{Y}}(0) &= n_0^{-1} \sum_{i=1}^{n} (1 - Z_i) Y_i, \end{aligned}$$

and the sample variances

$$\begin{aligned} \hat{S}^2(1) &= (n_1 - 1)^{-1} \sum_{i=1}^{n} Z_i \{Y_i - \hat{\bar{Y}}(1)\}^2, \\ \hat{S}^2(0) &= (n_0 - 1)^{-1} \sum_{i=1}^{n} (1 - Z_i) \{Y_i - \hat{\bar{Y}}(0)\}^2. \end{aligned}$$

But there are no sample versions of $S(1,0)$ and $S^2(\tau)$ because the potential outcomes $Y_i(1)$ and $Y_i(0)$ are never jointly observed for each unit i. Neyman (1923) proved the following theorem.

Theorem 4.1 *Under a CRE,*

1. *the difference-in-means estimator $\hat{\tau} = \hat{\bar{Y}}(1) - \hat{\bar{Y}}(0)$ is unbiased for τ:*

$$E(\hat{\tau}) = \tau;$$

2. *$\hat{\tau}$ has variance*

$$\begin{aligned} \operatorname{var}(\hat{\tau}) &= \frac{S^2(1)}{n_1} + \frac{S^2(0)}{n_0} - \frac{S^2(\tau)}{n} && (4.1) \\ &= \frac{n_0}{n_1 n} S^2(1) + \frac{n_1}{n_0 n} S^2(0) + \frac{2}{n} S(1,0); && (4.2) \end{aligned}$$

3. *the variance estimator*

$$\hat{V} = \frac{\hat{S}^2(1)}{n_1} + \frac{\hat{S}^2(0)}{n_0}$$

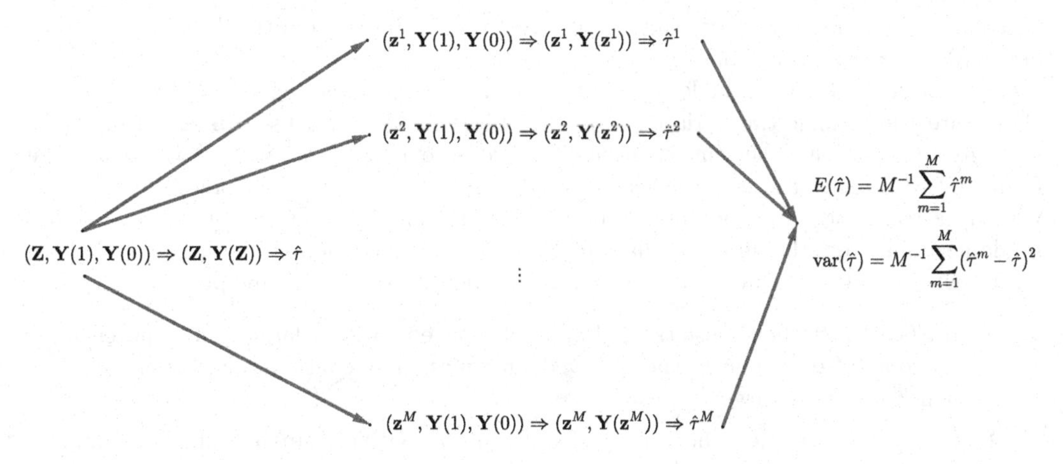

FIGURE 4.1: Illustration of Neyman's (1923) theorem, where $\boldsymbol{Y}(\boldsymbol{z}^m)$ is the observed outcome vector under the treatment vector $\boldsymbol{z}^m$.

is conservative for estimating $\text{var}(\hat\tau)$ *in the sense that*

$$E(\hat V) - \text{var}(\hat\tau) = \frac{S^2(\tau)}{n} \geq 0$$

with equality holding if and only if $\tau_i = \tau$ *for all units.*

I will present the proof of Theorem 4.1 in Section 4.3. Before proving Theorem 4.1, it is important to clarify the meanings of $E(\cdot)$ and $\text{var}(\cdot)$ in Theorem 4.1. The potential outcomes are all fixed numbers, and only the treatment indicators Z_i's are random. Therefore, the expectations and variances are all over the randomness of the Z_i's, which are random permutations of n_1 1's and n_0 0's. Figure 4.1 illustrates the randomness of $\hat\tau$, which is a discrete uniform distribution over $\{\hat\tau^1, \ldots, \hat\tau^M\}$ induced by $M = \binom{n}{n_1}$ possible treatment allocations. Compare Figure 4.1 with Figure 3.1 to see the key differences between the FRT and Neyman's (1923) theorem:

1. The FRT works for any test statistic but Neyman's (1923) theorem is only about the difference in means. Although we could derive the properties of other estimators similar to Neyman's (1923) theorem, this mathematical exercise is often quite challenging for general estimators.
2. In Figure 3.1, the observed outcome vector $\boldsymbol{Y}$ is fixed but in Figure 4.1, the observed outcome vector $\boldsymbol{Y}(\boldsymbol{z}^m)$ changes as $\boldsymbol{z}^m$ changes.
3. The $T(\boldsymbol{z}^m, \boldsymbol{Y})$'s In Figure 3.1 are all computable based on the observed data, but the $\hat\tau^m$'s in Figure 4.1 are hypothetical values because not all potential outcomes are known.

The point estimator $\hat\tau$ is standard but it has a non-trivial variance under the potential outcomes framework with a CRE. The variance formula (4.1) differs from the classic variance formula for the difference in means[2] because it not only depends on the finite population variances of the potential outcomes but also depends on the finite population variance of the individual effects, or, equivalently, the finite population covariance of the potential

[2] In the classic two-sample problem, the outcomes under the treatment are IID draws from a distribution with mean μ_1 and variance σ_1^2, and the outcomes under the control are IID draws from a distribution with

outcomes. Unfortunately, $S^2(\tau)$ and $S(1,0)$ are not identifiable from the data because $Y_i(1)$ and $Y_i(0)$ are never jointly observed.

The formula (4.1) is a little puzzling in that the more heterogeneous the individual effects are the smaller the variability of $\hat{\tau}$ is. Section 4.5.1 will use numerical examples to verify (4.1). What is the intuition here? I give an explanation based on the equivalent form (4.2). Compare the case with positively correlated potential outcomes and the case with negatively correlated potential outcomes. Although the treatment group is a simple random sample from the finite population of n units, it is possible to observe relatively large treatment potential outcomes in a realized experiment. Assume this happens.

1. If $S(1,0) > 0$, then those treated units also have relatively large control potential outcomes. Consequently, the observed outcomes under control are relatively small, which results in large $\hat{\tau}$.
2. If $S(1,0) < 0$, then those treated units have relatively small control potential outcomes. Consequently, the observed outcomes under control are relatively large, which results in small $\hat{\tau}$.

The reverse can also happen if we observe relatively small treatment potential outcomes in a realized experiment. Overall, although the unbiasedness of $\hat{\tau}$ does not depend on the correlation between the potential outcomes, it is more likely to observe more extreme $\hat{\tau}$ under $S(1,0) > 0$ than under $S(1,0) < 0$. So the variance of $\hat{\tau}$ is larger when the potential outcomes are positively correlated.

Li and Ding (2017, Theorem 5 and Proposition 3) further proved the following asymptotic Normality of $\hat{\tau}$ based on the finite population CLT.

Theorem 4.2 *Let $n \rightarrow \infty$ and $n_1 \rightarrow \infty$. If n_1/n has a limiting value in $(0,1)$, $\{S^2(1), S^2(0), S(1,0)\}$ have limiting values, and*

$$\begin{aligned} \max_{1\leq i\leq n} \{Y_i(1) - \bar{Y}(1)\}^2/n &\rightarrow 0, \\ \max_{1\leq i\leq n} \{Y_i(0) - \bar{Y}(0)\}^2/n &\rightarrow 0, \end{aligned}$$

then

$$\frac{\hat{\tau} - \tau}{\sqrt{\text{var}(\hat{\tau})}} \rightarrow \text{N}(0,1)$$

in distribution, and

$$\begin{aligned} \hat{S}^2(1) &\rightarrow S^2(1), \\ \hat{S}^2(0) &\rightarrow S^2(0) \end{aligned}$$

in probability.

The proof of Theorem 4.2 is technical and beyond the scope of this book. It ensures that the sampling distribution of $\hat{\tau}$ can be approximated by a Normal distribution with a large sample size and some regularity conditions. Moreover, it ensures that the sample variances of the outcomes are consistent for the population variances, which further ensures that the

mean μ_0 and variance σ_0^2. Under this assumption, we have

$$\text{var}(\hat{\tau}) = \frac{\sigma_1^2}{n_1} + \frac{\sigma_0^2}{n_0}.$$

Here, var$(\cdot)$ is over the randomness of the outcomes. This variance formula does not involve a third term that depends on the variance of the individual causal effects.

probability limit of Neyman (1923)'s variance estimator is larger than the true variance of $\hat{\tau}$. This justifies the Wald-type large-sample confidence interval (see Section A.2.4 for the terminology) for τ:

$$\hat{\tau} \pm z_{1-\alpha/2}\sqrt{\hat{V}},$$

where $z_{1-\alpha/2}$ is the $1-\alpha/2$ upper quantile of the standard Normal random variable. It is the same as the confidence interval for the standard two-sample problem asymptotically (see Section A.4.1). This confidence interval covers τ with probability at least as large as $1-\alpha$ when the sample size is large enough. By the duality between the confidence interval and hypothesis testing (see Section A.2.5 for a review), the confidence interval implies a test for

$$H_{0\mathrm{N}} : \tau = 0,$$

which is called the weak null hypothesis.

Due to the fundamental problem of missing one potential outcome, we can at most obtain a conservative variance estimator. In statistics, the definition of the confidence interval allows for over coverage and thus conservativeness in variance estimation (see Section A.2). Conservativeness is not a big problem if underreporting the treatment effect is not a big problem in practice. Sometimes, it can be problematic if the outcomes measure the side effects of a treatment. In medical experiments, underreporting the side effects of a new drug can have severe consequences on patients' health.

4.3 Proofs

In this section, I will prove Theorem 4.1.

4.3.1 Unbiasedness

First, we have the representation

$$\begin{aligned}
\hat{\tau} &= n_1^{-1}\sum_{i=1}^{n} Z_i Y_i - n_0^{-1}\sum_{i=1}^{n}(1-Z_i)Y_i \\
&= n_1^{-1}\sum_{i=1}^{n} Z_i Y_i(1) - n_0^{-1}\sum_{i=1}^{n}(1-Z_i)Y_i(0).
\end{aligned}$$

Then the unbiasedness of $\hat{\tau}$ follows from the linearity of the expectation:

$$\begin{aligned}
E(\hat{\tau}) &= E\left\{ n_1^{-1}\sum_{i=1}^{n} Z_i Y_i(1) - n_0^{-1}\sum_{i=1}^{n}(1-Z_i)Y_i(0) \right\} \\
&= n_1^{-1}\sum_{i=1}^{n} E(Z_i)Y_i(1) - n_0^{-1}\sum_{i=1}^{n} E(1-Z_i)Y_i(0) \\
&= n_1^{-1}\sum_{i=1}^{n} \frac{n_1}{n} Y_i(1) - n_0^{-1}\sum_{i=1}^{n} \frac{n_0}{n} Y_i(0) \\
&= n^{-1}\sum_{i=1}^{n} Y_i(1) - n^{-1}\sum_{i=1}^{n} Y_i(0) \\
&= \tau.
\end{aligned}$$

4.3.2 Variance formulas

First, we can write $\hat{\tau}$ as

$$\hat{\tau} = n_1 \times n_1^{-1} \sum_{i=1}^{n} Z_i c_i - n_0^{-1} \sum_{i=1}^{n} Y_i(0),$$

with $c_i = \frac{Y_i(1)}{n_1} + \frac{Y_i(0)}{n_0}$. Apply Lemma C.2 for simple random sampling to obtain

$$\begin{aligned} \text{var}(\hat{\tau}) &= n_1^2 \times \text{var}\left(n_1^{-1} \sum_{i=1}^{n} Z_i c_i \right) \\ &= n_1^2 \times \frac{n_0}{n n_1} S_c^2, \end{aligned}$$

where $S_c^2 = (n-1)^{-1} \sum_{i=1}^{n} (c_i - \bar{c})^2$ with $\bar{c} = n^{-1} \sum_{i=1}^{n} c_i$. Then the variance of $\hat{\tau}$ follows from

$$\begin{aligned} \text{var}(\hat{\tau}) &= \frac{n_1 n_0}{n(n-1)} \sum_{i=1}^{n} \left\{ \frac{Y_i(1)}{n_1} + \frac{Y_i(0)}{n_0} - \frac{\bar{Y}(1)}{n_1} - \frac{\bar{Y}(0)}{n_0} \right\}^2 \\ &= \frac{n_1 n_0}{n(n-1)} \left[\frac{1}{n_1^2} \sum_{i=1}^{n} \{Y_i(1) - \bar{Y}(1)\}^2 + \frac{1}{n_0^2} \sum_{i=1}^{n} \{Y_i(0) - \bar{Y}(0)\}^2 \right. \\ &\quad \left. + \frac{2}{n_1 n_0} \sum_{i=1}^{n} \{Y_i(1) - \bar{Y}(1)\}\{Y_i(0) - \bar{Y}(0)\} \right] \\ &= \frac{n_0}{n_1 n} S^2(1) + \frac{n_1}{n_0 n} S^2(0) + \frac{2}{n} S(1,0). \end{aligned}$$

From Lemma 4.1, we can also write the variance as

$$\begin{aligned} \text{var}(\hat{\tau}) &= \frac{n_0}{n_1 n} S^2(1) + \frac{n_1}{n_0 n} S^2(0) + \frac{1}{n} \{S^2(1) + S^2(0) - S^2(\tau)\} \\ &= \frac{S^2(1)}{n_1} + \frac{S^2(0)}{n_0} - \frac{S^2(\tau)}{n}. \end{aligned}$$

4.3.3 Unbiased variance estimation

Because the treatment group is a simple random sample of size n_1 from the n units, Lemma C.3 ensures that the sample variance of $Y_i(1)$'s is unbiased for its population variance:

$$E\{\hat{S}^2(1)\} = S^2(1).$$

Similarly, $E\{\hat{S}^2(0)\} = S^2(0)$. Therefore, $\hat{V}$ is unbiased for the first two terms in (4.1).

4.4 Regression analysis of the CRE

Practitioners often use regression-based inference for the average causal effect τ. A standard approach is to run the ordinary least squares (OLS) of the outcomes on the treatment indicators with an intercept

$$(\hat{\alpha}, \hat{\beta}) = \arg\min_{(a,b)} \sum_{i=1}^{n} (Y_i - a - bZ_i)^2,$$

and use the coefficient of the treatment $\hat{\beta}$ as the estimator for the average causal effect. We can show the coefficient $\hat{\beta}$ equals the difference in means:

$$\hat{\beta} = \hat{\tau}. \tag{4.3}$$

However, the usual variance estimator from the OLS (see (B.4) in Appendix B), e.g., the output from the `lm` function of `R`, equals

$$\begin{aligned} \hat{V}_{\text{OLS}} &= \frac{n(n_1-1)}{(n-2)n_1 n_0}\hat{S}^2(1) + \frac{n(n_0-1)}{(n-2)n_1 n_0}\hat{S}^2(0) \\ &\approx \frac{\hat{S}^2(1)}{n_0} + \frac{\hat{S}^2(0)}{n_1}, \end{aligned} \tag{4.4}$$

where the approximation holds with large n_1 and n_0. It differs from $\hat{V}$ even with large n_1 and n_0.

Fortunately, the Eicker–Huber–White (EHW) robust variance estimator (see (B.3) in Appendix B) is close to $\hat{V}$:

$$\begin{aligned} \hat{V}_{\text{EHW}} &= \frac{\hat{S}^2(1)}{n_1}\frac{n_1-1}{n_1} + \frac{\hat{S}^2(0)}{n_0}\frac{n_0-1}{n_0} \\ &\approx \frac{\hat{S}^2(1)}{n_1} + \frac{\hat{S}^2(0)}{n_0}, \end{aligned} \tag{4.5}$$

where the approximation holds with large n_1 and n_0. It is almost identical to $\hat{V}$. Moreover, the so-called HC2 variant of the EHW robust variance estimator is identical to $\hat{V}$. The `hccm` function in the `car` package returns the EHW robust variance estimator as well as its HC2 variant.

Problem 4.3 provides more technical details for (4.3)–(4.5).

4.5 Examples

4.5.1 Simulation

I first choose the sample size as $n = 100$ with 60 treated and 40 control units, and generate the potential outcomes with constant individual causal effects.

```
n  = 100
n1 = 60
n0 = 40
y0 = rexp(n)
y0 = sort(y0, decreasing = TRUE)
y1 = y0 + 1
```

With the Science Table fixed, I repeatedly generate CREs and apply Theorem 4.1 to obtain the point estimator, the conservative variance estimator, and the confidence interval based on the Normal approximation. The $(1,1)$th panel of Figure 4.2 shows the scatter plot of the potential outcomes, and the $(1,2)$th panel of Figure 4.2 shows the histogram of $\hat{\tau} - \tau$.

I then change the potential outcomes by sorting the control potential outcomes in reverse order

```
y0 = sort(y0, decreasing = FALSE)
```

and repeat the above simulation. The $(2,1)$th panel of Figure 4.2 shows the scatter plot of the potential outcomes, and the $(2,2)$th panel of Figure 4.2 shows the histogram of $\hat{\tau} - \tau$.

I finally permute the control potential outcomes randomly

```
y0 = sample(y0)
```

and repeat the above simulation. The $(3,1)$th panel of Figure 4.2 shows the scatter plot of the potential outcomes, and the $(2,2)$th panel of Figure 4.2 shows the histogram of $\hat{\tau} - \tau$.

Importantly, in the above three sets of simulations, the correlations between potential outcomes are different but the marginal distributions are the same. The following table compares the true variances, the average estimated variances, and the coverage rates of the 95% confidence intervals.

```
               constant effect negatively correlated uncorrelated
var                      0.059                 0.015        0.025
estimated var            0.059                 0.059        0.059
coverege rate            0.946                 1.000        0.995
```

The true variance depends on the correlation between the potential outcomes, with positively correlated potential outcomes corresponding to a larger sampling variance. This verifies (4.2). The estimated variances are almost identical because the formula of $\hat{V}$ depends only on the marginal distributions of the potential outcomes. Due to the discrepancy between the true and estimated variances, the coverage rates differ across the three sets of simulations. Only with constant causal effects, the estimated variance is identical to the true variance, verifying point 3 of Theorem 4.1.

Figure 4.2 also shows the Normal density curves based on the CLT for $\hat{\tau}$. They are very close to the histogram over simulations, verifying Theorem 4.2.

4.5.2 Heavy-tailed outcome and failure of Normal approximations

The CLT of $\hat{\tau}$ in Theorem 4.2 holds under some regularity conditions. Those conditions will be violated with heavy-tailed potential outcomes. We can modify the above simulation studies to illustrate this point. Assume the individual causal effects are constant but the control potential outcomes are contaminated by a Cauchy component with probability $0.1, 0.3$, or 0.5. The following code generates the potential outcomes with the probability of contamination being 0.1.

```
eps = rbinom(n, 1, 0.1)
y0 = (1 - eps)*rexp(n) + eps*rcauchy(n)
y1 = y0 + 1
```

Figures 4.3 and 4.4 show two realizations of the histograms of $\hat{\tau} - \tau$ with the corresponding Normal approximations. With heavy-tailed potential outcomes, the Normal approximations are quite poor. Moreover, unlike Figure 4.2, the histograms are quite sensitive to the random seed of the simulation.

4.5.3 Application

I analyzed the `lalonde` data in Section 3.4 to illustrate the idea of the FRT. Now I use the data to illustrate the theory in this chapter.

```
> library(Matching)
> data(lalonde)
> z = lalonde$treat
> y = lalonde$re78
```

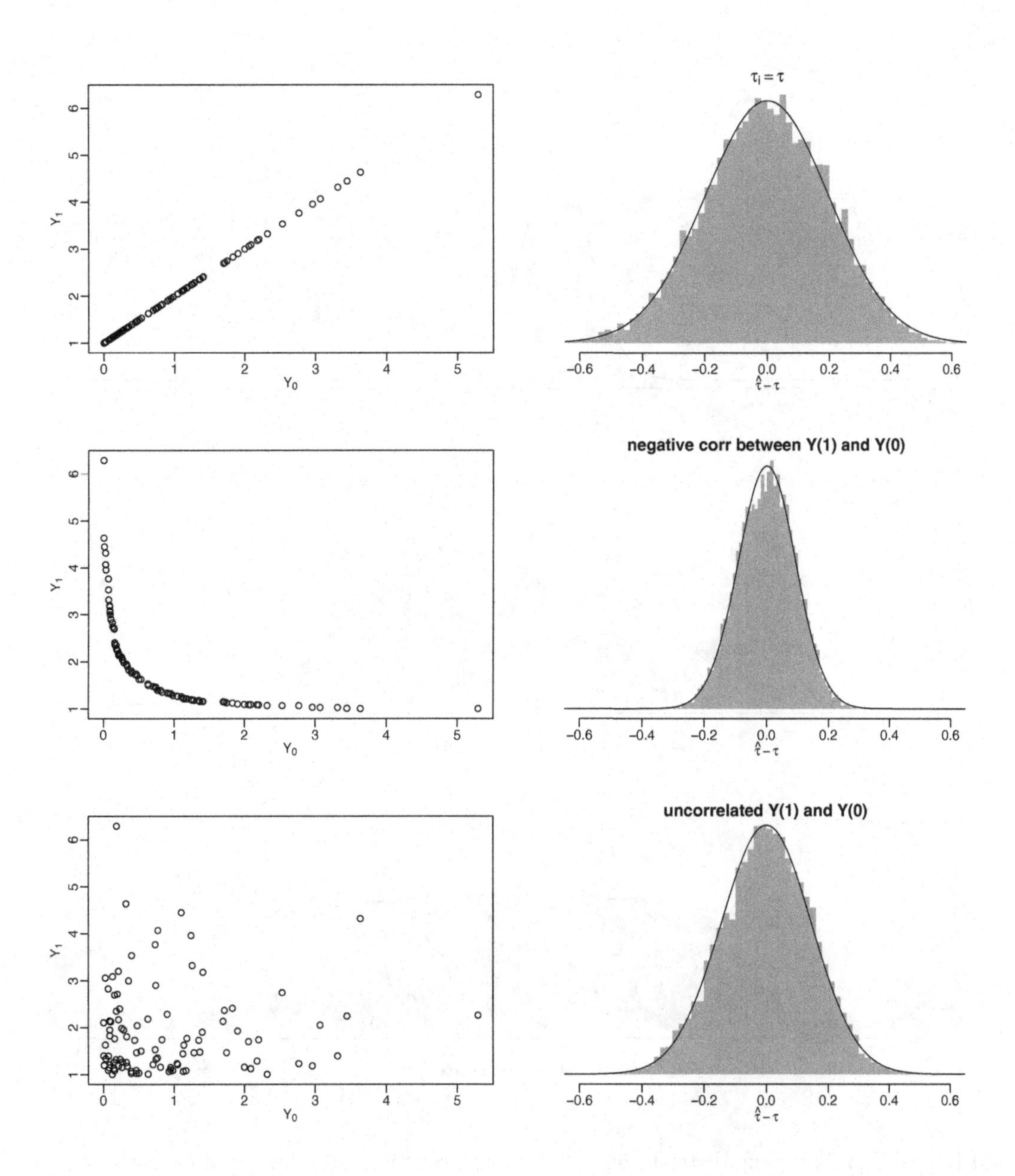

FIGURE 4.2: Left: joint distributions of the potential outcomes with the same marginal distributions. Right: sampling distribution of $\hat{\tau} - \tau$ over 10^4 simulations with different joint distributions of the potential outcomes.

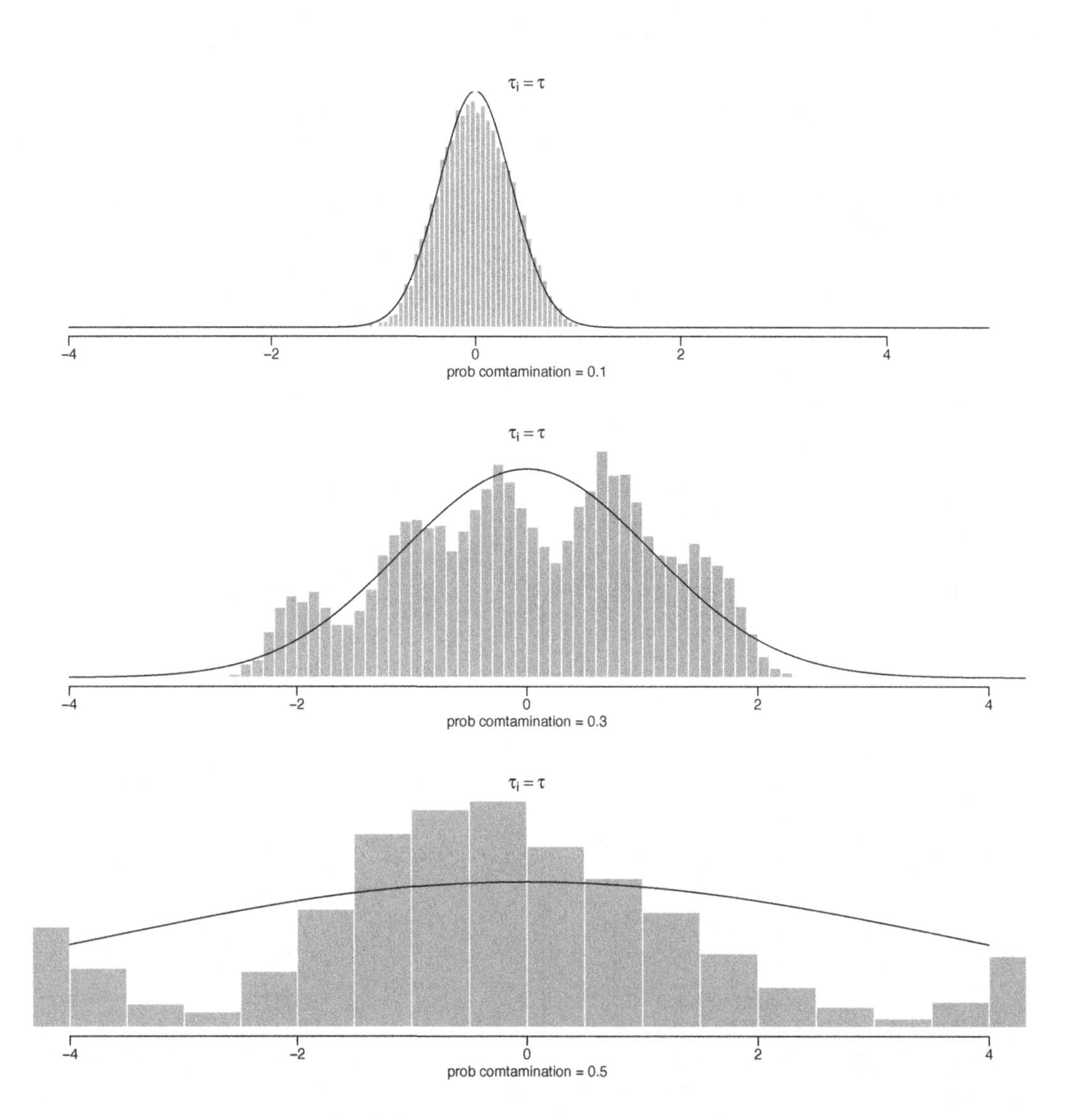

FIGURE 4.3: Sampling distribution of $\hat{\tau} - \tau$ with contaminated potential outcomes (with different contamination probabilities): realization one

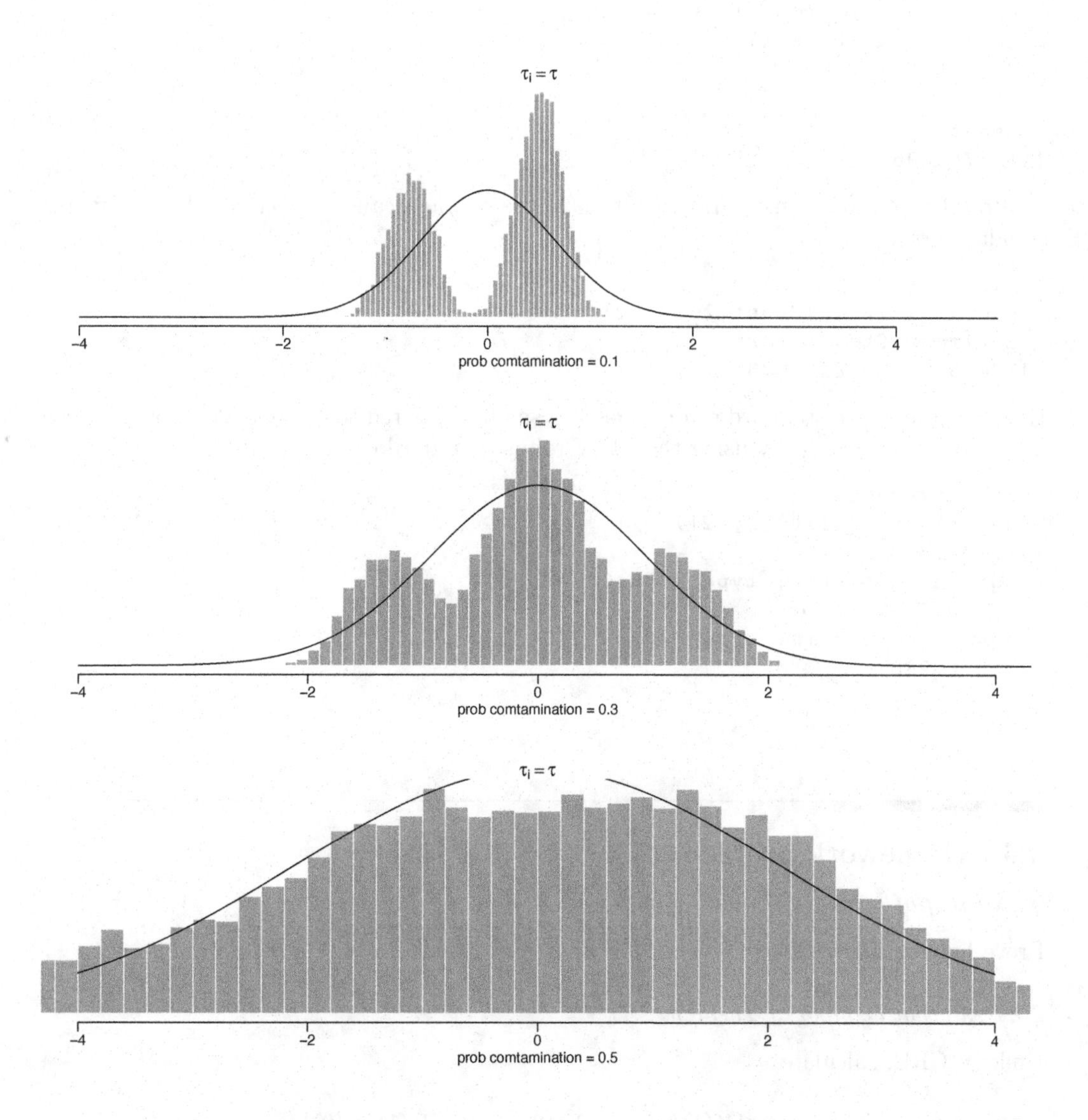

FIGURE 4.4: Sampling distribution of $\hat{\tau} - \tau$ with contaminated potential outcomes (with different contamination probabilities): realization two

We can easily calculate the point estimator and standard error based on the formulas in Theorem 4.1:

```
> n1= sum(z)
> n0= length(z) - n1
> tauhat = mean(y[z==1]) - mean(y[z==0])
> vhat   = var(y[z==1])/n1 + var(y[z==0])/n0
> sehat  = sqrt(vhat)
> tauhat
[1] 1794.343
> sehat
[1] 670.9967
```

Practitioners often use OLS to estimate the average causal effect which also gives a standard error.

```
> olsfit = lm(y ~ z)
> summary(olsfit)$coef[2, 1: 2]
  Estimate Std. Error
 1794.3431   632.8536
```

However, the above standard error seems too small compared to the one based on Theorem 4.1. This can be solved by using the EHW robust standard error.

```
> library(car)
> sqrt(hccm(olsfit)[2, 2])
[1] 672.6823
> sqrt(hccm(olsfit, type = "hc0")[2, 2])
[1] 669.3155
> sqrt(hccm(olsfit, type = "hc2")[2, 2])
[1] 670.9967
```

4.6 Homework problems

4.1 Proof of Lemma 4.1

Prove Lemma 4.1.

4.2 Alternative proof of Theorem 4.1

Under a CRE, calculate

$$\operatorname{var}\{\hat{\bar{Y}}(1)\}, \quad \operatorname{var}\{\hat{\bar{Y}}(0)\}, \quad \operatorname{cov}\{\hat{\bar{Y}}(1), \hat{\bar{Y}}(0)\}$$

and use these formulas to calculate $\operatorname{var}(\hat{\tau})$.

Remark: Use the results in Appendix C.

4.3 Neymanian inference and OLS

Prove (4.3)–(4.5). Moreover, prove that the HC2 variant of the EHW robust variance estimator recovers $\hat{V}$ exactly.

Remark: Appendix B reviews some important technical results about OLS.

4.4 Treatment effect heterogeneity

Show that $S^2(\tau) = 0$ implies that $S^2(1) = S^2(0)$. Give a counterexample with $S^2(1) = S^2(0)$ but $S^2(\tau) \neq 0$.

Show that $S^2(1) < S^2(0)$ implies that

$$S(Y(0), \tau) = (n-1)^{-1} \sum_{i=1}^{n} \{Y_i(0) - \bar{Y}(0)\}(\tau_i - \tau) < 0.$$

Give a counterexample with $S^2(1) > S^2(0)$ but $S(Y(0), \tau) < 0$.

Remark: The first result states that no treatment effect heterogeneity implies equal variances in the treated and control potential outcomes. But the converse is not true. The second result states that if the treated potential outcome has a smaller variance than the control potential outcome, then the individual treatment effect is negatively correlated with the control potential outcome. But the converse is not true. Gerber and Green (2012, page 293) and Ding et al. (2019, Section B.3) gave related discussions.

4.5 A better bound of the variance formula

Neyman (1923)'s conservative variance estimator essentially uses the following upper bound on the true variance:

$$\begin{aligned} \text{var}(\hat{\tau}) &= \frac{S^2(1)}{n_1} + \frac{S^2(0)}{n_0} - \frac{S^2(\tau)}{n} \\ &\leq \frac{S^2(1)}{n_1} + \frac{S^2(0)}{n_0}, \end{aligned}$$

which uses the trivial fact that $S^2(\tau) \geq 0$. Show the following upper bound

$$\text{var}(\hat{\tau}) \leq \frac{1}{n} \left\{ \sqrt{\frac{n_0}{n_1}} S(1) + \sqrt{\frac{n_1}{n_0}} S(0) \right\}^2. \tag{4.6}$$

When does the equality in (4.6) hold?

The upper bound (4.6) motivates another conservative variance estimator

$$\hat{V}' = \frac{1}{n} \left\{ \sqrt{\frac{n_0}{n_1}} \hat{S}(1) + \sqrt{\frac{n_1}{n_0}} \hat{S}(0) \right\}^2.$$

Section 4.5.1 used $\hat{V}$ in the simulation. Repeat the simulation with an additional comparison with the variance estimator $\hat{V}'$ and the associated confidence interval.

Remark: You may find Section A.1.4 useful for the proof. The upper bound (4.6) can be further improved. Aronow et al. (2014) derived the sharp upper bound for $\text{var}(\hat{\tau})$ using the Frechet–Hoeffding inequality. Those improvements are rarely used in practice mainly for two reasons. First, they are more complicated than $\hat{V}$ which can be conveniently implemented by OLS. Second, the confidence interval based on $\hat{V}$ also works under other formulations, for example, under a true linear model of the outcome on the treatment, but those improvements do not. Although they are theoretically interesting, those improvements have little practical impact.

4.6 Vector version of Neyman (1923)

The classic result of Neyman (1923) is about a scalar outcome. It is common to have multiple outcomes in practice. Therefore, we now extend the potential outcomes to vectors.

We consider the average causal effect on a vector outcome $\boldsymbol{V} \in \mathbb{R}^K$:

$$\tau_{\boldsymbol{V}} = \frac{1}{n}\sum_{i=1}^{n}\{\boldsymbol{V}_i(1) - \boldsymbol{V}_i(0)\},$$

where $\boldsymbol{V}_i(1)$ and $\boldsymbol{V}_i(0)$ are the potential outcomes of $\boldsymbol{V}$ for unit i. The Neyman-type estimator for $\tau_{\boldsymbol{V}}$ is the difference between the sample mean vectors of the observed outcomes under treatment and control:

$$\begin{aligned}\hat{\tau}_{\boldsymbol{V}} &= \bar{\boldsymbol{V}}_1 - \bar{\boldsymbol{V}}_0 \\ &= \frac{1}{n_1}\sum_{i=1}^{n} Z_i \boldsymbol{V}_i - \frac{1}{n_0}\sum_{i=1}^{n}(1 - Z_i)\boldsymbol{V}_i.\end{aligned}$$

Consider a CRE. Show that $\hat{\tau}_{\boldsymbol{V}}$ is unbiased for $\tau_{\boldsymbol{V}}$. Find the covariance matrix of $\hat{\tau}_{\boldsymbol{V}}$. Find a (possibly conservative) estimator for the variance.

4.7 Inference for the average treatment effect on the treated units

Under the CRE, we can also make inference of the average treatment effect on the treated units:

$$\tau_{\mathrm{T}} = \frac{1}{n_1}\sum_{i=1}^{n} Z_i\{Y_i(1) - Y_i(0)\}$$

which is a random variable depending on the treatment indicators.

Show that

$$E(\hat{\tau} - \tau_{\mathrm{T}}) = 0$$

and

$$\mathrm{var}(\hat{\tau} - \tau_{\mathrm{T}}) = \frac{n}{n_1 n_0} S^2(0).$$

Remark: A key feature of the variance formula above is that it depends on only the finite-population variance of the potential outcomes under the control. In particular, it does not involve $S^2(\tau)$. Therefore, we can easily construct an unbiased variance estimator

$$\hat{\mathrm{var}}(\hat{\tau} - \tau_{\mathrm{T}}) = \frac{n}{n_1 n_0} \hat{S}^2(0).$$

This nice feature comes at the price of estimating a random parameter τ_{T} which varies across different realizations of the treatment assignment. Hansen and Bowers (2009) and Ding and Miratrix (2019) discussed related results.

4.8 Inference in the BRE

In this book, I use the following definition for the BRE.

Definition 4.1 (BRE) *The treatment indicators Z_i's are IID Bernoulli(π) with $n_1 = \sum_{i=1}^{n} Z_i$ receiving the treatment and $n_0 = \sum_{i=1}^{n}(1 - Z_i)$ receiving the control, respectively.*

First, we can use the FRT to analyze the BRE. How do we test $H_{0\mathrm{F}}$ in the BRE? Can we use the same FRT procedure as in the CRE if the actual experiment is the BRE? If yes, give a justification; if no, explain why.

Second, we can obtain point estimators for τ and find the associated variance estimators, as Neyman (1923) did for the CRE.

1. Is $\hat{\tau}$ unbiased for τ? Is it consistent?

2. Find an unbiased estimator for τ.
3. Compare the variance of the above unbiased estimator and the asymptotic variance of $\hat{\tau}$.

Remark: Under the BRE, the estimator $\hat{\tau}$ does not have finite variance but the variance of its asymptotic distribution is finite.

4.9 Recommended reading

Ding (2016) compared the Fisherian approach and Neymanian approach to analyzing the CRE.

5

Stratification and Post-Stratification in Randomized Experiments

Block what you can and randomize what you cannot.
— Box et al. (1978, page 103)

This is the second most famous quote from George Box.[1] This chapter will explain its meaning.

5.1 Stratification

A CRE may generate an undesired treatment allocation by chance. Let us start with a CRE with a discrete covariate $X_i \in \{1, \ldots, K\}$. Define

$$n_{[k]} = \#\{i : X_i = k\} \quad \text{and} \quad \pi_{[k]} = n_{[k]}/n$$

as the number and proportion of units in stratum k ($k = 1, \ldots, K$), respectively. A CRE assigns n_1 units to the treatment group and n_0 units to the control group, which results in

$$\begin{aligned} n_{[k]1} &= \#\{i : X_i = k, Z_i = 1\}, \\ n_{[k]0} &= \#\{i : X_i = k, Z_i = 0\} \end{aligned}$$

units in the treatment and control groups, respectively, within stratum k. With positive probability, $n_{[k]1}$ or $n_{[k]0}$ is zero for some k, that is, it is possible that some strata only have treated or control units. Even if none of the $n_{[k]1}$'s or $n_{[k]0}$'s are zero, with high probability

$$\frac{n_{[k]1}}{n_1} - \frac{n_{[k]0}}{n_0} \neq 0, \tag{5.1}$$

and the magnitude can be quite large. So the proportions of units in stratum k are different across the treatment and control groups although on average their difference is zero (see Problem 5.1):

$$E\left(\frac{n_{[k]1}}{n_1} - \frac{n_{[k]0}}{n_0}\right) = 0. \tag{5.2}$$

When $n_{[k]1}/n_1 - n_{[k]0}/n_0$ is large for some strata k's, the treatment and control groups have undesirable covariate imbalance. Such covariate imbalance deteriorates the quality of the experiment, making it difficult to interpret the results of the experiment since the difference in the outcomes may be attributed to the treatment or the covariate imbalance.

How can we actively avoid covariate imbalance in the experiment? We can conduct stratified randomized experiments (SRE).

[1] His most famous quote is "all models are wrong but some are useful" (Box, 1979, page 202).

Definition 5.1 (SRE) *Fix the $n_{[k]1}$'s or $n_{[k]0}$'s. We conduct K independent CREs within the K strata of a discrete covariate X.*

In agricultural experiments, the SRE is also called the *randomized block design*, with the strata called the blocks. Analogously, *stratified randomization* is also called *block randomization*.[2] The total number of randomizations in an SRE equals

$$\prod_{k=1}^{K} \binom{n_{[k]}}{n_{[k]1}},$$

and each feasible randomization has equal probability. Within stratum k, the proportion of units receiving the treatment is

$$e_{[k]} = \frac{n_{[k]1}}{n_{[k]}},$$

which is also called the *propensity score*, a concept that will play a central role in Part III of this book (see Definition 11.1). An SRE is different from a CRE: first, all feasible randomizations in an SRE form a subset of all feasible randomizations in a CRE, so

$$\prod_{k=1}^{K} \binom{n_{[k]}}{n_{[k]1}} < \binom{n}{n_1};$$

second, $e_{[k]}$ is fixed in an SRE but random in a CRE.

For every unit i, we have potential outcomes $Y_i(1)$ and $Y_i(0)$, and individual causal effect $\tau_i = Y_i(1) - Y_i(0)$. For stratum k, we have the stratum-specific average causal effect

$$\tau_{[k]} = n_{[k]}^{-1} \sum_{X_i=k} \tau_i.$$

The average causal effect is

$$\begin{aligned} \tau &= n^{-1} \sum_{i=1}^{n} \tau_i \\ &= n^{-1} \sum_{k=1}^{K} \sum_{X_i=k} \tau_i \\ &= \sum_{k=1}^{K} \pi_{[k]} \tau_{[k]}, \end{aligned}$$

which is a weighted average of the stratum-specific average causal effects, with the weight $\pi_{[k]} = n_{[k]}/n$ being the proportion of units in stratum k.

If we are interested in $\tau_{[k]}$, then we can use the methods in Chapters 3 and 4 for the CRE within stratum k. Below I will discuss statistical inference for τ.

[2] Chapter 3.5.3 mentioned two Fisherian principles for experiments: *randomization* and *replication*. *Blocking* is the third Fisherian principle.

5.2 FRT

5.2.1 Theory

In parallel with the discussion of a CRE, I will start with the FRT in an SRE. The sharp null hypothesis is still

$$H_{0\mathrm{F}} : Y_i(1) = Y_i(0) \text{ for all units } i = 1, \ldots, n.$$

The fundamental idea of the FRT applies to any randomized experiment: we can use any test statistic $T = T(\boldsymbol{Z}, \boldsymbol{Y}, \boldsymbol{X})$, where $\boldsymbol{Z}, \boldsymbol{Y}$, and $\boldsymbol{X}$ are the treatment vector, observed outcome vector, and the covariate matrix, respectively. Under the SRE and $H_{0\mathrm{F}}$, the test statistic T has a known distribution because $\boldsymbol{Z}$ has a known distribution under the SRE. However, we must be careful with two subtle issues. First, when we simulate the treatment vector, we must permute the treatment indicators within strata of X according to Definition 5.1. The resulting FRT is sometimes called the *conditional randomization test* or *conditional permutation test.* Second, we should choose test statistics that can reflect the nature of the SRE. Below I give some canonical choices of the test statistic.

Example 5.1 (Stratified estimator) *Motivated by estimating τ (see Section 5.3 below), we can use the following stratified estimator in the FRT:*

$$\hat{\tau}_{\mathrm{S}} = \sum_{k=1}^{K} \pi_{[k]} \hat{\tau}_{[k]},$$

where

$$\hat{\tau}_{[k]} = n_{[k]1}^{-1} \sum_{i=1}^{n} I(X_i = k, Z_i = 1) Y_i - n_{[k]0}^{-1} \sum_{i=1}^{n} I(X_i = k, Z_i = 0) Y_i$$

is the stratum-specific difference-in-means within stratum k.

Example 5.2 (Studentized stratified estimator) *Motivated by the studentized statistic in the simple two-sample problem, we can use the following studentized statistic for the stratified estimator in the FRT:*

$$t_{\mathrm{S}} = \frac{\hat{\tau}_{\mathrm{S}}}{\sqrt{\hat{V}_{\mathrm{S}}}},$$

with

$$\hat{V}_{\mathrm{S}} = \sum_{k=1}^{K} \pi_{[k]}^2 \left(\frac{\hat{S}_{[k]}^2(1)}{n_{[k]1}} + \frac{\hat{S}_{[k]}^2(0)}{n_{[k]0}} \right)$$

where $\hat{S}_{[k]}^2(1)$ and $\hat{S}_{[k]}^2(0)$ are the stratum-specific sample variances of the outcomes under the treatment and control, respectively. The exact form of this statistic is motivated by the Neymanian perspective discussed in Section 5.3 below.

Example 5.3 (Combining Wilcoxon rank-sum statistics) *We first compute the Wilcoxon rank-sum statistic $W_{[k]}$ within stratum k (recall Example 3.3) and then combine them as*

$$W_{\mathrm{S}} = \sum_{k=1}^{K} c_{[k]} W_{[k]}.$$

Based on different asymptotic schemes and optimality criteria, Van Elteren (1960) proposed two weighting methods, one with

$$c_{[k]} = \frac{1}{n_{[k]1}n_{[k]0}},$$

and the other with

$$c_{[k]} = \frac{1}{n_{[k]} + 1}.$$

The motivations for these weights are quite technical, and other choices of weights may also be reasonable. For instance, I will use $c_{[k]} = 1/\pi_{[k]}$ in Section 5.2.2, which is close to the second weighting method proposed by Van Elteren (1960).

Example 5.4 (Aligned rank statistic of Hodges and Lehmann (1962)) *The statistic of Van Elteren (1960) works well with a few large strata. However, it does not work well with many small strata since it does not make enough comparisons, potentially losing information in the data. Hodges and Lehmann (1962) proposed a test statistic that makes more comparisons across strata after standardizing the outcomes. They suggested first centering the outcomes as*

$$\tilde{Y}_i = Y_i - \bar{Y}_{[k]}$$

with the stratum-specific mean $\bar{Y}_{[k]} = n_{[k]}^{-1} \sum_{X_i=k} Y_i$ if $X_i = k$, then obtaining the ranks $(\tilde{R}_1, \ldots, \tilde{R}_n)$ of the pooled outcomes $(\tilde{Y}_1, \ldots, \tilde{Y}_n)$, and finally constructing the test statistic

$$\tilde{W} = \sum_{i=1}^{n} Z_i \tilde{R}_i.$$

We can simulate the exact distributions of the above test statistics under the SRE. We can also calculate their means and variances and obtain the p-values based on Normal approximations.

After searching for a while, I failed to find a detailed discussion of the Kolmogorov–Smirnov statistic for the SRE. Below is my proposal.

Example 5.5 (Kolmogorov–Smirnov statistic) *We compute $D_{[k]}$, the maximum difference between the empirical distributions of the outcomes under treatment and control within stratum k. The final test statistic can be*

$$D_{\mathrm{S}} = \sum_{k=1}^{K} c_{[k]} D_{[k]}$$

or

$$D_{\max} = \max_{1 \le k \le K} c_{[k]} D_{[k]},$$

where $c_{[k]} = \sqrt{n_{[k]1}n_{[k]0}/n_{[k]}}$ is motivated by the limiting distribution of $D_{[k]}$ with $n_{[k]1}$ and $n_{[k]0}$ approaching infinity (see Example 3.4). The statistics D_{S} and $D_{\max}$ are more appropriate when all strata have large sample sizes. Another reasonable choice is

$$D = \max_{y} \left| \sum_{k=1}^{K} \pi_{[k]} \{\hat{F}_{[k]1}(y) - \hat{F}_{[k]0}(y)\} \right|,$$

where $\hat{F}_{[k]1}(y)$ and $\hat{F}_{[k]0}(y)$ are the stratum-specific empirical distribution functions of the outcomes under the treatment and control, respectively. The statistic D is appropriate in both the cases with large strata and the cases with many small strata.

5.2.2 An application

The Penn Bonus experiment is an example to illustrate the FRT in the SRE. The dataset used by Koenker and Xiao (2002) is from a job training program stratified on `quarter`, with the outcome being the duration before employment.

```
> penndata = read.table("Penn46_ascii.txt")
> z = penndata$treatment
> y = log(penndata$duration)
> block = penndata$quarter
> table(penndata$treatment, penndata$quarter)

      0   1   2   3   4   5
  0 234  41 687 794 738 860
  1  87  48 757 866 811 461
```

I will focus on $\hat{\tau}_S$ and W_S, and leave the FRT with other statistics as Problem 5.6. The following function computes $\hat{\tau}_S$ and W_S:

```
stat_SRE = function(z, y, x)
{
       xlevels = unique(x)
       K       = length(xlevels)
       PiK     = rep(0, K)
       TauK    = rep(0, K)
       WK      = rep(0, K)
       for(k in 1:K)
       {
             xk           = xlevels[k]
             zk           = z[x == xk]
             yk           = y[x == xk]
             PiK[k]       = length(zk)/length(z)
             TauK[k]      = mean(yk[zk==1]) - mean(yk[zk==0])
             WK[k]        = wilcox.test(yk[zk==1], yk[zk==0])$statistic
       }

       return(c(sum(PiK*TauK), sum(WK/PiK)))
}
```

The following function generates a random treatment assignment in the SRE based on the observed data:

```
zRandomSRE = function(z, x)
{
         xlevels = unique(x)
         K       = length(xlevels)
         zrandom = z
         for(k in 1:K)
         {
              xk = xlevels[k]
              zrandom[x == xk] = sample(z[x == xk])
         }

         return(zrandom)
}
```

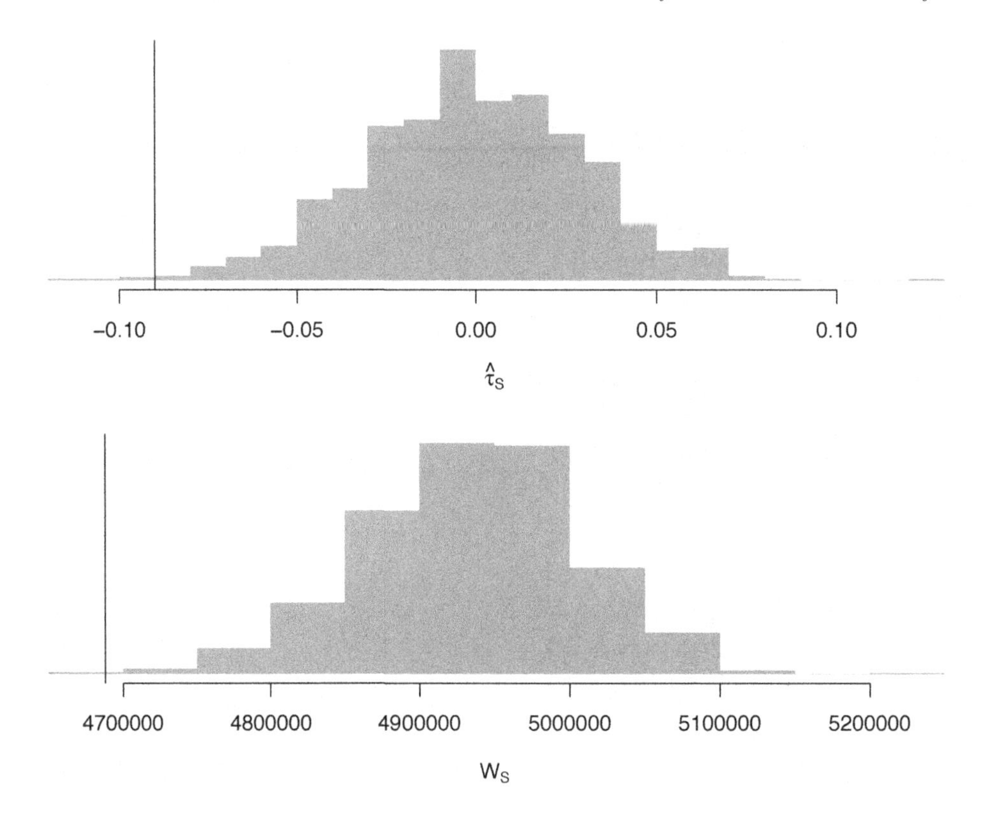

FIGURE 5.1: The randomization distributions of $\hat{\tau}_S$ and W_S based on the data from the Penn Bonus experiment, with 10^4 Monte Carlo draws

Based on the above data and functions, we can simulate the randomization distributions of the test statistics and compute the p-values.

```
> stat.obs = stat_SRE(z, y, block)
> MC = 10^3
> statSREMC = matrix(0, MC, 2)
> for(mc in 1:MC)
+ {
+     zrandom            = zRandomSRE(z, block)
+     statSREMC[mc, ] = stat_SRE(zrandom, y, block)
+ }
> mean(statSREMC[, 1] <= stat.obs[1])
[1] 0.002
> mean(statSREMC[, 2] <= stat.obs[2])
[1] 0.001
```

In the above, I calculate the p-values based on left-tail probabilities because the treatment has a negative effect on the outcome. See Figure 5.1 for more details about the observed test statistics and randomization distributions.

5.3 Neymanian inference

5.3.1 Point and interval estimation

Statistical inference for an SRE builds on the fact that it essentially consists of K independent CREs. Based on this, we can extend Neyman's (1923) results to the SRE. Within stratum k, the difference-in-means $\hat{\tau}_{[k]}$ is unbiased for $\tau_{[k]}$ with variance

$$\text{var}(\hat{\tau}_{[k]}) = \frac{S^2_{[k]}(1)}{n_{[k]1}} + \frac{S^2_{[k]}(0)}{n_{[k]0}} - \frac{S^2_{[k]}(\tau)}{n_{[k]}},$$

where $S^2_{[k]}(1), S^2_{[k]}(0)$, and $S^2_{[k]}(\tau)$ are the stratum-specific variances of potential outcomes and the individual causal effects, respectively. Therefore, the stratified estimator $\hat{\tau}_{\text{S}} = \sum_{k=1}^K \pi_{[k]}\hat{\tau}_{[k]}$ is unbiased for $\tau = \sum_{k=1}^K \pi_{[k]}\tau_{[k]}$ with variance

$$\text{var}(\hat{\tau}_{\text{S}}) = \sum_{k=1}^K \pi^2_{[k]}\text{var}(\hat{\tau}_{[k]}).$$

If $n_{[k]1} \geq 2$ and $n_{[k]0} \geq 2$, then we can obtain the sample variances $\hat{S}^2_{[k]}(1)$ and $\hat{S}^2_{[k]}(0)$ of the outcomes within stratum k and construct a conservative variance estimator

$$\hat{V}_{\text{S}} = \sum_{k=1}^K \pi^2_{[k]} \left(\frac{\hat{S}^2_{[k]}(1)}{n_{[k]1}} + \frac{\hat{S}^2_{[k]}(0)}{n_{[k]0}} \right),$$

where $\hat{S}^2_{[k]}(1)$ and $\hat{S}^2_{[k]}(0)$ are the stratum-specific sample variances of the outcomes under treatment and control, respectively. Based on a Normal approximation of $\hat{\tau}_{\text{S}}$, we can construct a Wald-type $1-\alpha$ confidence interval (see Section A.2.4 for the terminology) for τ as

$$\hat{\tau}_{\text{S}} \pm z_{1-\alpha/2}\sqrt{\hat{V}_{\text{S}}}.$$

From a hypothesis testing perspective, under $H_{0\text{N}} : \tau = 0$, we can compare

$$t_{\text{S}} = \hat{\tau}_{\text{S}} / \sqrt{\hat{V}_{\text{S}}}$$

with the standard Normal quantiles to obtain asymptotic p-values. The statistic t_{S} appears in Example 5.2 for the FRT. Chapter 8 later will show that using t_{S} in the FRT yields finite-sample exact p-value under $H_{0\text{F}}$ and asymptotically valid p-value under $H_{0\text{N}}$.

Here I omit the technical details for the CLT of $\hat{\tau}_{\text{S}}$. See Liu and Yang (2020) for a proof, which includes the two regimes with a few large strata and many small strata. I will illustrate this theoretical issue using numerical examples in Section 5.3.2 below.

5.3.2 Numerical examples

The following function computes the Neymanian point and variance estimators under the SRE:

```
Neyman_SRE = function(z, y, x)
{
        xlevels  = unique(x)
        K        = length(xlevels)
```

```
        PiK      = rep(0, K)
        TauK     = rep(0, K)
        varK     = rep(0, K)
        for(k in 1:K)
        {
                xk          = xlevels[k]
                zk          = z[x == xk]
                yk          = y[x == xk]
                PiK[k]      = length(zk)/length(z)
                TauK[k]     = mean(yk[zk==1]) - mean(yk[zk==0])
                varK[k]     = var(yk[zk==1])/sum(zk) +
                                   var(yk[zk==0])/sum(1 - zk)
        }

        return(c(sum(PiK*TauK), sum(PiK^2*varK)))
}
```

The first simulation setting has $K = 5$ and each stratum has 80 units. `TauHat` and `VarHat` are the point and variance estimators over 10^4 simulations.

```
> K   = 5
> n   = 80
> n1  = 50
> n0  = 30
> x   = rep(1:K, each = n)
> y0  = rexp(n*K, rate = x)
> y1  = y0 + 1
> zb  = c(rep(1, n1), rep(0, n0))
> MC  = 10^4
> TauHat = rep(0, MC)
> VarHat = rep(0, MC)
> for(mc in 1:MC)
+ {
+    z   = replicate(K, sample(zb))
+    z   = as.vector(z)
+    y   = z*y1 + (1-z)*y0
+    est = Neyman_SRE(z, y, x)
+    TauHat[mc]  = est[1]
+    VarHat[mc]  = est[2]
+ }
> var(TauHat)
[1] 0.002248925
> mean(VarHat)
[1] 0.002266396
```

The upper panel of Figure 5.2 shows the histogram of the point estimator, which is symmetric and bell-shaped around the true parameter. From the above, the average value of the variance estimator is almost identical to the variance of the estimators because the individual causal effects are constant.

The second simulation setting has $K = 50$ and each stratum has 8 units.

```
> K   = 50
> n   = 8
> n1  = 5
> n0  = 3
> x   = rep(1:K, each = n)
> y0  = rexp(n*K, rate = log(x + 1))
```

```
> y1 = y0 + 1
> zb = c(rep(1, n1), rep(0, n0))
> MC = 10^4
> TauHat = rep(0, MC)
> VarHat = rep(0, MC)
> for(mc in 1:MC)
+ {
+   z  = replicate(K, sample(zb))
+   z  = as.vector(z)
+   y  = z*y1 + (1-z)*y0
+   est = Neyman_SRE(z, y, x)
+   TauHat[mc] = est[1]
+   VarHat[mc] = est[2]
+ }
>
> hist(TauHat, xlab = expression(hat(tau)[S]),
+      ylab = "", main = "many small strata",
+      border = FALSE, col = "grey",
+      breaks = 30, yaxt = 'n',
+      xlim = c(0.8, 1.2))
> abline(v = 1)
>
> var(TauHat)
[1] 0.001443111
> mean(VarHat)
[1] 0.001473616
```

The lower panel of Figure 5.2 shows the histogram of the point estimator, which is symmetric and bell-shaped around the true parameter. Again, the average value of the variance estimator is almost identical to the variance of the estimators because the individual causal effects are constant.

We finally use the Penn Bonus Experiment to illustrate the Neymanian inference in an SRE. Applying the function `Neyman_SRE` to the dataset, we obtain:

```
> penndata = read.table("Penn46_ascii.txt")
> z = penndata$treatment
> y = log(penndata$duration)
> block = penndata$quarter
> est = Neyman_SRE(z, y, block)
> est[1]
[1] -0.08990646
> sqrt(est[2])
[1] 0.03079775
```

So the job training program significantly shortens the log of the duration time before employment.

5.3.3 Comparing the SRE and the CRE

What are the benefits of the SRE compared with the CRE? I have motivated the SRE from the covariate balance perspective. In addition, I will show that better covariate balance in turn results in better estimation precision of the average causal effect. To make a fair comparison, I assume that $e_{[k]} = e$ for all k which ensures that the difference in means equals the stratified estimator:

$$\hat{\tau} = \hat{\tau}_{\mathrm{S}}. \tag{5.3}$$

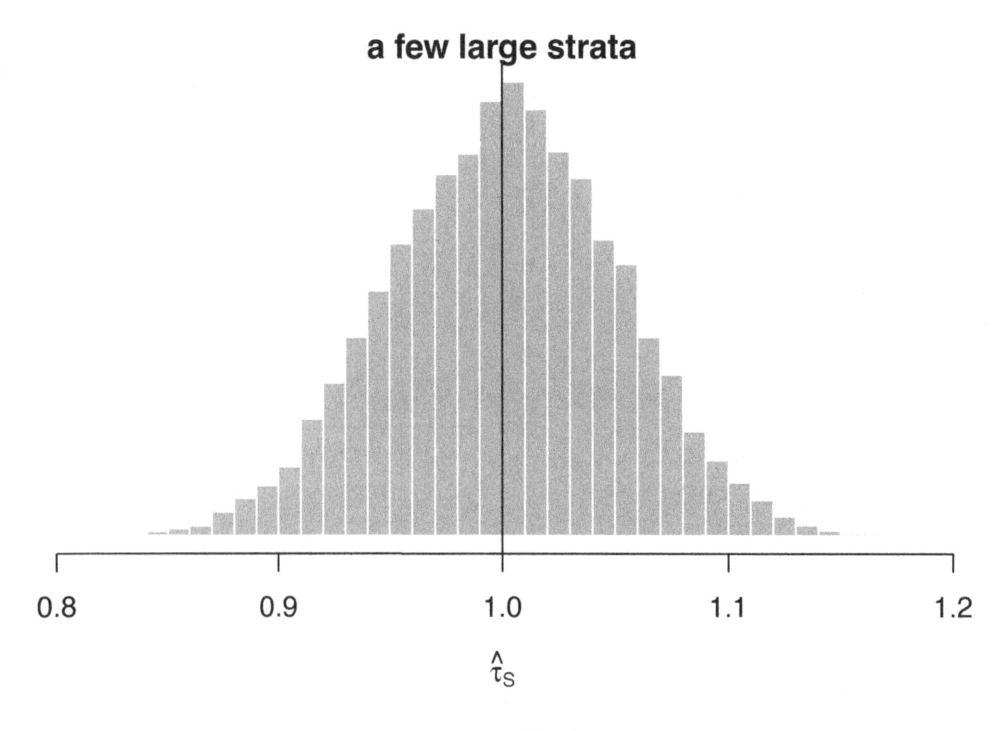

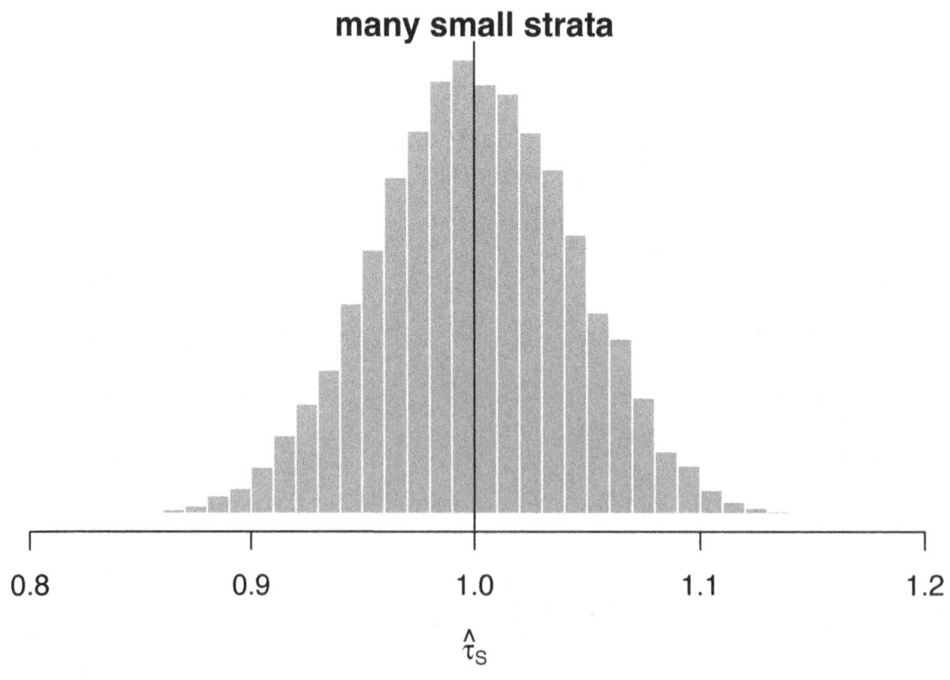

FIGURE 5.2: Normal approximations under two regimes

I leave the proof of (5.3) as Problem 5.2.

We now compare the sampling variances. The classic analysis of variance technique motivates the decomposition of the total variance into the summation of the within-strata and between-strata variances. For the variance of the potential outcomes under the treatment, we have

$$\begin{aligned} S^2(1) &= (n-1)^{-1}\sum_{i=1}^{n}\{Y_i(1)-\bar{Y}(1)\}^2 \\ &= (n-1)^{-1}\sum_{k=1}^{K}\sum_{X_i=k}\{Y_i(1)-\bar{Y}_{[k]}(1)+\bar{Y}_{[k]}(1)-\bar{Y}(1)\}^2 \\ &= (n-1)^{-1}\sum_{k=1}^{K}\sum_{X_i=k}\left[\{Y_i(1)-\bar{Y}_{[k]}(1)\}^2+\{\bar{Y}_{[k]}(1)-\bar{Y}(1)\}^2\right] \\ &= \sum_{k=1}^{K}\left[\frac{n_{[k]}-1}{n-1}S^2_{[k]}(1)+\frac{n_{[k]}}{n-1}\{\bar{Y}_{[k]}(1)-\bar{Y}(1)\}^2\right], \end{aligned}$$

where the key step three above holds because the cross term equals 0. Similarly, for the variance of the potential outcomes under the control, we have

$$S^2(0)=\sum_{k=1}^{K}\left[\frac{n_{[k]}-1}{n-1}S^2_{[k]}(0)+\frac{n_{[k]}}{n-1}\{\bar{Y}_{[k]}(0)-\bar{Y}(0)\}^2\right];$$

for the variance of the individual effects, we have

$$S^2(\tau)=\sum_{k=1}^{K}\left[\frac{n_{[k]}-1}{n-1}S^2_{[k]}(\tau)+\frac{n_{[k]}}{n-1}\{\tau_{[k]}-\tau\}^2\right].$$

The variance of the difference-in-means estimator under the CRE decomposes into

$$\begin{aligned} &\text{var}_{\text{CRE}}(\hat{\tau}) \\ =\ &\frac{S^2(1)}{n_1}+\frac{S^2(0)}{n_0}-\frac{S^2(\tau)}{n} \\ =\ &\sum_{k=1}^{K}\left[\frac{n_{[k]}-1}{(n-1)n_1}S^2_{[k]}(1)+\frac{n_{[k]}-1}{(n-1)n_0}S^2_{[k]}(0)-\frac{n_{[k]}-1}{(n-1)n}S^2_{[k]}(\tau)\right] \\ &+\sum_{k=1}^{K}\left[\frac{n_{[k]}-1}{(n-1)n_1}\{\bar{Y}_{[k]}(1)-\bar{Y}(1)\}^2+\frac{n_{[k]}-1}{(n-1)n_0}\{\bar{Y}_{[k]}(0)-\bar{Y}(0)\}^2\right. \\ &\left.-\frac{n_{[k]}-1}{(n-1)n}\{\tau_{[k]}-\tau\}^2\right]. \end{aligned}$$

With large $n_{[k]}$'s, it is approximately

$$\begin{aligned} &\text{var}_{\text{CRE}}(\hat{\tau}) \\ \approx\ &\sum_{k=1}^{K}\left[\frac{\pi_{[k]}}{n_1}S^2_{[k]}(1)+\frac{\pi_{[k]}}{n_0}S^2_{[k]}(0)-\frac{\pi_{[k]}}{n}S^2_{[k]}(\tau)\right] \\ &+\sum_{k=1}^{K}\left[\frac{\pi_{[k]}}{n_1}\{\bar{Y}_{[k]}(1)-\bar{Y}(1)\}^2+\frac{\pi_{[k]}}{n_0}\{\bar{Y}_{[k]}(0)-\bar{Y}(0)\}^2-\frac{\pi_{[k]}}{n}\{\tau_{[k]}-\tau\}^2\right]. \end{aligned}$$

The constant propensity scores assumption ensures

$$\pi_{[k]}/n_{[k]1} = 1/(ne), \quad \pi_{[k]}/n_{[k]0} = 1/\{n(1-e)\},$$

and by definition, $\pi_{[k]}/n_{[k]} = 1/n$, which together allow us to rewrite the variance of $\hat{\tau}_{\mathrm{S}}$ under the SRE as

$$\begin{aligned} \mathrm{var}_{\mathrm{SRE}}(\hat{\tau}_{\mathrm{S}}) &= \sum_{k=1}^{K} \pi_{[k]}^2 \left[\frac{S_{[k]}^2(1)}{n_{[k]1}} + \frac{S_{[k]}^2(0)}{n_{[k]0}} - \frac{S_{[k]}^2(\tau)}{n_{[k]}} \right] \\ &= \sum_{k=1}^{K} \left[\frac{\pi_{[k]}}{n_1} S_{[k]}^2(1) + \frac{\pi_{[k]}}{n_0} S_{[k]}^2(0) - \frac{\pi_{[k]}}{n} S_{[k]}^2(\tau) \right]. \end{aligned}$$

With large $n_{[k]}$'s, approximately, the difference between $\mathrm{var}_{\mathrm{CRE}}(\hat{\tau})$ and $\mathrm{var}_{\mathrm{SRE}}(\hat{\tau}_{\mathrm{S}})$ is

$$\sum_{k=1}^{K} \left[\frac{\pi_{[k]}}{n_1} \{\bar{Y}_{[k]}(1) - \bar{Y}(1)\}^2 + \frac{\pi_{[k]}}{n_0} \{\bar{Y}_{[k]}(0) - \bar{Y}(0)\}^2 - \frac{\pi_{[k]}}{n} (\tau_{[k]} - \tau)^2 \right]$$

which is non-negative because it equals (see Problem 5.4)

$$\sum_{k=1}^{K} \frac{\pi_{[k]}}{n} \left\{ \sqrt{\frac{n_0}{n_1}} \{\bar{Y}_{[k]}(1) - \bar{Y}(1)\} + \sqrt{\frac{n_1}{n_0}} \{\bar{Y}_{[k]}(0) - \bar{Y}(0)\} \right\}^2 \geq 0. \tag{5.4}$$

The difference in (5.4) is zero only in the extreme case that

$$\sqrt{\frac{n_0}{n_1}} \{\bar{Y}_{[k]}(1) - \bar{Y}(1)\} + \sqrt{\frac{n_1}{n_0}} \{\bar{Y}_{[k]}(0) - \bar{Y}(0)\} = 0$$

for $k = 1, \ldots, K$. When the covariate is predictive of the potential outcomes, the above quantities are usually not all zeros, which ensures the efficiency gain of the SRE compared with the CRE. Only in the extreme cases that the covariate is not predictive at all, the large-sample efficiency gain is zero. In those cases, the SRE can even result in less efficient estimators in finite samples. The above discussion corroborates the quote from George Box at the beginning of this chapter.

I will end this section with several remarks. First, the above comparison is based on the sampling variance, and we can also compare the estimated variances under the SRE and the CRE. The results are similar. Second, increasing K improves efficiency, but this argument depends on the large strata assumption. So we face a trade-off in practice. We cannot arbitrarily increase K, and the most extreme case is $n_{[k]1} = n_{[k]0} = 1$, which is called the matched-pairs experiment and will be discussed in Chapter 7.

5.4 Post-stratification in a CRE

In a CRE with a discrete covariate X, the numbers of units receiving the treatment and control are random within stratum k. In an SRE, these numbers are fixed. But if we conduct conditional inference given $\boldsymbol{n} = \{n_{[k]1}, n_{[k]0}\}_{k=1}^{K}$, then a CRE becomes an SRE. Mathematically, if none of the components of $\boldsymbol{n}$ are zero, then

$$\begin{aligned} \mathrm{pr}_{\mathrm{CRE}}(\boldsymbol{Z} = \boldsymbol{z} \mid \boldsymbol{n}) &= \frac{\mathrm{pr}_{\mathrm{CRE}}(\boldsymbol{Z} = \boldsymbol{z}, \boldsymbol{n})}{\mathrm{pr}_{\mathrm{CRE}}(\boldsymbol{n})} \\ &= \frac{1}{\prod_{k=1}^{K} \binom{n_{[k]}}{n_{[k]1}}}, \end{aligned} \tag{5.5}$$

that is, the conditional distribution of $\boldsymbol{Z}$ from a CRE given $\boldsymbol{n}$ is identical to the distribution of $\boldsymbol{Z}$ from an SRE. I leave the proof of (5.5) as Problem 5.5. So conditional on $\boldsymbol{n}$, we can analyze a CRE with a discrete covariate X in the same way as in an SRE. In particular, the FRT becomes a *conditional FRT*, and the Neymanian analysis becomes *post-stratification*:

$$\hat{\tau}_{\text{PS}} = \sum_{k=1}^{K} \pi_{[k]} \hat{\tau}_{[k]},$$

which has an identical form as $\hat{\tau}_{\text{S}}$. The variance of $\hat{\tau}_{\text{PS}}$ conditioning on $\boldsymbol{n}$ is identical to the variance of $\hat{\tau}_{\text{S}}$ under the SRE.

Hennessy et al. (2016) used simulation to show that the conditional FRT is often more powerful than the unconditional one. Miratrix et al. (2013) used theory to show that in many cases, post-stratification improves efficiency compared with $\hat{\tau}$. Both results hold if X is predictive of the outcome. However, the simulation is based on a limited number of data-generating processes, and the theory assumes all strata are large enough. We can not go too extreme in the conditional FRT or post-stratification because with a larger K it is more likely that some $n_{[k]1}$ or $n_{[k]0}$ become zero. Small or zero values of $n_{[k]1}$ or $n_{[k]0}$ greatly reduce the number of randomizations in the FRT, possibly reducing the power dramatically. The problem for the Neymanian counterpart is more salient because we cannot even define $\hat{\tau}_{\text{PS}}$ and the corresponding variance estimator.

Stratification uses X in the design stage and post-stratification uses X in the analysis stage. They are duals for using X. Asymptotically, their difference is small with large strata (Miratrix et al., 2013).

5.4.1 Reanalyzing the data from Meinert et al. (1970)

We use the data from a CRE reported in Meinert et al. (1970), which were also used by Rothman et al. (2008). The treatment is tolbutamide and the control is a placebo. The aggregate data are below:

	total	
	surviving	dead
$Z = 1$	174	30
$Z = 0$	184	21

Based on age, the stratum-specific data are below:

	age < 55			age ≥ 55	
	surviving	dead		surviving	dead
$Z = 1$	98	8	$Z = 1$	76	22
$Z = 0$	115	5	$Z = 0$	69	16

The following table shows the estimates for two strata separately, the post-stratified estimator, and the crude estimator ignoring the binary covariate, as well as the corresponding standard errors.

	stratum 1	stratum 2	post-stratification	crude
est	-0.034	-0.036	-0.035	-0.045
se	0.031	0.060	0.032	0.033

The crude estimator and the post-stratification estimator do not lead to fundamentally different results. However, the crude estimator is larger than both of the stratum-specific estimators, whereas the post-stratification estimator is within the range.

5.4.2 Reanalyzing the data from Chong et al. (2016)

Chong et al. (2016) conducted an SRE on 219 students of a rural secondary school in the Cajamarca district of Peru during the 2009 school year. They first provided the village clinic with iron supplements and trained the local staff to distribute one free iron pill to any adolescent who requested one in person. They then randomly assigned students to three arms with three different types of videos: in the first video, a popular soccer player was encouraging the use of iron supplements to maximize energy ("soccer" arm); in the second video, a physician was encouraging the use of iron supplements to improve overall health ("physician" arm); the third video did not mention iron at all ("control" arm). The experiment was stratified on the class level (1–5). The treatment and control group sizes within classes are shown below:

```
> library("foreign")
> dat_chong = read.dta("chong.dta")
> table(dat_chong$treatment, dat_chong$class_level)

                 1  2  3  4  5
  Soccer Player 16 19 15 10 10
  Physician     17 20 15 11 10
  Placebo       15 19 16 12 10
```

One outcome of interest is the average grades in the third and fourth quarters of 2009, and an important background covariate was the anemia status at baseline. I will only use a subset of the original data in this chapter.

```
> use.vars = c("treatment",
+              "gradesq34",
+              "class_level",
+              "anemic_base_re")
> dat_physician = subset(dat_chong,
+                        treatment != "Soccer Player",
+                        select = use.vars)
> dat_physician$z = (dat_physician$treatment=="Physician")
> dat_physician$y = dat_physician$gradesq34
> table(dat_physician$z,
+       dat_physician$class_level)

         1  2  3  4  5
  FALSE 15 19 16 12 10
  TRUE  17 20 15 11 10
> table(dat_physician$z,
+       dat_physician$class_level,
+       dat_physician$anemic_base_re)
, ,  = No


         1  2  3  4  5
  FALSE  6 14 12  7  4
  TRUE   8 12  9  5  6

, ,  = Yes


         1  2  3  4  5
  FALSE  9  5  4  5  6
```

```
TRUE    9  8  6  6  4
```

We can use the `Neyman_SRE` function defined before to compute the stratified estimator and its estimated variance.

```
tauS = with(dat_physician,
            Neyman_SRE(z, gradesq34, class_level))
```

An important additional covariate is the baseline anemic indicator which is quite important for predicting the outcome. Further conditioning the baseline anemic indicator, we have an experiment with $5 \times 2 = 10$ strata, with the treatment and control group sizes shown above. Again we can use the `Neyman_SRE` function defined before to compute the post-stratified estimator and its estimated variance.

```
> tauSPS = with(dat_physician, {
+   sps = interaction(class_level, anemic_base_re)
+   Neyman_SRE(z, gradesq34, sps)
+ })
```

The following table compares these two estimators. The post-stratified estimator yields a much smaller p-value.

	est	se	t.stat	p.value
stratify	0.406	0.202	2.005	0.045
stratify and post-stratify	0.463	0.190	2.434	0.015

This example illustrates that post-stratification can be used not only in the CRE but also in the SRE with additional discrete covariates.

5.5 Practical questions

I end this chapter with remarks on some practical questions about conducting SREs.

How do we choose X to conduct an SRE? Theoretically, X should be predictive of the potential outcomes. In some cases, the experimenter has enough background knowledge about the predictive covariates based on, for example, some pilot studies. Then the choice of X should be straightforward. In some other cases, this background knowledge may not be clear enough. Experimenters instead choose X based on logistical convenience, for example, X can be the indicator for the study areas or the cohort of students.

The choice of K is a related problem. Theoretically, more stratification increases the estimation efficiency if all strata are large enough. However, extremely large K may even decrease the estimation efficiency. In simulation studies, we observe diminishing marginal returns of increasing K. Anecdotally, $K = 5$ often suffices for efficiency gain (the magic number 5 will appear again in Chapter 11). Some experimenters prefer the most extreme version of the SRE with $K = n/2$. This results in the matched-pairs experiment, which will be discussed in Chapter 7 later.

Some experiments have multidimensional continuous covariates. Can the SRE still be used? If we have a pilot study, we can build a model for the potential outcome $Y(0)$ given those covariates, and then we can choose X as a discretized version of the predictor $\hat{Y}(0)$. In general, if we do not have such a pilot study or we do not want to make ad hoc discretizations, we can use a more general strategy called rerandomization, which will be the topic for Chapter 6.

5.6 Homework problems

5.1 Covariate balance in the CRE

Under the CRE, prove (5.2).

5.2 Consequence of the constant propensity score

Prove (5.3).

5.3 Consquence of constant individual causal effects

Assume that the individual causal effects are constant $\tau_i = \tau$ for all $i = 1, \ldots, n$. Consider the following class of weighted estimator for τ:

$$\hat{\tau}_w = \sum_{k=1}^{K} w_{[k]} \hat{\tau}_{[k]},$$

where the weights $w_{[k]}$'s are non-negative for all k.

Find the condition on the $w_{[k]}$'s such that $\hat{\tau}_w$ is unbiased for τ. Among all unbiased estimators, find the weights that give the $\hat{\tau}_w$ with the minimum variance.

5.4 Compare the CRE and SRE

Prove (5.4).

5.5 From the CRE to the SRE

Prove (5.5).

5.6 More FRTs for Section 5.2.2

Extend the analysis in Section 5.2.2 using FRTs with other test statistics.

5.7 FRT for an SRE in Imbens and Rubin (2015)

Imbens and Rubin (2015) discussed an SRE from the Student/Teacher Achievement Ratio experiment conducted in 1985–1986 in Tennessee. The kindergarten data are below:

```
treatment = list(c(1,1,0,0),
                 c(1,1,0,0),
                 c(1,1,1,0,0),
                 c(1,1,0,0),
                 c(1,1,0,0),
                 c(1,1,0,0),
                 c(1,1,0,0),
                 c(1,1,1,1,0,0),
                 c(1,1,0,0),
                 c(1,1,0,0),
                 c(1,1,0,0),
                 c(1,1,1,0,0),
                 c(1,1,0,0),
                 c(1,1,0,0),
                 c(1,1,0,0),
                 c(1,1,0,0))
```

```
outcome = list(c(0.165,0.321,-0.197,0.236),
               c(0.918,-0.202,1.19,0.117),
               c(0.341,0.561,-0.059,-0.496,0.225),
               c(-0.024,-0.450,-1.104,-0.956),
               c(-0.258,-0.083,-0.126,0.106),
               c(1.151,0.707,0.597,-0.495),
               c(0.077,0.371,0.685,0.270),
               c(-0.870,-0.496,-0.444,0.392,-0.934,-0.633),
               c(-0.568,-1.189,-0.891,-0.856),
               c(-0.727,-0.580,-0.473,-0.807),
               c(-0.533,0.458,-0.383,0.313),
               c(1.001,0.102,0.484,0.474,0.140),
               c(0.855,0.509,0.205,0.296),
               c(0.618,0.978,0.742,0.175),
               c(-0.545,0.234,-0.434,-0.293),
               c(-0.240,-0.150,0.355,-0.130))
```

The strata correspond to schools, and the unit of analysis is the teacher or class. The treatment equals 1 for small classes (13–17 students per teacher) and 0 for regular classes (22–25 students per teacher). The outcome is the standardized average mathematics score.

Reanalyze the Project STAR data below using the FRT. Use $\hat{\tau}_S$, W_S, and $\tilde{W}$ in the FRT. Compare the p-values.

Remark: This book uses Z for the treatment indicator but Imbens and Rubin (2015) use W.

5.8 A multi-center trial

Gould (1998, Table 1) reported the following data from a multi-center trial:

```
> multicenter = read.csv("multicenter.csv")
> multicenter
   center n0 mean0  sd0 n1 mean1  sd1 n5 mean5  sd5
1       1  7  0.43 4.58  7 -5.43 5.53  8 -2.63 3.38
2       2 11  0.10 4.21 11 -2.59 3.95 12 -2.21 4.14
3       3  6  2.58 4.80  6 -3.94 4.25  7  1.29 7.39
4       4 10 -2.30 3.86 10 -1.23 5.17 10 -1.40 2.27
5       5 10  2.08 6.46 10 -6.70 7.45 10 -5.13 3.91
6       6  6  1.13 3.24  5  3.40 8.17  5 -1.59 3.19
7       7  5  1.20 7.85  6 -3.67 4.89  5 -1.40 2.61
8       8 12 -1.21 2.66 13  0.18 3.81 12 -4.08 6.32
9       9  8  1.13 5.28  8 -2.19 5.17  9 -1.96 5.84
10     10  9 -0.11 3.62 10 -2.00 5.35 10  0.60 3.53
11     11 15 -4.37 6.12 14 -2.68 5.34 15 -2.14 4.27
12     12  8 -1.06 5.27  9  0.44 4.39  9 -2.03 5.76
13     13 12 -0.08 3.32 12 -4.60 6.16 11 -6.22 5.33
14     14  9  0.00 5.20  9 -0.25 8.23  7 -3.29 5.12
15     15  6  1.83 5.85  7 -1.23 4.33  6 -1.00 2.61
16     16 14 -4.21 7.53 14 -2.10 5.78 12 -5.75 5.63
17     17 13  0.76 3.82 13  0.55 2.53 13 -0.63 5.41
18     18 15 -1.05 4.54 13  2.54 4.16 14 -2.80 2.89
19     19 15  2.07 4.88 15 -1.67 4.95 15 -3.43 4.71
20     20 11 -1.46 5.48 10 -1.99 5.63 10 -6.77 5.19
21     21  5  0.80 4.21  5 -3.35 4.73  5 -0.23 4.14
22     22 11 -2.92 5.42 10 -1.22 5.95 11 -4.45 6.65
23     23  9 -3.37 4.73  9 -1.38 4.17  7  0.57 2.70
24     24 12 -1.92 2.91 12 -0.66 3.55 12 -2.39 2.27
```

```
25       25  9 -3.89 4.76  9 -3.22 5.54  8 -1.23 4.91
26       26 15 -3.48 5.98 15 -2.13 3.25 14 -3.71 5.30
27       27 11 -1.91 6.49 12 -1.33 4.40 11 -1.52 4.68
28       28 10 -2.66 3.80 10 -1.29 3.18 10 -4.70 3.43
29       29 13 -0.77 4.73 13 -2.31 3.88 13 -0.47 4.95
```

This is an SRE with centers being the strata. The trial was conducted to study the efficacy and tolerability of finasteride, a drug for treating benign prostatic hyperplasia. Within each of the 29 centers, patients were randomized into three arms: control, finasteride 1mg, and finasteride 5mg. The above dataset provides summary statistics for the outcome, which is the change from baseline in total symptom score. The total symptom score is the sum of the responses to nine questions (score 0 to 4) about symptoms of various aspects of impaired urinary ability. The meanings of the columns are:

1. `center`: ID of the centers;
2. `n0, n1, n5`: sample sizes under the three arms;
3. `mean0, mean1, mean5`: means of the outcomes under the three arms;
4. `sd0, sd1, sd5`: standard deviations of the outcomes under the three arms.

The individual-level outcomes are not reported so we cannot implement the FRT. However, the Neymanian inference only requires the summary statistics. Report the point estimators and variance estimators for comparing "finasteride 1mg" and "finasteride 5mg" to "control", separately.

5.9 Data reanalyses

Reanalyze the LaLonde data used in Section 4.5.3. Conduct both Fisherian and Neymanian inferences.

The original experiment is a CRE. Now we pretend that the original experiment is an SRE. First, reanalyze the data pretending that the experiment is stratified on the race (black, Hispanic, or other). Second, reanalyze the data pretending that the experiment is stratified on marital status. Third, reanalyze the data pretending that the experiment is stratified on the indicator of a high school diploma.

Compare with the results obtained under a CRE.

5.10 Recommended reading

Miratrix et al. (2013) provided a solid theory for post-stratification and compared it with stratification. A main theoretical result is that their difference is small asymptotically although they can differ in finite samples.

6

Rerandomization and Regression Adjustment

Stratification and post-stratification in Chapter 5 are duals for discrete covariates in the design and analysis of randomized experiments. How should we deal with multidimensional, possibly continuous, covariates? We can discretize continuous covariates, but this is not an ideal strategy with many covariates. Rerandomization and regression adjustment are duals for general covariates, which are the topics for this chapter.

The following table summarizes the topics of Chapters 5 and 6:

	design	analysis
discrete covariate	stratification	post-stratification
general covariate	rerandomization	regression adjustment

6.1 Rerandomization

6.1.1 Experimental design

Again we consider a finite population of n experimental units, where n_1 of them receive the treatment and n_0 of them receive the control. Let $\boldsymbol{Z} = (Z_1, \dots, Z_n)$ be the treatment vector for these units. Unit i has covariate $X_i \in \mathbb{R}^K$ which can have continuous or binary components. Concatenate them as an $n \times K$ covariate matrix $\boldsymbol{X} = (X_1, \dots, X_n)^{\mathsf{T}}$ and center them at mean zero $\bar{X} = n^{-1} \sum_{i=1}^{n} X_i = 0$ to simplify the presentation.

The CRE balances the covariates in the treatment and control groups on average, for instance, the difference in means of the covariates

$$\hat{\tau}_X = n_1^{-1} \sum_{i=1}^{n} Z_i X_i - n_0^{-1} \sum_{i=1}^{n} (1 - Z_i) X_i$$

has mean zero under the CRE. However, it can result in undesired covariate balance across the treatment and control groups in the realized treatment allocation, that is, the realized value of $\hat{\tau}_X$ is often not zero. Using the vector form of Neyman (1923) in Problem 4.6 before, we can show that

$$\begin{aligned} \operatorname{cov}(\hat{\tau}_X) &= \frac{1}{n_1} S_X^2 + \frac{1}{n_0} S_X^2 \\ &= \frac{n}{n_1 n_0} S_X^2, \end{aligned}$$

where $S_X^2 = (n-1)^{-1} \sum_{i=1}^{n} X_i X_i^{\mathsf{T}}$ is the finite-population covariance matrix of the covariates. The following Mahalanobis distance measures the difference between the treatment and control groups:

$$\begin{aligned} M &= \hat{\tau}_X^{\mathsf{T}} \operatorname{cov}(\hat{\tau}_X)^{-1} \hat{\tau}_X \\ &= \hat{\tau}_X^{\mathsf{T}} \left(\frac{n}{n_1 n_0} S_X^2 \right)^{-1} \hat{\tau}_X. \end{aligned} \tag{6.1}$$

Technically, the above formula for M in (6.1) is meaningful only if S_X^2 is invertible, which means that the columns of the covariate matrix $\boldsymbol{X}$ are linearly independent. If a column can be represented by a linear combination of other columns, it is redundant and should be dropped before the experiment. A nice feature of M is that it is invariant under non-degenerate linear transformations of X. Lemma 6.1 below summarizes the result with the proof relegated to Problem 6.2.

Lemma 6.1 *M in (6.1) remains the same if we transform X_i to $b_0 + BX_i$ for all units $i = 1, \ldots, n$ where $b_0 \in \mathbb{R}^K$ and $B \in \mathbb{R}^{K \times K}$ is invertible.*

The finite population CLT (Li and Ding, 2017) ensures that with large n, the Mahalanobis distance M is approximately χ_K^2 under the CRE. Therefore, it is likely that M has a large realized value under the CRE with asymptotic mean K and variance $2K$. Rerandomization avoids covariate imbalance by discarding the treatment allocations with large values of M. Below I give a formal definition of the rerandomization using the Mahalanobis distance (ReM), which was proposed by Cox (1982) and further studied by Morgan and Rubin (2012).

Definition 6.1 (ReM) *Draw $\boldsymbol{Z}$ from CRE and accept it if and only if*

$$M \leq a,$$

for some predetermined constant $a > 0$.

The problem of choosing a is similar to the problem of choosing the number of strata in the SRE, which is non-trivial in practice. At one extreme, $a = \infty$, we just conduct the CRE. At the other extreme, $a = 0$, there are very few feasible treatment allocations, and consequently, the experiment has little randomness, rendering randomization-based inference useless. As a compromise, we choose a small but not extremely small a, for example, $a = 0.001$ or some upper quantile of a χ_K^2 distribution.

ReM has many desirable properties. As mentioned above, it is invariant to linear transformations of the covariates. Moreover, it has nice geometric properties and elegant mathematical theory. This chapter will focus on ReM.

6.1.2 Statistical inference

An important question is how to analyze the data under ReM. Bruhn and McKenzie (2009) and Morgan and Rubin (2012) argued that we can always use the FRT as long as we simulate $\boldsymbol{Z}$ under the constraint $M \leq a$. This always yields finite-sample exact p-values under the sharp null hypothesis. See Problem 6.1.

It is a challenging problem to derive the finite sample properties of ReM without assuming the sharp null hypothesis. Instead, Li et al. (2018b) derived the asymptotic distribution of the difference in means of the outcome $\hat{\tau}$ under ReM and the regularity conditions below.

Condition 6.1 *As $n \to \infty$,*

1. *n_1/n and n_0/n have positive limits;*
2. *the finite population covariance of $\{X_i, Y_i(1), Y_i(0), \tau_i\}$ has a finite limit;*
3. *$\max_{1 \leq i \leq n}\{Y_i(1) - \bar{Y}(1)\}^2/n \to 0$, $\max_{1 \leq i \leq n}\{Y_i(0) - \bar{Y}(0)\}^2/n \to 0$, and $\max_{1 \leq i \leq n} X_i^{\top} X_i/n \to 0$.*

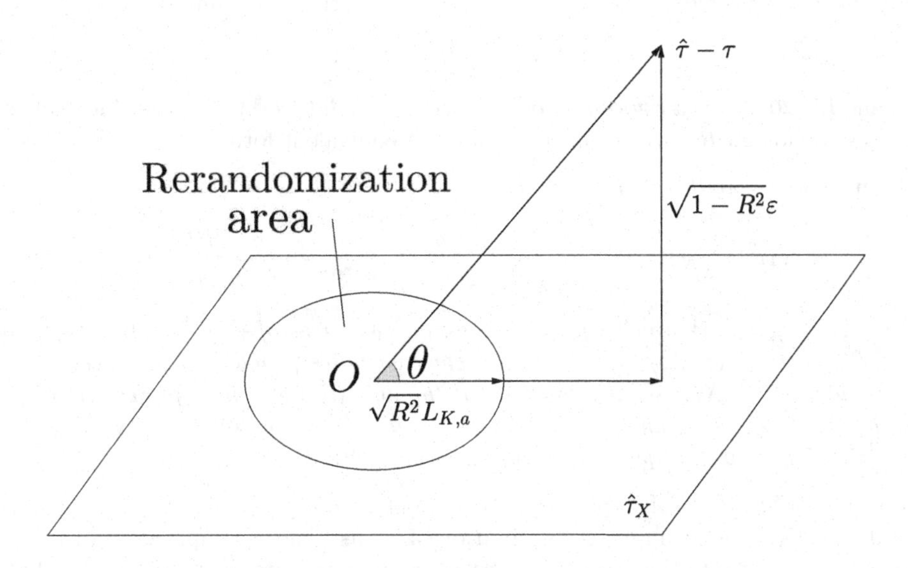

FIGURE 6.1: Geometry of the asymptotic distribution of $\hat{\tau}$ under ReM

Below is the main theorem for ReM, which relies on additional notation. Let

$$L_{K,a} \sim D_1 \mid \boldsymbol{D}^{\mathsf{T}}\boldsymbol{D} \le a$$

where $\boldsymbol{D} = (D_1, \ldots, D_K)$ follows a K-dimensional standard Normal distribution; let ε follow a univariate standard Normal distribution; $L_{K,a} \perp\!\!\!\perp \varepsilon$.

Theorem 6.1 *Under ReM with $M \le a$ and Condition 6.1, we have*[1]

$$\hat{\tau} - \tau \dot{\sim} \sqrt{\text{var}(\hat{\tau})}\left\{\sqrt{R^2}L_{K,a} + \sqrt{1-R^2}\varepsilon\right\},$$

where

$$\text{var}(\hat{\tau}) = \frac{S^2(1)}{n_1} + \frac{S^2(0)}{n_0} - \frac{S^2(\tau)}{n}$$

is Neyman's (1923) variance formula proved in Chapter 4, and

$$R^2 = \text{corr}^2(\hat{\tau}, \hat{\tau}_X)$$

is the squared multiple correlation coefficient (see Section A.1.1 for the definition) between $\hat{\tau}$ and $\hat{\tau}_X$ under the CRE.

Although the proof of Li et al. (2018b) is technical, the asymptotic distribution in Theorem 6.1 has a clear geometric interpretation, as shown in Figure 6.1. It shows that $\hat{\tau}$ decomposes into a component that is a linear combination of $\hat{\tau}_X$ and a component that is orthogonal to $\hat{\tau}_X$. Geometrically, $\cos^2\theta = R^2$, where θ is the angle between $\hat{\tau}$ and $\hat{\tau}_X$. ReM affects the first component but does not change the second component. The truncated Normal distribution $L_{K,a}$ is due to the restriction of ReM on the first component.

When $a = \infty$, the asymptotic distribution simplifies to the one under the CRE:

$$\hat{\tau} - \tau \dot{\sim} \sqrt{\text{var}(\hat{\tau})}\varepsilon.$$

[1] The notation "$A \dot{\sim} B$" means that A and B have the same asymptotic distributions.

When the threshold a is close to zero, the asymptotic distribution simplifies to

$$\hat{\tau} - \tau \dot{\sim} \sqrt{\text{var}(\hat{\tau})(1 - R^2)}\varepsilon;$$

see Wang and Li (2022) for a rigorous proof. So with a small threshold a, the efficiency gain due to ReM depends on R^2, which has the following equivalent form.

Proposition 6.1 *Under the CRE, we have*

$$R^2 = \text{corr}^2(\hat{\tau}, \hat{\tau}_X) = \frac{n_1^{-1} S^2(1 \mid X) + n_0^{-1} S^2(0 \mid X) - n^{-1} S^2(\tau \mid X)}{n_1^{-1} S^2(1) + n_0^{-1} S^2(0) - n^{-1} S^2(\tau)},$$

where $\{S^2(1), S^2(0), S^2(\tau)\}$ *are the finite population variances of* $\{Y_i(1), Y_i(0), \tau_i\}_{i=1}^n$, *and* $\{S^2(1 \mid x), S^2(0 \mid x), S^2(\tau \mid x)\}$ *are the corresponding finite population variances of their linear projections on* $(1, X_i)$; *see Section B.2 for the definition of linear projections. Under the constant causal effect assumption with* $\tau_i = \tau$, *the* R^2 *reduces to* $S^2(0 \mid X)/S^2(0)$, *the finite population squared multiple correlation between* $Y_i(0)$ *and* X_i.

I leave the proof of Proposition 6.1 to Problem 6.4.

When $0 < a < \infty$, the asymptotic distribution of $\hat{\tau}$ has a more complicated form and is more concentrated at τ and thus the difference in means $\hat{\tau}$ is more precise under ReM than under the CRE.

If we ignore the design of ReM and still use the confidence interval based on Neyman's (1923) variance formula and the Normal approximation, it is overly conservative and over-covers τ even if the individual causal effects are constant. Li et al. (2018b) proposed to construct confidence intervals based on Theorem 6.1. We omit the discussion here but will come back to the inference issue in Section 6.3.

6.2 Regression adjustment

What if we do not conduct rerandomization in the design stage but want to adjust for covariate imbalance in the analysis stage of the CRE? We will discuss several regression adjustment strategies.

6.2.1 Covariate-adjusted FRT

The covariates $\boldsymbol{X}$ are all fixed, and furthermore, under $H_{0\text{F}}$, the observed outcomes are all fixed. Therefore, we can simulate the distribution of any test statistic $T = T(\boldsymbol{Z}, \boldsymbol{Y}, \boldsymbol{X})$ and calculate the p-value. The basic idea of the FRT remains the same in the presence of additional covariates.

There are two general strategies to construct the test statistic. Problem 3.6 before hints at both of them. I summarize them below, using the terminology from Zhao and Ding (2021).

Definition 6.2 (pseudo-outcome strategy for covariate-adjusted FRT) *We can construct the test statistic based on residuals from fitted statistical models. We can regress* Y_i *on* X_i *to obtain residual* $\hat{\varepsilon}_i$, *and then treat* $\hat{\varepsilon}_i$ *as the pseudo outcome to construct test statistics.*

Definition 6.3 (model-output strategy for covariate-adjusted FRT) *We can use a regression coefficient as a test statistic. We can regress* Y_i *on* (Z_i, X_i) *to obtain the coefficient of* Z_i *as the test statistic.*

In the pseudo-outcome strategy in Definition 6.2, the regression of Y_i on X_i should not include the treatment Z_i because we want to ensure that the pseudo outcome satisfies H_{0F} if the original outcome satisfies H_{0F}. In the model-output strategy in Definition 6.3, the regression of Y_i on (Z_i, X_i) includes the treatment Z_i because we want to use the coefficient of Z_i to measure the deviation from H_{0F}. Computationally, in strategy one, we only need to run regression once, whereas in strategy two, we need to run regression many times.

In the above, "regression" is a generic term, which can be linear regression, logistic regression, or even machine learning algorithms. The FRT with any test statistics from these two strategies will be finite-sample exact under H_{0F} although they differ under alternative hypotheses. The rest of this section will review some test statistics based on OLS.

6.2.2 Analysis of covariance and extensions

Now we turn to direct estimation of the average causal effect τ that adjusts for the observed covariates.

Historically, Fisher (1925) proposed to use the analysis of covariance (ANCOVA) to improve estimation efficiency. This remains a standard strategy in many fields. He suggested running the OLS of Y_i on $(1, Z_i, X_i)$ and obtaining the coefficient of Z_i as an estimator for τ. Let $\hat{\tau}_F$ denote Fisher's ANCOVA estimator.

A former Berkeley Statistics Professor, David A. Freedman, reanalyzed Fisher's ANCOVA under Neyman's (1923) potential outcomes framework. Freedman (2008a,b) found the following negative results:

1. $\hat{\tau}_F$ is biased, but the simple difference in means $\hat{\tau}$ is unbiased.
2. The asymptotic variance of $\hat{\tau}_F$ may be even larger than that of $\hat{\tau}$ when $n_1 \neq n_0$.
3. The standard error from the OLS is inconsistent for the true standard error of $\hat{\tau}_F$ under the CRE.

A former Berkeley Ph.D. student, Winston Lin, wrote a thesis in response to Freedman's critiques. Lin (2013) found the following positive results:

1. The bias of $\hat{\tau}_F$ is small in large samples, and it goes to zero as the sample size approaches infinity.
2. We can improve the asymptotic efficiency of both $\hat{\tau}$ and $\hat{\tau}_F$ by using the coefficient of Z_i in the OLS of Y_i on $(1, Z_i, X_i, Z_i \times X_i)$. Let $\hat{\tau}_L$ denote Lin's (2013) estimator. Moreover, the EHW standard error is a conservative estimator for the true standard error of $\hat{\tau}_L$ under the CRE.
3. The EHW standard error[2] for $\hat{\tau}_F$ in the OLS fit of Y_i on $(1, Z_i, X_i)$ is a conservative estimator for the true standard error of $\hat{\tau}_F$ under the CRE.

6.2.2.1 Some heuristics for Lin's (2013) results

Neyman's (1923) result demonstrates that the variance of the difference-in-means estimator $\hat{\tau}$ depends on the variances of the potential outcomes. Intuitively, we can reduce the variance

[2] Without covariates, the HC2 correction yields identical variance estimator as Neyman's (1923) classic one; see Problem 4.3. For coherence, we can also use the HC2 correction for Lin's (2013) estimator with covariate adjustment. When the number of covariates is small compared with the sample size and the covariates do not contain outliers, the variants of the EHW standard error perform similarly to the original one. When the number of covariates is large compared with the sample size or the covariates contain outliers, the variants can outperform the original one. In those cases, Lei and Ding (2021) recommend using the HC3 variant of the EHW standard error. Under a different model, Long and Ervin (2000) also recommended the HC3 correction based on finite-sample simulation of OLS.

of the estimator by reducing the variances of the potential outcomes. A family of linearly adjusted estimators is

$$\begin{aligned} \hat{\tau}(\beta_1, \beta_0) &= n_1^{-1} \sum_{i=1}^{n} Z_i (Y_i - \beta_1^{\mathsf{T}} X_i) - n_0^{-1} \sum_{i=1}^{n} (1 - Z_i)(Y_i - \beta_0^{\mathsf{T}} X_i) \quad (6.2) \\ &= \left\{ \hat{\bar{Y}}(1) - \beta_1^{\mathsf{T}} \hat{\bar{X}}(1) \right\} - \left\{ \hat{\bar{Y}}(0) - \beta_0^{\mathsf{T}} \hat{\bar{X}}(0) \right\}, \quad (6.3) \end{aligned}$$

where $\{\hat{\bar{Y}}(1), \hat{\bar{Y}}(0)\}$ are the sample means of the outcomes, and $\{\hat{\bar{X}}(1), \hat{\bar{X}}(0)\}$ are the sample means of the covariates. This covariate-adjusted estimator $\hat{\tau}(\beta_1, \beta_0)$ tries to reduce the variance of $\hat{\tau}$ by residualizing the potential outcomes. It reduces to $\hat{\tau}$ with $\beta_1 = \beta_0 = 0$. It has mean τ for any fixed values of β_1 and β_0 because $\bar{X} = 0$. We are interested in finding the (β_1, β_0) that minimizes the variance of $\hat{\tau}(\beta_1, \beta_0)$. This estimator is essentially the difference in means of the adjusted potential outcomes $\{Y_i(1) - \beta_1^{\mathsf{T}} X_i, Y_i(0) - \beta_0^{\mathsf{T}} X_i\}_{i=1}^{n}$. Applying Neyman's (1923) result, this estimator has variance

$$\operatorname{var}\{\hat{\tau}(\beta_1, \beta_0)\} = \frac{S^2(1; \beta_1)}{n_1} + \frac{S^2(0; \beta_0)}{n_0} - \frac{S^2(\tau; \beta_1, \beta_0)}{n},$$

where $S^2(z; \beta_z)$ $(z = 1, 0)$ and $S^2(\tau; \beta_1, \beta_0)$ are the finite population variances of the adjusted potential outcomes and individual causal effects, respectively; moreover, a conservative variance estimate is

$$\hat{V}(\beta_1, \beta_0) = \frac{\hat{S}^2(1; \beta_1)}{n_1} + \frac{\hat{S}^2(0; \beta_0)}{n_0},$$

where

$$\begin{aligned} \hat{S}^2(1; \beta_1) &= (n_1 - 1)^{-1} \sum_{i=1}^{n} Z_i (Y_i - \gamma_1 - \beta_1^{\mathsf{T}} X_i)^2, \\ \hat{S}^2(0; \beta_0) &= (n_0 - 1)^{-1} \sum_{i=1}^{n} (1 - Z_i)(Y_i - \gamma_0 - \beta_0^{\mathsf{T}} X_i)^2 \end{aligned}$$

are the sample variances of the adjusted potential outcomes with γ_1 and γ_0 being the sample means of $Y_i - \beta_1^{\mathsf{T}} X_i$ under treatment and $Y_i - \beta_0^{\mathsf{T}} X_i$ under control. To minimize $\hat{V}(\beta_1, \beta_0)$, we need to solve two OLS problems:[3]

$$\begin{aligned} &\min_{\gamma_1, \beta_1} \sum_{i=1}^{n} Z_i (Y_i - \gamma_1 - \beta_1^{\mathsf{T}} X_i)^2, \\ &\min_{\gamma_0, \beta_0} \sum_{i=1}^{n} (1 - Z_i)(Y_i - \gamma_0 - \beta_0^{\mathsf{T}} X_i)^2. \end{aligned}$$

We run OLS of Y_i on X_i for the treatment and control groups separately and obtain $(\hat{\gamma}_1, \hat{\beta}_1)$ and $(\hat{\gamma}_0, \hat{\beta}_0)$. The final estimator is

$$\begin{aligned} \hat{\tau}(\hat{\beta}_1, \hat{\beta}_0) &= n_1^{-1} \sum_{i=1}^{n} Z_i (Y_i - \hat{\beta}_1^{\mathsf{T}} X_i) - n_0^{-1} \sum_{i=1}^{n} (1 - Z_i)(Y_i - \hat{\beta}_0^{\mathsf{T}} X_i) \\ &= \left\{ \hat{\bar{Y}}(1) - \hat{\beta}_1^{\mathsf{T}} \hat{\bar{X}}(1) \right\} - \left\{ \hat{\bar{Y}}(0) - \hat{\beta}_0^{\mathsf{T}} \hat{\bar{X}}(0) \right\}. \end{aligned}$$

[3]We can also minimize the true variance of $\hat{\tau}(\beta_1, \beta_0)$. See Problem 6.6 for more details.

From the properties of the OLS fits (see (B.5)), we know

$$\begin{aligned}\hat{\bar{Y}}(1) &= \hat{\gamma}_1 + \hat{\beta}_1^{\mathsf{T}} \hat{\bar{X}}(1),\\ \hat{\bar{Y}}(0) &= \hat{\gamma}_0 + \hat{\beta}_0^{\mathsf{T}} \hat{\bar{X}}(0).\end{aligned}$$

Therefore, we can rewrite the estimator as

$$\hat{\tau}(\hat{\beta}_1, \hat{\beta}_0) = \hat{\gamma}_1 - \hat{\gamma}_0. \tag{6.4}$$

The equivalent form in (6.4) suggests that we can obtain $\hat{\tau}(\hat{\beta}_1, \hat{\beta}_0)$ from a single OLS fit below.

Proposition 6.2 *The estimator $\hat{\tau}(\hat{\beta}_1, \hat{\beta}_0)$ in (6.4) equals the coefficient of Z_i in the OLS fit of Y_i on $(1, Z_i, X_i, Z_i \times X_i)$, which is Lin's (2013) estimator $\hat{\tau}_{\mathrm{L}}$ introduced before.*

I leave the proof of Proposition 6.2 to Problem 6.7, which is a pure linear algebra fact. Based on the discussion above, a conservative variance estimator for $\hat{\tau}_{\mathrm{L}}$ is

$$\begin{aligned}\hat{V}(\hat{\beta}_1, \hat{\beta}_0) &= \frac{1}{n_1(n_1-1)} \sum_{i=1}^{n} Z_i (Y_i - \hat{\gamma}_1 - \hat{\beta}_1^{\mathsf{T}} X_i)^2 \\ &\quad + \frac{1}{n_0(n_0-1)} \sum_{i=1}^{n} (1-Z_i)(Y_i - \hat{\gamma}_0 - \hat{\beta}_0^{\mathsf{T}} X_i)^2.\end{aligned}$$

Based on quite technical calculations, Lin (2013) further showed that the EHW standard error from the OLS in Proposition 6.2 is almost identical to $\hat{V}(\hat{\beta}_1, \hat{\beta}_0)$ which is a conservative estimator of the true standard error of $\hat{\tau}_{\mathrm{L}}$ under the CRE. Intuitively, this is because we do not assume that the linear model is correctly specified, and the EHW standard error is robust to model misspecification.

There is a subtle issue with the discussion above. The variance formula $\mathrm{var}\{\hat{\tau}(\beta_1, \beta_0)\}$ works for fixed (β_1, β_0), but the estimator $\hat{\tau}(\hat{\beta}_1, \hat{\beta}_0)$ uses two estimated coefficients $(\hat{\beta}_1, \hat{\beta}_0)$. The additional uncertainty in the estimated coefficients may cause finite-sample bias in the final estimator. Lin (2013) showed that this issue goes away asymptotically because $\hat{\tau}(\hat{\beta}_1, \hat{\beta}_0)$ behaves similarly to $\hat{\tau}(\tilde{\beta}_1, \tilde{\beta}_0)$, where $\tilde{\beta}_1$ and $\tilde{\beta}_0$ are the limit of $\hat{\beta}_1$ and $\hat{\beta}_0$, respectively. Heuristically, the difference between $\hat{\tau}(\hat{\beta}_1, \hat{\beta}_0)$ and $\hat{\tau}(\tilde{\beta}_1, \tilde{\beta}_0)$ depends on

$$(\hat{\beta}_z - \tilde{\beta}_z)^{\mathsf{T}} \hat{\bar{X}}(z), \quad (z = 0, 1).$$

The central limit theorem ensures that $\hat{\beta}_z - \tilde{\beta}_z$ behaves like a term of order $n^{-1/2}$, and $\hat{\bar{X}}(z)$ also behaves like a term of order $n^{-1/2}$ because $\bar{X}$ is centered at 0. Therefore, $(\hat{\beta}_z - \tilde{\beta}_z)^{\mathsf{T}} \hat{\bar{X}}(z)$ is a product of two terms of small order, and therefore, we can ignore the difference between $\hat{\tau}(\hat{\beta}_1, \hat{\beta}_0)$ and $\hat{\tau}(\tilde{\beta}_1, \tilde{\beta}_0)$ when the large size n is large. As a warning, the asymptotic theory requires a large sample size and some regularity conditions on the potential outcomes and covariates. In finite samples, regression adjustment can be even harmful due to the additional uncertainty in the estimated coefficients $\hat{\beta}_1$ and $\hat{\beta}_0$.

6.2.2.2 Understanding Lin's (2013) estimator via predicting the potential outcomes

We can view Lin's (2013) estimator as a *predictive estimator* based on OLS fits of the potential outcomes on the covariates. We build a prediction model for $Y(1)$ based on X using the data from the treatment group:

$$\hat{\mu}_1(X) = \hat{\gamma}_1 + \hat{\beta}_1^{\mathsf{T}} X. \tag{6.5}$$

TABLE 6.1: Predicting the potential outcomes

X	Z	$Y(1)$	$Y(0)$	$\hat{Y}(1)$	$\hat{Y}(0)$
X_1	1	$Y_1(1)$	?	$\hat{\mu}_1(X_1)$	$\hat{\mu}_0(X_1)$
$\vdots$					
X_{n_1}	1	$Y_{n_1}(1)$	?	$\hat{\mu}_1(X_{n_1})$	$\hat{\mu}_0(X_{n_1})$
X_{n_1+1}	0	?	$Y_{n_1+1}(0)$	$\hat{\mu}_1(X_{n_1+1})$	$\hat{\mu}_0(X_{n_1+1})$
$\vdots$					
X_n	0	?	$Y_n(0)$	$\hat{\mu}_1(X_n)$	$\hat{\mu}_0(X_n)$

Similarly, we build a prediction model for $Y(0)$ based on X using the data from the control group:

$$\hat{\mu}_0(X) = \hat{\gamma}_0 + \hat{\beta}_0^{\mathsf{T}} X. \tag{6.6}$$

Table 6.1 illustrates the prediction of the potential outcomes. If we predict the missing potential outcomes, then we have the following predictive estimator:

$$\hat{\tau}_{\text{pred}} = n^{-1} \left\{ \sum_{Z_i=1} Y_i + \sum_{Z_i=0} \hat{\mu}_1(X_i) - \sum_{Z_i=1} \hat{\mu}_0(X_i) - \sum_{Z_i=0} Y_i \right\}. \tag{6.7}$$

We can verify that with (6.5) and (6.6), the predictive estimator equals Lin's (2013) estimator:

$$\hat{\tau}_{\text{pred}} = \hat{\tau}_{\text{L}}. \tag{6.8}$$

If we predict all potential outcomes even if they are observed, we have the following *projective estimator*:

$$\hat{\tau}_{\text{proj}} = n^{-1} \sum_{i=1}^{n} \{\hat{\mu}_1(X_i) - \hat{\mu}_0(X_i)\}. \tag{6.9}$$

We can verify that with (6.5) and (6.6), the projective estimator equals Lin's (2013) estimator:

$$\hat{\tau}_{\text{proj}} = \hat{\tau}_{\text{L}}. \tag{6.10}$$

I leave the proofs of (6.8) and (6.10) to Problem 6.8.

The terminology "predictive" and "projective" is from the survey sampling literature (Firth and Bennett, 1998; Ding and Li, 2018). The more general formulas (6.7) and (6.9) are well-defined with other predictors of the potential outcomes. To make connections with Lin's (2013) estimator, I focus on the linear predictors here. The predictors $\hat{\mu}_1(X)$ and $\hat{\mu}_0(X)$ can be quite general, including much more complicated machine learning algorithms. However, constructing a point estimator is just the first step in analyzing the CRE. A more important second step is to quantify the uncertainty associated with the estimator, which depends on the properties of the predictors of the potential outcomes. Nevertheless, without doing additional theoretical analysis, we can always use (6.7) and (6.9) as the test statistics in the FRT.

6.2.2.3 Understanding Lin's (2013) estimator via adjusting for covariate imbalance

The linearly adjusted estimator has an equivalent form

$$\hat{\tau}(\beta_1, \beta_0) = \hat{\tau} - \gamma^{\mathsf{T}} \hat{\tau}_X \tag{6.11}$$

where $\gamma = \frac{n_0}{n}\beta_1 + \frac{n_1}{n}\beta_0$, so we can also write it as $\hat{\tau}(\gamma) = \hat{\tau}(\beta_1, \beta_0)$. Similarly, Lin's (2013) estimator has an equivalent form

$$\hat{\tau}_{\mathrm{L}} = \hat{\tau} - \hat{\gamma}^{\mathsf{T}} \hat{\tau}_X, \tag{6.12}$$

where $\hat{\gamma} = \frac{n_0}{n}\hat{\beta}_1 + \frac{n_1}{n}\hat{\beta}_0$. I leave the proofs of (6.11) and (6.12) to Problem 6.9. The forms (6.11) and (6.12) are the mathematical statements of "adjusting for the covariate imbalance." They essentially subtract some linear combinations of the difference in means of the covariates. Since $\hat{\tau}$ and $\hat{\tau}_X$ are correlated under the CRE, the covariate adjustment with an appropriately chosen γ reduces the variance of $\hat{\tau}$. Another interesting feature of (6.11) and (6.12) is that the final estimators depend only on γ or $\hat{\gamma}$, so the choice of the β-coefficients is not unique. Li and Ding (2020) pointed out this simple fact. Therefore, Lin's (2013) estimator is just one of the optimal estimators. However, it can be easily implemented via the standard OLS with the EHW standard error. That's why this book focuses on it.

6.2.3 Some additional remarks on regression adjustment

6.2.3.1 Equivalence of regression adjustment and post-stratification

If we have discrete covariate C_i with K categories, we can create $K-1$ centered dummy variables

$$X_i = (I(C_i = 1) - \pi_{[1]}, \ldots, I(C_i = K-1) - \pi_{[K-1]})$$

where $\pi_{[k]}$ equals the proportion of units with $C_i = k$. In this case, Lin's (2013) regression adjustment is equivalent to post-stratification, as summarized by the following proposition.

Proposition 6.3 *$\hat{\tau}_{\mathrm{L}}$ based in X_i is numerically identical to the post-stratification estimator $\hat{\tau}_{\mathrm{PS}}$ based on C_i (see Section 5.4).*

I leave the proof of Proposition 6.3 as Problem 6.11.

6.2.3.2 Difference-in-differences as a special case of covariate adjustment $\hat{\tau}(\beta_1, \beta_0)$

An important covariate X in many studies is the lagged outcome before the treatment. For instance, the covariate X is the pre-test score if the outcome Y is the post-test score in educational research; the covariate X is the log wage before the job training program if the outcome Y is the log wage after the job training program. With the lagged outcome X as a covariate, a popular estimator is the *gain score* or *difference-in-differences* estimator with $\beta_1 = \beta_0 = 1$ in (6.2) and (6.3):

$$\begin{aligned}
\hat{\tau}(1,1) &= n_1^{-1}\sum_{i=1}^{n} Z_i(Y_i - X_i) - n_0^{-1}\sum_{i=1}^{n}(1-Z_i)(Y_i - X_i) \\
&= \left\{\hat{\bar{Y}}(1) - \hat{\bar{Y}}(0)\right\} - \left\{\hat{\bar{X}}(1) - \hat{\bar{X}}(0)\right\}.
\end{aligned}$$

The first form of $\hat{\tau}(1,1)$ justifies the name *gain score* because it is essentially the difference in means of the gain score $g_i = Y_i - X_i$. The second form of $\hat{\tau}(1,1)$ justifies the name *difference-in-differences* because it is the difference between two differences in means. This estimator is different from Lin's (2013) estimator: it fixes $\beta_1 = \beta_0 = 1$ in advance while Lin's (2013) estimator involves two estimated β's. The $\hat{\tau}(1,1)$ is unbiased with a conservative variance estimator

$$\begin{aligned}\hat{V}(1,1) &= \frac{1}{n_1(n_1-1)}\sum_{i=1}^{n} Z_i\{g_i - \hat{\bar{g}}(1)\}^2 \\ &\quad + \frac{1}{n_0(n_0-1)}\sum_{i=1}^{n}(1-Z_i)\{g_i - \hat{\bar{g}}(0)\}^2,\end{aligned}$$

where $\hat{\bar{g}}(1)$ and $\hat{\bar{g}}(0)$ are the sample means of the gain score $g_i = Y_i - X_i$ under the treatment and control, respectively. When the lagged outcome is a strong predictor of the outcome, the gain score $g_i = Y_i - X_i$ often has a much smaller variance than the outcome itself. In this case, $\hat{\tau}(1,1)$ often greatly reduces the variance of the simple difference in means of the outcome. See Problem 6.12 for more details.

In theory, Lin's (2013) estimator is always more efficient than $\hat{\tau}(1,1)$ with large samples. However, Lin's (2013) estimator is biased in finite samples, whereas $\hat{\tau}(1,1)$ is always unbiased.

6.2.4 Extension to the SRE

It is possible that we have an experiment stratified on a discrete variable C and we also observe additional covariates X. If all strata are large, then we can obtain Lin's (2013) estimators within strata $\hat{\tau}_{\mathrm{L},[k]}$ and obtain the final estimator as

$$\hat{\tau}_{\mathrm{L,S}} = \sum_{k=1}^{K} \pi_{[k]}\hat{\tau}_{\mathrm{L},[k]}.$$

A conservative variance estimator is

$$\hat{V}_{\mathrm{L,S}} = \sum_{k=1}^{K} \pi_{[k]}^2 \hat{V}_{\mathrm{EHW},[k]},$$

where $\hat{V}_{\mathrm{EHW},[k]}$ is the EHW variance estimator from the OLS fit of the outcome on the intercept, the treatment indicator, the covariates, and their interactions within stratum k. Importantly, we need to center covariates by their stratum-specific means.

6.3 Discussion

6.3.1 Duality between ReM and regression adjustment

Li et al. (2018b) pointed out that ReM and Lin's (2013) regression adjustment are duals in using covariates in the design and analysis stages of the experiment. To be more specific, when a is small, the asymptotic distribution of $\hat{\tau}$ under ReM is almost identical to the asymptotic distribution of $\hat{\tau}_{\mathrm{L}}$ under the CRE. So ReM uses covariates in the design stage and Lin's (2013) regression adjustment uses covariates in the analysis stage, achieving nearly the same asymptotic efficiency gain when a is small.

TABLE 6.2: Design and analysis of experiments

		analysis		
design	CRE	$\hat{\tau}$ (Neyman, 1923)	$\xrightarrow{1}$	$\hat{\tau}_{\mathrm{L}}$ (Lin, 2013)
		$2\downarrow$		$\downarrow 4$
	ReM	$\hat{\tau}$ (Li et al., 2018b)	$\xrightarrow{3}$	$\hat{\tau}_{\mathrm{L}}$ (Li and Ding, 2020)

6.3.2 Unification of ReM and regression adjustment

Li and Ding (2020) unified the literature of ReM and regression adjustment. Table 6.2 summarizes the literature from Neyman (1923) to Li and Ding (2020). Arrow 1 illustrates the efficiency gain of covariate adjustment in the CRE: asymptotically, $\hat{\tau}_{\mathrm{L}}$ has a smaller variance than $\hat{\tau}$. Arrow 2 illustrates the efficiency gain of ReM: asymptotically, $\hat{\tau}$ has narrower quantile ranges under the ReM than under the CRE. Arrows 3 and 4 illustrate the benefits of the combination of ReM and the CRE.

6.3.3 Final recommendation

Based on theoretical analysis, Li and Ding (2020) recommended that we can combine rerandomization and regression adjustment. That is, if we rerandomize in the design stage, we can use Lin's (2013) estimator with the EHW standard error in the analysis stage. The combination of rerandomization and regression adjustment improves covariate balance in the design stage and estimation efficiency in the analysis stage.

6.4 Simulation

Angrist et al. (2009) conducted an experiment to evaluate different strategies to improve academic performance among college freshmen in a Canadian university. Here I use a subset of the original data, focusing on the control group and the treatment group offered academic support services and financial incentives for good grades. The outcome is the GPA at the end of the first year. I impute the missing outcomes with the observed average which is somewhat arbitrary (see Problem 6.14). Two covariates are the gender and baseline GPA.

```
> library("car")
> library("foreign")
> ## Angrist 2009 data: Canadian university - Section 6.4
> angrist   = read.dta("star.dta")
> angrist2  = subset(angrist, control == 1|sfsp == 1)
> ## imputing missing outcomes
> y = angrist2$GPA_year1
> meany    = mean(y, na.rm = TRUE)
> y = ifelse(is.na(y), meany, y)
> z = angrist2$sfsp
> x = angrist2[, c("female", "gpa0")]
```

The following code gives the results based on the unadjusted and adjusted estimators.

```
> ## unadjusted estimator
> fit_unadj  = lm(y ~ z)
> ace_unadj  = coef(fit_unadj)[2]
```

```
> se_unadj   = sqrt(hccm(fit_unadj, type = "hc2")[2, 2])
>
> ## regression adjustment
> x          = scale(x)
> fit_adj    = lm(y ~ z*x)
> ace_adj    = coef(fit_adj)[2]
> se_adj     = sqrt(hccm(fit_adj, type = "hc2")[2, 2])
>
> res = c(ace_unadj, ace_adj, se_unadj, se_adj)
> dim(res) = c(2, 2)
> t.stat   = res[, 1]/res[, 2]
> p.value  = 2*pnorm(abs(t.stat), lower.tail = FALSE)
> res      = cbind(res, t.stat, p.value)
> rownames(res) = c("Neyman", "Lin")
> colnames(res) = c("estimate", "s.e.", "t-stat", "p-value")
> round(res, 3)
       estimate  s.e. t-stat p-value
Neyman    0.052 0.078  0.669   0.504
Lin       0.068 0.074  0.925   0.355
```

The adjusted estimator has a smaller standard error although it gives the same insignificant result as the unadjusted estimator.

I also use this dataset to conduct simulation studies to evaluate the four design and analysis strategies summarized in Table 6.2. I fit quadratic functions of the outcome on the covariates and use them to impute all the missing potential outcomes, separately for the treated and control groups. To show the improvement of ReM and regression adjustment, I also rescale the error terms by 0.1 and 0.25 to increase the signal-to-noise ratio. With the imputed Science Table, I generate 2000 treatments, obtain the observed data, and calculate the estimators. In the simulation, the "true" outcome model is nonlinear, but we still use linear adjustment for estimation. By doing this, we can evaluate the properties of the estimators even when the linear model is misspecified.

Figure 6.2 shows the violin plots[4] of the four combinations, subtracting the true τ from the estimates. As predicted by the theory, all estimators are nearly unbiased, and both ReM and regression adjustment improve efficiency. They are more effective when the noise level is smaller.

6.5 Final remarks

ReM uses the Mahalanobis distance as the balance criterion. We can consider general rerandomization with the balance criterion defined as a function of $\boldsymbol{Z}$ and $\boldsymbol{X}$. For example, we can use the following criterion based on marginal tests for all coordinates of $X_i = (x_{i1}, \dots, x_{iK})^\top$. We accept $\boldsymbol{Z}$ if and only if

$$\left| \frac{\hat{\tau}_{xk}}{\sqrt{\frac{n}{n_1 n_0} S^2_{xk}}} \right| \leq a \quad (k = 1, \dots, K) \tag{6.13}$$

[4]A violin plot is similar to a box plot, with the addition of a rotated kernel density plot on each side.

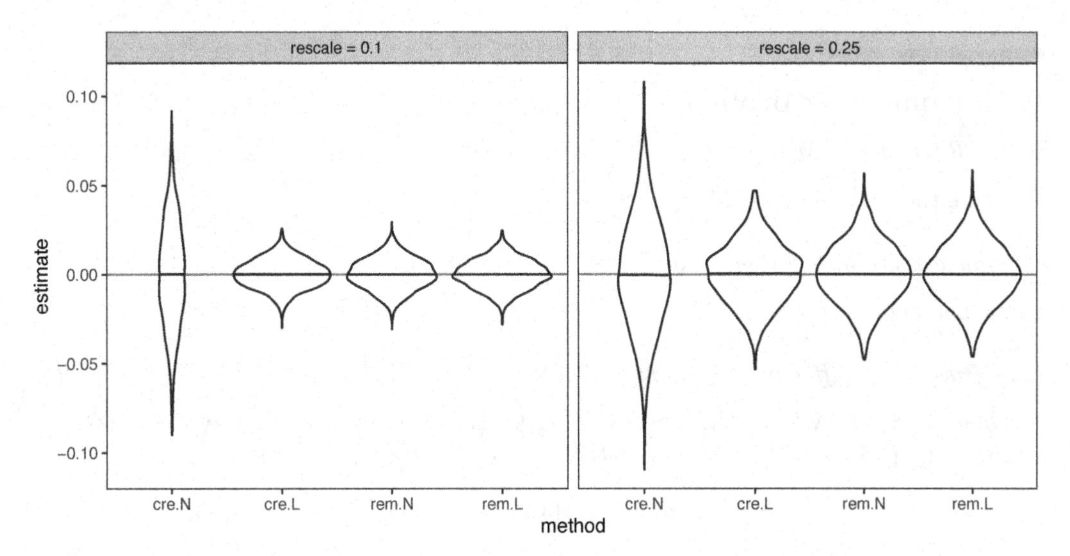

FIGURE 6.2: Simulation with 2000 Monte Carlo replicates and $a = 0.05$ for ReM

for some predetermined constant $a > 0$, where S^2_{xk} is the finite-population variance of covariate x_{ik}. For example, a is some upper quantile of a standard Normal distribution. See Zhao and Ding (2024) for the theory for the rerandomization based on criterion (6.13) as well as other criteria.

With a continuous outcome, Fisher's ANCOVA has been the standard approach for many years. Lin's (2013) improvement has better theoretical properties even when the linear model is misspecified. With a binary outcome, it is common to use the coefficient of the treatment in the logistic regression of the observed outcome on the intercept, the treatment indicator, and covariates to estimate the causal effects. However, Freedman (2008c) showed that this logistic regression does not have nice properties under the potential outcomes framework. Even if the logistic model is correct, the coefficient estimates the conditional odds ratio (see Section B.6) which may not be the parameter of interest; when the logistic model is incorrect, it is even harder to interpret the coefficient. From the discussion above, if the parameter of interest is the average causal effect, we can still use Lin's (2013) estimator to analyze the binary outcome data in the CRE. Guo and Basse (2023) extend Lin's (2013) theory to allow for using generalized linear models to construct estimators for the average causal effect under the potential outcomes framework.

Other extensions of Lin's (2013) theory focus on high dimensional covariates. Bloniarz et al. (2016) focus on the regime with many covariates than the sample size, and suggest replacing the OLS fits with the least absolute shrinkage and selection operator (LASSO) fits (Tibshirani, 1996) of the outcome on the intercept, the treatment, covariates and their interactions. Lei and Ding (2021) focus on the regime with a diverging number of covariates without assuming sparsity, and under certain regularity conditions, they show that Lin's (2013) estimator is still consistent and asymptotically Normal. Wager et al. (2016) propose to use machine learning methods to analyze high dimensional experimental data.

6.6 Homework problems

6.1 FRT under ReM

Describe the FRT under ReM.

6.2 Invariance of the Mahalanobis Distance

Prove Lemma 6.1.

6.3 Bias of the difference-in-means estimator under rerandomization

Assume that we draw $\boldsymbol{Z} = (Z_1, \ldots, Z_n)$ from a CRE and accept it if and only if $\phi(\boldsymbol{Z}, \boldsymbol{X}) = 1$, where ϕ is a predetermined balance criterion. Show that if $n_1 = n_0$ and

$$\phi(\boldsymbol{Z}, \boldsymbol{X}) = \phi(\boldsymbol{1}_n - \boldsymbol{Z}, \boldsymbol{X}), \tag{6.14}$$

then $\hat{\tau}$ is unbiased for τ. Verify that rerandomization using the Mahalanobis distance satisfies (6.14) if $n_1 = n_0$. Give a counterexample that $\hat{\tau}$ is biased for τ when these two conditions do not hold.

Remark: $\phi(\boldsymbol{Z}, \boldsymbol{X})$ can be a general balance criterion in this problem. ReM is a special case with $\phi(\boldsymbol{Z}, \boldsymbol{X}) = I(M \leq a)$.

6.4 Equivalent form of R^2 in the CRE

Prove Proposition 6.1.

6.5 More on linear projections of the potential outcomes onto covariates

Show that

$$S^2(1 \mid X) = S_{Y(1)X}(S_X^2)^{-1}S_{XY(1)}$$

where $S_{XY(1)}$ is the finite population covariance between $Y_i(1)$'s and X_i's, $S_{Y(1)X} = S_{XY(1)}^{\mathrm{T}}$, and S_X^2 is the finite population covariance matrix of X_i's. Find the analogous formulas for $S^2(0 \mid X)$ and $S^2(\tau \mid X)$.

6.6 Comparing the true variances within the family of linearly adjusted estimator

Show that the variance $\hat{\tau}(\beta_1, \beta_0)$ decomposes into

$$\mathrm{var}\{\hat{\tau}(\beta_1, \beta_0)\} = \mathrm{var}\{\hat{\tau}(\tilde{\beta}_1, \tilde{\beta}_0)\} + \mathrm{var}\{\hat{\tau}(\beta_1, \beta_0) - \hat{\tau}(\tilde{\beta}_1, \tilde{\beta}_0)\}$$

where $\tilde{\beta}_1$ and $\tilde{\beta}_0$ are the coefficients of X_i in the OLS projection of $Y_i(1)$'s and $Y_i(0)$'s onto $(1, X_i)$'s, respectively.

Remark: Li and Ding (2017, Example 9) derived the decomposition. It shows that based on the true variance, $\hat{\tau}(\tilde{\beta}_1, \tilde{\beta}_0)$ is an optimal choice within the family of linearly adjusted estimator because $\mathrm{var}\{\hat{\tau}(\beta_1, \beta_0) - \hat{\tau}(\tilde{\beta}_1, \tilde{\beta}_0)\}$ is always non-negative.

6.7 Lin's (2013) estimator for covariate adjustment

Prove Proposition 6.2.

6.8 Predictive and projective estimators

Prove (6.8) and (6.10).

6.9 Equivalent form of the covariate-adjusted estimator

Prove (6.11) and (6.12).

6.10 ANCOVA also adjusts for covariate imbalance

This problem gives a result for ANCOVA that is similar to (6.12).

Show that

$$\hat{\tau}_{\text{F}} = \hat{\tau} - \hat{\gamma}_{\text{F}}^{\text{T}} \hat{\tau}_X,$$

where $\hat{\gamma}_{\text{F}}$ is the coefficient of X_i in the OLS fit of Y_i on $(1, Z_i, X_i)$.

6.11 Regression adjustment and post-stratification of CRE

Prove Proposition 6.3.

Remark: Sometimes $\hat{\tau}_{\text{PS}}$ or $\hat{\tau}_{\text{L}}$ may not be well-defined. In those cases, we treat $\hat{\tau}_{\text{PS}}$ and $\hat{\tau}_{\text{L}}$ as equal. You can ignore this complexity in the proof.

6.12 More on the difference-in-differences estimator in the CRE

This problem gives more details for the difference-in-differences estimator in the CRE in Section 6.2.3.2.

Show that $\hat{\tau}(1,1)$ is unbiased for τ, calculate its variance, and show that $\hat{V}(1,1)$ is a conservative estimator for the true variance of $\hat{\tau}(1,1)$. When does $E\{\hat{V}(1,1)\} = \text{var}\{\hat{\tau}(1,1)\}$ hold?

Compare the variances of $\hat{\tau}(0,0)$ and $\hat{\tau}(1,1)$ to show that

$$\text{var}\{\hat{\tau}(0,0)\} \geq \text{var}\{\hat{\tau}(1,1)\}$$

if and only if

$$2\frac{n_0}{n}\beta_1 + 2\frac{n_1}{n}\beta_0 \geq 1,$$

where

$$\begin{aligned} \beta_1 &= \frac{\sum_{i=1}^n (X_i - \bar{X})\{Y_i(1) - \bar{Y}(1)\}}{\sum_{i=1}^n (X_i - \bar{X})^2}, \\ \beta_0 &= \frac{\sum_{i=1}^n (X_i - \bar{X})\{Y_i(0) - \bar{Y}(0)\}}{\sum_{i=1}^n (X_i - \bar{X})^2} \end{aligned}$$

are the coefficients of X_i in the OLS fits of $Y_i(1)$ and $Y_i(0)$ on $(1, X_i)$, respectively.

Remark: Gerber and Green (2012, page 28) discussed a special case of this problem with $n_1 = n_0$.

6.13 Data reanalyses of the Penn Bonus Experiment

Reanalyze the Penn Bonus Experiment data. The analysis in Chapter 5 uses the treatment indicator, the outcome, and the block indicator. Now we want to use all other covariates.

Conduct regression adjustments within the strata of the experiment, and then combine these adjusted estimators to estimate the average causal effect. Report the point estimator, estimated standard error, and 95% confidence interval. Compare them with those without regression adjustment.

6.14 Missing outcomes in randomized experiments

The data analysis in Section 6.4 uses a naive imputation method to deal with missing outcomes. It is somewhat arbitrary.

Impute the missing outcomes under the treatment and control, respectively, based on the observed means. Do the results change?

Impute the missing outcomes under the treatment and control, respectively, based on linear regressions of the observed outcomes on the covariates. Do the results change?

Do you have other ways to deal with missing outcomes? Justify them and implement them to analyze the dataset.

6.15 Recommended reading

The title of this chapter is the same as that of Li and Ding (2020), which studied the roles of rerandomization and regression adjustment in the design and analysis stages of randomized experiments, respectively.

7

Matched-Pairs Experiment

The matched-pairs experiment (MPE) is the most extreme version of the SRE with only one treated unit and one control unit within each stratum. In this case, the strata are also called pairs. Although this type of experiment is a special case of the SRE discussed in Chapter 5, it has its own estimation and inference strategy. Moreover, it has many new features and it is closely related to the "matching" strategy in observational studies which will be covered in Chapter 15 later. So we discuss the MPE here, in its own chapter.

7.1 Design of the experiment and potential outcomes

Consider an experiment with $2n$ units. If we have predictive covariates to the outcomes, we can pair units based on the similarity of covariates. With a scalar covariate, we can order units based on this covariate and then form pairs based on the adjacent units. With many covariates, we can define pairwise distances between units and then form pairs based on these distances. In this case, pair matching can be done using a greedy algorithm or an optimal nonbipartite matching algorithm. The greedy algorithm pairs the two units with the smallest distance, drops them from the pool of units, pairs the two remaining units with the smallest distance, etc. The optimal nonbipartite matching algorithm divides the $2n$ units into n pairs of two units to minimize the sum of the within-pair distances. See Greevy et al. (2004) for more details of the computational aspect of the MPE. In this chapter, we assume that the pairs are formed based on the covariates, and discuss the subsequent design and analysis issues.

Let (i, j) index the unit j in pair i, where $i = 1, \ldots, n$ and $j = 1, 2$. Unit (i, j) has potential outcomes $Y_{ij}(1)$ and $Y_{ij}(0)$ under the treatment and control, respectively. Within each pair, we randomly assign one unit to receive the treatment and the other to receive the control. Let

$$Z_i = \begin{cases} 1, & \text{if the first unit receives the treatment,} \\ 0, & \text{if the second unit receives the treatment.} \end{cases}$$

We can formally define MPE based on the treatment assignment mechanism.

Definition 7.1 (MPE) *We have*

$$(Z_i)_{i=1}^n \overset{\text{IID}}{\sim} \textit{Bernoulli}(1/2). \tag{7.1}$$

The observed outcomes within pair i are

$$\begin{aligned} Y_{i1} &= Z_i Y_{i1}(1) + (1 - Z_i) Y_{i1}(0) \\ &= \begin{cases} Y_{i1}(1), & \text{if } Z_i = 1; \\ Y_{i1}(0), & \text{if } Z_i = 0; \end{cases} \end{aligned}$$

and

$$\begin{aligned} Y_{i2} &= Z_i Y_{i2}(0) + (1 - Z_i) Y_{i2}(1) \\ &= \begin{cases} Y_{i2}(0), & \text{if } Z_i = 1; \\ Y_{i2}(1), & \text{if } Z_i = 0. \end{cases} \end{aligned}$$

So the observed data are $(Z_i, Y_{i1}, Y_{i2})_{i=1}^n$.

7.2 FRT

Similar to the discussion before, we can always use the FRT to test the sharp null hypothesis:

$$H_{0\mathrm{F}} : Y_{ij}(1) = Y_{ij}(0) \text{ for all } i = 1, \dots n \text{ and } j = 1, 2.$$

When conducting the FRT, we need to simulate the distribution of $(Z_1, \dots, Z_n)$ from (7.1). I will discuss some canonical choices of test statistics based on the within-pair differences between the treated and control outcomes:

$$\begin{aligned} \hat{\tau}_i &= \text{outcome under treatment} - \text{outcome under control (within pair } i) \\ &= (2Z_i - 1)(Y_{i1} - Y_{i2}) \\ &= S_i(Y_{i1} - Y_{i2}), \end{aligned}$$

where the $S_i = 2Z_i - 1$ are IID random signs with mean 0 and variance 1, for $i = 1, \dots, n$. Since the pairs with zero $\hat{\tau}_i$'s do not contribute to the randomization distribution, we drop those pairs in the discussion of the FRT.

Example 7.1 (paired t statistic) *The average of the within-pair differences is*

$$\hat{\tau} = n^{-1} \sum_{i=1}^n \hat{\tau}_i.$$

Under $H_{0\mathrm{F}}$, $\hat{\tau}$ has mean

$$E(\hat{\tau}) = 0$$

and variance

$$\begin{aligned} \mathrm{var}(\hat{\tau}) &= n^{-2} \sum_{i=1}^n \mathrm{var}(\hat{\tau}_i) \\ &= n^{-2} \sum_{i=1}^n \mathrm{var}(S_i)(Y_{i1} - Y_{i2})^2 \\ &= n^{-2} \sum_{i=1}^n \hat{\tau}_i^2. \end{aligned}$$

Based on the CLT for the sum of independent random variables, we have the Normal approximation:

$$\frac{\hat{\tau}}{\sqrt{n^{-2} \sum_{i=1}^n \hat{\tau}_i^2}} \to \mathrm{N}(0, 1)$$

in distribution. We can use this Normal approximation to construct an asymptotic test. Many standard textbooks suggest using the following paired t statistic in the MPE:

$$t_{pair} = \frac{\hat{\tau}}{\sqrt{\{n(n-1)\}^{-1}\sum_{i=1}^{n}(\hat{\tau}_i - \hat{\tau})^2}},$$

which is almost identical to $\hat{\tau}$ with large n and small $\hat{\tau}$ under H_{0F}.

In classic statistics, the motivation for using t_{pair} is under a different framework. When $\hat{\tau}_i \stackrel{\text{IID}}{\sim} \mathrm{N}(0, \sigma^2)$, we can show that $t_{\text{pair}} \sim t(n-1)$, i.e., the exact distribution of t_{pair} is t with degrees of freedom $n-1$, which is close to $\mathrm{N}(0,1)$ with a large n. The `R` function `t.test` with `paired=TRUE` can implement this test. With a large n, these procedures give similar results. The discussion in Example 7.1 gives another justification of the classic paired t test without assuming the Normality of the data.

Example 7.2 (Wilcoxon sign-rank statistic) *Based on the ranks $(R_1, \ldots, R_n)$ of $(|\hat{\tau}_1|, \ldots, |\hat{\tau}_n|)$, we can define a test statistic*

$$W = \sum_{i=1}^{n} I(\hat{\tau}_i > 0) R_i.$$

Under H_{0F}, the $|\hat{\tau}_i|$'s are fixed so the R_i's are also fixed, which implies that W has mean

$$\begin{aligned} E(W) &= \frac{1}{2}\sum_{i=1}^{n} R_i \\ &= \frac{1}{2}\sum_{i=1}^{n} i \\ &= \frac{n(n+1)}{4} \end{aligned}$$

and variance

$$\begin{aligned} \mathrm{var}(W) &= \frac{1}{4}\sum_{i=1}^{n} R_i^2 \\ &= \frac{1}{4}\sum_{i=1}^{n} i^2 \\ &= \frac{n(n+1)(2n+1)}{24}. \end{aligned}$$

The CLT for the sum of independent random variables ensures the following Normal approximation:

$$\frac{W - n(n+1)/4}{\sqrt{n(n+1)(2n+1)/24}} \to \mathrm{N}(0,1)$$

in distribution. We can use this Normal approximation to construct an asymptotic test. The `R` function `wilcox.test` with `paired=TRUE` can implement both the exact and asymptotic tests.

Example 7.3 (Kolmogorov–Smirnov-type statistic) *Under H_{0F}, the absolute values $(|\hat{\tau}_1|, \ldots, |\hat{\tau}_n|)$ are fixed but their signs are random. So $(\hat{\tau}_1, \ldots, \hat{\tau}_n)$ and $-(\hat{\tau}_1, \ldots, \hat{\tau}_n)$ should have the same distribution. Let*

$$\hat{F}(t) = n^{-1}\sum_{i=1}^{n} I(\hat{\tau}_i \le t)$$

be the empirical distribution of $(\hat{\tau}_1, \ldots, \hat{\tau}_n)$, *and*

$$1 - \hat{F}(-t-) = n^{-1} \sum_{i=1}^{n} I(-\hat{\tau}_i \leq t)$$

be the empirical distribution of $(\hat{\tau}_1, \ldots, \hat{\tau}_n)$, *where* $\hat{F}(-t-)$ *is the left limit of the function* $\hat{F}(\cdot)$ *at* $-t$. *A Kolmogorov–Smirnov-type statistic is then*

$$D = \max_t |\hat{F}(t) + \hat{F}(-t-) - 1|.$$

Butler (1969) proposed this test statistic and derived its exact and asymptotic distributions. Unfortunately, this is not implemented in standard software packages. Nevertheless, we can simulate its exact distribution and compute the p-value based on the FRT. [1]

Example 7.4 (sign statistic) *The sign statistic uses only the signs of the within-pair differences*

$$\Delta = \sum_{i=1}^{n} I(\hat{\tau}_i > 0).$$

Under H_{0F},

$$I(\hat{\tau}_i > 0) \overset{\text{IID}}{\sim} \text{Bernoulli}(1/2)$$

and therefore

$$\Delta \sim \text{Binomial}(n, 1/2).$$

Based on this we have an exact Binomial test, which is implemented in the `R` *function* `binom.test` *with* `p=1/2`. *Using the CLT, we can also conduct a test based on the following Normal approximation of the Binomial distribution:*

$$\frac{\Delta - n/2}{\sqrt{n/4}} \to \text{N}(0, 1)$$

in distribution.

Example 7.5 (McNemar's statistic for a binary outcome) *If the outcome is binary, we can summarize the observed data from the MPE in a more compact way. Given a pair, the treated outcome can be either* 1 *or* 0 *and the control outcome can be either* 1 *or* 0, *yielding a two-by-two table as in Table 7.1.*

Under H_{0F}, *the numbers of concordant pairs* m_{11} *and* m_{00} *are fixed, and* $m_{10} + m_{01}$ *is also fixed. So the only random component is* m_{10} *which has distribution*

$$m_{10} \sim \text{Binomial}(m_{10} + m_{01}, 1/2).$$

[1] Butler (1969) proposed this test statistic under a slightly different framework. Given IID draws of $(\hat{\tau}_1, \ldots, \hat{\tau}_n)$ from a distribution $F(y)$, if they are symmetrically distributed around 0, then

$$F(t) = \text{pr}(\hat{\tau}_i \leq t) = \text{pr}(-\hat{\tau}_i \leq t) = 1 - \text{pr}(\hat{\tau}_i < -t) = 1 - F(-t-).$$

Therefore, $\hat{F}(t) + \hat{F}(-t-) - 1$ measures the deviation from the null hypothesis of symmetry, which motivates the definition of D in Example 7.3. Rothman and Woodroofe (1972) also attributed the test statistic D in Example 7.3 to Smirnov.

A naive definition of the Kolmogorov–Smirnov-type statistic is to compare the empirical distributions of the outcomes under treatment and control as in Example 3.4. Using that definition, we effectively break the pairs. Although it can still be used in the FRT for the MPE, it does not capture the matched-pairs structure of the experiment.

TABLE 7.1: Counts of four types of pairs

	control outcome 1	control outcome 0
treated outcome 1	m_{11}	m_{10}
treated outcome 0	m_{01}	m_{00}

This implies an exact test based on the Binomial distribution. The `R` *function* `mcnemar.test` *gives an asymptotic test based on the Normal approximation of the Binomial distribution:*

$$\frac{m_{10} - (m_{10} + m_{01})/2}{\sqrt{(m_{10} + m_{01})/4}} = \frac{m_{10} - m_{01}}{\sqrt{m_{10} + m_{01}}} \to \mathrm{N}(0,1)$$

in distribution. Both the exact FRT and the asymptotic test do not depend on m_{11} *or* m_{00}*. Only the numbers of discordant pairs matter in these tests.*

7.3 Neymanian inference

The average causal effect within pair i is

$$\tau_i = \frac{1}{2}\{Y_{i1}(1) + Y_{i2}(1) - Y_{i1}(0) - Y_{i2}(0)\},$$

and the average causal effect for all units is

$$\begin{aligned} \tau &= n^{-1}\sum_{i=1}^{n}\tau_i \\ &= (2n)^{-1}\sum_{i=1}^{n}\sum_{j=1}^{2}\{Y_{ij}(1) - Y_{ij}(0)\}. \end{aligned}$$

It is intuitive that $\hat{\tau}_i$ is unbiased for τ_i, so $\hat{\tau}$ is unbiased for τ. We can also calculate the variance of $\hat{\tau}$. I relegate the exact formula to Problem 7.1 because the MPE is just a special case of the SRE.

However, the variance calculation in Problem 7.1 is of pure theoretical interest, because we cannot follow the same strategy under the SRE to estimate the variance of $\hat{\tau}$ under the MPE. The within-pair sample variances of the outcomes are not well defined because, within each pair, we have only one treated and one control unit. The data do not allow us to estimate the variance of $\hat{\tau}_i$ within pair i.

Is it possible to estimate the variance of $\hat{\tau}$ in the MPE? Let us forget about the MPE and change the perspective to the classic IID sampling. If the $\hat{\tau}_i$'s are IID with mean μ and σ^2, then the variance of $\hat{\tau} = n^{-1}\sum_{i=1}^{n}\hat{\tau}_i$ is σ^2/n. An unbiased estimator for σ^2 is the sample variance $(n-1)^{-1}\sum_{i=1}^{n}(\hat{\tau}_i - \hat{\tau})^2$, so an unbiased estimator for $\mathrm{var}(\hat{\tau})$ is

$$\hat{V} = \{n(n-1)\}^{-1}\sum_{i=1}^{n}(\hat{\tau}_i - \hat{\tau})^2. \tag{7.2}$$

The discussion also extends to the independent but not IID setting; see Problem A.1 in Appendix A. The above discussion seems a digression from the MPE which has completely different statistical assumptions. But at least it motivates a variance estimator $\hat{V}$, which uses the between-pair variance of $\hat{\tau}_i$ to estimate the variance of $\hat{\tau}$. Of course, it is derived under different assumptions. Does it also work for the MPE? Theorem 7.1 below is a positive result.

Theorem 7.1 *Under the MPE, $\hat{V}$ defined in (7.2) is a conservative estimator for the true variance of $\hat{\tau}$:*

$$E(\hat{V}) - \text{var}(\hat{\tau}) = \{n(n-1)\}^{-1} \sum_{i=1}^{n} (\tau_i - \tau)^2 \geq 0.$$

If the τ_i's are constant across pairs, then $E(\hat{V}) = \text{var}(\hat{\tau})$.

Theorem 7.1 states that under the MPE, $\hat{V}$ is a conservative variance estimator in general and becomes unbiased if the average causal effects are constant across pairs. It is somewhat surprising because $\hat{V}$ depends on the between-pair variance of the $\hat{\tau}_i$'s whereas $\text{var}(\hat{\tau})$ depends on the within-pair variance of each of $\hat{\tau}_i$. The proof below might provide some insights into this surprising result.

Proof of Theorem 7.1: Recall the basic algebraic fact that $\sum_{i=1}^{n}(a_i - \bar{a})^2 = \sum_{i=1}^{n} a_i^2 - n\bar{a}^2$ which parallels the fact that $\text{var}(W) = E(W^2) - (EW)^2$ for a random variable W. Using it in the following steps 2 and 5, we have

$$\begin{aligned}
n(n-1)E(\hat{V}) &= E\left\{\sum_{i=1}^{n}(\hat{\tau}_i - \hat{\tau})^2\right\} \\
&= E\left(\sum_{i=1}^{n}\hat{\tau}_i^2 - n\hat{\tau}^2\right) \\
&= \sum_{i=1}^{n}\{\text{var}(\hat{\tau}_i) + \tau_i^2\} - n\{\text{var}(\hat{\tau}) + \tau^2\} \\
&= \sum_{i=1}^{n}\text{var}(\hat{\tau}_i) - n\text{var}(\hat{\tau}) + \sum_{i=1}^{n}\tau_i^2 - n\tau^2 \\
&= n^2\text{var}(\hat{\tau}) - n\text{var}(\hat{\tau}) + \sum_{i=1}^{n}(\tau_i - \tau)^2.
\end{aligned}$$

Therefore,

$$E(\hat{V}) = \text{var}(\hat{\tau}) + \{n(n-1)\}^{-1}\sum_{i=1}^{n}(\tau_i - \tau)^2 \geq \text{var}(\hat{\tau}).$$

□

Similar to the discussions for other experiments, the Neymanian approach relies on the large-sample approximation:

$$\frac{\hat{\tau} - \tau}{\sqrt{\text{var}(\hat{\tau})}} \to \text{N}(0, 1)$$

in distribution if $n \to \infty$ and some regularity conditions hold. Due to the over-estimation of the variance, the Wald-type confidence interval

$$\hat{\tau} \pm z_{1-\alpha/2}\sqrt{\hat{V}}$$

covers τ with probability at least $1 - \alpha$.

Both the point estimator $\hat{\tau}$ and the variance estimator $\hat{V}$ can be conveniently obtained by OLS, as shown in the proposition below.

Proposition 7.1 *$\hat{\tau}$ and $\hat{V}$ are identical to the coefficient and variance estimator of the intercept from the OLS fit of the vector $(\hat{\tau}_1, \ldots, \hat{\tau}_n)^{\mathsf{T}}$ on the intercept only.*

I leave the proof of Proposition 7.1 as Problem 7.3.

7.4 Covariate adjustment

Although we have matched on covariates in the design stage, it is possible that the matching is not perfect ($X_{i1} \neq X_{i2}$) and sometimes we have additional covariates beyond those used in the pair-matching stage. In those cases, we can adjust for the covariates to further improve estimation efficiency. Assume that each unit (i, j) has covariates X_{ij}. Similar to the discussion in the CRE, there are two general strategies of covariate adjustment in the MPE.

7.4.1 FRT

I start with the covariate-adjusted FRT in the MPE. In parallel with Definition 6.2, we can construct test statistics based on the residuals from a model fitting of the outcome on the covariates, since those residuals are fixed numbers under the sharp null hypothesis. A canonical choice is to fit OLS of all observed Y_{ij}'s on X_{ij}'s to obtain the residuals $\hat{\varepsilon}_{ij}$'s. We can then construct test statistics pretending that the $\hat{\varepsilon}_{ij}$'s are the observed outcomes. Rosenbaum (2002a) advocated this strategy in particular to the MPE.

In parallel with Definition 6.3, we can directly use some coefficients from model fitting as the test statistics. The discussion in the next subsection will suggest a choice of the test statistic for this strategy.

7.4.2 Regression adjustment

I now focus on estimating τ. We can compute the within-pair differences in covariates $\hat{\tau}_{X,i}$ and their average $\hat{\tau}_X$ in the same way as the outcome. We can show that

$$E(\hat{\tau}_{X,i}) = 0, \quad E(\hat{\tau}_X) = 0,$$

and

$$\text{cov}(\hat{\tau}_X) = n^{-2} \sum_{i=1}^{n} \hat{\tau}_{X,i} \hat{\tau}_{X,i}^{\mathsf{T}}.$$

In a realized MPE, $\text{cov}(\hat{\tau}_X)$ is not zero unless all the $\hat{\tau}_{X,i}$'s are zero. With an unlucky draw of $(Z_1, \ldots, Z_n)$, it is possible that $\hat{\tau}_X$ differs substantially from zero. Similar to the discussion under the CRE in Section 6.2.2.3, adjusting for the imbalance of the covariate means is likely to improve estimation efficiency.

Similar to the discussion in Section 6.2.2.3, we can consider a class of estimators indexed by γ:

$$\hat{\tau}(\gamma) = \hat{\tau} - \gamma^{\mathsf{T}} \hat{\tau}_X$$

which has mean 0 for any fixed γ. We want to choose γ to minimize the variance of $\hat{\tau}(\gamma)$. Its variance is a quadratic function of γ:

$$\begin{aligned} \text{var}\{\hat{\tau}(\gamma)\} &= \text{var}(\hat{\tau} - \gamma^{\mathsf{T}} \hat{\tau}_X) \\ &= \text{var}(\hat{\tau}) + \gamma^{\mathsf{T}} \text{cov}(\hat{\tau}_X)\gamma - 2\gamma^{\mathsf{T}} \text{cov}(\hat{\tau}_X, \hat{\tau}), \end{aligned}$$

which is minimized at

$$\tilde{\gamma} = \text{cov}(\hat{\tau}_X)^{-1}\text{cov}(\hat{\tau}_X, \hat{\tau}).$$

We have obtained the formula for $\text{cov}(\hat{\tau}_X)$ in the above, which can also be written as

$$\text{cov}(\hat{\tau}_X) = n^{-2}\sum_{i=1}^{n} |\hat{\tau}_{X,i}||\hat{\tau}_{X,i}|^{\mathsf{T}},$$

where $|\cdot|$ denotes the component-wise absolute value of a vector. So $\text{cov}(\hat{\tau}_X)$ is fixed and known from the observed data. However, $\text{cov}(\hat{\tau}_X, \hat{\tau})$ depends on unknown potential outcomes. Fortunately, we can obtain an unbiased estimator for it, as shown in Theorem 7.2 below.

Theorem 7.2 *An unbiased estimator for* $\text{cov}(\hat{\tau}_X, \hat{\tau})$ *is*

$$\hat{\theta} = \{n(n-1)\}^{-1}\sum_{i=1}^{n}(\hat{\tau}_{X,i} - \hat{\tau}_X)(\hat{\tau}_i - \hat{\tau}).$$

The proof of Theorem 7.2 is similar to that of Theorem 7.1. I leave it to Problem 7.2.

Based on Theorem 7.2, we can estimate the optimal coefficient $\tilde{\gamma} = \text{cov}(\hat{\tau}_X)^{-1}\text{cov}(\hat{\tau}_X, \hat{\tau})$ by

$$\begin{aligned}
\hat{\gamma} &= \text{cov}(\hat{\tau}_X)^{-1}\hat{\theta} \\
&= \left(n^{-2}\sum_{i=1}^{n}\hat{\tau}_{X,i}\hat{\tau}_{X,i}^{\mathsf{T}}\right)^{-1}\left\{\{n(n-1)\}^{-1}\sum_{i=1}^{n}(\hat{\tau}_{X,i} - \hat{\tau}_X)(\hat{\tau}_i - \hat{\tau})\right\} \\
&\approx \left(\sum_{i=1}^{n}(\hat{\tau}_{X,i} - \hat{\tau}_X)(\hat{\tau}_{X,i} - \hat{\tau}_X)^{\mathsf{T}}\right)^{-1}\sum_{i=1}^{n}(\hat{\tau}_{X,i} - \hat{\tau}_X)(\hat{\tau}_i - \hat{\tau}),
\end{aligned}$$

which is approximately the coefficient of the $\hat{\tau}_{X,i}$ in the OLS fit of the $\hat{\tau}_i$'s on the $\hat{\tau}_{X,i}$'s with an intercept. The final estimator is

$$\hat{\tau}_{\text{adj}} = \hat{\tau}(\hat{\gamma}) = \hat{\tau} - \hat{\gamma}^{\mathsf{T}}\hat{\tau}_X,$$

which, by the property of OLS, is approximately the intercept in the OLS fit of the $\hat{\tau}_i$'s on the $\hat{\tau}_{X,i}$'s with an intercept.

A conservative variance estimator for $\hat{\tau}_{\text{adj}}$ is then

$$\begin{aligned}
\hat{V}_{\text{adj}} &= \hat{V} + \hat{\gamma}^{\mathsf{T}}\text{cov}(\hat{\tau}_X)\hat{\gamma} - 2\hat{\gamma}^{\mathsf{T}}\hat{\theta} \\
&= \hat{V} - \hat{\theta}^{\mathsf{T}}\text{cov}(\hat{\tau}_X)^{-1}\hat{\theta}.
\end{aligned}$$

A subtle technical issue is whether $\hat{\tau}(\hat{\gamma})$ has the same optimality as $\hat{\tau}(\tilde{\gamma})$. We have encountered a similar issue in the discussion of Lin's (2013) estimator. With large samples, we can show $\hat{\tau}(\hat{\gamma}) - \hat{\tau}(\tilde{\gamma}) = -(\hat{\gamma} - \tilde{\gamma})^{\mathsf{T}}\hat{\tau}_X$ is of higher order since it is the product of two "small" terms $\hat{\gamma} - \tilde{\gamma}$ and $\hat{\tau}_X$.

Moreover, Fogarty (2018b) discussed the asymptotically equivalent regression formulation of the above covariate-adjusted procedure and gave a rigorous proof for the associated CLT. I summarize the regression formulation below without giving the regularity conditions.

Proposition 7.2 *Under the MPE, the covariate-adjusted estimator* $\hat{\tau}_{\text{adj}}$ *and the associated variance estimator* $\hat{V}_{\text{adj}}$ *can be conveniently approximated by the intercept and the associated variance estimator from the OLS fit of the vector of the* $\hat{\tau}_i$*'s on the* 1*'s and the matrix of the* $\hat{\tau}_{X,i}$*'s.*

I leave the proof of Proposition 7.2 as Problem 7.3. Interestingly, neither Proposition 7.1 nor 7.2 requires the EHW correction of the variance estimator. See Fogarty (2018b) for more technical details. Because we reduce the data from the MPE to the within-pair differences, it is unnecessary to center the covariates, which is different from the implementation of Lin's (2013) estimator under the CRE.

7.5 Examples

7.5.1 Darwin's data comparing cross-fertilizing and self-fertilizing on the height of corns

This is a classical example from Fisher (1935). It contains 15 pairs of corns with either cross-fertilizing or self-fertilizing, with the height being the outcome. The `R` package `HistData` provides the original data, where `cross` and `self` are the heights under cross-fertilizing and self-fertilizing, respectively, and `diff` denotes their difference.

```
> library("HistData")
> ZeaMays
   pair pot  cross   self   diff
1     1   1 23.500 17.375  6.125
2     2   1 12.000 20.375 -8.375
3     3   1 21.000 20.000  1.000
4     4   2 22.000 20.000  2.000
5     5   2 19.125 18.375  0.750
6     6   2 21.500 18.625  2.875
7     7   3 22.125 18.625  3.500
8     8   3 20.375 15.250  5.125
9     9   3 18.250 16.500  1.750
10   10   3 21.625 18.000  3.625
11   11   3 23.250 16.250  7.000
12   12   4 21.000 18.000  3.000
13   13   4 22.125 12.750  9.375
14   14   4 23.000 15.500  7.500
15   15   4 12.000 18.000 -6.000
```

In total, the MPE has $2^{15} = 32768$ possible treatment assignments, which is a tractable number in `R`. The following function can enumerate all possible treatment assignments for the MPE:

```
MP_enumerate = function(i, n.pairs)
{
 if(i > 2^n.pairs)  print("i is too large.")
 a = 2^((n.pairs-1):0)
 b = 2*a
 2*sapply(i-1,
          function(x)
            as.integer((x %% b)>=a)) - 1
}
```

So we enumerate all the treatment assignments and calculate the corresponding $\hat{\tau}$'s and the one-sided exact p-value.

```
> difference = ZeaMays$diff
```

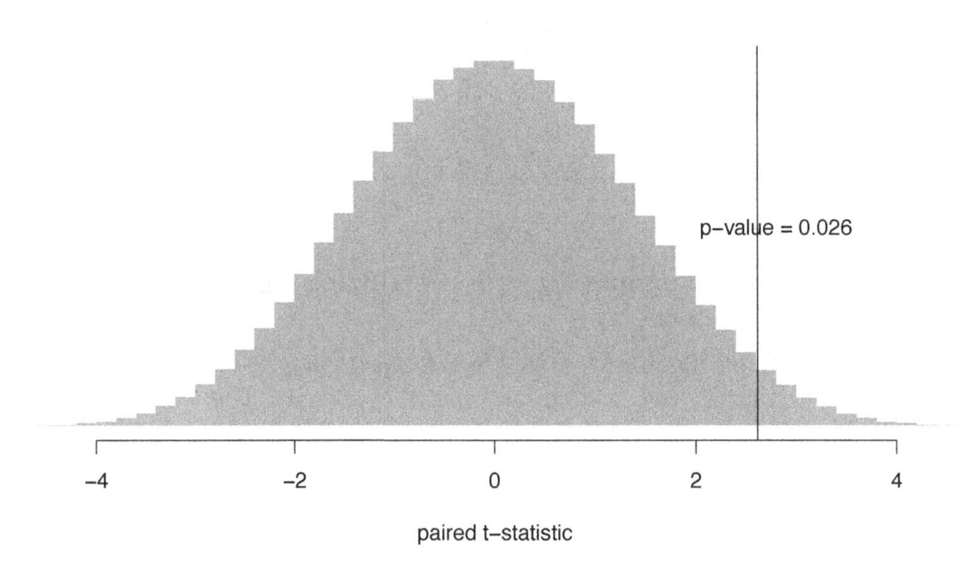

FIGURE 7.1: Exact randomization distribution of $\hat{\tau}$ using Darwin's data

```
> n.pairs     = length(difference)
> abs.diff    = abs(difference)
> t.obs       = mean(difference)
> t.ran       = sapply(1:2^15,
+                      function(x){
+                        sum(MP_enumerate(x, 15)*abs.diff)
+                        })/n.pairs
> pvalue      = mean(t.ran>=t.obs)
> pvalue
[1] 0.02633667
```

Figure 7.1 shows the exact randomization distribution of $\hat{\tau}$ under $H_{0\text{F}}$.

7.5.2 Children's Television Workshop experiment data

I also reanalyze a subset of the data from the Children's Television Workshop experiment from Ball et al. (1973) which was also analyzed by Imbens and Rubin (2015, Chapter 10). It contains eight pairs of classes, with one of the two classes randomly assigned to be shown *The Electric Company* show during the standard reading-class period. It contains the pre-test score (covariate) and the post-test score (outcome). The following table summarizes the within-pair covariates and outcomes, as well as their differences:

```
> dataxy = c(12.9, 12.0, 54.6, 60.6,
+            15.1, 12.3, 56.5, 55.5,
+            16.8, 17.2, 75.2, 84.8,
+            15.8, 18.9, 75.6, 101.9,
+            13.9, 15.3, 55.3, 70.6,
+            14.5, 16.6, 59.3, 78.4,
+            17.0, 16.0, 87.0, 84.2,
```

```
+              15.8, 20.1, 73.7, 108.6)
> dataxy = matrix(dataxy, 8, 4,   byrow = TRUE)
> diffx = dataxy[, 2] - dataxy[, 1]
> diffy = dataxy[, 4] - dataxy[, 3]
> dataxy = cbind(dataxy, diffx, diffy)
> rownames(dataxy) = 1:8
> colnames(dataxy) = c("x.control", "x.treatment",
+                     "y.control", "y.treatment",
+                     "diffx", "diffy")
> dataxy = as.data.frame(dataxy)
> dataxy
  x.control x.treatment y.control y.treatment diffx diffy
1      12.9        12.0      54.6        60.6  -0.9   6.0
2      15.1        12.3      56.5        55.5  -2.8  -1.0
3      16.8        17.2      75.2        84.8   0.4   9.6
4      15.8        18.9      75.6       101.9   3.1  26.3
5      13.9        15.3      55.3        70.6   1.4  15.3
6      14.5        16.6      59.3        78.4   2.1  19.1
7      17.0        16.0      87.0        84.2  -1.0  -2.8
8      15.8        20.1      73.7       108.6   4.3  34.9
```

The following R code calculates $\hat{\tau}$ and $\hat{V}$.

```
> n      = dim(dataxy)[1]
> tauhat = mean(dataxy[, "diffy"])
> vhat   = var(dataxy[, "diffy"])/n
> tauhat
[1] 13.425
> sqrt(vhat)
[1] 4.636337
```

By Propositions 7.1 and 7.2, we can use the OLS to obtain the point estimators and standard errors. Adjusting for the covariates, we have

```
> unadj = summary(lm(diffy ~ 1, data = dataxy))$coef
> round(unadj, 3)
            Estimate Std. Error t value Pr(>|t|)
(Intercept)   13.425      4.636   2.896    0.023
```

Adjusting for the covariates, we have

```
> adj = summary(lm(diffy ~ diffx, data = dataxy))$coef
> round(adj, 3)
            Estimate Std. Error t value Pr(>|t|)
(Intercept)    8.994      1.410   6.381    0.001
diffx          5.371      0.599   8.964    0.000
```

The above results assume large n, and p-values are justified if we believe the large-n approximation. However, $n = 8$ is not large. In total, we have $2^8 = 256$ possible treatment assignments, so the smallest possible p-value is $1/256 = 0.0039$, which is much larger than the p-value based on the Normal approximation of the covariate-adjusted estimator. In this example, it will be more reasonable to use the FRT with the studentized statistic, which is the `t value` from the `lm` function, to calculate exact p-values.[2] The following R code calculates the p-values based on two t-statistics.

```
> t.ran = sapply(1:2^8, function(x){
```

[2]See Chapter 8 for a justification.

```
+     z.mpe = MP_enumerate(x, 8)
+     diffy.mpe = diffy*z.mpe
+     diffx.mpe = diffx*z.mpe
+
+     c(summary(lm(diffy.mpe ~ 1))$coef[1, 3],
+       summary(lm(diffy.mpe ~ diffx.mpe))$coef[1, 3])
+ })
> p.unadj = mean(abs(t.ran[1, ]) >= abs(unadj[1, 3]))
> p.unadj
[1] 0.03125
> p.adj = mean(abs(t.ran[2, ]) >= abs(adj[1, 3]))
> p.adj
[1] 0.0078125
```

Figure 7.2 shows the exact distributions of the two studentized statistics, as well as the two-sided p-values. The figure highlights the fact that the randomization distributions of the test statistics are discrete, taking at most 256 possible values. The Normal approximations are unlikely to be accurate, especially at the tails. We should report the p-values based on the FRT.

7.6 Comparing the MPE and CRE

Imai (2008b) compared the MPE and CRE. Heuristically, the conclusion is that the MPE gives more precise estimators if the matching is well done and the covariates are predictive of the outcome. However, without the outcome data in the design stage, it is hard to decide whether this is true or not. In the FRT, if covariates are predictive of the outcome, the MPE usually gives more powerful tests compared with the CRE. Greevy et al. (2004) illustrated this using simulation based on the Wilcoxon sign-rank statistic. However, this can be a subtle issue with finite samples. Consider an experiment with $2n$ units, with n units receiving the

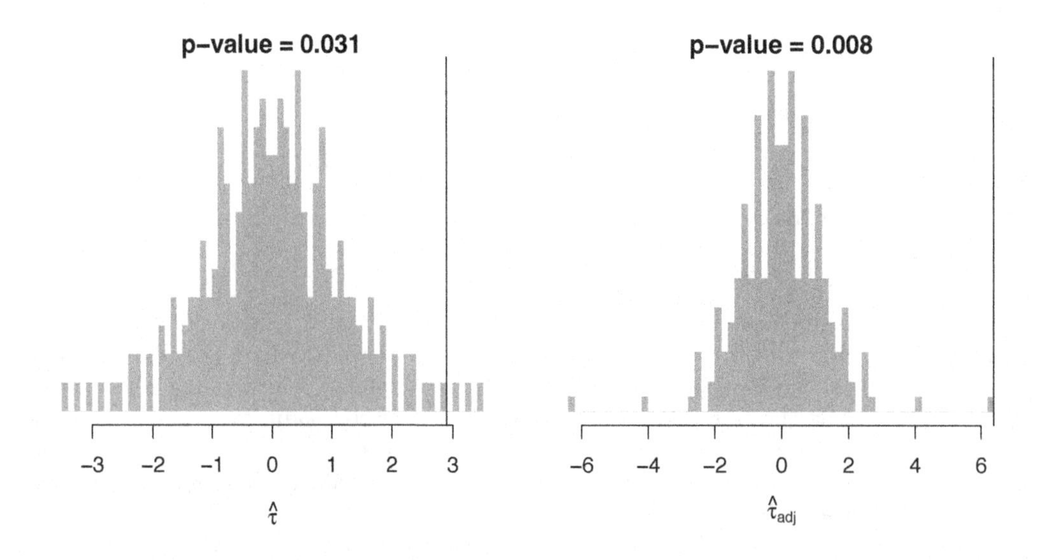

FIGURE 7.2: Exact randomization distributions of the studentized statistics in Section 7.5.2

treatment and n units receiving the control. If we test the sharp null hypothesis at level 0.05, then in the MPE, we need at least $2 \times 5 = 10$ units since the smallest p-value is $1/2^5 = 1/32 < 0.05$ but $1/2^4 = 1/16 > 0.05$, but in the CRE, we need at least $2 \times 4 = 8$ units since the smallest p-value is $1/\binom{8}{4} = 1/70 < 0.05$ but $1/\binom{6}{3} = 1/20 = 0.05$. So with 8 units, it is impossible to reject the sharp null hypothesis in the MPE but it is possible in the CRE. Even if the covariates are perfect predictors of the outcome, the MPE is not superior to the CRE based on the FRT.

7.7 Extension to the general matched experiment

It is straightforward to extend the MPE to the general matched experiment with varying numbers of control units. Assume that we have n matched sets indexed by $i = 1, \ldots, n$. For the matched set i, we have $1 + M_i$ units. The M_i's can vary. The total number of experimental units is $N = n + \sum_{i=1}^n M_i$. Let ij index the unit j within matched set i ($i = 1, \ldots, n$ and $j = 1, \ldots, M_i + 1$). Unit ij has potential outcomes $Y_{ij}(1)$ and $Y_{ij}(0)$ under the treatment and control, respectively.

Within the matched set i ($i = 1, \ldots, n$), the experimenter randomly selects exactly one unit to receive the treatment with the rest M_i units receiving the control. This general matched experiment is also a special case of the SRE with n strata of size $1 + M_i$ ($i = 1, \ldots, n$). Let Z_{ij} be the treatment indicator for unit ij, which reveals one of the potential outcomes as

$$Y_{ij} = Z_{ij} Y_{ij}(1) + (1 - Z_{ij}) Y_{ij}(0).$$

The average causal effect within the matched set i equals

$$\tau_i = (M_i + 1)^{-1} \sum_{j=1}^{1+M_i} \{Y_{ij}(1) - Y_{ij}(0)\}.$$

Since the general matched experiment is an SRE, an unbiased estimator of τ_i is

$$\hat{\tau}_i = \sum_{j=1}^{M_i+1} Z_{ij} Y_{ij} - M_i^{-1} \sum_{j=1}^{M_i+1} (1 - Z_{ij}) Y_{ij}$$

which is the difference in means of the outcomes within matched set i.

Below we discuss the statistical inference with the general matched experiment.

7.7.1 FRT

As usual, we can always use the FRT to test the sharp null hypothesis

$$H_{0\text{F}} : Y_{ij}(1) = Y_{ij}(0) \text{ for all } i = 1, \ldots, n; j = 1, \ldots, M_i + 1.$$

Because the general matched experiment is a special case of the SRE with many small strata, we can use the test statistics defined in Examples 5.4, 5.5, 7.2, 7.3, 7.4, as well as the estimators and the corresponding t-statistics from the following two subsections.

7.7.2 Estimating the average of the within-strata effects

We first focus on estimating the average of the within-strata effects:

$$\tau = n^{-1} \sum_{i=1}^{n} \tau_i.$$

It has an unbiased estimator

$$\hat{\tau} = n^{-1} \sum_{i=1}^{n} \hat{\tau}_i.$$

Interestingly, we can show that Theorem 7.1 holds for the general matched experiment, and so are other results for the MPE. In particular, we can use the OLS fit of the $\hat{\tau}_i$'s on the intercept to obtain the point and variance estimators for τ. With covariates, we can use the OLS fit of the $\hat{\tau}_i$'s on the intercept and the $\hat{\tau}_{X,i}$'s, where

$$\hat{\tau}_{X,i} = \sum_{j=1}^{M_i+1} Z_{ij} X_{ij} - M_i^{-1} \sum_{j=1}^{M_i+1} (1 - Z_{ij}) X_{ij}$$

is the corresponding difference in means of the covariates within matched set i.

7.7.3 A more general causal estimand

Importantly, the τ above is the average of the τ_i's, which does not equal the average causal effect for the N units in the experiment when the M_i's vary. The average causal effect equals

$$\begin{aligned} \tau' &= N^{-1} \sum_{i=1}^{n} \sum_{j=1}^{1+M_i} \{Y_{ij}(1) - Y_{ij}(0)\} \\ &= \sum_{i=1}^{n} \frac{1 + M_i}{N} \tau_i. \end{aligned}$$

To unify the discussion, I consider the weighted causal effect

$$\tau_w = \sum_{i=1}^{n} w_i \tau_i$$

with $\sum_{i=1}^{n} w_i = 1$, which includes τ as a special case with $w_i = n^{-1}$ and τ' as a special case with $w_i = (1 + M_i)/N$ for $i = 1, \dots, n$. It is straightforward to obtain an unbiased estimator

$$\hat{\tau}_w = \sum_{i=1}^{n} w_i \hat{\tau}_i,$$

and calculate its variance

$$\text{var}(\hat{\tau}_w) = \sum_{i=1}^{n} w_i^2 \text{var}(\hat{\tau}_i).$$

However, estimating the variance of $\hat{\tau}_w$ is quite tricky because the $\hat{\tau}_i$'s are independent random variables without any replicates. This is a famous problem in theoretical statistics studied by Hartley et al. (1969) and Rao (1970). Fogarty (2018a) also discussed this problem without recognizing these previous works. I will give the final form of the variance estimator without detailing the motivation:

$$\hat{V}_w = \sum_{i=1}^{n} c_i(\hat{\tau}_i - \hat{\tau}_w)^2$$

where

$$c_i = \frac{\frac{w_i^2}{1-2w_i}}{1 + \sum_{i=1}^{n} \frac{w_i^2}{1-2w_i}}.$$

As a sanity check, c_i reduces to $\{n(n-1)\}^{-1}$ in the MPE with $M_i = 1$ and $w_i = n^{-1}$. For simplicity, we focus on the case with $w_i < 1/2$ for all i's, that is, there is no matched set containing more than half of the total weights. The following theorem extends Theorem 7.1.

Theorem 7.3 *Under the general matched experiment with varying M_i's, if $w_i < 1/2$ for all i's, then*

$$E(\hat{V}_w) - \text{var}(\hat{\tau}_w) = \sum_{i=1}^{n} c_i(\tau_i - \tau_w)^2 \geq \text{var}(\hat{\tau}_w) \geq 0$$

with equality holding if the τ_i's are constant.

Although the theoretical motivation for $\hat{V}_w$ is quite complicated, it is not too difficult to verify Theorem 7.3 directly. I relegate the proof to Problem 7.9.

7.8 Homework problems

7.1 The true variance of $\hat{\tau}$ in the MPE

Express $\text{var}(\hat{\tau})$ in terms of the finite-population variances of the potential outcomes.

7.2 A covariance estimator

Prove Theorem 7.2.

7.3 Variance estimators via OLS

Prove Propositions 7.1 and 7.2.

7.4 Point and variance estimator with binary outcome

This problem extends Example 7.5 to Neymanian inference.

Express $\hat{\tau}$ and $\hat{V}$ in terms of the counts in Table 7.1. Do they depend on m_{11} and m_{00}?

7.5 Minimum sample size for the FRT

Extend the discussion in Section 7.6. Consider an experiment with $2n$ units, with n units receiving the treatment and n units receiving the control, and test the sharp null hypothesis at level 0.001. What is the minimum value of n for an MPE so that the smallest p-value does not exceed than 0.001, and what is the corresponding minimum value of n for a CRE?

7.6 Reanalyzing Darwin's data

Section 7.5.1 analyzed Darwin's data using the FRT based on the test statistic $\hat{\tau}$.

Reanalyze this dataset using the FRT with the Wilcoxon sign-rank statistic.

Reanalyze this dataset based on the Neymanian inference: report the unbiased point estimator, conservative variance estimator, and 95% confidence interval.

7.7 Reanalyzing Children's Television Workshop experiment data

Section 7.5.2 analyzed the data from based on Neymanian inference.

Reanalyze this dataset using the FRT with different test statistics.

Reanalyze this dataset using the FRT with covariate adjustment.

7.8 Reanalyzing the data from Angrist and Lavy (2009)

The original analysis of Angrist and Lavy (2009) was quite complicated. For this problem, please focus only on Table A1 of the original paper and view the schools as the experimental units. Angrist and Lavy (2009) essentially conducted an MPE on the schools. Dropping their pair 6 and all the pairs with noncompliance results in 14 complete pairs, with data shown below and also in `AL2009.csv`:

	pair	z	pr99	pr00	pr01	pr02
1	1	0	0.046	0.000	0.091	0.185
2	1	1	0.036	0.051	0.000	0.047
3	2	0	0.054	0.094	0.184	0.034
4	2	1	0.050	0.108	0.110	0.095
5	3	0	0.114	0.000	0.056	0.075
6	3	1	0.098	0.054	0.030	0.068
7	4	0	0.148	0.162	0.082	0.075
8	4	1	0.134	0.390	0.339	0.458
9	5	0	0.152	0.105	0.083	0.129
10	5	1	0.145	0.077	0.579	0.167
11	6	0	0.188	0.214	0.375	0.545
12	6	1	0.179	0.165	0.483	0.444
13	7	0	0.193	0.771	0.328	0.583
14	7	1	0.189	0.186	0.168	0.368
15	8	0	0.197	0.350	0.000	0.383
16	8	1	0.200	0.071	0.667	0.429
17	9	0	0.213	0.176	0.164	0.172
18	9	1	0.209	0.165	0.092	0.151
19	10	0	0.211	0.667	0.250	0.617
20	10	1	0.219	0.250	0.500	0.350
21	11	0	0.219	0.153	0.185	0.219
22	11	1	0.224	0.363	0.372	0.342
23	12	0	0.255	0.226	0.213	0.327
24	12	1	0.257	0.098	0.107	0.095
25	13	0	0.261	0.071	0.000	NA
26	13	1	0.263	0.441	0.448	0.435
27	14	0	0.286	0.161	0.126	0.181
28	14	1	0.285	0.389	0.353	0.309

The outcomes are the Bagrut passing rates in the years 2001 and 2002, with the Bagrut passing rates in 1999 and 2000 as pretreatment covariates. Reanalyze the data based on the Neymanian inference with and without adjusting for the covariates. In particular, how do you deal with the missing outcome in pair 25?

7.9 Variance estimation in the general matched experiment

This problem contains more details for Section 7.7.

First, prove Theorem 7.1 for the general matched experiment. Second, prove Theorem 7.3.

Remark: For the second part, we can first verify that $\hat{\tau}_i - \hat{\tau}_w$ has mean $\tau_i - \tau_w$ and variance

$$\text{var}(\hat{\tau}_i - \hat{\tau}_w) = \text{var}(\hat{\tau}_w) + (1 - 2w_i)\text{var}(\hat{\tau}_i).$$

See Pashley and Miratrix (2021) for the discussion of various variance estimators in blocked and matched experiments.

7.10 Recommended reading

Greevy et al. (2004) provided an algorithm to form matched pairs based on covariates. Imai (2008b) discussed the estimation of the average causal effect without covariates, and Fogarty (2018b) discussed covariate adjustment in MPEs.

8

Unification of the Fisherian and Neymanian Inferences in Randomized Experiments

Chapters 3–7 cover both the Fisherian and Neymanian inferences for different types of experiments. The Fisherian perspective focuses on the finite-sample exact p-values for testing the strong null hypothesis of no causal effects for any units whatsoever, whereas the Neymanian perspective focuses on unbiased estimation with a conservative large-sample confidence interval for the average causal effect. Both of them are justified by the physical randomization which is ensured by the design of the experiments. Because of this, they are both called *randomization-based inference* or *design-based inference*. Because they concern a finite population of units in the experiments, they are also called *finite-population inference*. They are related but also have distinct features.

In 1935, Neyman presented his seminal paper on randomization-based inference to the Royal Statistical Society. His paper (Neyman, 1935) was attacked by Fisher in the discussion session. Sabbaghi and Rubin (2014) reviewed this famous Neyman–Fisher controversy and presented some new results for this old problem. Instead of going to the philosophical issues, this chapter tries to provide a unified discussion.

8.1 Testing strong and weak null hypotheses in the CRE

8.1.1 CRE without covariates

Let us revisit the treatment-control CRE. The Fisherian perspective focuses on testing the strong null hypothesis

$$H_{0\mathrm{F}} : Y_i(1) = Y_i(0) \text{ for all units } i = 1, \ldots, n.$$

The FRT delivers a finite-sample exact p_{FRT} defined in (3.2).

By duality of the confidence interval and hypothesis testing, the Neymanian perspective gives a test for the weak null hypothesis

$$H_{0\mathrm{N}} : \tau = 0$$

or, equivalently,

$$H_{0\mathrm{N}} : \bar{Y}(1) = \bar{Y}(0)$$

based on

$$t = \frac{\hat{\tau}}{\sqrt{\hat{V}}}.$$

By the CLT of $\hat{\tau}$ and the conservativeness of the variance estimator, we have

$$t = \sqrt{\frac{\mathrm{var}(\hat{\tau})}{\hat{V}}} \times \frac{\hat{\tau}}{\sqrt{\mathrm{var}(\hat{\tau})}} \to C \times \mathrm{N}(0, 1)$$

in distribution, where C is smaller than or equal to 1 but depends on the unknown potential outcomes. Using $\text{N}(0,1)$ quantiles for the studentized statistic t, we have a conservative large-sample test for $H_{0\text{N}}$.

Furthermore, Ding and Dasgupta (2017) show that the FRT with the studentized statistic t has the dual guarantees:

1. p_{FRT} is finite-sample exact under $H_{0\text{F}}$;
2. p_{FRT} is asymptotically conservative under $H_{0\text{N}}$.

Importantly, this is a feature of the studentized statistic t. Ding and Dasgupta (2017) showed that the FRT with other test statistics may not have the dual guarantees. In particular, the FRT with $\hat{\tau}$ may be asymptotically anti-conservative under $H_{0\text{N}}$. I give some heuristics below to illustrate the importance of studentization in the FRT.

Under $H_{0\text{N}}$, we have

$$\hat{\tau} \dot{\sim} \text{N}\left(0, \frac{S^2(1)}{n_1} + \frac{S^2(0)}{n_0} - \frac{S^2(\tau)}{n}\right).$$

The FRT pretends that the Science Table is $(Y_i, Y_i)_{i=1}^n$ induced by $H_{0\text{F}}$, so the randomization distribution of $\hat{\tau}$ is

$$(\hat{\tau})^\pi \dot{\sim} \text{N}\left(0, \frac{s^2}{n_1} + \frac{s^2}{n_0}\right),$$

where $(\cdot)^\pi$ denotes the randomization distribution[1] and s^2 is the sample variance of the observed outcomes. Based on (3.7) in Chapter 3, we can approximate the asymptotic variance of $(\hat{\tau})^\pi$ under $H_{0\text{F}}$ as

$$\begin{aligned}
\frac{s^2}{n_1} + \frac{s^2}{n_0} &= \frac{n}{n_1 n_0}\left\{\frac{n_1 - 1}{n-1}\hat{S}^2(1) + \frac{n_0 - 1}{n-1}\hat{S}^2(0) + \frac{n_1 n_0}{n(n-1)}\hat{\tau}^2\right\} \\
&\approx \frac{\hat{S}^2(1)}{n_0} + \frac{\hat{S}^2(0)}{n_1} \\
&\approx \frac{S^2(1)}{n_0} + \frac{S^2(0)}{n_1},
\end{aligned}$$

which does not match the asymptotic variance of $\hat{\tau}$. Ideally, we should compute the p-value under $H_{0\text{N}}$ based on the true distribution of $\hat{\tau}$, which, however, depends on the unknown potential outcomes. In contrast, we use the FRT to compute the p_{FRT} based on the permutation distribution $(\hat{\tau})^\pi$, which does not match the true distribution of $\hat{\tau}$ under $H_{0\text{N}}$ even with large samples. Therefore, the FRT with $\hat{\tau}$ may not control the type one error rate under $H_{0\text{N}}$ even with large samples.

Fortunately, the undesired property of the FRT with $\hat{\tau}$ goes away if we replace the test statistic $\hat{\tau}$ with the studentized version t. Under $H_{0\text{N}}$, we have

$$t \dot{\sim} \text{N}(0, C^2)$$

where $C^2 \leq 1$ with equality holding if $Y_i(1) - Y_i(0) = \tau$ for all units $i = 1, \ldots, n$. The FRT pretends that $Y_i(1) = Y_i(0) = Y_i$ for all i's and generates the permutation distribution

$$t^\pi \dot{\sim} \text{N}(0, 1)$$

where the variance equals 1 because the Science Table used by the FRT has zero individual causal effects. Under $H_{0\text{N}}$, because the true distribution of t is more dispersed than the corresponding permutation distribution, the p_{FRT} based on t is asymptotically conservative.

[1] π is a standard notation for a random permutation.

8.1.2 Covariate-adjusted FRTs in the CRE

Extending the discussion in Section 8.1 to the case with covariates, Zhao and Ding (2021) recommend using the FRT with the studentized Lin's (2013) estimator:

$$t_{\mathrm{L}} = \frac{\hat{\tau}_{\mathrm{L}}}{\sqrt{\hat{V}_{\mathrm{L}}}},$$

which is the robust t-statistic for the coefficient of Z_i in the OLS fit of Y_i on $(1, Z_i, X_i, Z_i X_i)$. They show that the FRT with t_{L} has multiple guarantees:

1. p_{FRT} is finite-sample exact under $H_{0\mathrm{F}}$;
2. p_{FRT} is asymptotically conservative under $H_{0\mathrm{N}}$;
3. p_{FRT} is asymptotically more powerful than the FRT with t when $H_{0\mathrm{N}}$ does not hold and the covariates are predictive to the outcomes;
4. the above properties hold even when the linear outcome model is misspecified.

Similarly, this is a feature of the studentized statistic t_{L}. Zhao and Ding (2021) show that other covariate-adjusted FRTs reviewed in Section 6.2.1 may be either anti-conservative under $H_{0\mathrm{N}}$ or less powerful than the FRT with t_{L} when $H_{0\mathrm{N}}$ does not hold.

8.1.3 General recommendations

The recommendations for the SRE parallel those for the CRE if both the strong and weak null hypotheses are of interest. Recall the estimators in Chapters 5 and 6.2.4. Without additional covariates, Zhao and Ding (2021) recommend using the FRT with

$$t_{\mathrm{S}} = \frac{\hat{\tau}_{\mathrm{S}}}{\sqrt{\hat{V}_{\mathrm{S}}}};$$

with additional covariates, they recommend using the FRT with

$$t_{\mathrm{L,S}} = \frac{\hat{\tau}_{\mathrm{L,S}}}{\sqrt{\hat{V}_{\mathrm{L,S}}}}.$$

The analysis of ReM is trickier. Zhao and Ding (2021) show that the FRT with t does not have the dual guarantees in Section 8.1, but the FRT with t_{L} still has the guarantees in Section 8.1.2. This highlights the importance of both covariate adjustment and studentization in ReM.

Similar results hold for the MPE. Without covariates, we recommend using the FRT with the t-statistic for the intercept in the OLS fit of $\hat{\tau}_i$ on 1; with covariates, we recommend using the FRT with the t-statistic for the intercept in the OLS fit of $\hat{\tau}_i$ on 1 and $\hat{\tau}_{X,i}$. Figure 7.2 in Chapter 7 is based on these recommended FRTs.

Overall, the FRTs with studentized statistics are safer choices. When the large-sample Normal approximations to the studentized statistics are accurate, the p_{FRT}'s are almost identical to the p-values based on Normal approximations. When the large-sample approximations are inaccurate, the FRTs at least guarantee valid p-values under the strong null hypotheses. This is the recommendation of this book.

8.2 A simulation study

Now I use simulation to evaluate the finite-sample properties of the p_{FRT}'s under the weak null hypothesis. I will use the following twelve test statistics.

1. The first three test statistics are from the OLS implementation of Neyman (1923), including the coefficient of the treatment, the t-statistic based on the classic standard error, and the t-statistic based on the EHW standard error.
2. The next three test statistics are based on the pseudo-outcome strategy in Definition 6.2, advocated by Rosenbaum (2002a). We first residualize the outcomes by regressing Y_i on $(1, X_i)$, and then obtain the three test statistics similar to the first three.
3. The next three test statistics are based on the OLS implementation of Fisher (1925), as discussed in Section 6.2.2.
4. The final three test statistics are based on the OLS implementation of Lin (2013), as discussed in Section 6.2.2.

Consider a finite population of $n = 100$ units subjected to a CRE of size $(n_1, n_0) = (20, 80)$. For each i, we draw a univariate covariate X_i from Unif$(-1, 1)$ and generate potential outcomes as $Y_i(1) \sim \mathrm{N}(X_i^3, 1)$ and $Y_i(0) \sim \mathrm{N}(-X_i^3, 0.5^2)$. Center the $Y_i(1)$'s and $Y_i(0)$'s to ensure $\tau = 0$. Fix $\{Y_i(1), Y_i(0), X_i\}_{i=1}^n$ in simulation. We draw a random permutation of n_1 1's and n_0 0's to obtain the observed outcomes and conduct FRTs. The procedure is repeated 500 times, with the p-values approximated by 500 independent permutations of the treatment vector in each replication.

Figure 8.1 shows the p-values under the weak null hypothesis. The four robust t-statistics, as shown in the last row, are the only ones that preserve the correct type one error rates. In fact, they are conservative, which is coherent with the theory. All the other eight statistics yield type one error rates greater than the nominal levels and are thus not proper for testing the weak null hypothesis.

8.3 A case study

Recall the SRE in Chong et al. (2016) analyzed in Section 5.4.2. I compare the "soccer" arm versus the "control" arm and the "physician" arm versus the "control" arm. We also compare the FRTs with and without using the covariate indicating the baseline anemia status. We use their dataset to illustrate the FRTs in the CRE and SRE. The ten subgroup analyses within the same class levels use the FRTs with t and t_{L} for the CRE and the two overall analyses averaging over all class levels use the FRTs with t_{S} and $t_{\mathrm{L,S}}$ for the SRE.

Table 8.1 shows the point estimators, standard errors, the p-value based on the Normal approximation of the robust t-statistics, and the p-value based on the FRTs. In most strata, covariate adjustment decreases the standard error since the baseline anemia status

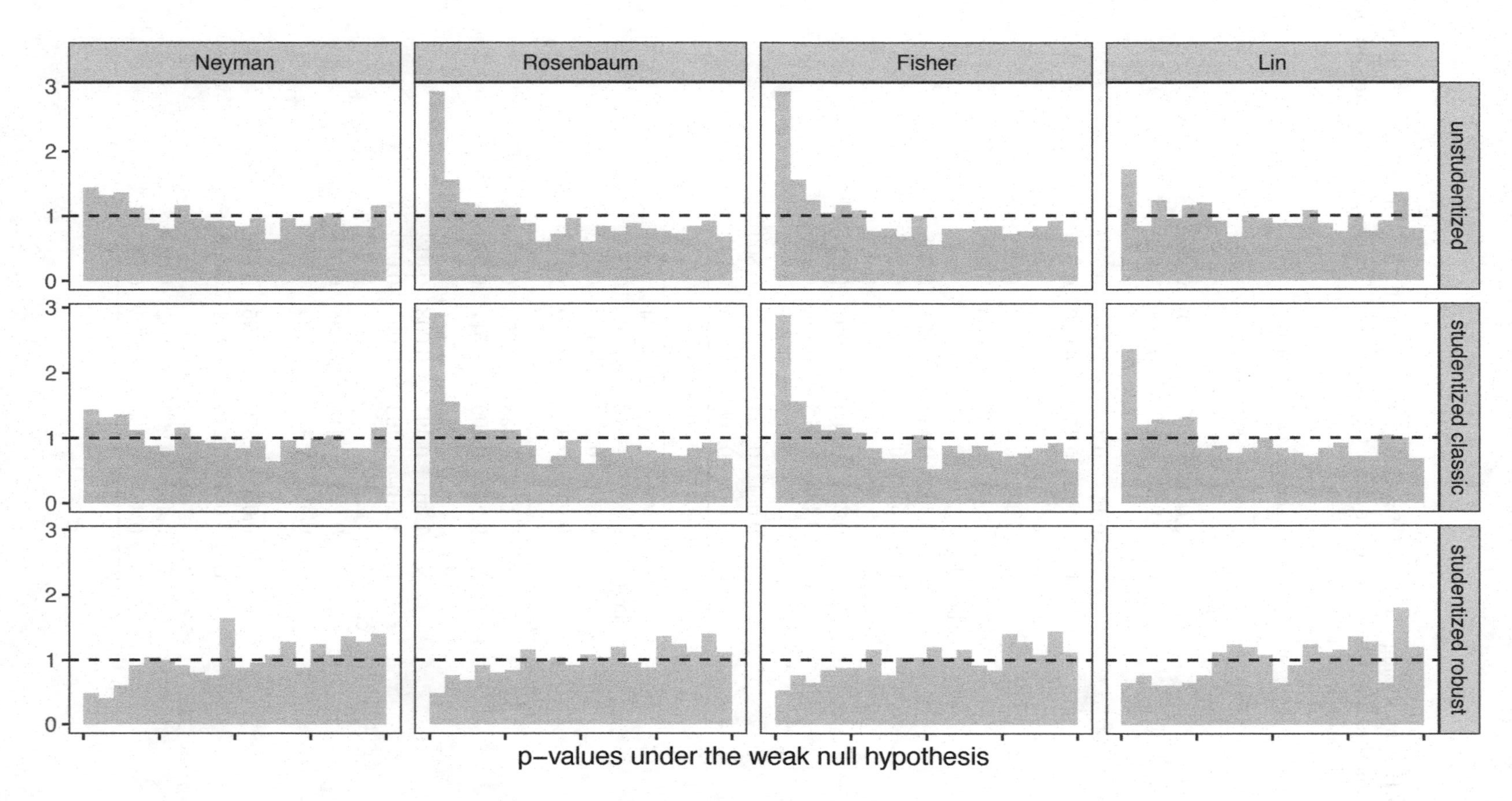

FIGURE 8.1: Histograms of p_{FRT} under the weak null hypothesis

TABLE 8.1: Reanalyzing the data from Chong et al. (2016). "N" corresponds to the unadjusted estimators and tests due to Neyman (1923), and "L" corresponds to the covariate-adjusted estimators and tests due to Lin (2013).

(a) soccer versus control

	est	s.e.	p_{Normal}	p_{FRT}
class 1				
N	0.051	0.502	0.919	0.924
L	0.050	0.489	0.919	0.929
class 2				
N	-0.158	0.451	0.726	0.722
L	-0.176	0.452	0.698	0.700
class 3				
N	0.005	0.403	0.990	0.989
L	-0.096	0.385	0.803	0.806
class 4				
N	-0.492	0.447	0.271	0.288
L	-0.511	0.447	0.253	0.283
class 5				
N	0.390	0.369	0.291	0.314
L	0.443	0.318	0.164	0.186
all				
N	-0.051	0.204	0.802	0.800
L	-0.074	0.200	0.712	0.712

(b) physician versus control

	est	s.e.	p_{Normal}	p_{FRT}
class 1				
N	0.567	0.426	0.183	0.192
L	0.588	0.418	0.160	0.174
class 2				
N	0.193	0.438	0.659	0.666
L	0.265	0.409	0.517	0.523
class 3				
N	1.305	0.494	0.008	0.012
L	1.501	0.462	0.001	0.003
class 4				
N	-0.273	0.413	0.508	0.515
L	-0.313	0.417	0.454	0.462
class 5				
N	-0.050	0.379	0.895	0.912
L	-0.067	0.279	0.811	0.816
all				
N	0.406	0.202	0.045	0.047
L	0.463	0.190	0.015	0.017

is predictive of the outcome. Table 8.1 also exhibits two exceptions: within class 2, covariate adjustment increases the standard error when comparing "soccer" and "control"; in class 4, covariate adjustment increases the standard error when comparing "physician" and "control". This is due to the small group sizes within these strata, causing the asymptotic approximation inaccurate. Nevertheless, in these two scenarios, the differences in the standard error are in the third digit. The p-values from the Normal approximation and the FRT are close with the latter being slightly larger in most cases. Based on the theory, the p-values based on the FRT should be trusted since it has an additional guarantee of being finite-sample exact under the sharp null hypothesis. This becomes important in this example since the group sizes are quite small within strata.

Figure 8.2 compares the histograms of the randomization distributions of the robust t-statistics with the asymptotic approximations. In the subgroup analysis, we can observe discrepancies between the randomization distributions and $\mathrm{N}(0,1)$; averaged over all class levels, the discrepancy becomes unnoticeable. Overall, in this application, the p-values based on the Normal approximation do not differ substantially from those based on the FRTs. Two approaches yield coherent conclusions: the video with a physician telling the benefits of iron supplements improved academic performance and the effect was most significant among students in class 3; in contrast, the video with a famous soccer player telling the benefits of the iron supplements did not have any significant effect.

8.4 Homework problems

8.1 Reanalyzing the data from Angrist and Lavy (2009)

This is the Fisherian counterpart of Problem 7.8. Report the p_{FRT}'s from the FRTs with studentized statistics.

8.2 Replication Figure 1 of Zhao and Ding (2021)

Zhao and Ding (2021) use simulation to evaluate the finite-sample properties of the p_{FRT}'s from the FRTs with various test statistics. Based on their Figure 1, they recommend using the FRT with $t_{\mathrm{L,S}}$ to analyze the SRE.

Read their Section 5 and replicate their Figure 1.

8.3 Recommended reading

Using the studentized statistics in permutation tests is not a new idea; see Janssen (1997) and Chung and Romano (2013). Under the design-based framework, Ding and Dasgupta (2017) and Wu and Ding (2021) made the recommendation for multiarmed experiments, and Zhao and Ding (2021) extended the recommendation to covariate adjustment.

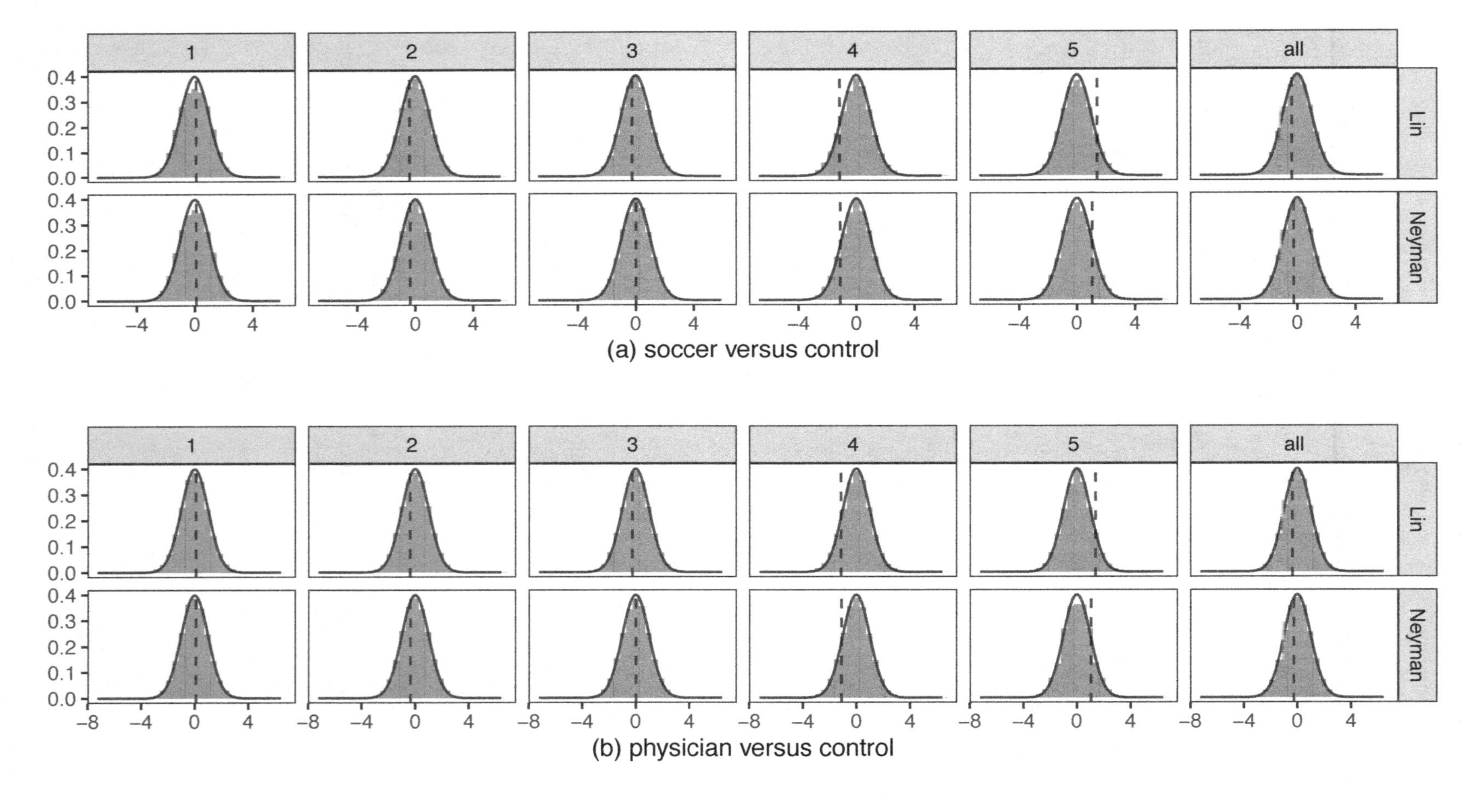

FIGURE 8.2: Reanalyzing the data from Chong et al. (2016): randomization distributions with 5×10^4 Monte Carlo draws and the $\mathrm{N}(0,1)$ approximations

9

Bridging Finite and Super Population Causal Inference

We have focused on the finite population perspective in randomized experiments. It treats all the potential outcomes as fixed numbers. Even if the potential outcomes are random variables, we can condition on them under the finite population perspective. The advantages of this perspective are

1. it focuses on the design of the experiments;
2. it requires minimal assumptions on the data-generating process of the outcomes.

However, it is often criticized for having only *internal validity* but not necessarily *external validity*, with informal definitions below.

Definition 9.1 (internal validity) *The statistical analysis is valid for the study samples at hand.*

Definition 9.2 (external validity) *The statistical analysis is valid for a broader population beyond the study samples.*

Obviously, all experimenters want not only the internal validity but also the external validity of their experiments. Since all statistical properties are conditional on the potential outcomes for the units we have, the results are only about the observed units under the finite population perspective. Then a natural question arises: do the finite population results generalize to a bigger population?

This is a fair critique of the finite population framework conditional on the potential outcomes. However, this can be a philosophical question. What we observed is a finite population, so any experimental design and analysis directly give us information about this finite population. Randomization only ensures internal validity given the potential outcomes of these units. The external validity of the results depends on the sampling process of the units. If the finite population is a representative sample of a larger population we are interested in, then of course the experimental results also have external validity. Otherwise, the results based on randomization inference may not generalize. To rigorously discuss this issue, we need to have a framework with two levels of randomness, one due to the sampling process of the units and the other due to the random assignment of the treatment. See Miratrix et al. (2018) and Abadie et al. (2020) for formal discussions of this idea.[1]

For some statisticians, this is just a technical problem. We can change the statistical framework, assuming that the units are sampled from a superpopulation. Then all the statements are about the population of interest. This is a convenient framework, although it does not really solve the problem mentioned above. Below, I will introduce this framework for two purposes:

1. it gives a different perspective on randomized experiments;

[1] Pearl and Bareinboim (2014) discussed the *transportability* problem from the perspective of causal diagrams.

2. it serves as a bridge between Parts II and III of this book.

The latter purpose is more important since the superpopulation framework allows us to derive more fruitful results for observational studies in which the treatment is not randomly assigned.

9.1 CRE

Assume

$$\{Z_i, Y_i(1), Y_i(0), X_i\}_{i=1}^n \overset{\text{IID}}{\sim} \{Z, Y(1), Y(0), X\}$$

from a superpopulation. So we can drop the subscript i for quantities of this population. With a little abuse of notation, we define the population average causal effect as

$$\tau = E\{Y(1) - Y(0)\} = E\{Y(1)\} - E\{Y(0)\}.$$

Under the superpopulation framework, we can formulate the CRE as below.

Definition 9.3 (CRE under the superpopulation framework) *We have*

$$Z \perp\!\!\!\perp \{Y(1), Y(0), X\}.$$

Under Definition 9.3, the average causal effect can be written as

$$\begin{aligned} \tau &= E\{Y(1) \mid Z = 1\} - E\{Y(0) \mid Z = 0\} \\ &= E(Y \mid Z = 1) - E(Y \mid Z = 0), \end{aligned} \tag{9.1}$$

which equals the difference in expectations of the outcomes. The first line in (9.1) is the key step that leverages the value of randomization. Since τ can be expressed as a function of the distributions of the observables, it is *nonparametrically identifiable*[2]. The formula (9.1) immediately suggests a moment estimator $\hat{\tau}$, which is the difference in means of the outcomes defined before. Conditioning on $\boldsymbol{Z}$, this is then a standard two-sample problem comparing the means of two independent samples. We have

$$E(\hat{\tau} \mid \boldsymbol{Z}) = \tau$$

and

$$\text{var}(\hat{\tau} \mid \boldsymbol{Z}) = \frac{\text{var}\{Y(1)\}}{n_1} + \frac{\text{var}\{Y(0)\}}{n_0}.$$

Under IID sampling, the sample variances are unbiased for the population variances, so Neyman's (1923) variance estimator is unbiased for $\text{var}(\hat{\tau} \mid \boldsymbol{Z})$. The conservativeness problem goes away under this superpopulation framework.

We can also discuss covariate adjustment. Based on the OLS decompositions (see Appendix B)

$$Y(1) = \gamma_1 + \beta_1^\top X + \varepsilon(1), \tag{9.2}$$

$$Y(0) = \gamma_0 + \beta_0^\top X + \varepsilon(0), \tag{9.3}$$

[2]In causal inference, we say that a parameter is nonparametrically identifiable if it can be determined by the distribution of the observed variables without imposing further parametric assumptions. See Definition 10.1 later for a more formal discussion.

we have

$$\begin{aligned} \tau &= E\{Y(1) - Y(0)\} \\ &= \gamma_1 - \gamma_0 + (\beta_1 - \beta_0)^\intercal E(X), \end{aligned}$$

since the residuals $\varepsilon(1)$ and $\varepsilon(0)$ have mean zero due to the inclusion of the intercepts. We can use the OLS with the treated and control data to estimate the coefficients in (9.2) and (9.3), respectively. The sample versions of the coefficients are $\hat{\gamma}_1, \hat{\beta}_1, \hat{\gamma}_0, \hat{\beta}_0$, so a covariate-adjusted estimator for τ is

$$\hat{\tau}_{\text{adj}} = \hat{\gamma}_1 - \hat{\gamma}_0 + (\hat{\beta}_1 - \hat{\beta}_0)^\intercal \bar{X}.$$

If we center covariates with $\bar{X} = 0$, the above estimator reduces to Lin's (2013) estimator

$$\hat{\tau}_{\text{L}} = \hat{\gamma}_1 - \hat{\gamma}_0,$$

which equals the coefficient of Z in the pooled regression with treatment-covariates interactions; see Proposition 6.2.

Unfortunately, the EHW variance estimator does not work for $\hat{\tau}_{\text{L}}$ because of the additional uncertainty in the sample mean of covariates $\bar{X}$ under the super population framework. Imbens and Wooldridge (2009), Berk et al. (2013), Negi and Wooldridge (2021), and Zhao and Ding (2021) proposed a correction of the EHW variance estimator by adding an extra term

$$(\hat{\beta}_1 - \hat{\beta}_0)^\intercal S_X^2 (\hat{\beta}_1 - \hat{\beta}_0)/n, \tag{9.4}$$

where $\hat{\beta}_1 - \hat{\beta}_0$ equals the coefficient of the interaction $Z_i X_i$ in obtaining Lin's (2013) estimator and S_X^2 is the finite-population covariance matrix of the covariates. A conceptually simpler yet computationally intensive approach is to use the bootstrap to estimate the variance; see Section A.6 and Problem 9.2.

9.2 Simulation under the CRE: the super population perspective

The following `linestimator` function can compute Lin's (2013) estimator, the EHW standard error, and the corrected standard error based on (9.4) for the super population.

```
library("car")
linestimator = function(Z, Y, X){
  ## standardize X
  X        = scale(X)
  n        = dim(X)[1]
  p        = dim(X)[2]

  ## fully interacted OLS
  linreg = lm(Y ~ Z*X)
  est    = coef(linreg)[2]
  vehw   = hccm(linreg)[2, 2]

  ## super population correction
  inter  = coef(linreg)[(p+3):(2*p+2)]
  vsuper = vehw + sum(inter*(cov(X)%*%inter))/n

  c(est, sqrt(vehw), sqrt(vsuper))
}
```

I then use simulation to compare the EHW standard error and the corrected standard error. I choose sample size $n = 500$ and use 2000 Monte Carlo repetitions. The potential outcomes are nonlinear in covariates and the error terms are not Normal. The true average causal effect equals 0.

```
res = replicate(2000, {
  n  = 500
  X  = matrix(rnorm(n*2), n, 2)
  Y1 = X[, 1] + X[, 1]^2 + runif(n, -0.5, 0.5)
  Y0 = X[, 2] + X[, 2]^2 + runif(n, -1, 1)
  Z  = rbinom(n, 1, 0.6)
  Y  = Z*Y1 + (1-Z)*Y0
  linestimator(Z, Y, X)
})
```

The following results confirm the theory. First, Lin's (2013) estimator is nearly unbiased. Second, the average EHW standard error is smaller than the empirical standard derivation, resulting in under coverage of the 95% confidence intervals. Third, the average corrected standard error for super population is nearly identical to the empirical standard derivation, resulting in more accurate coverage of the 95% confidence intervals.

```
> ## bias
> mean(res[1, ])
[1] -0.0001247585
> ## empirical standard deviation
> sd(res[1, ])
[1] 0.1507773
> ## estimated EHW standard error
> mean(res[2, ])
[1] 0.1388657
> ## coverage based on EHW standard error
> mean((res[1, ]-1.96*res[2, ])*(res[1, ]+1.96*res[2, ])<=0)
[1] 0.927
> ## estimated super population standard error
> mean(res[3, ])
[1] 0.1531519
> ## coverage based on population standard error
> mean((res[1, ]-1.96*res[3, ])*(res[1, ]+1.96*res[3, ])<=0)
[1] 0.9525
```

9.3 Extension to the SRE

We can extend the discussion in Section 9.1 to the SRE since it is equivalent to independent CREs within strata. The notation below will be slightly different from that in Chapter 5.

Assume that

$$\{Z_i, Y_i(1), Y_i(0), X_i\} \overset{\text{IID}}{\sim} \{Z, Y(1), Y(0), X\}.$$

With a discrete covariate $X_i \in \{1, \dots, K\}$, we can formulate the SRE as below.

Definition 9.4 (SRE under the super population framework) *We have*

$$Z \perp\!\!\!\perp \{Y(1), Y(0)\} \mid X.$$

Under Definition 9.4, the conditional average causal effect can be rewritten as

$$\begin{aligned} \tau_{[k]} &= E\{Y(1) - Y(0) \mid X = k\} \\ &= E(Y \mid Z = 1, X = k) - E(Y \mid Z = 0, X = k), \end{aligned}$$

so the average causal effect can be rewritten as

$$\begin{aligned} \tau &= E\{Y(1) - Y(0)\} \\ &= \sum_{k=1}^{K} \text{pr}(X = k) E\{Y(1) - Y(0) \mid X = k\} \\ &= \sum_{k=1}^{K} \text{pr}(X = k) \tau_{[k]}. \end{aligned}$$

The discussion in Section 9.1 holds within all strata, so we can derive the superpopulation analog for the SRE. When there are more than two treatment and control units within each stratum, we can use $\hat{V}_{\text{S}}$ as the variance estimator for $\text{var}(\hat{\tau}_{\text{S}})$.

We will see the exact form of Definition 9.4 in Part III later.

9.4 Homework problems

9.1 OLS decomposition of the observed outcome under the CRE

Based on (9.2) and (9.3), show that the OLS decomposition of the observed outcome on the treatment, covariates, and their interaction is

$$Y = \alpha_0 + \alpha_Z Z + \alpha_X^{\mathsf{T}} X + \alpha_{ZX}^{\mathsf{T}} XZ + \varepsilon$$

where

$$\alpha_0 = \gamma_0, \quad \alpha_Z = \gamma_1 - \gamma_0, \quad \alpha_X = \beta_0, \quad \alpha_{ZX} = \beta_1 - \beta_0$$

and

$$\varepsilon = Z\varepsilon(1) + (1 - Z)\varepsilon(0).$$

That is,

$$(\alpha_0, \alpha_Z, \alpha_X, \alpha_{ZX}) = \arg \min_{a_0, a_Z, a_X, a_{ZX}} E(Y - a_0 - a_Z Z - a_X^{\mathsf{T}} X - a_{ZX}^{\mathsf{T}} XZ)^2.$$

9.2 Variance estimation of Lin's (2013) estimator under the super population framework

Under the super population framework, simulate $\{X_i, Z_i, Y_i(1), Y_i(0)\}_{i=1}^n$ with $\beta_1 \neq \beta_0$ in (9.2) and (9.3). Calculate Lin's (2013) estimator in each simulated dataset. Compare the true variance and the following estimated variances:

1. the EHW robust variance estimator;
2. the EHW robust variance estimator with the correction term defined in (9.4);
3. the bootstrap variance estimator, with covariates centered at the beginning but not recentered for each bootstrap sample;

4. the bootstrap variance estimator, with covariates recentered for each bootstrap sample.

9.3 Recommended reading

Ding et al. (2017a) provide a unified discussion of the finite population and superpopulation inferences for the average causal effect.

Part III

Observational studies

10

Observational Studies, Selection Bias, and Nonparametric Identification of Causal Effects

Cochran (1965) summarized two common characteristics of observational studies:

1. the objective is to elucidate cause-and-effect relationships;
2. it is not feasible to use controlled experimentation.

The first characteristic is identical to that of randomized experiments discussed in Part II of this book, but the second differs fundamentally from randomized experiments.

Dorn (1953) suggested that the planner of an observational study should always ask the following question:

> How would the study be conducted if it were possible to do it by controlled experimentation?

It is always helpful to follow the suggestion from Dorn (1953) because the potential outcomes framework has an intrinsic link to an experiment, either a real experiment or a thought experiment. Part III of this book will discuss causal inference with observational studies. It will clarify the fundamental differences between observational studies and randomized experiments. Nevertheless, many ideas of causal inference with observational studies are deeply connected to those with randomized experiments.

10.1 Motivating examples

Example 10.1 (job training program) *LaLonde (1986) was interested in the causal effect of a job training program on earnings. He compared the results based on a randomized experiment to the results based on observational studies. We have used the experimental data before, which is the* `lalonde` *dataset in the* `Matching` *package. We have also used an observational counterpart* `cps1re74.csv` *in Section 1.2.1 and Problem 1.4. LaLonde (1986) found that many traditional statistical or econometric methods for observational studies gave quite different estimates compared with the estimates based on the experimental data. Dehejia and Wahba (1999) reanalyzed the data using methods motivated by causal inference, and found that those methods can recover the experimental gold standard well. Since then, their study has become a canonical example in causal inference with observational studies.*

Example 10.2 (smoking and homocysteine) *Bazzano et al. (2003) compared the homocysteine*[1] *levels in daily smokers and non-smokers based on the data from the National*

[1] Homocysteine is a type of amino acid. A high level of homocysteine in the blood is regarded as a marker of cardiovascular disease.

Health and Nutrition Examination Survey (NHANES) 2005–2006. Rosenbaum (2018) documented the data as `homocyst` *in the package* `senstrat`*. The dataset has the following important covariates:*

female	*1=female, 0=male*
age3	*three age categories: 20–39, 40–50, ≥60*
ed3	*three education categories: < high school, high school, some college*
bmi3	*three BMI categories: <30, [30, 35), ≥ 35*
pov2	*TRUE=income at least twice the poverty level, FALSE otherwise*

Example 10.3 (school meal program and body mass index) *Chan et al. (2016) used a subsample of the data from NHANES 2007–2008 to study whether participation in school meal programs led to an increase in BMI for school children. They documented the data as* `nhanes_bmi` *in the package* `ATE`*. The dataset has the following important covariates:*

age	*Age*
ChildSex	*Sex (1: male, 0: female)*
black	*Race (1: black, 0: otherwise)*
mexam	*Race (1: Hispanic: 0 otherwise)*
pir200_plus	*Family above 200% of the federal poverty level*
WIC	*Participation in the special supplemental nutrition program*
Food_Stamp	*Participation in food stamp program*
fsdchbi	*Childhood food security*
AnyIns	*Any insurance*
RefSex	*Sex of the adult respondent (1: male, 0: female)*
RefAge	*Age of the adult respondent*

A common feature of Examples 10.1–10.3 is that we are interested in estimating the causal effect of a nonrandomized treatment on an outcome. They are all from observational studies.

10.2 Causal effects and selection bias under the potential outcomes framework

For unit i ($i = 1, \ldots, n$), we have pretreatment covariates X_i, a binary treatment indicator Z_i, and an observed outcome Y_i with two potential outcomes $Y_i(1)$ and $Y_i(0)$ under the treatment and control, respectively. For simplicity, we assume

$$\{X_i, Z_i, Y_i(1), Y_i(0)\}_{i=1}^n \stackrel{\text{IID}}{\sim} \{X, Z, Y(1), Y(0)\}.$$

So we can drop the subscript i for quantities of this population. The causal effects of interest are the average causal effect

$$\tau = E\{Y(1) - Y(0)\},$$

the average causal effect on the treated units

$$\tau_{\text{T}} = E\{Y(1) - Y(0) \mid Z = 1\},$$

and the average causal effect on the control units:

$$\tau_{\text{C}} = E\{Y(1) - Y(0) \mid Z = 0\}.$$

By the linearity of the expectation, we have

$$\begin{aligned}\tau_{\mathrm{T}} &= E\{Y(1) \mid Z=1\} - E\{Y(0) \mid Z=1\} \\ &= E(Y \mid Z=1) - E\{Y(0) \mid Z=1\}\end{aligned}$$

and

$$\begin{aligned}\tau_{\mathrm{C}} &= E\{Y(1) \mid Z=0\} - E\{Y(0) \mid Z=0\} \\ &= E\{Y(1) \mid Z=0\} - E(Y \mid Z=0).\end{aligned}$$

In the above two formulas of τ_{T} and τ_{C}, the quantities $E(Y \mid Z=1)$ and $E(Y \mid Z=0)$ are directly estimable from the data, but the quantities $E\{Y(0) \mid Z=1\}$ and $E\{Y(1) \mid Z=0\}$ are not. The latter two are *counterfactuals* because they are the means of the potential outcomes corresponding to the treatment level that is the opposite of the actual received treatment.

The simple difference in means, also known as the *prima facie*[2] causal effect,

$$\begin{aligned}\tau_{\mathrm{PF}} &= E(Y \mid Z=1) - E(Y \mid Z=0) \\ &= E\{Y(1) \mid Z=1\} - E\{Y(0) \mid Z=0\}\end{aligned}$$

is generally biased for the causal effects defined above. For example,

$$\tau_{\mathrm{PF}} - \tau_{\mathrm{T}} = E\{Y(0) \mid Z=1\} - E\{Y(0) \mid Z=0\}$$

and

$$\tau_{\mathrm{PF}} - \tau_{\mathrm{C}} = E\{Y(1) \mid Z=1\} - E\{Y(1) \mid Z=0\}$$

are not zero in general, and they quantify the *selection bias.*[3] They measure the differences in the means of the potential outcomes across the treatment and control groups.

Why is randomization so important? Rubin (1978) first used potential outcomes to quantify the benefit of randomization. We have used the fact in Chapter 9 that

$$Z \perp\!\!\!\perp \{Y(1), Y(0)\} \tag{10.1}$$

in the CRE, which implies that the selection bias terms are both zero:

$$\begin{aligned}\tau_{\mathrm{PF}} - \tau_{\mathrm{T}} &= E\{Y(0) \mid Z=1\} - E\{Y(0) \mid Z=0\} \\ &= 0\end{aligned}$$

and

$$\begin{aligned}\tau_{\mathrm{PF}} - \tau_{\mathrm{C}} &= E\{Y(1) \mid Z=1\} - E\{Y(1) \mid Z=0\} \\ &= 0.\end{aligned}$$

So under the CRE in (10.1), we have

$$\tau = \tau_{\mathrm{T}} = \tau_{\mathrm{C}} = \tau_{\mathrm{PF}}. \tag{10.2}$$

[2]It is a Latin phrase, which means "based on the first impression" or "accepted as correct until proved otherwise." Holland (1986) used this phrase.

[3]Unfortunately, "selection bias" is abused across various research areas. The selection bias here corresponds to the selection of the treatment groups which is more commonly used in social sciences. In epidemiology, selection bias is often referred to as the bias in the sampling process. For instance, whether a unit is sampled or not can depend on its outcome status; see Section B.6.3 for a brief discussion of the case-control studies.

From the above discussion, the fundamental benefit of randomization is to balance the distributions of the potential outcomes across the treatment and control groups. This is a much stronger guarantee than balancing the distributions of the observed covariates.

Without randomization, the selection bias terms can be arbitrarily large, especially for unbounded outcomes. This highlights the fundamental difficulty of causal inference with observational studies.

10.3 Sufficient conditions for nonparametric identification of causal effects

10.3.1 Identification

Causal inference with observational studies is challenging. It relies on strong assumptions. A strategy is to use the information from the pretreatment covariates and assume that conditioning on the observed covariates X, the selection bias terms are zero, that is,

$$E\{Y(0) \mid Z=1, X\} = E\{Y(0) \mid Z=0, X\}, \tag{10.3}$$
$$E\{Y(1) \mid Z=1, X\} = E\{Y(1) \mid Z=0, X\}. \tag{10.4}$$

The assumptions in (10.3) and (10.4) state that the differences in the means of the potential outcomes across the treatment and control groups are entirely due to the difference in the observed covariates. So given the same value of the covariates, the potential outcomes have the same means across the treatment and control groups. Mathematically, (10.3) and (10.4) ensure that the conditional versions of the effects in (10.2) are identical:

$$\tau(X) = \tau_{\mathrm{T}}(X) = \tau_{\mathrm{C}}(X) = \tau_{\mathrm{PF}}(X),$$

where

$$\begin{aligned} \tau(X) &= E\{Y(1) - Y(0) \mid X\}, \\ \tau_{\mathrm{T}}(X) &= E\{Y(1) - Y(0) \mid Z=1, X\}, \\ \tau_{\mathrm{C}}(X) &= E\{Y(1) - Y(0) \mid Z=0, X\}, \\ \tau_{\mathrm{PF}}(X) &= E(Y \mid Z=1, X) - E(Y \mid Z=0, X). \end{aligned}$$

In particular, $\tau(X)$ is often called the *conditional average causal effect* or *conditional average treatment effect* (CATE).

A key result in this chapter is that the average causal effect τ is *nonparametrically identifiable* under (10.3) and (10.4). The notion of nonparametric identifiability does not appear frequently in classic statistics, but it is key to causal inference with observational studies. I first give an abstract definition below.

Definition 10.1 (identification) *A parameter θ is identifiable if it can be written as a function of the distribution of the observed data under certain model assumptions. A parameter θ is nonparametrically identifiable if it can be written as a function of the distribution of the observed data without any parametric model assumptions.*

Definition 10.1 is too abstract at the moment. I will use more concrete examples in later chapters to illustrate its meaning. It is often neglected in classic statistics problems. For instance, the mean $\theta = E(Y)$ is nonparametrically identifiable if we have IID draws of Y_i's; the Pearson correlation coefficient $\theta = \rho_{YX}$ is nonparametrically identifiable if we have IID draws of the pairs (X_i, Y_i)'s. In those examples, the parameters are nonparametrically identifiable automatically. However, Definition 10.1 is fundamental in causal inference with observational studies. In particular, the parameter of interest $\tau = E\{Y(1) - Y(0)\}$ depends on some unobserved random variables, so it is unclear whether it is nonparametrically identifiable based on observed data. Under the assumptions in (10.3) and (10.4), it is nonparametrically identifiable, with details below.

Because $\tau_{\text{PF}}(X)$ depends only on the observables, it is nonparametrically identified by definition. Moreover, (10.3) and (10.4) ensure that the three causal effects are the same as $\tau_{\text{PF}}(X)$, so $\tau(X)$, $\tau_{\text{T}}(X)$, and $\tau_{\text{C}}(X)$ are all nonparametrically identified. Consequently, the unconditional versions are also nonparametrically identified under (10.3) and (10.4) due to the law of total expectation:

$$\tau = E\{\tau(X)\}, \quad \tau_{\text{T}} = E\{\tau_{\text{T}}(X) \mid Z = 1\}, \quad \tau_{\text{C}} = E\{\tau_{\text{C}}(X) \mid Z = 0\}.$$

From now on, I will focus on τ unless stated otherwise (Chapter 13 will be an exception). The following theorem summarizes the identification formulas of τ.

Theorem 10.1 *Under* (10.3) *and* (10.4), *the average causal effect τ is identified by*

$$\begin{aligned} \tau &= E\{\tau(X)\} && (10.5) \\ &= E\{E(Y \mid Z = 1, X) - E(Y \mid Z = 0, X)\} && (10.6) \\ &= \int \{E(Y \mid Z = 1, X = x) - E(Y \mid Z = 0, X = x)\} f(x) \mathrm{d}x. && (10.7) \end{aligned}$$

The formula (10.6) was formally established by Rosenbaum and Rubin (1983b), which is also called the g-formula by Robins when it is extended to the setting with time-varying treatments. See Hernán and Robins (2020) and Chapter 29.

With a discrete covariate, we can write the identification formula in Theorem 10.1 as

$$\begin{aligned} \tau &= \sum_x E(Y \mid Z = 1, X = x) \text{pr}(X = x) \\ &\quad - \sum_x E(Y \mid Z = 0, X = x) \text{pr}(X = x), && (10.8) \end{aligned}$$

and also the simple difference in means as

$$\begin{aligned} \tau_{\text{PF}} &= \sum_x E(Y \mid Z = 1, X = x) \text{pr}(X = x \mid Z = 1) \\ &\quad - \sum_x E(Y \mid Z = 0, X = x) \text{pr}(X = x \mid Z = 0) && (10.9) \end{aligned}$$

by the law of total probability. Comparing (10.8) and (10.9), we can see that although both formulas compare the conditional expectations $E(Y \mid Z = 1, X = x)$ and $E(Y \mid Z = 0, X = x)$, they average over different distributions of the covariates. The causal parameter τ averages the conditional expectations over the common distribution of the covariates, but the difference in means τ_{PF} averages the conditional expectations over two different distributions of the covariates in the treated and control groups.

Usually, we impose a stronger assumption.

Assumption 10.1 (ignorability) *We have*

$$Y(z) \perp\!\!\!\perp Z \mid X \quad (z = 0, 1). \qquad (10.10)$$

Assumption 10.1 has many names:

1. *ignorability* due to Rubin (1978);[4]
2. *unconfoundedness* which is popular among epidemiologists;
3. *selection on observables* which is popular among social scientists;
4. *conditional independence* which is merely a description of the notation $\perp\!\!\!\perp$ in the assumption.

Sometimes, we impose an even stronger assumption.

Assumption 10.2 (strong ignorability) *We have*

$$\{Y(1), Y(0)\} \perp\!\!\!\perp Z \mid X. \tag{10.11}$$

Assumption 10.2 is called *strong ignorability* (Rubin, 1978; Rosenbaum and Rubin, 1983b). If the parameter of interest is τ, then the stronger Assumptions 10.1 and 10.2 are just imposed for notational simplicity thanks to the conditional independence notation $\perp\!\!\!\perp$. They are not necessary for identifying τ. However, they can not be relaxed if the parameter of interest is the causal effects on other scales (for example, distribution, quantile, or some transformation of the outcome). The *strong ignorability* assumption requires that the potential outcomes vector be independent of the treatment given the covariates, but the *ignorability* assumption only requires each potential outcome to be independent of the treatment given covariates. The former is stronger than the latter. However, their difference is rather technical and of pure theoretical interests; see Problem 10.5. In most reasonable statistical models, they both hold; see Section 10.3.2 below. We will not distinguish them in this book and will simply use *ignorability* to refer to both.

10.3.2 Plausibility of the ignorability assumption

A fundamental problem of causal inference with observational studies is the plausibility of the ignorability assumption. The discussion in Section 10.3.1 may seem too mathematical in the sense that the ignorability assumption serves as a sufficient condition to ensure the nonparametric identification of the average causal effect. What is its scientific meaning? Intuitively, it rules out all unmeasured covariates that affect the treatment and outcome simultaneously. Those "common causes" of the treatment and outcome are called confounders. That is why the ignorability assumption is also called the unconfoundedness assumption. More transparently, we can interpret the ignorability assumption based on the outcome data-generating process. If

$$\begin{aligned} Y(1) &= g_1(X, V_1), \\ Y(0) &= g_0(X, V_0), \\ Z &= I\{g(X, V) \geq 0\} \end{aligned}$$

where the g's are general functions and the random error terms satisfy $(V_1, V_0) \perp\!\!\!\perp V$, then both Assumptions 10.1 and 10.2 hold. In the above data-generating process, the "common causes" X of the treatment and the outcome are all observed, and the remaining random components are independent. If the data-generating process changes to

$$\begin{aligned} Y(1) &= g_1(X, U, V_1), \\ Y(0) &= g_0(X, U, V_0), \\ Z &= I\{g(X, U, V) \geq 0\} \end{aligned}$$

[4]The reason for using this name is unclear based on the discussion in this book. In fact, it is from the Bayesian perspective of causal inference. It is beyond the scope of this book.

where the g's are general functions and the random error terms satisfy $(V_1, V_0) \perp\!\!\!\perp V$, then Assumptions 10.1 and 10.2 do not hold in general. The unmeasured "common cause" U induces dependence between the treatment and the potential outcomes even conditional on the observed covariates X. If we do not have access to U and analyze the data based only on (Z, X, Y), the final estimator will be biased for the causal parameter in general. This type of bias is called the *omitted variable bias* in econometrics; see Problem 16.2 in a later chapter.

The ignorability assumption can be reasonable if we observe a rich set of covariates X that affect the treatment and the outcome simultaneously. I start with this assumption to discuss identification and estimation strategies in Part III of this book. However, it is fundamentally untestable. We may justify it based on the scientific background knowledge, but we are often not sure whether it holds or not. Parts IV and V of this book will discuss other strategies when the ignorability assumption is not plausible.

10.4 Two simple estimation strategies and their limitations

10.4.1 Stratification or standardization based on discrete covariates

If the covariate $X_i \in \{1, \ldots, K\}$ is discrete, then ignorability (10.10) reads as

$$Y(z) \perp\!\!\!\perp Z \mid X = k \quad (z = 0, 1; k = 1, \ldots, K),$$

which essentially assumes that the observational study is an SRE under the superpopulation framework in Chapter 9. Therefore, we can use the estimator

$$\hat{\tau} = \sum_{k=1}^{K} \pi_{[k]} \left\{ \hat{\bar{Y}}_{[k]}(1) - \hat{\bar{Y}}_{[k]}(0) \right\},$$

which is identical to the stratified or post-stratified estimator discussed in Chapter 5.

This method is still widely used in practice. Example 10.2 contains discrete covariates, and I relegate the analysis to Problem 10.4. However, there are several obvious difficulties in implementing this method. First, it works well for the case with small K. For large K, it is very likely that many strata have $n_{[k]1} = 0$ or $n_{[k]0} = 0$, leading to the poorly defined $\hat{\tau}_{[k]}$'s for those strata. This is related to the issue of overlap which will be discussed in Chapter 20 later. Second, it is not obvious how to apply this stratification method to multidimensional continuous or mixed covariates X. A standard method is to create strata based on the initial covariates and then apply the stratification method. This may result in arbitrariness in the analysis due to the non-uniqueness of the created strata.

10.4.2 Outcome regression

A commonly used method based on outcome regression is to run the OLS with an additive model of the observed outcome on the treatment indicator and covariates, which assumes

$$E(Y \mid Z, X) = \beta_0 + \beta_z Z + \beta_x^{\mathrm{T}} X.$$

If the above linear model is correct, then we have

$$\begin{aligned} \tau(X) &= E(Y \mid Z = 1, X) - E(Y \mid Z = 0, X) \\ &= (\beta_0 + \beta_z + \beta_x^{\mathrm{T}} X) - (\beta_0 + \beta_x^{\mathrm{T}} X) \\ &= \beta_z, \end{aligned}$$

which implies that the causal effect is homogeneous with respect to the covariates. This, coupled with ignorability, implies that

$$\tau = E\{\tau(X)\} = \beta_z.$$

Therefore, if ignorability holds and the outcome model is linear, then the average causal effect equals the coefficient of Z. This is one of the most important applications of the linear model. However, the causal interpretation of the coefficient of Z is valid only under two strong assumptions: ignorability and the linear model.

We have discussed in Chapter 6 that the above procedure is suboptimal even in CREs because it ignores the treatment effect heterogeneity induced by the covariates. If we assume

$$E(Y \mid Z, X) = \beta_0 + \beta_z Z + \beta_x^{\mathsf{T}} X + \beta_{zx}^{\mathsf{T}} XZ,$$

we have

$$\begin{aligned} \tau(X) &= E(Y \mid Z=1, X) - E(Y \mid Z=0, X) \\ &= (\beta_0 + \beta_z + \beta_x^{\mathsf{T}} X + \beta_{zx}^{\mathsf{T}} X) - (\beta_0 + \beta_x^{\mathsf{T}} X) \\ &= \beta_z + \beta_{zx}^{\mathsf{T}} X, \end{aligned}$$

which, coupled with ignorability, implies that

$$\tau = E\{\tau(X)\} = E(\beta_z + \beta_{zx}^{\mathsf{T}} X) = \beta_z + \beta_{zx}^{\mathsf{T}} E(X).$$

The estimator for τ is then $\hat\beta_z + \hat\beta_{zx}^{\mathsf{T}} \bar X$, where $\hat\beta_z$ is the regression coefficient of Z and $\bar X$ is the sample mean of X. If we center the covariates to ensure $\bar X = 0$, then the estimator is simply the regression coefficient of Z. To simplify the procedure, we usually center the covariates at the beginning; also recall Lin's (2013) estimator introduced in Chapter 6. Rosenbaum and Rubin (1983b) and Hirano and Imbens (2001) discussed this estimator.

In general, we can use other more complex models to estimate the causal effects. For example, if we build two predictors $\hat\mu_1(X)$ and $\hat\mu_0(X)$ based on the treated and control data, respectively, then we have an estimator for the conditional average causal effect

$$\hat\tau(X) = \hat\mu_1(X) - \hat\mu_0(X)$$

and an estimator for the average causal effect:

$$\hat\tau^{\text{reg}} = n^{-1} \sum_{i=1}^{n} \{\hat\mu_1(X_i) - \hat\mu_0(X_i)\}.$$

The estimator $\hat\tau$ above has the same form as the projective estimator discussed in Chapter 6. It is sometimes called the *outcome regression* estimator. The OLS-based estimators above are special cases. We can use the nonparametric bootstrap (see Section A.6) to estimate the standard error of the outcome regression estimator.

I give another example for binary outcomes below.

Example 10.4 (outcome regression estimator for binary outcomes) *With a binary outcome, we may model Y using a logistic model*

$$E(Y \mid Z, X) = \text{pr}(Y = 1 \mid Z, X) = \frac{e^{\beta_0 + \beta_z Z + \beta_x^{\mathsf{T}} X}}{1 + e^{\beta_0 + \beta_z Z + \beta_x^{\mathsf{T}} X}},$$

then based on the estimators of the coefficients $\hat\beta_0, \hat\beta_z, \hat\beta_x$, we have the following estimator for the average causal effect:

$$\hat\tau = n^{-1}\sum_{i=1}^{n}\left\{\frac{e^{\hat\beta_0+\hat\beta_z+\hat\beta_x^{\mathrm{T}}X_i}}{1+e^{\hat\beta_0+\hat\beta_z+\hat\beta_x^{\mathrm{T}}X_i}} - \frac{e^{\hat\beta_0+\hat\beta_x^{\mathrm{T}}X_i}}{1+e^{\hat\beta_0+\hat\beta_x^{\mathrm{T}}X_i}}\right\}.$$

This estimator is not simply the coefficient of the treatment in the logistic model.[5] *It is a nonlinear function of all the coefficients as well as the empirical distribution of the covariates. In econometrics, this estimator is called the "average partial effect" or "average marginal effect" of the treatment in the logistic model. Some econometric software packages report this estimator associated with the standard error. Similarly, we can also derive the corresponding estimator based on a fully interacted logistic model; see Problem 10.3 for more details.*

The above predictors for the conditional means of the outcome can also be obtained from other machine learning tools. In particular, Hill (2011) championed the use of tree methods for estimating τ, and Wager and Athey (2018) proposed to use them also for estimating $\tau(X)$. Wager and Athey (2018) also combined the tree methods with the idea of the *propensity score* in the next chapter. Since then, the intersection of machine learning and causal inference has been an active research area (e.g., Hahn et al., 2020; Künzel et al., 2019).

The biggest problem with the above approach based on outcome regressions is its sensitivity to the specification of the outcome model. Problem 1.4 gave such an example. Depending on the incentive of empirical research and publications, people might report their favorable causal effects estimates after searching over a wide set of candidate models, without confessing this model searching process. This is a major source of p-hacking in causal inference. Problem 1.4 hinted at this issue. Leamer (1978) criticized this approach in empirical research.

10.5 Homework problems

10.1 A simple identity

Show that $\tau = \mathrm{pr}(Z=1)\tau_{\mathrm{T}} + \mathrm{pr}(Z=0)\tau_{\mathrm{C}}$.

10.2 Nonparametric identification of other causal effects

Under ignorability, show that

1. the distributional causal effect

$$\mathrm{DCE}_y = \mathrm{pr}\{Y(1) > y\} - \mathrm{pr}\{Y(0) > y\}$$

is nonparametrically identifiable for all y;

[5] If the logistic outcome model is correct, then $\hat\beta_z$ estimates the conditional odds ratio of the treatment on the outcome given covariates, which does not equal τ. Freedman (2008c) gave a warning about using the logistic regression coefficient to estimate τ in CREs. See Appendix B for more details of the logistic regression.

2. the quantile causal effect

$$\text{QCE}_q = \text{quantile}_q\{Y(1)\} - \text{quantile}_q\{Y(0)\},$$

is nonparametrically identifiable for all q, where $\text{quantile}_q\{\cdot\}$ is the qth quantile of a random variable.

Remark: In probability theory, $\text{pr}\{Y(z) \leq y\}$ is the cumulative distribution function and $\text{pr}\{Y(z) > y\}$ is the survival function of the potential outcome $Y(z)$. The distributional causal effect compares the survival functions of the potential outcomes under treatment and control.

The quantile causal effect compares the marginal quantiles of the potential outcomes, which is different from the quantile of the individual causal effect

$$\tau_q = \text{quantile}_q\{Y(1) - Y(0)\}.$$

In fact, τ_q is not identifiable in the sense that the marginal distributions $\text{pr}\{Y(1) \leq y\}$ and $\text{pr}\{Y(0) \leq y\}$ can not uniquely determine τ_q.

10.3 Outcome imputation estimator in the fully interacted logistic model

This problem extends Example 10.4.

Assume that a binary outcome follows a logistic model

$$E(Y \mid Z, X) = \text{pr}(Y = 1 \mid Z, X) = \frac{e^{\beta_0 + \beta_z Z + \beta_x^{\mathrm{T}} X + \beta_{xz}^{\mathrm{T}} XZ}}{1 + e^{\beta_0 + \beta_z Z + \beta_x^{\mathrm{T}} X + \beta_{xz}^{\mathrm{T}} XZ}}.$$

What is the corresponding outcome regression estimator for the average causal effect?

10.4 Data analysis: stratification and regression

Use the dataset `homocyst` in the package `senstrat`. The outcome is `homocysteine`, the homocysteine level, and the treatment is `z`, where $z = 1$ for a daily smoker and $z = 0$ for a never smoker. Covariates are `female, age3, ed3, bmi3, pov2` with detailed explanations in the package, and `st` is a stratum indicator, defined by all the combinations of the discrete covariates.

1. How many strata have only treated or control units? What is the proportion of the units in these strata? Drop these strata and perform a stratified analysis of the observational study. Report the point estimator, variance estimator, and 95% confidence interval for the average causal effect.
2. Run the OLS of the outcome on the treatment indicator and covariates without interactions. Report the coefficient of the treatment and the robust standard error.

   Drop the strata with only treated or control units. Rerun the OLS and report the result.
3. Apply Lin's (2013) estimator of the average causal effect. Report the coefficient of the treatment and the robust standard error.

   If you do not drop the strata with only treated or control units, what will happen?
4. Compare the results in the above three analyses. Which one is more credible?

10.5 Ignorability versus strong ignorability

Give an example such that the ignorability holds but the strong ignorability does not hold.

Remark: This is related to a classic probability problem of finding three random variables A, B, C such that

$$A \perp\!\!\!\perp C \text{ and } B \perp\!\!\!\perp C \text{ but } (A, B) \not\!\perp\!\!\!\perp C.$$

10.6 Recommended reading

Cochran (1965) is a classic reference on observational studies. It contains many useful insights but does not use the formal potential outcomes framework. Dylan Small founded a new journal called *Observational Studies* in 2015 (Small, 2015) and reprinted an old article "Observational Studies" by W. G. Cochran as Cochran (2015). Many leading researchers also made insightful comments on Cochran (2015).

11

The Central Role of the Propensity Score in Observational Studies for Causal Effects

Rosenbaum and Rubin (1983b) proposed the key concept of *propensity score* and discussed its role in causal inference with observational studies. It is one of the most cited papers in statistics, and Titterington (2013) listed it as the second most cited paper published in *Biometrika* during the past 100 years. Its number of citations has grown very fast in recent years.

Under the IID sampling assumption, each unit has four random variables $\{X, Z, Y(1), Y(0)\}$. Following the basic probability rules, we can factorize the joint distribution as

$$\begin{aligned} & \text{pr}\{X, Z, Y(1), Y(0)\} \\ = \ & \text{pr}(X) \times \text{pr}\{Y(1), Y(0) \mid X\} \times \text{pr}\{Z \mid X, Y(1), Y(0)\}, \end{aligned}$$

where

1. $\text{pr}(X)$ is the covariate distribution,
2. $\text{pr}\{Y(1), Y(0) \mid X\}$ is the outcome distribution conditional on the covariates X,
3. $\text{pr}\{Z \mid X, Y(1), Y(0)\}$ is the treatment distribution conditional on the covariates X, also known as the treatment assignment mechanism.

Usually, we do not want to model the covariates because they are background information happening before the treatment and outcome. If we want to go beyond the outcome model, then we must focus on the treatment assignment mechanism, which leads to the definition of the propensity score.

Definition 11.1 (propensity score) *Define*

$$e(X, Y(1), Y(0)) = \text{pr}\{Z = 1 \mid X, Y(1), Y(0)\}$$

as the propensity score. Under strong ignorability, we have

$$\begin{aligned} e(X, Y(1), Y(0)) &= \text{pr}\{Z = 1 \mid X, Y(1), Y(0)\} \\ &= \text{pr}(Z = 1 \mid X), \end{aligned}$$

so the propensity score reduces to

$$e(X) = \text{pr}(Z = 1 \mid X),$$

the conditional probability of receiving the treatment given the observed covariates.

Rosenbaum and Rubin (1983b) used $e(X) = \text{pr}(Z = 1 \mid X)$ as the definition of the propensity score because they focused on observational studies under strong ignorability. It is sometimes helpful to view $e(X, Y(1), Y(0)) = \text{pr}\{Z = 1 \mid X, Y(1), Y(0)\}$ as the general definition of the propensity score even when strong ignorability fails. See Problem 11.1 for more details.

Following Rosenbaum and Rubin (1983b), this chapter will demonstrate that $e(X)$ is a key quantity in causal inference with observational studies under ignorability.

11.1 The propensity score as a dimension reduction tool

11.1.1 Theory

Theorem 11.1 *If* $Z \perp\!\!\!\perp \{Y(1), Y(0)\} \mid X$, *then* $Z \perp\!\!\!\perp \{Y(1), Y(0)\} \mid e(X)$.

Theorem 11.1 states that, if strong ignorability holds conditional on covariates X, then it also holds conditional on the scalar propensity score $e(X)$. The ignorability requires conditioning on many background characteristics X of the units, but Theorem 11.1 implies that conditioning on the propensity score $e(X)$ removes all confounding induced by covariates X. The original covariates X can be general and have many dimensions, but the propensity score $e(X)$ is a one-dimensional scalar variable bounded between 0 and 1. Therefore, the propensity score reduces the dimension of the original covariates but still maintains the ignorability. As a statistical terminology, we can view the propensity score as a *dimensional reduction* tool. We will first prove Theorem 11.1 below and then give an application of the dimension reduction property of the propensity score.

Proof of Theorem 11.1: By the definition of conditional independence, we need to show that

$$\mathrm{pr}\{Z = 1 \mid Y(1), Y(0), e(X)\} = \mathrm{pr}\{Z = 1 \mid e(X)\}. \tag{11.1}$$

The left-hand side of (11.1) equals

$$\begin{aligned}
& \mathrm{pr}\{Z = 1 \mid Y(1), Y(0), e(X)\} \\
= \ & E\{Z \mid Y(1), Y(0), e(X)\} \\
= \ & E\Big[E\{Z \mid Y(1), Y(0), e(X), X\} \mid Y(1), Y(0), e(X)\Big] \\
& \qquad\qquad \text{(tower property; see Section A.1.5)} \\
= \ & E\Big[E\{Z \mid Y(1), Y(0), X\} \mid Y(1), Y(0), e(X)\Big] \\
= \ & E\Big\{E(Z \mid X) \mid Y(1), Y(0), e(X)\Big\} \quad \text{(strong ignorability)} \\
= \ & E\Big\{e(X) \mid Y(1), Y(0), e(X)\Big\} \\
= \ & e(X).
\end{aligned}$$

The right-hand side of (11.1) equals

$$\begin{aligned}
& \mathrm{pr}\{Z = 1 \mid e(X)\} \\
= \ & E\{Z \mid e(X)\} \\
= \ & E\Big[E\{Z \mid e(X), X\} \mid e(X)\Big] \quad \text{(tower property)} \\
= \ & E\Big\{E(Z \mid X) \mid e(X)\Big\} \\
= \ & E\Big\{e(X) \mid e(X)\Big\} \\
= \ & e(X).
\end{aligned}$$

So the left-hand side of (11.1) equals the right-hand side of (11.1). □

11.1.2 Propensity score stratification

Theorem 11.1 motivates a simple method for estimating causal effects: propensity score stratification. Starting from the simple case, we assume that the propensity score is known and only takes K possible values $\{e_1, \ldots, e_K\}$ with K being much smaller than the sample size n. Theorem 11.1 reduces to

$$Z \perp\!\!\!\perp \{Y(1), Y(0)\} \mid e(X) = e_k \quad (k = 1, \ldots, K).$$

Therefore, we have an SRE, that is, we have K independent CREs within strata of the propensity score. We can analyze the observational data in the same way as the SRE stratified on $e(X)$.

In general, the propensity score $e(X)$ is not known and moreover, it is not a discrete random variable. We often fit a statistical model for $\text{pr}(Z = 1 \mid X)$ (for example, a logistic model of the binary Z given X) to obtain the estimated propensity score $\hat{e}(X)$. This estimated propensity score can take as many values as the sample size, but we can discretize it to approximate the simple case above. For example, we can discretize the estimated propensity score by its K quantiles to obtain $\hat{e}'(X)$: $\hat{e}'(X_i) = e_k$, the k/K-th quantile of $\hat{e}(X)$, if $\hat{e}(X_i)$ is between the $(k-1)/K$-th and k/K-th quantiles of the $\hat{e}(X_i)$'s. Then

$$Z \perp\!\!\!\perp \{Y(1), Y(0)\} \mid \hat{e}'(X) = e_k \quad (k = 1, \ldots, K)$$

holds approximately. So we can analyze the observational data in the same way as the SRE stratified on $\hat{e}'(X)$. The ignorability holds only approximately given $\hat{e}'(X)$. We can further use regression adjustment based on covariates to remove bias and improve efficiency. To be more specific, we can obtain Lin's (2013) estimator within each stratum and construct the final estimator by a weighted average (see Section 6.2.4).

With an unknown propensity score, we need to fit a statistical model to obtain the estimated propensity score $\hat{e}(X)$. This makes the final estimator dependent on the model specification. However, the propensity score stratification estimator only requires the correct ordering of the estimated propensity scores rather than their exact values, which makes it relatively robust compared with other methods. This robustness property of propensity score stratification appeared in many numerical examples but its rigorous quantification is still missing in the literature.

An important practical question is how to choose K? If K is too small, then ignorability does not hold conditional on $\hat{e}'(X)$ even approximately. If K is too large, then we do not have enough units within each stratum of the estimated propensity score and many strata have only treated or control units. Therefore, we face a trade-off. Rosenbaum and Rubin (1983b) and Rosenbaum and Rubin (1984) suggested $K = 5$ which removes a large amount of bias in many settings, following the heuristic argument from Cochran (1968). However, with extremely large datasets, propensity score stratification leads to biased estimators with a fixed K (Lunceford and Davidian, 2004). It is thus reasonable to increase K as long as each stratum still has enough treated and control units. Wang et al. (2020) suggested a greedy choice of K, which is the maximum number of strata such that the stratified estimator is well-defined. However, the rigorous theory for this procedure is not fully established.

Another important practical question is how to compute the standard errors of the estimators based on propensity score stratification. Some researchers have conditioned on the discretized propensity scores $\hat{e}'(X)$ and reported standard errors based on the SRE. This effectively ignores the uncertainty in the estimated propensity scores. This often leads to an overestimation of the true variance since a surprising result in the literature is that using the estimated propensity scores decreases the asymptotic variance for estimating the average causal effect (Su et al., 2023). Other researchers accounted for the full uncertainty

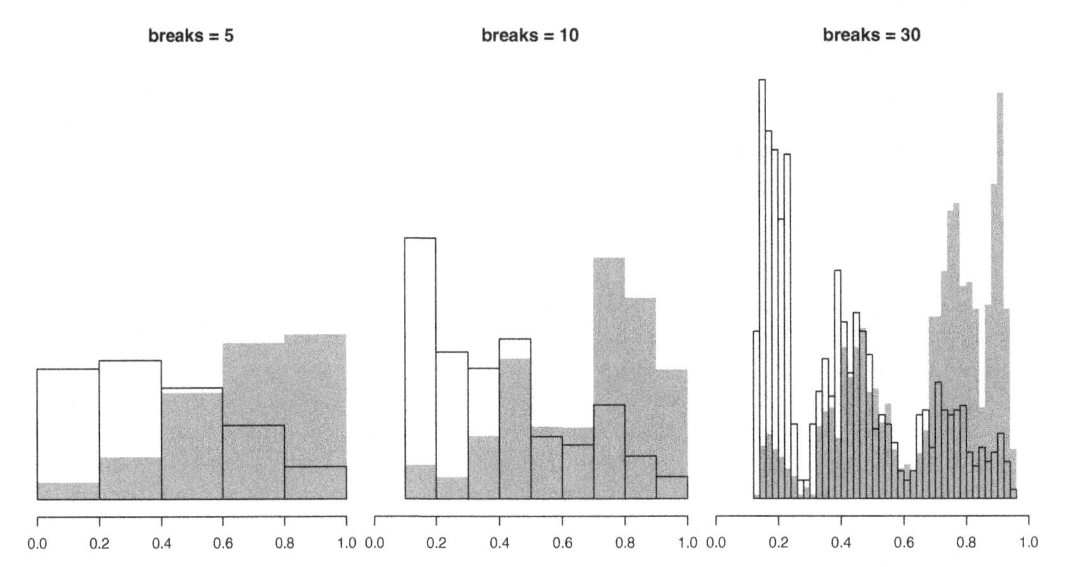

FIGURE 11.1: Histograms of the estimated propensity scores based on the `nhanes_bmi` data: white for the control group and grey for the treatment group

by bootstrapping the whole procedure including the first stage of estimating the propensity scores. However, the theory for the bootstrap is still unclear due to the discreteness of this estimator.

11.1.3 Application

To illustrate the propensity score stratification method, I revisited Example 10.3.

```
> nhanes_bmi = read.csv("nhanes_bmi.csv")[, -1]
> z = nhanes_bmi$School_meal
> y = nhanes_bmi$BMI
> x = as.matrix(nhanes_bmi[, -c(1, 2)])
> x = scale(x)
```

Example 10.3 introduced the covariates in the dataset. The treatment `School_meal` is the indicator for participation in the school meal plan, and the outcome `BMI` is the BMI.

Figure 11.1 shows the histograms of the estimated propensity scores with different numbers of bins ($K = 5, 10, 30$). Based on propensity score stratification, we can calculate the point estimators and the standard errors for difference choice of $K \in \{5, 10, 20, 50, 80\}$ as follows (with the function `Neyman_SRE` defined in Chapter 5 for analyzing the SRE):

```
> pscore = glm(z ~ x, family = binomial)$fitted.values
> n.strata = c(5, 10, 20, 50, 80)
> strat.res = sapply(n.strata, FUN = function(nn){
+   q.pscore = quantile(pscore, (1:(nn-1))/nn)
+   ps.strata = cut(pscore, breaks = c(0,q.pscore,1),
+                   labels = 1:nn)
+   Neyman_SRE(z, y, ps.strata)})
>
> rownames(strat.res) = c("est", "se")
> colnames(strat.res) = n.strata
> round(strat.res, 3)
         5     10     20     50     80
```

```
est  -0.116  -0.178  -0.200  -0.265  -0.204
se    0.283   0.282   0.279   0.272      NA
```

Increasing K from 5 to 50 reduces the standard error. However, we cannot go as extreme as $K = 80$ because the standard error is not well-defined in some strata with only one treated or one control unit. The above estimators show a negative but insignificant effect of the meal program on BMI.

We can also compare the above estimator with the three simple regression estimators: the one without adjusting for any covariates, and Fisher's and Lin's estimators.

```
> DiM = lm(y ~ z)
> Fisher = lm(y ~ z + x)
> Lin = lm(y ~ z + x + z*x)
> res.regression = c(coef(DiM)[2], hccm(DiM)[2, 2]^0.5,
+                    coef(Fisher)[2], hccm(Fisher)[2, 2]^0.5,
+                    coef(Lin)[2], hccm(Lin)[2, 2]^0.5)
> res.regression = matrix(res.regression,
+                         nrow = 2, ncol = 3)
> rownames(res.regression) = c("est", "se")
> colnames(res.regression) = c("naive", "fisher", "lin")
> round(res.regression, 3)
    naive fisher    lin
est 0.534  0.061 -0.017
se  0.225  0.227  0.226
```

The naive difference in means differs greatly from the other methods due to the large imbalance in covariates. Fisher's estimator still gives a positive point estimate although it is not significant. Lin's estimator and the propensity score stratification estimators give qualitatively the same results. The propensity score stratification estimators are stable across different choices of K.

11.2 Propensity score weighting

11.2.1 Theory

Theorem 11.2 *If* $Z \perp\!\!\!\perp \{Y(1), Y(0)\} \mid X$ *and* $0 < e(X) < 1$, *then*

$$E\{Y(1)\} = E\left\{\frac{ZY}{e(X)}\right\}, \qquad E\{Y(0)\} = E\left\{\frac{(1-Z)Y}{1-e(X)}\right\},$$

and

$$\tau = E\{Y(1) - Y(0)\} = E\left\{\frac{ZY}{e(X)} - \frac{(1-Z)Y}{1-e(X)}\right\}.$$

Before proving Theorem 11.2, it is important to note the additional assumption $0 < e(X) < 1$. It is called the *overlap* or *positivity* condition. The formulas in Theorem 11.2 become infinity if $e(X) = 0$ or 1 for some values of X. It is a requirement not only for the identification formulas based on propensity score weighting but also for those based on outcome modeling. Although it was not stated explicitly in Theorem 10.1, the conditional expectations $E(Y \mid Z = 1, X)$ and $E(Y \mid Z = 0, X)$ in the identification formula of τ

in (10.6) is well defined only if $0 < e(X) < 1$. The overlap condition can be viewed as a technical condition to ensure that the formulas in Theorems 10.1 and 11.2 are well defined. It can also cause some philosophical issues for causal inference with observational studies. When unit i has $e(X_i) = 1$, we always observe its potential outcome under the treatment, $Y_i(1)$, but can never observe its potential outcome under the control, $Y_i(0)$. In this case, the potential outcome $Y_i(0)$ may not even be well defined, making the definition of the causal effect ambiguous for unit i. King and Zeng (2006) called $Y_i(0)$ an *extreme counterfactual* when $e(X_i) = 1$, and discussed their dangers in causal inference. A similar problem arises if unit i has $e(X_i) = 0$.

In sum, $Z \perp\!\!\!\perp \{Y(1), Y(0)\} \mid X$ requires adequate covariates to ensure the conditional independence of the treatment and potential outcomes, and $0 < e(X) < 1$ requires residual randomness in the treatment conditional on the covariates. In fact, the definition of strong ignorability in Rosenbaum and Rubin (1983b) includes both of these conditions. In modern literature, they are often stated separately.

Proof of Theorem 11.2: I only prove the result for $E\{Y(1)\}$ because the proof of the result for $E\{Y(0)\}$ is similar. We have

$$
\begin{aligned}
& E\left\{\frac{ZY}{e(X)}\right\} \\
= \quad & E\left\{\frac{ZY(1)}{e(X)}\right\} \\
= \quad & E\left[E\left\{\frac{ZY(1)}{e(X)} \mid X\right\}\right] \quad \text{(tower property)} \\
= \quad & E\left[\frac{1}{e(X)} E\{ZY(1) \mid X\}\right] \\
= \quad & E\left[\frac{1}{e(X)} E(Z \mid X) E\{Y(1) \mid X\}\right] \quad \text{(strong ignorability)} \\
= \quad & E\left[\frac{1}{e(X)} e(X) E\{Y(1) \mid X\}\right] \\
= \quad & E[E\{Y(1) \mid X\}] \\
= \quad & E\{Y(1)\}.
\end{aligned}
$$

□

11.2.2 Inverse propensity score weighting estimators

Theorem 11.2 motivates the following moment estimator for the average causal effect:

$$
\hat{\tau}^{\mathrm{ht}} = \frac{1}{n} \sum_{i=1}^{n} \frac{Z_i Y_i}{\hat{e}(X_i)} - \frac{1}{n} \sum_{i=1}^{n} \frac{(1 - Z_i) Y_i}{1 - \hat{e}(X_i)},
$$

where $\hat{e}(X_i)$ is the estimated propensity score. This is the inverse propensity score weighting (IPW) estimator, which is also called the Horvitz–Thompson (HT) estimator. Horvitz and Thompson (1952) proposed it in survey sampling and Rosenbaum (1987a) used in causal inference with observational studies.

However, the estimator $\hat{\tau}^{\mathrm{ht}}$ has many problems. In particular, it is not invariant to the location transformation of the outcome. Proposition 11.1 states this problem precisely, with the proof relegated to Problem 11.3.

Proposition 11.1 (lack of invariance for the HT estimator) *If we change Y_i to $Y_i + c$ with a constant c, then the HT estimator $\hat{\tau}^{\text{ht}}$ becomes $\hat{\tau}^{\text{ht}} + c(\hat{1}_{\text{T}} - \hat{1}_{\text{C}})$, where*

$$\hat{1}_{\text{T}} = \frac{1}{n}\sum_{i=1}^{n} \frac{Z_i}{\hat{e}(X_i)},$$

$$\hat{1}_{\text{C}} = \frac{1}{n}\sum_{i=1}^{n} \frac{(1 - Z_i)}{1 - \hat{e}(X_i)}$$

can be viewed as two different estimates of the constant 1.

In Proposition 11.1, I use the funny notation $\hat{1}_{\text{T}}$ and $\hat{1}_{\text{C}}$ because with the true propensity score these two terms both have expectation 1; see Problem 11.3 for more details. In general, $\hat{1}_{\text{T}} - \hat{1}_{\text{C}}$ is not zero in finite samples. Since adding a constant to every outcome should not change the average causal effect, the HT estimator is not reasonable because of its dependence on c. A simple fix to the problem is to normalize the weights by $\hat{1}_{\text{T}}$ and $\hat{1}_{\text{C}}$ respectively, resulting in the following estimator

$$\hat{\tau}^{\text{hajek}} = \frac{\sum_{i=1}^{n} \frac{Z_i Y_i}{\hat{e}(X_i)}}{\sum_{i=1}^{n} \frac{Z_i}{\hat{e}(X_i)}} - \frac{\sum_{i=1}^{n} \frac{(1-Z_i) Y_i}{1-\hat{e}(X_i)}}{\sum_{i=1}^{n} \frac{1-Z_i}{1-\hat{e}(X_i)}}.$$

This is the Hajek estimator due to Hájek (1971) in the context of survey sampling with varying probabilities. We can verify that the Hajek estimator is invariant to the location transformation. That is, if we replace Y_i by $Y_i + c$, then $\hat{\tau}^{\text{hajek}}$ remains the same; see Problem 11.3. Moreover, many numerical studies have found that $\hat{\tau}^{\text{hajek}}$ is much more stable than $\hat{\tau}^{\text{ht}}$ in finite samples.

11.2.3 A problem of IPW and a fundamental problem of causal inference

Many asymptotic analyses require a *strong overlap* condition

$$0 < \alpha_{\text{L}} \le e(X) \le \alpha_{\text{U}} < 1,$$

that is, the true propensity score is bounded away from 0 and 1. However, D'Amour et al. (2021) pointed out that this is a rather strong assumption, especially with many covariates. Chapter 20 will discuss this problem in detail.

Even if the strong overlap condition holds for the true propensity score, the estimated propensity scores can be close to 0 or 1. When this happens, the weighting estimators blow up to infinity, which results in extremely unstable behavior in finite samples. We can either truncate the estimated propensity score by changing it to

$$\max\left[\alpha_{\text{L}}, \min\{\hat{e}(X_i), \alpha_{\text{U}}\}\right],$$

or trim the observations by dropping units with $\hat{e}(X_i)$ outside the interval $[\alpha_{\text{L}}, \alpha_{\text{U}}]$. Crump et al. (2009) suggested $\alpha_{\text{L}} = 0.1$ and $\alpha_{\text{U}} = 0.9$, and Kurth et al. (2005) suggested $\alpha_{\text{L}} = 0.05$ and $\alpha_{\text{U}} = 0.95$. Yang and Ding (2018b) established some asymptotic theory for trimming. Overall, although trimming often stabilizes the IPW estimators, it also injects additional arbitrariness into the procedure.

11.2.4 Application

The following functions can compute the IPW estimators and their bootstrap standard errors.

```
ipw.est = function(z, y, x, truncps = c(0, 1))
{
  ## fitted propensity score
  pscore   = glm(z ~ x, family = binomial)$fitted.values
  pscore   = pmax(truncps[1], pmin(truncps[2], pscore))

  y.treat     = mean(z*y/pscore)
  y.control   = mean((1 - z)*y/(1 - pscore))
  one.treat   = mean(z/pscore)
  one.control = mean((1 - z)/(1 - pscore))
  ace.ipw0    = y.treat - y.control
  ace.ipw     = y.treat/one.treat - y.control/one.control

  return(c(ace.ipw0, ace.ipw))
}


ipw.boot = function(z, y, x, n.boot = 500, truncps = c(0, 1))
{
  point.est  = ipw.est(z, y, x, truncps)

  ## nonparametric bootstrap
  n.sample   = length(z)
  x          = as.matrix(x)
  boot.est   = replicate(n.boot, {
    id.boot = sample(1:n.sample, n.sample, replace = TRUE)
    ipw.est(z[id.boot], y[id.boot], x[id.boot, ], truncps)
  })
  boot.se    = apply(boot.est, 1, sd)

  res = cbind(point.est, boot.se)
  colnames(res) = c("est", "se")
  rownames(res) = c("HT", "Hajek")

  return(res)
}
```

Revisiting Example 10.3, we can obtain the IPW estimators based on different truncations of the estimated propensity scores. The following results are the two weighting estimators with the bootstrap standard errors, with truncations at $(0, 1)$, $(0.01, 0.99)$, $(0.05, 0.95)$, and $(0.1, 0.9)$:

```
> trunc.list = list(trunc0 = c(0,1),
+                   trunc.01 = c(0.01, 0.99),
+                   trunc.05 = c(0.05, 0.95),
+                   trunc.1 = c(0.1, 0.9))
> trunc.est = lapply(trunc.list,
+                    function(t){
+                      est = ipw.boot(z, y, x, truncps = t)
+                      round(est, 3)
+                    })
> trunc.est
```

```
$trunc0
          est    se
HT     -1.516 0.496
Hajek  -0.156 0.258

$trunc.01
          est    se
HT     -1.516 0.501
Hajek  -0.156 0.254

$trunc.05
          est    se
HT     -1.499 0.501
Hajek  -0.152 0.255

$trunc.1
          est    se
HT     -0.713 0.425
Hajek  -0.054 0.246
```

The HT estimator gives results far away from all other estimators we discussed so far. The point estimates seem too large and they are negatively significant unless we truncate the estimated propensity scores at $(0.1, 0.9)$. This is an example showing the instability of the HT estimator.

11.3 The balancing property of the propensity score

11.3.1 Theory

Theorem 11.3 *The propensity score satisfies*

$$Z \perp\!\!\!\perp X \mid e(X).$$

Moreover, for any function $h(\cdot)$, we have

$$E\left\{\frac{Zh(X)}{e(X)}\right\} = E\left\{\frac{(1-Z)h(X)}{1-e(X)}\right\} \tag{11.2}$$

provided the existence of the moments on both sides of (11.2).

Theorem 11.3 does not require the ignorability assumption. It is about the treatment Z and covariates X only. The first part of Theorem 11.3 states that conditional on the propensity score, the treatment indicator, and the covariates are independent. Therefore, within the same level of the propensity score, the covariate distributions are balanced across the treatment and control groups. The second part of Theorem 11.3 states an equivalent form of covariate balance based on IPW. I give a proof of Theorem 11.3 below.

Proof of Theorem 11.3: First, we show $Z \perp\!\!\!\perp X \mid e(X)$, that is,

$$\text{pr}\{Z = 1 \mid X, e(X)\} = \text{pr}\{Z = 1 \mid e(X)\}. \tag{11.3}$$

Following similar steps as the proof of Theorem 11.1, we can show that the left-hand side of (11.3) equals

$$\begin{aligned} \text{pr}\{Z=1 \mid X, e(X)\} &= \text{pr}(Z=1 \mid X) \\ &= e(X), \end{aligned}$$

and the right-hand side of (11.3) equals

$$\begin{aligned} \text{pr}\{Z=1 \mid e(X)\} &= E\{Z \mid e(X)\} \\ &= E\Big[E\{Z \mid X, e(X)\} \mid e(X)\Big] \\ &= E\Big[E\{Z \mid X\} \mid e(X)\Big] \\ &= E\Big[e(X) \mid e(X)\Big] \\ &= e(X). \end{aligned}$$

Therefore, (11.3) holds.

Second, we show (11.2). We can use similar steps as the proof of Theorem 11.2. But given Theorem 11.2, we have a simpler proof. If we view $h(X)$ as an outcome, then its two potential outcomes are identical and ignorability holds: $Z \perp\!\!\!\perp \{h(X), h(X)\} \mid X$. The difference between the left-hand and right-hand sides of (11.2) is the average causal effect of Z on $h(X)$, which is zero. □

11.3.2 Covariate balance check

The proof of Theorem 11.3 is simple. Nevertheless, Theorem 11.3 has useful implications for the statistical analysis. Before getting access to the outcome data, we can check whether the propensity score model is specified well enough to ensure the covariate balance in the data. Rubin (2007) viewed this as the design stage of the observational study, and Rubin (2008) argued that this can result in more objective causal inference because the design stage does not involve the values of the outcomes.[1]

In the propensity score stratification, we have the discretized estimated propensity score $\hat{e}'(X)$ and approximately

$$Z \perp\!\!\!\perp X \mid \hat{e}'(X) = e_k \quad (k = 1, \ldots, K).$$

Therefore, we can check whether the covariate distributions are the same across the treatment and control groups within each stratum of the discretized estimated propensity score.

In propensity score weighting, we can view $h(X)$ as a pseudo outcome and estimate the average causal effect on $h(X)$. Because the true average causal effect on $h(X)$ is 0, the estimate should not be significantly different from 0. A canonical choice of $h(X)$ is X.

Let us revisit Example 10.3 again. Based on propensity score stratification with $K = 5$, all the covariates are well-balanced across the treatment and control groups. Similar results hold for the Hajek estimator. The only exception is `Food_Stamp`, the seventh covariate in Figure 11.2. Figure 11.2 shows the balance-checking results.

[1]While this is a useful recommendation in practice, it is not entirely clear how to quantify the objectiveness.

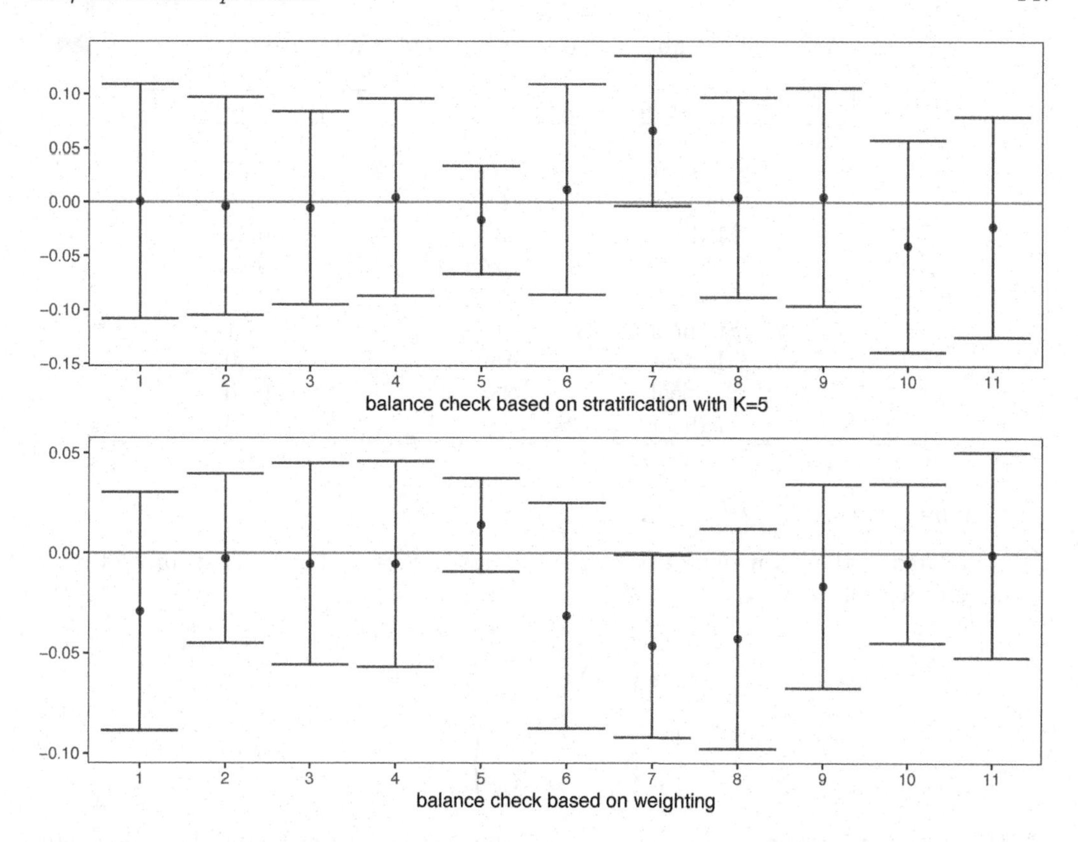

FIGURE 11.2: Balance check: point estimates and 95% confidence intervals of the average causal effect on covariates

11.4 Homework problems

11.1 Another version of Theorem 11.1

Prove that

$$Z \perp\!\!\!\perp \{Y(1), Y(0), X\} \mid e(X, Y(1), Y(0)). \tag{11.4}$$

Remark: This result holds without assuming strong ignorability. It implies that

$$Z \perp\!\!\!\perp \{Y(1), Y(0)\} \mid \{X, e(X, Y(1), Y(0)\}.$$

Rosenbaum (2020) and Rosenbaum and Rubin (2023) pointed out the result in (11.4) and called $e(X, Y(1), Y(0))$ the *principal unobserved covariate*.

11.2 Another version of Theorem 11.1

Theorem 11.1 states a result under strong ignorability. An analogous result also holds under ignorability. That is, if ignorability holds conditional on covariates X, then it also holds conditional on the scalar propensity score $e(X)$.

Theorem 11.4 *If* $Z \perp\!\!\!\perp Y(z) \mid X$ *for* $z = 0, 1$, *then* $Z \perp\!\!\!\perp Y(z) \mid e(X)$ *for* $z = 0, 1$.

Prove Theorem 11.4.

TABLE 11.1: Table 1 of Rosenbaum and Rubin (1983a)

stratum by $\hat{e}(X)$	treatment	number of patients	proportion improved
1	Surgical	26	0.54
	Medical	277	0.35
2	Surgical	68	0.70
	Medical	235	0.40
3	Surgical	98	0.70
	Medical	205	0.35
4	Surgical	164	0.71
	Medical	139	0.30
5	Surgical	234	0.70
	Medical	69	0.39

11.3 More results on the IPW estimators

This is related to the discussion of the HT estimator in Section 11.2.2. First, prove Proposition 11.1. Second, prove

$$
\begin{aligned}
E\left\{\frac{1}{n}\sum_{i=1}^{n}\frac{Z_i}{e(X_i)}\right\} &= 1, \\
E\left\{\frac{1}{n}\sum_{i=1}^{n}\frac{(1-Z_i)}{1-e(X_i)}\right\} &= 1.
\end{aligned}
$$

Third, prove that if we add a constant c to every observed outcome Y_i, the Hajek estimator $\hat{\tau}^{\text{hajek}}$ remains the same.

11.4 Reanalyzing the data from Rosenbaum and Rubin (1983a)

Table 11.1 is from Rosenbaum and Rubin (1983a), which concerned the causal effect of the coronary artery bypass surgery compared with the medical therapy on the functional improvement 6 months after cardiac catheterization. They first estimated the propensity score based on 74 observed covariates and then formed 5 strata based on the discretized estimated propensity score. Because the treatment is binary and the outcome is also binary, they represented the data in a table. Based on Table 11.1, estimate the average causal effect, and report the 95% confidence interval of the average causal effect.

Remark: If you are interested, you can read the whole paper of Rosenbaum and Rubin (1983a) after reading Part IV of the book. It is a canonical paper on sensitivity analysis in causal inference.

11.5 Balancing score and propensity score: more theoretical results

Rosenbaum and Rubin (1983b) also introduced the notion of *balancing score*.

Definition 11.2 (balancing score) $b(X)$ *is a balancing score if*

$$
Z \perp\!\!\!\perp X \mid b(X).
$$

In Definition 11.2, $b(X)$ can be a scalar or a vector. An obvious balancing score is $b(X) = X$, but it is not a useful one without any simplification of the original covariates. By Theorem 11.3, the propensity score is a special balancing score. More interestingly,

Rosenbaum and Rubin (1983b) showed that the propensity score is the coarsest balancing score, as in Theorem 11.5 below which includes Theorem 11.3 as a special case.

Theorem 11.5 *$b(X)$ is a balancing score if and only if $b(X)$ is finer than $e(X)$ in the sense that $e(X) = f(b(X))$ for some function $f(\cdot)$.*

Theorem 11.5 implies the following corollary.

Corollary 11.1 *$Z \perp\!\!\!\perp X \mid \{e(X), g(X)\}$, where $e(X)$ is the propensity score and $g(X)$ is an arbitrary function of X.*

Prove Theorem 11.5 and Corollary 11.1.

Remark: Corollary 11.1 is relevant in subgroup analysis; see Problem 11.6 below. In particular, we may be interested in not only the average causal effect τ but also the subgroup effects. For instance, we may want to estimate the average causal effects among boys and girls, respectively. Without loss of generality, assume the first component of X is a discrete random variable indicating the subgroups of interest, e.g. a binary indicator for girls, and we are interested in estimating

$$\tau(x_1) = E\{Y(1) - Y(0) \mid X_1 = x_1\}.$$

Corollary 11.1 implies that

$$Z \perp\!\!\!\perp X \mid \{e(X), X_1\},$$

Therefore, within each level of X_1, the treatment and covariates are independent conditional on the propensity score. See Problems 11.6 and 11.7 below for more discussion on subgroup effects.

11.6 Some basics of subgroup effects

This problem is related to Problem 11.5, but you can work on it independently.

Consider a standard observational study with covariates $X = (X_1, X_2)$, where X_1 denotes a discrete random variable indicating the subgroups of interest (e.g., a binary indicator for statistics major or not statistics major) and X_2 contains the rest of the covariates. The parameter of interest is the subgroup causal effect

$$\tau(x_1) = E\{Y(1) - Y(0) \mid X_1 = x_1\}.$$

We have Theorem 11.6 below.

Theorem 11.6 *Assume ignorability that $Z \perp\!\!\!\perp \{Y(1), Y(0)\} \mid X$.*

1. *Based on outcome modeling, we have*

$$\tau(x_1) = E\{E(Y \mid Z = 1, X) \mid X_1 = x_1\} - E\{E(Y \mid Z = 0, X) \mid X_1 = x_1\}.$$

2. *Based on the propensity score, we have*

$$\tau(x_1) = E\left\{\frac{1(X_1 = x_1)ZY}{e(X)} - \frac{1(X_1 = x_1)(1 - Z)Y}{1 - e(X)}\right\} \Big/ \mathrm{pr}(X_1 = x_1).$$

Prove Theorem 11.6. Based on part 1, construct an estimator for $\tau(x_1)$ based on outcome modeling. Based on part 2, construct the corresponding HT and Hajek estimators for $\tau(x_1)$.

11.7 More results on the propensity score for subgroup effects

Inherit the setting in Problem 11.6. Theorem 11.7 extends Theorem 11.1.

Theorem 11.7 *If* $Z \perp\!\!\!\perp \{Y(1), Y(0)\} \mid X$, *then* $Z \perp\!\!\!\perp \{Y(1), Y(0)\} \mid \{e(X), g(X)\}$, *where* $e(X)$ *is the propensity score and* $g(X)$ *is an arbitrary function of* X.

Prove Theorem 11.7.

Remark: Theorem 11.7 implies that under ignorability, we have

$$Z \perp\!\!\!\perp \{Y(1), Y(0)\} \mid \{e(X), X_1\}.$$

Recall Problem 11.5 that $b(X) = \{e(X), X_1\}$ is finer than $e(X)$ and thus a balancing score. Theorem 11.7 further ensures that ignorability holds given the propensity score, within each level of X_1. Therefore, we can perform the same analysis based on the propensity score, within each level of X_1.

11.8 Recommended reading

The title of this chapter is the same as the title of the classic paper by Rosenbaum and Rubin (1983b). Most results in this chapter are directly drawn from their original paper.

Rubin (2007) and Rubin (2008) highlighted the importance of the design stage of observational studies for more objective causal inference.

12

The Doubly Robust or the Augmented Inverse Propensity Score Weighting Estimator for the Average Causal Effect

Under ignorability $Z \perp\!\!\!\perp \{Y(1), Y(0)\} \mid X$ and overlap $0 < e(X) < 1$, Chapter 11 has shown two identification formulas of the average causal effect $\tau = E\{Y(1) - Y(0)\}$. First, the outcome regression formula is

$$\tau = E\{\mu_1(X)\} - E\{\mu_0(X)\} \tag{12.1}$$

where

$$\begin{aligned} \mu_1(X) &= E\{Y(1) \mid X\} = E(Y \mid Z = 1, X), \\ \mu_0(X) &= E\{Y(0) \mid X\} = E(Y \mid Z = 0, X) \end{aligned}$$

are the two conditional mean functions of the outcome given covariates under the treatment and control, respectively. Second, the IPW formula is

$$\tau = E\left\{\frac{ZY}{e(X)}\right\} - E\left\{\frac{(1-Z)Y}{1-e(X)}\right\} \tag{12.2}$$

where

$$e(X) = \text{pr}(Z = 1 \mid X)$$

is the propensity score introduced in Chapter 11.

The outcome regression estimator requires fitting a model for the outcome given the treatment and covariates. It is consistent if the outcome model is correctly specified. The IPW estimator requires fitting a model for the treatment given the covariates. It is consistent if the propensity score model is correctly specified.

Mathematically, we have many combinations of (12.1) and (12.2) that lead to different identification formulas of the average causal effect. Below I will discuss a particular combination that has appealing theoretical properties. This combination motivates an estimator that is consistent if either the propensity score or the outcome model is correctly specified. It is called the *doubly robust* estimator, championed by James Robins (Scharfstein et al., 1999; Bang and Robins, 2005).

12.1 The doubly robust estimator

12.1.1 Population version

We posit a model for the conditional means of the outcome $\mu_1(X, \beta_1)$ and $\mu_0(X, \beta_0)$, indexed by the parameters β_1 and β_0. For example, if the conditional means are linear or logistic under the working model, then the parameters are just the regression coefficients. If the outcome model is correctly specified, then $\mu_1(X,\beta_1) = \mu_1(X)$ and $\mu_0(X,\beta_0) = \mu_0(X)$. We posit a working model for the propensity score $e(X,\alpha)$, indexed by the parameter α. For example, if the working model is logistic, then α is the regression coefficient. If the propensity score model is correctly specified, then $e(X,\alpha) = e(X)$. In practice, both models may be misspecified. Sometimes, we call them *working models* due to the possibility of misspecification.

Define

$$\tilde{\mu}_1^{\mathrm{dr}} = E\left[\frac{Z\{Y-\mu_1(X,\beta_1)\}}{e(X,\alpha)} + \mu_1(X,\beta_1)\right], \tag{12.3}$$

$$\tilde{\mu}_0^{\mathrm{dr}} = E\left[\frac{(1-Z)\{Y-\mu_0(X,\beta_0)\}}{1-e(X,\alpha)} + \mu_0(X,\beta_0)\right], \tag{12.4}$$

which can also be written as

$$\tilde{\mu}_1^{\mathrm{dr}} = E\left[\frac{ZY}{e(X,\alpha)} - \frac{Z-e(X,\alpha)}{e(X,\alpha)}\mu_1(X,\beta_1)\right], \tag{12.5}$$

$$\tilde{\mu}_0^{\mathrm{dr}} = E\left[\frac{(1-Z)Y}{1-e(X,\alpha)} - \frac{e(X,\alpha)-Z}{1-e(X,\alpha)}\mu_0(X,\beta_0)\right]. \tag{12.6}$$

The formulas in (12.3) and (12.4) augment the outcome regression estimator by inverse propensity score weighting terms of the residuals. The formulas in (12.5) and (12.6) augment the IPW estimator by the imputed outcomes. For this reason, the doubly robust estimator is also called the *augmented inverse propensity score weighting* (AIPW) estimator.

The augmentation strengthens the theoretical properties in the following sense.

Theorem 12.1 *Assume ignorability* $Z \perp\!\!\!\perp \{Y(1), Y(0)\} \mid X$ *and overlap* $0 < e(X) < 1$.

1. *If either* $e(X,\alpha) = e(X)$ *or* $\mu_1(X,\beta_1) = \mu_1(X)$, *then* $\tilde{\mu}_1^{\mathrm{dr}} = E\{Y(1)\}$.
2. *If either* $e(X,\alpha) = e(X)$ *or* $\mu_0(X,\beta_0) = \mu_0(X)$, *then* $\tilde{\mu}_0^{\mathrm{dr}} = E\{Y(0)\}$.
3. *If either* $e(X,\alpha) = e(X)$ *or* $\{\mu_1(X,\beta_1) = \mu_1(X), \mu_0(X,\beta_0) = \mu_0(X)\}$, *then* $\tilde{\mu}_1^{\mathrm{dr}} - \tilde{\mu}_0^{\mathrm{dr}} = \tau$.

By Theorem 12.1, $\tilde{\mu}_1^{\mathrm{dr}} - \tilde{\mu}_0^{\mathrm{dr}}$ equals τ if either the propensity score model or the outcome model is correctly specified. That's why it is called the doubly robust estimator.

Proof of Theorem 12.1: I only prove the result for $\mu_1 = E\{Y(1)\}$. The proof for the result for $\mu_0 = E\{Y(0)\}$ is similar.

We have the decomposition

$$\begin{aligned}
& \tilde{\mu}_1^{\mathrm{dr}} - E\{Y(1)\} \\
= \; & E\left[\frac{Z\{Y(1)-\mu_1(X,\beta_1)\}}{e(X,\alpha)} - \{Y(1)-\mu_1(X,\beta_1)\}\right] && \text{(by definition)} \\
= \; & E\left[\frac{Z-e(X,\alpha)}{e(X,\alpha)}\{Y(1)-\mu_1(X,\beta_1)\}\right] && \text{(combining terms)} \\
= \; & E\left(E\left[\frac{Z-e(X,\alpha)}{e(X,\alpha)}\{Y(1)-\mu_1(X,\beta_1)\} \mid X\right]\right) && \text{(tower property)} \\
= \; & E\left[E\left\{\frac{Z-e(X,\alpha)}{e(X,\alpha)} \mid X\right\} \times E\left\{Y(1)-\mu_1(X,\beta_1) \mid X\right\}\right] && \text{(ignorability)} \\
= \; & E\left[\frac{e(X)-e(X,\alpha)}{e(X,\alpha)} \times \{\mu_1(X)-\mu_1(X,\beta_1)\}\right].
\end{aligned}$$

Therefore, $\tilde{\mu}_1^{\mathrm{dr}} - E\{Y(1)\} = 0$ if either $e(X,\alpha) = e(X)$ or $\mu_1(X,\beta_1) = \mu_1(X)$. □

12.1.2 Sample version

Based on the population versions of $\tilde{\mu}_1^{\text{dr}}$ and $\tilde{\mu}_0^{\text{dr}}$, we can obtain their sample analogs to construct a doubly robust estimator for τ.

Definition 12.1 (doubly robust estimator for the average causal effect) *Based on the data $(X_i, Z_i, Y_i)_{i=1}^n$, we can obtain a doubly robust estimator for τ by the following steps:*[1]

1. *obtain the fitted values of the propensity scores: $e(X_i, \hat{\alpha})$;*
2. *obtain the fitted values of the outcome means: $\mu_1(X_i, \hat{\beta}_1)$ and $\mu_0(X_i, \hat{\beta}_0)$;*
3. *construct the doubly robust estimator: $\hat{\tau}^{\text{dr}} = \hat{\mu}_1^{\text{dr}} - \hat{\mu}_0^{\text{dr}}$, where*

$$\hat{\mu}_1^{\text{dr}} = \frac{1}{n}\sum_{i=1}^n \left[\frac{Z_i\{Y_i - \mu_1(X_i, \hat{\beta}_1)\}}{e(X_i, \hat{\alpha})} + \mu_1(X_i, \hat{\beta}_1)\right]$$

and

$$\hat{\mu}_0^{\text{dr}} = \frac{1}{n}\sum_{i=1}^n \left[\frac{(1 - Z_i)\{Y_i - \mu_0(X_i, \hat{\beta}_0)\}}{1 - e(X_i, \hat{\alpha})} + \mu_0(X_i, \hat{\beta}_0)\right].$$

By Definition 12.1, we can also write the doubly robust estimator as

$$\hat{\tau}^{\text{dr}} = \hat{\tau}^{\text{reg}} + \frac{1}{n}\sum_{i=1}^n \frac{Z_i\{Y_i - \mu_1(X_i, \hat{\beta}_1)\}}{e(X_i, \hat{\alpha})} - \frac{1}{n}\sum_{i=1}^n \frac{(1 - Z_i)\{Y_i - \mu_0(X_i, \hat{\beta}_0)\}}{1 - e(X_i, \hat{\alpha})}.$$

Analogous to (12.5) and (12.6), we can also rewrite it as

$$\hat{\tau}^{\text{dr}} = \hat{\tau}^{\text{ipw}} - \frac{1}{n}\sum_{i=1}^n \frac{Z_i - e(X_i, \hat{\alpha})}{e(X_i, \hat{\alpha})}\mu_1(X_i, \hat{\beta}_1) + \frac{1}{n}\sum_{i=1}^n \frac{e(X_i, \hat{\alpha}) - Z_i}{1 - e(X_i, \hat{\alpha})}\mu_0(X_i, \hat{\beta}_0).$$

Funk et al. (2011) suggested to approximate the variance of $\hat{\tau}^{\text{dr}}$ via the nonparametric bootstrap by resampling from $(Z_i, X_i, Y_i)_{i=1}^n$.

12.2 More intuition and theory for the doubly robust estimator

Although the beginning of this chapter claims that the basic identification formulas based on outcome regression and IPW immediately yield infinitely many other identification formulas, the particular forms of the doubly robust estimators in (12.3) and (12.4) are not obvious to come up with. The original motivation for (12.3) and (12.4) was quite theoretical, and relied on something called the *semiparametric efficiency theory* in advanced mathematical statistics (Bickel et al., 1993). It is beyond the level of this book. Below I will give two intuitive perspectives to construct (12.3) and (12.4). Both Sections 12.2.1 and 12.2.2 below focus on the estimation of $E\{Y(1)\}$ since the estimation of $E\{Y(0)\}$ is similar by symmetry.

[1] I used $\hat{e}(X_i)$ for $e(X_i, \hat{\alpha})$ and $\hat{\mu}_z(X_i)$ for $\mu_z(X_i, \hat{\beta}_1)$ before when I did not want to emphasize the parameters in the working models.

12.2.1 Reducing the variance of the IPW estimator

The IPW estimator for μ_1 based on

$$\mu_1 = E\left\{\frac{ZY}{e(X)}\right\}$$

completely ignores the outcome model of Y. It has the advantage of being consistent without assuming any outcome model. However, if the covariates are predictive of the outcome, the residual based on a working outcome model usually has a smaller variance than the outcome even when this working outcome model is wrong. With a possibly misspecified outcome model $\mu_1(X, \beta_1)$, a trivial decomposition holds:

$$\begin{aligned} \mu_1 &= E\{Y(1)\} \\ &= E\{Y(1) - \mu_1(X, \beta_1)\} + E\{\mu_1(X, \beta_1)\}. \end{aligned}$$

If we apply the IPW formula to the first term in the above formula viewing $Y(1) - \mu_1(X, \beta_1)$ as a "pseudo potential outcome" under the treatment, we can rewrite the above formula as

$$\mu_1 = E\left\{\frac{Z\{Y - \mu_1(X, \beta_1)\}}{e(X)}\right\} + E\{\mu_1(X, \beta_1)\} \tag{12.7}$$

$$= E\left\{\frac{Z\{Y - \mu_1(X, \beta_1)\}}{e(X)} + \mu_1(X, \beta_1)\right\}, \tag{12.8}$$

which holds if the propensity score model is correct without assuming that the outcome model is correct. Using a working model to improve efficiency is an old idea from survey sampling. Little and An (2004) and Lumley et al. (2011) pointed out its connection with the doubly robust estimator.

12.2.2 Reducing the bias of the outcome regression estimator

The discussion in Section 12.2.1 starts with the IPW estimator and improves its efficiency based on a working outcome model. Alternatively, we can also start with an outcome regression estimator based on

$$\tilde{\mu}_1 = E\{\mu_1(X, \beta_1)\}$$

which may not be the same as μ_1 since the outcome model may be wrong. The bias of this estimator is $E\{\mu_1(X, \beta_1) - Y(1)\}$, which can be estimated by an IPW estimator

$$B = E\left\{\frac{Z\{\mu_1(X, \beta_1) - Y\}}{e(X)}\right\}$$

if the propensity score model is correct. So a de-biased estimator is $\tilde{\mu}_1 - B$, which is identical to (12.8).

12.3 Examples

12.3.1 Summary of some canonical estimators for the average causal effect

The following `R` code implements the outcome regression, HT, Hajek, and doubly robust estimators for τ. These estimators can be conveniently implemented based on the fitted values of the `glm` function. The default choice for the propensity score model is the logistic model, and the default choice for the outcome model is the linear model with `out.family = gaussian`.[2] For binary outcomes, we can also specify `out.family = binomial` to fit the logistic model.

```
OS_est = function(z, y, x, out.family = gaussian,
                  truncps = c(0, 1))
{
    ## fitted propensity score
    pscore   = glm(z ~ x, family = binomial)$fitted.values
    pscore   = pmax(truncps[1], pmin(truncps[2], pscore))

    ## fitted potential outcomes
    outcome1 = glm(y ~ x, weights = z,
                   family = out.family)$fitted.values
    outcome0 = glm(y ~ x, weights = (1 - z),
                   family = out.family)$fitted.values

    ## outcome regression estimator
    ace.reg  = mean(outcome1 - outcome0)
    ## IPW estimators
    y.treat     = mean(z*y/pscore)
    y.control   = mean((1 - z)*y/(1 - pscore))
    one.treat   = mean(z/pscore)
    one.control = mean((1 - z)/(1 - pscore))
    ace.ipw0    = y.treat - y.control
    ace.ipw     = y.treat/one.treat - y.control/one.control
    ## doubly robust estimator
    res1      = y - outcome1
    res0      = y - outcome0
    r.treat   = mean(z*res1/pscore)
    r.control = mean((1 - z)*res0/(1 - pscore))
    ace.dr    = ace.reg + r.treat - r.control

    return(c(ace.reg, ace.ipw0, ace.ipw, ace.dr))
}
```

It is tedious to calculate the analytic formulas for the variances of the above estimators. The bootstrap provides convenient approximations to the variances based on resampling from $\{Z_i, X_i, Y_i\}_{i=1}^n$. Building upon the function `OS_est` above, the following function returns point estimators as well as the bootstrap standard errors.

```
OS_ATE = function(z, y, x, n.boot = 2*10^2,
                  out.family = gaussian, truncps = c(0, 1))
{
    point.est  = OS_est(z, y, x, out.family, truncps)

    ## nonparametric bootstrap
    n          = length(z)
    x          = as.matrix(x)
    boot.est   = replicate(n.boot, {
      id.boot = sample(1:n, n, replace = TRUE)
      OS_est(z[id.boot], y[id.boot], x[id.boot, ],
             out.family, truncps)
    })
```

[2]The `glm` function is more general than the `lm` function. With `out.family = gaussian`, `glm` is identical to `lm`.

```
    boot.se      = apply(boot.est, 1, sd)

    res          = rbind(point.est, boot.se)
    rownames(res) = c("est", "se")
    colnames(res) = c("reg", "HT", "Hajek", "DR")

    return(res)
}
```

12.3.2 Simulation

I will use simulation to evaluate the finite-sample properties of the estimators under four scenarios:

1. both the propensity score and outcome models are correct;
2. the propensity score model is wrong but the outcome model is correct;
3. the propensity score model is correct but the outcome model is wrong;
4. both the propensity score and outcome models are wrong.

I will report the average bias, the true standard error, and the average estimated standard error of the estimators over simulation.

In case 1, the data generating process is

```
x       = matrix(rnorm(n*2), n, 2)
x1      = cbind(1, x)
beta.z  = c(0, 1, 1)
pscore  = 1/(1 + exp(- as.vector(x1%*%beta.z)))
z       = rbinom(n, 1, pscore)
beta.y1 = c(1, 2, 1)
beta.y0 = c(1, 2, 1)
y1      = rnorm(n, x1%*%beta.y1)
y0      = rnorm(n, x1%*%beta.y0)
y       = z*y1 + (1 - z)*y0
```

In case 2, I modify the propensity score model to be nonlinear:

```
x1      = cbind(1, x, exp(x))
beta.z  = c(-1, 0, 0, 1, -1)
pscore  = 1/(1 + exp(- as.vector(x1%*%beta.z)))
```

In case 3, I modify the outcome model to be nonlinear:

```
beta.y1 = c(1, 0, 0, 0.2, -0.1)
beta.y0 = c(1, 0, 0, -0.2, 0.1)
y1      = rnorm(n, x1%*%beta.y1)
y0      = rnorm(n, x1%*%beta.y0)
```

In case 4, I modify both the propensity score and the outcome model.

We set the sample size to be $n = 500$ and generate 500 independent data sets according to the data-generating processes above. In case 1,

```
         reg   HT Hajek   DR
ave.bias 0.00 0.02  0.03 0.01
true.se  0.11 0.28  0.26 0.13
est.se   0.10 0.25  0.23 0.12
```

All estimators are nearly unbiased. The two weighting estimators have larger variances. In case 2,

	reg	HT	Hajek	DR
ave.bias	0.00	-0.76	-0.75	-0.01
true.se	0.12	0.59	0.47	0.18
est.se	0.13	0.50	0.38	0.18

The two weighting estimators are severely biased due to the misspecification of the propensity score model. The outcome regression and doubly robust estimators are nearly unbiased.

In case 3,

	reg	HT	Hajek	DR
ave.bias	-0.05	0.00	-0.01	0.00
true.se	0.11	0.15	0.14	0.14
est.se	0.11	0.14	0.13	0.14

The outcome regression estimator has a larger bias than the other three estimators due to the misspecification of the outcome model. The weighting and doubly robust estimators are nearly unbiased.

In case 4,

	reg	HT	Hajek	DR
ave.bias	-0.08	0.11	-0.07	0.16
true.se	0.13	0.32	0.20	0.41
est.se	0.13	0.25	0.16	0.26

All estimators are biased because both the propensity score and outcome models are wrong. The HT and doubly robust estimator has the largest bias. When both models are wrong, the doubly robust estimator appears to be doubly fragile.

In all the cases above, the bootstrap standard errors are close to the true ones when the estimators are nearly unbiased for the true average causal effect.

12.3.3 Applications

Revisiting Example 10.3, we obtain the following estimators and bootstrap standard errors:

	reg	HT	Hajek	DR
est	-0.017	-1.516	-0.156	-0.019
se	0.230	0.492	0.246	0.233

The two weighting estimators are much larger than the other two estimators. Truncating the estimated propensity score at $[0.1, 0.9]$, we obtain the following estimators and bootstrap standard errors:

	reg	HT	Hajek	DR
est	-0.017	-0.713	-0.054	-0.043
se	0.223	0.422	0.235	0.231

The Hajek estimator becomes much closer to the outcome regression and doubly robust estimators, while the HT estimator is still an outlier.

12.4 Some further discussion

Recall the proof of Theorem 12.1, the key for the double robustness property is the product structure in

$$\tilde{\mu}_1^{\mathrm{dr}} - E\{Y(1)\} = E\left[\frac{e(X) - e(X,\alpha)}{e(X,\alpha)} \times \{\mu_1(X) - \mu_1(X,\beta_1)\}\right],$$

which ensures that the estimation error is zero if either $e(X) = e(X,\alpha)$ or $\mu_1(X) = \mu_1(X,\beta_1)$. This delicate structure renders the doubly robust estimator possibly doubly fragile when both the propensity score and the outcome models are misspecified. The product of two errors multiply to yield potentially much larger errors. The simulation in Section 12.3.2 confirms this point.

Kang and Schafer (2007) criticized the doubly robust estimator based on simulation studies. They found that the finite-sample performance of the doubly robust estimator can be even wilder than the simple outcome regression and IPW estimators. Despite the critique from Kang and Schafer (2007), the doubly robust estimator has been a standard strategy in causal inference since the seminal work of Scharfstein et al. (1999). Recently, it was resurrected in the theoretical statistics and econometrics literature with a fancier name "double machine learning" (Chernozhukov et al., 2018). The basic idea is to replace the working models for the propensity score and outcome with machine learning tools which can be viewed as more flexible models than the traditional parametric models.

12.5 Homework problems

12.1 A sanity check

Consider the case in which the covariate is discrete $X \in \{1,\dots,K\}$ and the parameter of interest is τ. Without imposing any model assumptions, the estimated propensity score $\hat{e}(X)$ equals $\hat{e}_{[k]} = \hat{\mathrm{pr}}(Z = 1 \mid X = k)$, the proportion of units receiving the treatment, and the estimated outcome means are the sample means of the outcomes $\hat{\bar{Y}}_{[k]}(1) = \hat{E}(Y \mid Z = 1, X = k)$ and $\hat{\bar{Y}}_{[k]}(0) = \hat{E}(Y \mid Z = 0, X = k)$ under treatment, within stratum $X = k$ $(k = 1,\dots,K)$. Show that the stratified estimator, outcome regression estimator, HT estimator, Hajek estimator, and doubly robust estimator are all identical numerically.

12.2 Double robustness for estimating CATE

Theorem 12.1 states the double robustness for estimating the average causal effect τ. Theorem 12.2 below states the double robustness for estimating the conditional average treatment effect $\tau(X) = E\{Y(1) - Y(0) \mid X\}$.

Define

$$\tilde{\tau}^{\mathrm{dr}}(X,Z,Y) = \tilde{\mu}_1^{\mathrm{dr}}(X,Z,Y) - \tilde{\mu}_0^{\mathrm{dr}}(X,Z,Y),$$

where

$$\tilde{\mu}_1^{\mathrm{dr}}(X,Z,Y) = \frac{Z\{Y - \mu_1(X,\beta_1)\}}{e(X,\alpha)} + \mu_1(X,\beta_1), \tag{12.9}$$

$$\tilde{\mu}_0^{\mathrm{dr}}(X,Z,Y) = \frac{(1-Z)\{Y - \mu_0(X,\beta_0)\}}{1 - e(X,\alpha)} + \mu_0(X,\beta_0) \tag{12.10}$$

are functions of the observables X, Z, Y. Prove Theorem 12.2 below.

Theorem 12.2 *Assume ignorability* $Z \perp\!\!\!\perp \{Y(1), Y(0)\} \mid X$ *and overlap* $0 < e(X) < 1$. *If either* $e(X, \alpha) = e(X)$ *or* $\mu_1(X, \beta_1) = \mu_1(X)$, *then* $E\{\tilde{\tau}^{\text{dr}}(X, Z, Y) - \tau(X) \mid X\} = 0$.

Remark: To prove Theorem 12.2, you can revisit the proof of Theorem 12.1.

12.3 An alternative form of the doubly robust estimator for τ

Motivated by (12.7), we have an alternative form of the doubly robust estimator for $\mu_1 = E\{Y(1)\}$:

$$\tilde{\mu}_1^{\text{dr2}} = \frac{E\left[\frac{Z\{Y-\mu_1(X,\beta_1)\}}{e(X,\alpha)}\right]}{E\left[\frac{Z}{e(X,\alpha)}\right]} + E\{\mu_1(X, \beta_1)\}.$$

Show that $\tilde{\mu}_1^{\text{dr2}} = \mu_1$ if either $e(X, \alpha) = e(X)$ or $\mu_1(X, \beta_1) = \mu_1(X)$. Give the analogous formula for estimating μ_0. Give the sample analog of the doubly robust estimator for τ based on these formulas.

Remark: This form of doubly robust estimator appeared in Robins et al. (2007).

12.4 An upper bound of the bias of the doubly robust estimator

Consider the population version of the doubly robust estimator $\tilde{\mu}_1^{\text{dr}}$ for $E\{Y(1)\}$. Show that

$$|\tilde{\mu}_1^{\text{dr}} - E\{Y(1)\}| \leq \sqrt{E\left[\frac{\{e(X) - e(X, \alpha)\}^2}{e(X, \alpha)^2}\right] \times E\left[\{\mu_1(X) - \mu_1(X, \beta_1)\}^2\right]}.$$

Find the analogous upper bound for the bias of $\tilde{\mu}_0^{\text{dr}}$ for $E\{Y(0)\}$.

Remark: You may find Section A.1.4 useful for the proof.

12.5 Data analysis of Example 10.1

Analyze the dataset `cps1re74.csv` using the methods discussed so far.

12.6 Analyzing a dataset from the Karolinska Institute

Rubin (2008) used the dataset `karolinska.txt` to illustrate the ideas of causal inference in observational studies. The dataset has 158 cardia cancer patients diagnosed between 1988 and 1995 in Central and Northern Sweden, 79 diagnosed at large volume hospitals, defined as treating more than ten patients with cardia cancer during that period, and 79 diagnosed at the remaining small volume hospitals. The treatment `z` is the indicator of whether a patient was diagnosed at a large volume hospital. The outcome `y` is whether the patient survived longer than 1 year after the diagnosis. The covariates `x` contain information about age, whether a patient was from a rural area, and whether a patient was male.

```
karolinska = read.table("karolinska.txt", header = TRUE)
z = karolinska$hvdiag
y = 1 - (karolinska$year.survival == 1)
x = as.matrix(karolinska[, c(3, 4, 5)])
```

Analyze the dataset using the methods discussed so far.

12.7 Recommended reading

Lunceford and Davidian (2004) gave a review and comparison of many methods discussed in Chapters 11 and 12.

13

The Average Causal Effect on the Treated Units and Other Estimands

Chapters 10–12 focused on the identification and estimation of the average causal effect $\tau = E\{Y(1) - Y(0)\}$ under the ignorability and overlap assumptions. Conceptually, it is straightforward to extend the discussion to the average causal effects on the treated and control units:

$$\begin{aligned} \tau_{\mathrm{T}} &= E\{Y(1) - Y(0) \mid Z = 1\}, \\ \tau_{\mathrm{C}} &= E\{Y(1) - Y(0) \mid Z = 0\}. \end{aligned}$$

If τ_{T} and τ_{C} differ from τ, then the average causal effects are heterogeneous across the treatment and control groups. Whether we should estimate τ_{T}, τ_{C}, or τ depends on the practical question of interest.

Because of the symmetry, this chapter focuses on τ_{T}. Section 13.4 also discusses extensions to other estimands.

13.1 Nonparametric identification of τ_{T}

The average causal effect on the treated units equals

$$\tau_{\mathrm{T}} = E(Y \mid Z = 1) - E\{Y(0) \mid Z = 1\},$$

where the first term $E(Y \mid Z = 1)$ is directly identifiable from the data and the second term $E\{Y(0) \mid Z = 1\}$ is counterfactual. The key assumption to identify the second term is the following ignorability and overlap assumptions.

Assumption 13.1 *$Z \perp\!\!\!\perp Y(0) \mid X$ and $e(X) < 1$.*

Because the key is to identify $E\{Y(0) \mid Z = 1\}$, we only need the "one-sided" ignorability and overlap assumptions. Under Assumption 13.1, we have the following identification result for τ_{T}.

Theorem 13.1 *Under Assumption 13.1, we have*

$$\begin{aligned} E\{Y(0) \mid Z = 1\} &= E\{E(Y \mid Z = 0, X) \mid Z = 1\} \\ &= \int E(Y \mid Z = 0, X = x) f(x \mid Z = 1) \mathrm{d}x. \end{aligned}$$

By Theorem 13.1 the counterfactual mean $E\{Y(0) \mid Z = 1\}$ equals the conditional mean of the observed outcomes under the control, averaged over the distribution of the covariates under the treatment. It implies that τ_{T} is nonparametrically identified by

$$\tau_{\mathrm{T}} = E(Y \mid Z = 1) - E\{E(Y \mid Z = 0, X) \mid Z = 1\} \tag{13.1}$$

Proof of Theorem 13.1: We have

$$\begin{aligned} E\{Y(0) \mid Z=1\} &= E\big[E\{Y(0) \mid Z=1, X\} \mid Z=1\big] \\ &= E\big[E\{Y(0) \mid Z=0, X\} \mid Z=1\big] \\ &= E\left\{E(Y \mid Z=0, X) \mid Z=1\right\} \\ &= \int E(Y \mid Z=0, X=x) f(x \mid Z=1) \mathrm{d}x. \end{aligned}$$

□

With a discrete X, the identification formula in Theorem 13.1 reduces to

$$E\{Y(0) \mid Z=1\} = \sum_{k=1}^{K} E(Y \mid Z=0, X=k) \mathrm{pr}(X=k \mid Z=1),$$

motivating the following stratified estimator for τ_{T}:

$$\hat{\tau}_{\mathrm{T}} = \hat{\bar{Y}}(1) - \sum_{k=1}^{K} \hat{\pi}_{[k]|1} \hat{\bar{Y}}_{[k]}(0),$$

where $\hat{\pi}_{[k]|1} = n_{[k]1}/n_1$ is the proportion of category k of X among the treated units.

For continuous X, we need to fit an outcome model for $E(Y \mid Z=0, X)$ using the control units. If the fitted values for the control potential outcomes are $\hat{\mu}_0(X_i)$, then the outcome regression estimator is

$$\begin{aligned} \hat{\tau}_{\mathrm{T}}^{\mathrm{reg}} &= \hat{\bar{Y}}(1) - n_1^{-1} \sum_{i=1}^{n} Z_i \hat{\mu}_0(X_i) \\ &= n_1^{-1} \sum_{i=1}^{n} Z_i \{Y_i - \hat{\mu}_0(X_i)\}. \end{aligned}$$

Example 13.1 *If we specify a linear model for all units*

$$E(Y \mid Z, X) = \beta_0 + \beta_z Z + \beta_x^{\mathsf{T}} X,$$

then

$$\begin{aligned} \tau_{\mathrm{T}} &= E\{E(Y \mid Z=1, X) - E(Y \mid Z=0, X) \mid Z=1\} \\ &= \beta_z. \end{aligned}$$

If we run OLS to obtain $(\hat{\beta}_0, \hat{\beta}_z, \hat{\beta}_x)$, then we can use $\hat{\beta}_z$ to estimate τ_{T}. Section 10.4.2 shows that $\hat{\beta}_z$ is an estimator for τ, and this example further shows that $\hat{\beta}_z$ is an estimator for τ_{T}. This is not surprising because the linear model assumes constant causal effects across units.

Example 13.2 *The identification formula depends only on $E(Y \mid Z=0, X)$, so we need only to specify a model for the control units. When this model is linear,*

$$E(Y \mid Z=0, X) = \beta_{0|0} + \beta_{x|0}^{\mathsf{T}} X,$$

we have

$$\begin{aligned} \tau_{\mathrm{T}} &= E(Y \mid Z=1) - E(\beta_{0|0} + \beta_{x|0}^{\mathsf{T}} X \mid Z=1) \\ &= E(Y \mid Z=1) - \beta_{0|0} - \beta_{x|0}^{\mathsf{T}} E(X \mid Z=1). \end{aligned}$$

If we run OLS with only the control units to obtain $(\hat{\beta}_{0|0}, \hat{\beta}_{x|0})$, then the estimator is

$$\hat{\tau}_{\mathrm{T}} = \hat{\bar{Y}}(1) - \hat{\beta}_{0|0} - \hat{\beta}_{x|0}^{\mathrm{T}} \hat{\bar{X}}(1).$$

Using the property of the OLS (see B.5), we have

$$\hat{\bar{Y}}(0) = \hat{\beta}_{0|0} + \hat{\beta}_{x|0}^{\mathrm{T}} \hat{\bar{X}}(0).$$

Therefore, the above estimator reduces to

$$\hat{\tau}_{\mathrm{T}} = \left\{ \hat{\bar{Y}}(1) - \hat{\bar{Y}}(0) \right\} - \hat{\beta}_{x|0}^{\mathrm{T}} \left\{ \hat{\bar{X}}(1) - \hat{\bar{X}}(0) \right\}.$$

As an algebraic fact, we can show that this estimator equals the coefficient of Z in the OLS fit of the outcome on the treatment, covariates, and their interactions, with the covariates centered as $X_i - \hat{\bar{X}}(1)$. See Problem 13.3 for more details.

13.2 Inverse propensity score weighting and doubly robust estimation of τ_{T}

Theorem 13.2 *Under Assumption 13.1, we have*

$$E\{Y(0) \mid Z = 1\} = E\left\{ \frac{e(X)}{e} \frac{1-Z}{1-e(X)} Y \right\} \tag{13.2}$$

and

$$\tau_{\mathrm{T}} = E(Y \mid Z = 1) - E\left\{ \frac{e(X)}{e} \frac{1-Z}{1-e(X)} Y \right\}, \tag{13.3}$$

where $e = \mathrm{pr}(Z = 1)$ is the marginal probability of the treatment.

Proof of Theorem 13.2: The left-hand side of (13.2) equals

$$\begin{aligned} E\{Y(0) \mid Z = 1\} &= E\{ZY(0)\}/e \\ &= E\big[E(Z \mid X) E\{Y(0) \mid X\}\big]/e \\ &= E\big[e(X) E\{Y(0) \mid X\}\big]/e. \end{aligned}$$

The right-hand side of (13.2) equals

$$\begin{aligned} E\left\{ \frac{e(X)}{e} \frac{1-Z}{1-e(X)} Y \right\} &= E\left[E\left\{ \frac{e(X)}{e} \frac{1-Z}{1-e(X)} Y(0) \mid X \right\} \right] \\ &= E\left[\frac{e(X)}{e\{1-e(X)\}} E\left\{(1-Z) Y(0) \mid X\right\} \right] \\ &= E\left[\frac{e(X)}{e\{1-e(X)\}} E(1-Z \mid X) E\left\{Y(0) \mid X\right\} \right] \\ &= E\big[e(X) E\{Y(0) \mid X\}\big]/e. \end{aligned}$$

So (13.2) holds. □

We have two IPW estimators

$$\hat{\tau}_{\mathrm{T}}^{\mathrm{ht}} = \hat{\bar{Y}}(1) - n_1^{-1}\sum_{i=1}^{n} \hat{o}(X_i)(1-Z_i)Y_i$$

and

$$\hat{\tau}_{\mathrm{T}}^{\mathrm{hajek}} = \hat{\bar{Y}}(1) - \frac{\sum_{i=1}^{n} \hat{o}(X_i)(1-Z_i)Y_i}{\sum_{i=1}^{n} \hat{o}(X_i)(1-Z_i)},$$

where $\hat{o}(X_i) = \hat{e}(X_i)/\{1-\hat{e}(X_i)\}$ is the fitted odds of the treatment given covariates.

We also have a doubly robust estimator for $E\{Y(0) \mid Z=1\}$ which combines the propensity score and the outcome models. Define

$$\tilde{\mu}_{0\mathrm{T}}^{\mathrm{dr}} = E\left[o(X,\alpha)(1-Z)\{Y-\mu_0(X,\beta_0)\} + Z\mu_0(X,\beta_0)\right]/e, \tag{13.4}$$

where $o(X,\alpha) = e(X,\alpha)/\{1-e(X,\alpha)\}$.

Theorem 13.3 *Under Assumption 13.1, if either $e(X,\alpha)=e(X)$ or $\mu_0(X,\beta_0)=\mu_0(X)$, then $\tilde{\mu}_{0\mathrm{T}}^{\mathrm{dr}} = E\{Y(0) \mid Z=1\}$.*

Proof of Theorem 13.3: We have the decomposition

$$\begin{aligned}
& e\left[\tilde{\mu}_{0\mathrm{T}}^{\mathrm{dr}} - E\{Y(0) \mid Z=1\}\right] \\
= & E\left[o(X,\alpha)(1-Z)\{Y(0)-\mu_0(X,\beta_0)\} + Z\mu_0(X,\beta_0)\right] - E\{ZY(0)\} \\
= & E\left[o(X,\alpha)(1-Z)\{Y(0)-\mu_0(X,\beta_0)\} - Z\{Y(0)-\mu_0(X,\beta_0)\}\right] \\
= & E\left[\{o(X,\alpha)(1-Z)-Z\}\{Y(0)-\mu_0(X,\beta_0)\}\right] \\
= & E\left[\frac{e(X,\alpha)-Z}{1-e(X,\alpha)}\{Y(0)-\mu_0(X,\beta_0)\}\right] \\
= & E\left[E\left\{\frac{e(X,\alpha)-Z}{1-e(X,\alpha)} \mid X\right\} \times E\{Y(0)-\mu_0(X,\beta_0) \mid X\}\right] \\
= & E\left[\frac{e(X,\alpha)-e(X)}{1-e(X,\alpha)} \times \{\mu_0(X)-\mu_0(X,\beta_0)\}\right].
\end{aligned}$$

Therefore, $\tilde{\mu}_{0\mathrm{T}}^{\mathrm{dr}} - E\{Y(0) \mid Z=1\} = 0$ if either $e(X,\alpha)=e(X)$ or $\mu_0(X,\beta_0)=\mu_0(X)$. □

Based on the population versions of $\tilde{\mu}_{0\mathrm{T}}^{\mathrm{dr}}$ in (13.4), we can obtain its sample version to construct a doubly robust estimator for τ_{T}.

Definition 13.1 (doubly robust estimator for τ_{T}) *Based on the data $(X_i,Z_i,Y_i)_{i=1}^n$, we can obtain a doubly robust estimator for τ_{T} by the following steps:*

1. *obtain the fitted values of the propensity scores $e(X_i,\hat{\alpha})$ and then obtain the fitted values of the odds $o(X_i,\hat{\alpha}) = e(X_i,\hat{\alpha})/(1-e(X_i,\hat{\alpha}))$;*
2. *obtain the fitted values of the outcome mean under control $\mu_0(X_i,\hat{\beta}_0)$;*
3. *construct the doubly robust estimator: $\hat{\tau}_{\mathrm{T}}^{\mathrm{dr}} = \hat{\bar{Y}}(1) - \hat{\mu}_{0\mathrm{T}}^{\mathrm{dr}}$, where*

$$\hat{\mu}_{0\mathrm{T}}^{\mathrm{dr}} = \frac{1}{n_1}\sum_{i=1}^{n}\left[o(X_i,\hat{\alpha})(1-Z_i)\{Y_i-\mu_0(X_i,\hat{\beta}_0)\} + Z_i\mu_0(X_i,\hat{\beta}_0)\right].$$

By Definition 13.1, we can rewrite $\hat{\tau}_{\mathrm{T}}^{\mathrm{dr}}$ as

$$\hat{\tau}_{\mathrm{T}}^{\mathrm{dr}} = \hat{\tau}_{\mathrm{T}}^{\mathrm{reg}} - \frac{1}{n_1}\sum_{i=1}^{n} o(X_i,\hat{\alpha})(1-Z_i)\{Y_i-\mu_0(X_i,\hat{\beta}_0)\}$$

or

$$\hat{\tau}_{\mathrm{T}}^{\mathrm{dr}} = \hat{\tau}_{\mathrm{T}}^{\mathrm{ht}} - \frac{1}{n_1}\sum_{i=1}^{n}\{o(X_i,\hat{\alpha})(1-Z_i)+Z_i\}\mu_0(X_i,\hat{\beta}_0).$$

Similar to the discussion of $\hat{\tau}^{\mathrm{dr}}$, we can estimate the variance of $\hat{\tau}_{\mathrm{T}}^{\mathrm{dr}}$ via the bootstrap by resampling from $(Z_i, X_i, Y_i)_{i=1}^n$. Hahn (1998), Mercatanti and Li (2014), Shinozaki and Matsuyama (2015), and Yang and Ding (2018b) are references on the estimation of τ_{T}.

13.3 An example

The following R function implements two outcome regression estimators, two IPW estimators, and the doubly robust estimator for τ_{T}. To avoid extreme estimated propensity scores, we can also truncate them from the above.

```
ATT.est = function(z, y, x, out.family = gaussian, Utruncps = 1)
{
  ## sample size
  nn  = length(z)
  nn1 = sum(z)

  ## fitted propensity score
  pscore    = glm(z ~ x, family = binomial)$fitted.values
  pscore    = pmin(Utruncps, pscore)
  odds      = pscore/(1 - pscore)

  ## fitted potential outcomes
  outcome0 = glm(y ~ x, weights = (1 - z),
                 family = out.family)$fitted.values

  ## outcome regression estimator
  ace.reg0 = lm(y ~ z + x)$coef[2]
  ace.reg  = mean(y[z==1]) - mean(outcome0[z==1])
  ## propensity score weighting estimator
  ace.ipw0 = mean(y[z==1]) -
               mean(odds*(1 - z)*y)*nn/nn1
  ace.ipw  = mean(y[z==1]) -
               mean(odds*(1 - z)*y)/mean(odds*(1 - z))
  ## doubly robust estimator
  res0     = y - outcome0
  ace.dr   = ace.reg - mean(odds*(1 - z)*res0)*nn/nn1

  return(c(ace.reg0, ace.reg, ace.ipw0, ace.ipw, ace.dr))
}
```

The following R function further implements the bootstrap variance estimators.

```
OS_ATT = function(z, y, x, n.boot = 10^2,
                  out.family = gaussian, Utruncps = 1)
{
  point.est  = ATT.est(z, y, x, out.family, Utruncps)

  ## nonparametric bootstrap
  n   = length(z)
```

```
  x             = as.matrix(x)
  boot.est      = replicate(n.boot, {
    id.boot = sample(1:n, n, replace = TRUE)
    ATT.est(z[id.boot], y[id.boot], x[id.boot, ],
            out.family, Utruncps)
  })

  boot.se       = apply(boot.est, 1, sd)

  res           = rbind(point.est, boot.se)
  rownames(res) = c("est", "se")
  colnames(res) = c("reg0", "reg", "HT", "Hajek", "DR")

  return(res)
}
```

Now we reanalyze the data in Example 10.3 to estimate τ_T. We obtain

```
      reg0     reg      HT   Hajek      DR
est  0.061  -0.351  -1.992  -0.351  -0.187
se   0.227   0.258   0.705   0.328   0.287
```

without truncating the estimated propensity scores, and

```
      reg0     reg      HT   Hajek      DR
est  0.061  -0.351  -0.597  -0.192  -0.230
se   0.223   0.255   0.579   0.302   0.276
```

by truncating the estimated propensity scores from the above at 0.9. The HT estimator is sensitive to the truncation as expected. The regression estimator in Example 13.1 is quite different from other estimators. It imposes an unnecessary assumption that the regression functions in the treatment and control group share the same coefficient of X. The regression estimator in Example 13.2 is much closer to the Hajek and doubly robust estimators. The estimates above are slightly different from those in Section 12.3.3, suggesting the existence of treatment effect heterogeneity across τ_T and τ.

13.4 Other estimands

Li et al. (2018a) gave a unified discussion of the causal estimands in observational studies. Starting from the conditional average causal effect $\tau(X)$, they proposed a general class of estimands

$$\tau^h = \frac{E\{h(X)\tau(X)\}}{E\{h(X)\}}$$

indexed by a weighting function $h(X)$ with $E\{h(X)\} \neq 0$. The normalization in the denominator is to ensure that a constant causal effect $\tau(X) = \tau$ averages to the same τ.

Under ignorability,

$$\tau^h = \frac{E[h(X)\{\mu_1(X) - \mu_0(X)\}]}{E\{h(X)\}}$$

which motivates the outcome regression estimator

$$\hat{\tau}^h = \frac{\sum_{i=1}^n h(X_i)\{\hat{\mu}_1(X_i) - \hat{\mu}_0(X_i)\}}{\sum_{i=1}^n h(X_i)}.$$

Moreover, we can show that τ^h has the following weighting form:

Theorem 13.4 *Under the ignorability and overlap assumption, we have*

$$\tau^h = E\left\{\frac{ZYh(X)}{e(X)} - \frac{(1-Z)Yh(X)}{1-e(X)}\right\}/E\{h(X)\}.$$

The proof of Theorem 13.4 is similar to those of Theorems 11.2 and 13.2 which is relegated to Problem 13.10. Based on Theorem 13.4, we can construct the corresponding IPW estimator for τ^h.

By Theorem 13.4, each unit is associated with the weight due to the definition of the estimand as well as the weight due to the inverse of the propensity score. Finally, the treated units are weighted by $h(X)/e(X)$ and the control units are weighted by $h(X)/\{1-e(X)\}$. Li et al. (2018a, Table 1) summarized several estimands, and I present a part of it below:

population	$h(X)$	estimand	weights
combined	1	τ	$1/e(X)$ and $1/\{1-e(X)\}$
treated	$e(X)$	τ_T	1 and $e(X)/\{1-e(X)\}$
control	$1-e(X)$	τ_C	$\{1-e(X)\}/e(X)$ and 1
overlap	$e(X)\{1-e(X)\}$	τ_O	$1-e(X)$ and $e(X)$

The overlap population and the corresponding estimand

$$\tau_\text{O} = \frac{E\left[e(X)\{1-e(X)\}\tau(X)\right]}{E\left[e(X)\{1-e(X)\}\right]}$$

is new to us. This estimand has the largest weight for units with $e(X) = 1/2$ and down-weights the units with extreme propensity scores. A nice feature of this estimand is that its IPW estimator is stable without the possibly extremely small values of $e(X)$ and $1-e(X)$ in the denominator. If $e(X) \perp\!\!\!\perp \tau(X)$ including the special case of $\tau(X) = \tau$, the parameter τ_O reduces to τ. In general, however, the estimand τ_O may cause controversy because it changes the initial population and depends on the propensity score which may be misspecified in practice. Li et al. (2018a) and Li et al. (2019) gave some justifications and numerical evidence. This estimand will appear again in Chapter 14.

We can also construct the doubly robust estimator for τ^h. I relegate the details to Problem 13.11.

13.5 Homework problems

13.1 Comparing $\tau_\text{T}, \tau_\text{C}$, and τ

Assume $Z \perp\!\!\!\perp \{Y(1), Y(0)\} \mid X$. Recall $e(X) = \text{pr}(Z = 1 \mid X)$ is the propensity score, $e = \text{pr}(Z = 1)$ is the marginal probability of the treatment, and $\tau(X) = E\{Y(1) - Y(0) \mid X\}$ is the CATE. Show that

$$\begin{aligned} \tau_\text{T} - \tau &= \frac{\text{cov}\{e(X), \tau(X)\}}{e}, \\ \tau_\text{C} - \tau &= -\frac{\text{cov}\{e(X), \tau(X)\}}{1-e}. \end{aligned}$$

Remark: The results above also imply that

$$\tau_\text{T} - \tau_\text{C} = \frac{\text{cov}\{e(X), \tau(X)\}}{e(1-e)}.$$

So the differences among τ_T, τ_C, τ depend on the covariance between the propensity score and CATE. When $e(X)$ and $\tau(X)$ are uncorrelated, their differences are 0.

13.2 Decomposing the bias of the difference in means for estimating τ_T

Define the bias of the difference in means for estimating τ_T as

$$B = E(Y \mid Z=1) - E(Y \mid Z=0) - \tau_T.$$

For simplicity, assume the covariate vector X is continuous and has density, with $f(x \mid Z=1)$ and $f(x \mid Z=0)$ denoting the densities of X under the treatment and control, respectively. Show that B decomposes into two components

$$B = B_{\text{obs}} + B_{\text{unobs}},$$

where

$$B_{\text{obs}} = \int E\{Y(0) \mid Z=0, X=x\}\{f(x \mid Z=1) - f(x \mid Z=0)\}\text{d}x$$

and

$$B_{\text{unobs}} = \int [E\{Y(0) \mid Z=1, X=x\} - E\{Y(0) \mid Z=0, X=x\}] f(x \mid Z=1)\text{d}x.$$

Remark: The bias term B_{obs} is due to the difference between the observed distributions $f(x \mid Z=1)$ and $f(x \mid Z=0)$. The bias term B_{unobs} is due to the difference between $E\{Y(0) \mid Z=1, X=x\}$ which is unobservable and $E\{Y(0) \mid Z=0, X=x\}$ which is observable. Heckman et al. (1997, Section 9) presented a more general decomposition which allows for the violation of the overlap between the distributions $f(x \mid Z=1)$ and $f(x \mid Z=0)$.

13.3 An algebraic fact about a regression estimator for τ_T

This problem provides more details for Example 13.2.

Show that if we center the covariates by $X_i - \hat{\bar{X}}(1)$ for all units, then $\hat{\tau}_T$ equals the coefficient of Z in the OLS fit of the outcome on the intercept, the treatment, covariates, and their interactions.

13.4 Simulation for the average causal effect on the treated units

Chapter 12 ran some simulation studies for τ. Run similar simulation studies for τ_T with either correct or incorrect propensity score or outcome models.

You can choose different model parameters, a larger number of simulation settings, and a larger number of bootstrap replicates. Report your findings, including at least the bias, variance, and variance estimator via the bootstrap. You can also report other properties of the estimators, for example, the asymptotic Normality and the coverage rates of the confidence intervals.

13.5 An alternative form of the doubly robust estimator for τ_T

Motivated by (13.4), we have an alternative form of doubly robust estimator for $E\{Y(0) \mid Z=1\}$:

$$\tilde{\mu}_{0\text{T}}^{\text{dr2}} = \frac{E\left[o(X,\alpha)(1-Z)\{Y - \mu_0(X,\beta_0)\}\right]}{E\left[o(X,\alpha)(1-Z)\right]} + E\{Z\mu_0(X,\beta_0)\}/e.$$

Show that under Assumption 13.1, $\tilde{\mu}_{0\text{T}}^{\text{dr2}} = E\{Y(0) \mid Z=1\}$ if either $e(X,\alpha) = e(X)$ or $\mu_0(X,\beta_0) = \mu_0(X)$. Give the sample analog of the doubly robust estimator for τ_T.

13.6 Average causal effect on the control units

Prove the identification formulas for τ_{C}, analogous to (13.1) and (13.3). Propose the doubly robust estimator for τ_{C}.

13.7 Estimating individual effect and conditional average causal effect

Assume that $\{Z_i, X_i, Y_i(1), Y_i(0)\}_{i=1}^n \stackrel{\text{IID}}{\sim} \{Z, X, Y(1), Y(0)\}$. The individual effect is $\tau_i = Y_i(1) - Y_i(0)$ and the conditional average causal effect is $\tau(X_i) = E\{Y_i(1) - Y_i(0) \mid X_i\}$. Since we will discuss the individual effect, we do not drop the subscript i since τ means the average causal effect, not the population version of $Y(1) - Y(0)$.

1. Under randomization with $Z_i \perp\!\!\!\perp \{Y_i(1), Y_i(0)\}$ and $e = \text{pr}(Z_i = 1)$, show that
$$\delta_i = \frac{Z_i Y_i}{e} - \frac{(1 - Z_i) Y_i}{1 - e}$$
is an unbiased predictor of the individual effect in the sense that
$$E(\delta_i - \tau_i) = 0 \quad (i = 1, \ldots, n).$$
Further show that $E(\delta_i) = \tau$ for all $i = 1, \ldots, n$.
2. Under ignorability with $Z_i \perp\!\!\!\perp \{Y_i(1), Y_i(0)\} \mid X_i$ and $e(X_i) = \text{pr}(Z_i = 1 \mid X_i)$, show that
$$\delta_i = \frac{Z_i Y_i}{e(X_i)} - \frac{(1 - Z_i) Y_i}{1 - e(X_i)}$$
is an unbiased predictor of the individual effect and the conditional average causal effect in the sense that
$$E(\delta_i - \tau_i) = 0, \quad E\{\delta_i - \tau(X_i)\} = 0, \quad (i = 1, \ldots, n).$$
Further show that $E(\delta_i) = \tau$ for all $i = 1, \ldots, n$.

13.8 General estimand and $(\tau_{\text{T}}, \tau_{\text{C}})$

Assume unconfoundedness. Show that $\tau^h = \tau_{\text{T}}$ if $h(X) = e(X)$, and $\tau^h = \tau_{\text{C}}$ if $h(X) = 1 - e(X)$.

13.9 More on τ_{O}

Show that
$$\tau_{\text{O}} = \frac{E[\{1 - e(X)\}\tau(X) \mid Z = 1]}{E\{1 - e(X) \mid Z = 1\}} = \frac{E\{e(X)\tau(X) \mid Z = 0\}}{E\{e(X) \mid Z = 0\}}.$$

13.10 IPW for the general estimand

Prove Theorem 13.4.

13.11 Doubly robust estimation for general estimand

For a given $h(X)$, we have the following formulas for constructing the doubly robust estimator for τ^h:

$$\begin{aligned} \tilde{\mu}_1^{h,\text{dr}} &= E\left[\frac{Z h(X)\{Y - \mu_1(X, \beta_1)\}}{e(X, \alpha)} + h(X)\mu_1(X, \beta_1)\right], \\ \tilde{\mu}_0^{h,\text{dr}} &= E\left[\frac{(1 - Z) h(X)\{Y - \mu_0(X, \beta_0)\}}{1 - e(X, \alpha)} + h(X)\mu_0(X, \beta_0)\right]. \end{aligned}$$

Show that under ignorability and overlap,

1. if either $e(X, \alpha) = e(X)$ or $\mu_1(X, \beta_1) = \mu_1(X)$, then $\tilde{\mu}_1^{h,\mathrm{dr}} = E\{h(X)Y(1)\}$;
2. if either $e(X, \alpha) = e(X)$ or $\mu_0(X, \beta_0) = \mu_0(X)$, then $\tilde{\mu}_0^{h,\mathrm{dr}} = E\{h(X)Y(0)\}$;
3. if either $e(X, \alpha) = e(X)$ or $\{\mu_1(X, \beta_1) = \mu_1(X), \mu_0(X, \beta_0) = \mu_0(X)\}$, then

$$\frac{\tilde{\mu}_1^{h,\mathrm{dr}} - \tilde{\mu}_0^{h,\mathrm{dr}}}{E\{h(X)\}} = \tau^h.$$

Remark: Tao and Fu (2019) proved the above results. However, they hold only for a given $h(X)$. The most interesting cases of $\tau_{\mathrm{T}}, \tau_{\mathrm{C}}$, and τ_{O} all have weight depending on the propensity score $e(X)$, which must be estimated in the first place. The above formulas do not apply to constructing the doubly robust estimators for τ_{T} and τ_{C}; there does not exist a doubly robust estimator for τ_{O}.

13.12 Analyzing a dataset from the Karolinska Institute

Revisit Problem 12.6. Estimate τ_{T} based on the methods introduced in this chapter.

13.13 Recommended reading

Shinozaki and Matsuyama (2015) focused on τ_{T}, and Li et al. (2018a) discussed general τ^h.

14

Using the Propensity Score in Regressions for Causal Effects

Since the publication of the seminal paper by Rosenbaum and Rubin (1983b), many creative uses of the propensity score have appeared in the literature (e.g., Bang and Robins, 2005; Robins et al., 2007; Van der Laan and Rose, 2011; Vansteelandt and Daniel, 2014). This chapter discusses two simple methods to use the propensity score:

1. including the propensity score as a covariate in regressions;
2. running regressions weighted by the inverse of the propensity score.

I choose to focus on these two methods because of the following reasons:

1. they are easy to implement, and involve only standard statistical software packages for regressions;
2. their properties are comparable to many more complex methods;
3. they can be easily extended to allow for flexible statistical models including machine learning algorithms.

14.1 Regressions with the propensity score as a covariate

By Theorem 11.1, if ignorability holds conditioning on X, then it also holds conditioning on $e(X)$:

$$Z \perp\!\!\!\perp \{Y(1), Y(0)\} \mid e(X).$$

Analogous to (10.6), τ is also nonparametrically identified by

$$\tau = E\Big[E\{Y \mid Z=1, e(X)\} - E\{Y \mid Z=0, e(X)\}\Big],$$

which motivates methods based on regressions of Y on Z and $e(X)$.

The simplest regression specification is the OLS fit of Y on $\{1, Z, e(X)\}$, with the coefficient of Z as an estimator, denoted by τ_e. For simplicity, I will discuss the population OLS:

$$\arg\min_{a,b,c} E\{Y - a - bZ - ce(X)\}^2$$

with τ_e defined as the coefficient of Z. It is consistent for τ if we have a correct propensity score model and the outcome model is indeed linear in Z and $e(X)$. The more interesting result is that τ_e estimates $\tau_{\rm O}$ introduced in Section 13.4 if we have a correct propensity score model even if the outcome model is completely misspecified.

Theorem 14.1 *If $Z \perp\!\!\!\perp \{Y(1), Y(0)\} \mid X$, then the coefficient of Z in the population OLS fit of Y on $\{1, Z, e(X)\}$ equals*

$$\tau_e = \tau_O = \frac{E\{h_O(X)\tau(X)\}}{E\{h_O(X)\}},$$

recalling that $h_O(X) = e(X)\{1 - e(X)\}$ and $\tau(X) = E\{Y(1) - Y(0) \mid X\}$.

An unusual feature of Theorem 14.1 is that the overlap condition is not needed anymore. Even if some units have propensity score $e(X)$ equaling 0 or 1, their associated weight $e(X)\{1 - e(X)\}$ is zero so they do not contribute anything to the final parameter τ_O.
Proof of Theorem 14.1: I will use the FWL theorem reviewed in Section B.3 to prove Theorem 14.1. By the FWL theorem, we can obtain τ_e in two steps:

1. we obtain the residual $\tilde{Z}$ from the OLS fit of Z on $\{1, e(X)\}$;
2. we obtain τ_e from the OLS fit of Y on $\tilde{Z}$.

We can use the result on covariance in Section A.1.5 to simplify the coefficient of $e(X)$ in the OLS fit of Z on $\{1, e(X)\}$ is

$$\begin{aligned} \frac{\mathrm{cov}\{Z, e(X)\}}{\mathrm{var}\{e(X)\}} &= \frac{E[\mathrm{cov}\{Z, e(X) \mid X\}] + \mathrm{cov}\{E(Z \mid X), e(X)\}}{\mathrm{var}\{e(X)\}} \\ &= \frac{0 + \mathrm{var}\{e(X)\}}{\mathrm{var}\{e(X)\}} = 1, \end{aligned}$$

so the intercept is $E(Z) - E\{e(X)\} = 0$ and the residual is $\tilde{Z} = Z - e(X)$. This makes sense since $Z - e(X)$ is uncorrelated with any function of X.

Therefore, we can obtain τ_e from the univariate OLS fit of Y on a centered variable $Z - e(X)$:

$$\tau_e = \frac{\mathrm{cov}\{Z - e(X), Y\}}{\mathrm{var}\{Z - e(X)\}}.$$

The denominator simplifies to

$$\begin{aligned} \mathrm{var}\{Z - e(X)\} &= E\{Z - e(X)\}^2 \\ &= E\{Z + e(X)^2 - 2Ze(X)\} \\ &= E\{e(X) + e(X)^2 - 2e(X)^2\} \\ &= E\{h_O(X)\}. \end{aligned}$$

The numerator simplifies to

$$\begin{aligned} & \mathrm{cov}\{Z - e(X), Y\} \\ = \; & E[\{Z - e(X)\}Y] \\ = \; & E[\{Z - e(X)\}ZY(1)] + E[\{Z - e(X)\}(1 - Z)Y(0)] \\ & \qquad (\text{by } Y = ZY(1) + (1 - Z)Y(0)) \\ = \; & E[\{Z - Ze(X)\}Y(1)] - E[e(X)(1 - Z)Y(0)] \\ = \; & E[Z\{1 - e(X)\}Y(1)] - E[e(X)(1 - Z)Y(0)] \\ = \; & E[e(X)\{1 - e(X)\}\mu_1(X)] - E[e(X)\{1 - e(X)\}\mu_0(X)] \\ & \qquad (\text{tower property and ignorability}) \\ = \; & E\{h_O(X)\tau(X)\}. \end{aligned}$$

The conclusion follows. □

From the proof of Theorem 14.1, we can simply run the OLS of Y on the centered treatment $Z - e(X)$. Lee (2018) proposed this procedure. Moreover, we can also include X in the OLS fit which may improve efficiency in finite samples. However, this does not change the estimand, which is still τ_O. I summarize these two results in the corollary below.

Corollary 14.1 *If $Z \perp\!\!\!\perp \{Y(1), Y(0)\} \mid X$, then*

(1) the coefficient of $Z - e(X)$ in the population OLS fit of Y on $Z - e(X)$ or $\{1, Z - e(X)\}$ equals τ_O;

(2) the coefficient of Z in the population OLS fit of Y on $\{1, Z, e(X), X\}$ equals τ_O.

Proof of Corollary 14.1: (1) The first result is an intermediate step in the proof of Theorem 14.1. The second result holds because regressing Y on $Z - e(X)$ or $\{1, Z - e(X)\}$ does not change the coefficient of $Z - e(X)$ since it has mean zero.

(2) We can use the FWL theorem again. We can first obtain the residual from the population OLS of Z on $\{1, e(X), X\}$, which is $Z - e(X)$ because

$$Z - e(X) = Z - 0 - 1 \cdot e(X) - 0^{\mathsf{T}} X$$

and $Z - e(X)$ is uncorrelated with any functions of X. Then the coefficient of Z in the population OLS fit of Y on $\{1, Z, e(X), X\}$ equals the coefficient of Z in the population OLS fit of Y on $Z - e(X)$. □

Theorem 14.1 motivates a two-step estimator for τ_O:

1. fit a propensity score model to obtain $\hat{e}(X_i)$;
2. run OLS of Y_i on $(1, Z_i, \hat{e}(X_i))$ to obtain the coefficient of Z_i.

Corollary 14.1(1) motivates a two-step estimator for τ_O:

1. fit a propensity score model to obtain $\hat{e}(X_i)$;
2. run OLS of Y_i on $Z_i - \hat{e}(X_i)$ to obtain the coefficient of $Z_i - \hat{e}(X_i)$.

Corollary 14.1(2) motivates another two-step estimator for τ_O:

1. fit a propensity score model to obtain $\hat{e}(X_i)$;
2. run OLS of Y_i on $(1, Z_i, \hat{e}(X_i), X_i)$ to obtain the coefficient of Z_i.

Although OLS is convenient for obtaining point estimators, the corresponding standard errors are incorrect due to the uncertainty in the first step estimation of the propensity score. We can use the bootstrap to approximate the standard errors.

Robins et al. (1992) discussed many OLS estimators based on the propensity score. The above results seem special cases of their general theory although they did not point out the connection with the estimand under the overlap weight, which was resurrected by Li et al. (2018a). Lee (2018) proposed to regress Y on $Z - e(X)$ from a different perspective without making connections to the existing results in Robins et al. (1992) and Li et al. (2018a).

Rosenbaum and Rubin (1983b) proposed to estimate the average causal effect based on the OLS fit of Y on $\{1, Z, e(X), Ze(X)\}$. When this outcome model is correct, their estimator is consistent for the average causal effect. However, when the model is incorrect, the corresponding estimator has a much more complicated interpretation. Little and An (2004) suggested constructing estimators based on the OLS of Y on Z and a flexible function[1] of

[1] For example, we can include polynomial terms of $e(X)$ in the regression. Little and An (2004) suggested using splines.

$e(X)$ and showed it enjoys certain double robustness property. Due to the complexity of implementation, I omit the discussion.

14.2 Regressions weighted by the inverse of the propensity score

14.2.1 Average causal effect

We first reexamine the Hajek estimator of τ:

$$\hat{\tau}^{\text{hajek}} = \frac{\sum_{i=1}^n \frac{Z_i Y_i}{\hat{e}(X_i)}}{\sum_{i=1}^n \frac{Z_i}{\hat{e}(X_i)}} - \frac{\sum_{i=1}^n \frac{(1-Z_i)Y_i}{1-\hat{e}(X_i)}}{\sum_{i=1}^n \frac{1-Z_i}{1-\hat{e}(X_i)}},$$

which equals the difference between the weighted means of the outcomes in the treatment and control groups. Numerically, it is identical to the coefficient of Z_i in the following weighted least squares (WLS) of Y_i on $(1, Z_i)$.

Proposition 14.1 $\hat{\tau}^{\text{hajek}}$ *equals* $\hat{\beta}$ *from the following WLS:*

$$(\hat{\alpha}, \hat{\beta}) = \arg\min_{\alpha,\beta} \sum_{i=1}^n w_i (Y_i - \alpha - \beta Z_i)^2$$

with weights

$$\begin{aligned} w_i &= \frac{Z_i}{\hat{e}(X_i)} + \frac{1-Z_i}{1-\hat{e}(X_i)} \\ &= \begin{cases} \frac{1}{\hat{e}(X_i)} & \textit{if } Z_i = 1; \\ \frac{1}{1-\hat{e}(X_i)} & \textit{if } Z_i = 0. \end{cases} \end{aligned} \tag{14.1}$$

Imbens (2004) pointed out the result in Proposition 14.1. I leave it as Problem 14.1. By Proposition 14.1, it is convenient to obtain $\hat{\tau}^{\text{hajek}}$ based on WLS. However, due to the uncertainty in the estimated propensity score, the standard error reported by WLS is incorrect for the true standard error of $\hat{\tau}^{\text{hajek}}$. The bootstrap provides a convenient approximation to the true standard error.

Why does the WLS give a consistent estimator for τ? Recall that in the CRE with a constant propensity score, we can simply use the coefficient of Z_i in the OLS fit of Y_i on $(1, Z_i)$ to estimate τ. In observational studies, units have different probabilities of receiving the treatment and control, respectively. If we weight the treated units by $1/e(X_i)$ and the control units by $1/\{1 - e(X_i)\}$, then both treated and control groups can represent the whole population. Thus, by weighting, we effectively have a pseudo-randomized experiment. Consequently, the difference between the weighted means is consistent for τ. The numerical equivalence of $\hat{\tau}^{\text{hajek}}$ and WLS is not only a fun numerical fact itself but also useful for motivating more complex estimators with covariate adjustment. I give one extension below.

Recall that in the CRE, we can use the coefficient of Z_i in the OLS fit of Y_i on $(1, Z_i, X_i, Z_i X_i)$ to estimate τ, where the covariates are centered with $\bar{X} = 0$. This is Lin's (2013) estimator which uses covariates to improve efficiency. A natural extension to observational studies is to estimate τ using the coefficient of Z_i in the WLS fit of Y_i on $(1, Z_i, X_i, Z_i X_i)$ with weights defined in (14.1). Hirano and Imbens (2001) used this estimator in an application. The fully interacted linear model is equivalent to two separate linear models for the treated and control groups. If the linear models

$$\begin{aligned} E(Y \mid Z=1, X) &= \beta_{10} + \beta_{1x}^{\mathsf{T}} X, \\ E(Y \mid Z=0, X) &= \beta_{00} + \beta_{0x}^{\mathsf{T}} X, \end{aligned}$$

are correctly specified, then both OLS and WLS give consistent estimators for the coefficients and the estimators of the coefficient of Z are consistent for τ. More interestingly, the estimator of the coefficient of Z based on WLS is also consistent for τ if the propensity score model is correct and the outcome model is incorrect. That is, the estimator based on WLS is doubly robust. Robins et al. (2007) discussed this property and attributed this result to M. Joffe's unpublished paper. I will give more details below.

Let $e(X_i, \hat{\alpha})$ be the fitted propensity score and $(\mu_1(X_i, \hat{\beta}_1), \mu_0(X_i, \hat{\beta}_0))$ be the fitted values of the outcome means based on the WLS. The outcome regression estimator is

$$\hat{\tau}_{\text{wls}}^{\text{reg}} = \frac{1}{n} \sum_{i=1}^{n} \mu_1(X_i, \hat{\beta}_1) - \frac{1}{n} \sum_{i=1}^{n} \mu_0(X_i, \hat{\beta}_0)$$

and the doubly robust estimator for τ is

$$\hat{\tau}_{\text{wls}}^{\text{dr}} = \hat{\tau}_{\text{wls}}^{\text{reg}} + \frac{1}{n} \sum_{i=1}^{n} \frac{Z_i\{Y_i - \mu_1(X_i, \hat{\beta}_1)\}}{e(X_i, \hat{\alpha})} - \frac{1}{n} \sum_{i=1}^{n} \frac{(1-Z_i)\{Y_i - \mu_0(X_i, \hat{\beta}_0)\}}{1 - e(X_i, \hat{\alpha})}.$$

An interesting result is that this doubly robust estimator equals the outcome regression estimator, which reduces to the coefficient of Z_i in the WLS fit of Y_i on $(1, Z_i, X_i, Z_i X_i)$ if we use weights (14.1).

Theorem 14.2 *If $\bar{X} = 0$ and $(\mu_1(X_i, \hat{\beta}_1), \mu_0(X_i, \hat{\beta}_0)) = (\hat{\beta}_{10} + \hat{\beta}_{1x}^{\mathsf{T}} X_i, \hat{\beta}_{00} + \hat{\beta}_{0x}^{\mathsf{T}} X_i)$ based on the WLS fit of Y_i on $(1, Z_i, X_i, Z_i X_i)$ with weights* (14.1), *then*

$$\hat{\tau}_{\text{wls}}^{\text{dr}} = \hat{\tau}_{\text{wls}}^{\text{reg}} = \hat{\beta}_{10} - \hat{\beta}_{00},$$

which is the coefficient of Z_i in the WLS fit.

Proof of Theorem 14.2: The WLS fit of Y_i on $(1, Z_i, X_i, Z_i X_i)$ is equivalent to two WLS fits based on the treated and control data. Both WLS fits include intercepts, so the weighted residuals have mean 0 (see (B.5)):

$$\sum_{i=1}^{n} \frac{Z_i(Y_i - \hat{\beta}_{10} - \hat{\beta}_{1x}^{\mathsf{T}} X_i)}{\hat{e}(X_i)} = 0$$

and

$$\sum_{i=1}^{n} \frac{(1-Z_i)(Y_i - \hat{\beta}_{00} - \hat{\beta}_{0x}^{\mathsf{T}} X_i)}{1 - \hat{e}(X_i)} = 0.$$

So the difference between $\hat{\tau}^{\text{dr}}$ and $\hat{\tau}^{\text{reg}}$ is exactly zero. Both reduce to

$$\begin{aligned} \frac{1}{n} \sum_{i=1}^{n} (\hat{\beta}_{10} + \hat{\beta}_{1x}^{\mathsf{T}} X_i) - \frac{1}{n} \sum_{i=1}^{n} (\hat{\beta}_{00} + \hat{\beta}_{0x}^{\mathsf{T}} X_i) &= \hat{\beta}_{10} - \hat{\beta}_{00} + (\hat{\beta}_{1x} - \hat{\beta}_{0x})^{\mathsf{T}} \bar{X} \\ &= \hat{\beta}_{10} - \hat{\beta}_{00} \end{aligned}$$

with centered covariates. So they both equal the coefficient of Z_i in the WLS fit of Y_i on $(1, Z_i, X_i, Z_i X_i)$. □

TABLE 14.1: Regression estimators in CREs and unconfounded observational studies. The weights w_i's are defined in (14.1). Assume covariates are centered at $\bar{X} = 0$.

	CRE	unconfounded observational studies
without X	$Y_i \sim (1, Z_i)$	$Y_i \sim (1, Z_i)$ with weights w_i
with X	$Y_i \sim (1, Z_i, X_i, Z_i X_i)$	$Y_i \sim (1, Z_i, X_i, Z_i X_i)$ with weights w_i

Freedman and Berk (2008) discouraged the use of the WLS estimator above based on some simulation studies. They showed that when the outcome model is correct, the WLS estimator is worse than the OLS estimator since the WLS estimator has large variability in their simulation setting with homoskedastic outcomes. This may not be true in general. When the errors have variance proportional to the inverse of the propensity scores, the WLS estimator will be more efficient than the OLS estimator.

Freedman and Berk (2008) also showed that the estimated standard error based on the WLS fit is not consistent for the true standard error because it ignores the uncertainty in the estimated propensity score. This can be easily fixed by using the bootstrap to approximate the variance of the WLS estimator.

Nevertheless, Freedman and Berk (2008) found that "weighting may help under some circumstances" because when the outcome model is incorrect, the WLS estimator is still consistent if the propensity score model is correct.

I end this section with Table 14.1 summarizing the regression estimators for causal effects in both randomized experiments and observational studies.

14.2.2 Average causal effect on the treated units

The results for τ_{T} parallel those for τ. First, the Hajek estimator for τ_{T}

$$\hat{\tau}_{\mathrm{T}}^{\text{hajek}} = \hat{\bar{Y}}(1) - \frac{\sum_{i=1}^{n} \hat{o}(X_i)(1 - Z_i)Y_i}{\sum_{i=1}^{n} \hat{o}(X_i)(1 - Z_i)},$$

with $\hat{o}(X_i) = \hat{e}(X_i)/\{1 - \hat{e}(X_i)\}$, equals the coefficient of Z_i in the following WLS fit Y_i on $(1, Z_i)$.

Proposition 14.2 *$\hat{\tau}_{\mathrm{T}}^{hajek}$ is numerically identical to $\hat{\beta}$ in the following WLS:*

$$(\hat{\alpha}, \hat{\beta}) = \arg\min_{\alpha,\beta} \sum_{i=1}^{n} w_{\mathrm{T}i}(Y_i - \alpha - \beta Z_i)^2$$

with weights

$$\begin{aligned} w_{\mathrm{T}i} &= Z_i + (1 - Z_i)\hat{o}(X_i) \\ &= \begin{cases} 1 & \textit{if } Z_i = 1; \\ \hat{o}(X_i) & \textit{if } Z_i = 0. \end{cases} \end{aligned} \qquad (14.2)$$

Similar to Proposition 14.1, Proposition 14.2 is a pure linear algebra result. I relegate its proof to Problem 14.1.

Second, if we center covariates at $\hat{\bar{X}}(1) = 0$, then we can estimate τ_{T} using the coefficient of Z_i in the WLS fit of Y_i on $(1, Z_i, X_i, Z_i X_i)$ with weights defined in (14.2). Similarly, this estimator equals the regression estimator

$$\hat{\tau}_{\mathrm{T,wls}}^{\text{reg}} = \hat{\bar{Y}}(1) - \frac{1}{n_1} \sum_{i=1}^{n} Z_i \mu_0(X_i, \hat{\beta}_0),$$

which also equals the doubly robust estimator

$$\hat{\tau}^{\text{dr}}_{\text{T,wls}} = \hat{\tau}^{\text{reg}}_{\text{T,wls}} - \frac{1}{n_1}\sum_{i=1}^{n} \hat{o}(X_i)(1-Z_i)\{Y_i - \mu_0(X_i, \hat{\beta}_0)\}.$$

Theorem 14.3 *If $\hat{\bar{X}}(1) = 0$ and $\mu_0(X_i, \hat{\beta}_0) = \hat{\beta}_{00} + \hat{\beta}_{0x}^{\text{T}} X_i$ based on the WLS fit of Y_i on $(1, Z_i, X_i, Z_i X_i)$ with weights (14.2), then*

$$\hat{\tau}^{\text{dr}}_{\text{T,wls}} = \hat{\tau}^{\text{reg}}_{\text{T,wls}} = \hat{\beta}_{10} - \hat{\beta}_{00},$$

which is the coefficient of Z_i in the WLS fit.

Proof of Theorem 14.3: Based on the WLS fits in the treatment and control groups, we have

$$\sum_{i=1}^{n} Z_i(Y_i - \hat{\beta}_{10} - \hat{\beta}_{1x}^{\text{T}} X_i) = 0, \quad (14.3)$$

$$\sum_{i=1}^{n} \hat{o}(X_i)(1-Z_i)(Y_i - \hat{\beta}_{00} - \hat{\beta}_{0x}^{\text{T}} X_i) = 0. \quad (14.4)$$

The second result (14.4) ensures that $\hat{\tau}^{\text{dr}}_{\text{T,wls}} = \hat{\tau}^{\text{reg}}_{\text{T,wls}}$. Both reduce to

$$\hat{\bar{Y}}(1) - \frac{1}{n_1}\sum_{i=1}^{n} Z_i(\hat{\beta}_{00} + \hat{\beta}_{0x}^{\text{T}} X_i) = \frac{1}{n_1}\sum_{i=1}^{n} Z_i(Y_i - \hat{\beta}_{00} - \hat{\beta}_{0x}^{\text{T}} X_i).$$

With covariates centered at $\hat{\bar{X}}(1) = 0$, the first result (14.3) implies that $\hat{\bar{Y}}(1) = \hat{\beta}_{10}$ which further simplifies the estimator to $\hat{\beta}_{10} - \hat{\beta}_{00}$. □

14.3 Homework problems

14.1 Hajek estimators as WLS estimators

Prove Propositions 14.1 and 14.2.

Remark: These are special cases of Problem B.3 on the univariate WLS.

14.2 Predictive estimator and doubly robust estimator

Another outcome regression estimator is the predictive estimator

$$\hat{\tau}^{\text{pred}} = \hat{\mu}_1^{\text{pred}} - \hat{\mu}_0^{\text{pred}}$$

where

$$\hat{\mu}_1^{\text{pred}} = \frac{1}{n}\sum_{i=1}^{n}\left\{Z_i Y_i + (1-Z_i)\mu_1(X_i, \hat{\beta}_1)\right\}$$

and

$$\hat{\mu}_0^{\text{pred}} = \frac{1}{n}\sum_{i=1}^{n}\left\{Z_i \mu_0(X_i, \hat{\beta}_1) + (1-Z_i)Y_i\right\}.$$

It differs from the outcome regression estimator discussed before in that it only predicts the counterfactual outcomes but not the observed outcomes.

Show that the doubly robust estimator equals $\hat{\tau}^{\text{pred}}$ if $(\mu_1(X_i, \hat{\beta}_1), \mu_0(X_i, \hat{\beta}_1)) = (\hat{\beta}_{10} + \hat{\beta}_{1x}^{\mathsf{T}} X_i, \hat{\beta}_{00} + \hat{\beta}_{0x}^{\mathsf{T}} X_i)$ are from the WLS fits of Y_i on $(1, X_i)$ based on the treated and control data, respectively, with weights

$$\begin{aligned} w_i &= Z_i/\hat{o}(X_i) + (1 - Z_i)\hat{o}(X_i) \\ &= \begin{cases} \frac{1}{\hat{o}(X_i)} = \frac{1-\hat{e}(X_i)}{\hat{e}(X_i)} & \text{if } Z_i = 1; \\ \hat{o}(X_i) = \frac{\hat{e}(X_i)}{1-\hat{e}(X_i)} & \text{if } Z_i = 0. \end{cases} \end{aligned}$$

Remark: Cao et al. (2009) and Vermeulen and Vansteelandt (2015) motivated the weights in (14.5) from other more theoretical perspectives.

14.3 Weighted logistic regression with a binary outcome

With a binary outcome, we can replace linear outcome models with logistic outcome models. Show that with weights in the logistic regressions, the doubly robust estimators equal the outcome regression estimator. The result holds for both τ and τ_{T}.

14.4 Causal inference with a misspecified linear regression

Define the population OLS of Y on $(1, Z, X)$ as

$$(\beta_0, \beta_1, \beta_2) = \arg\min_{b_0, b_1, b_2} E(Y - b_0 - b_1 Z - b_2^{\mathsf{T}} X)^2.$$

Recall that $e(X) = \text{pr}(Z = 1 \mid X)$ is the propensity score, and define $\tilde{e}(X) = \gamma_0 + \gamma_1^{\mathsf{T}} X$ as the OLS projection of Z on X with

$$(\gamma_0, \gamma_1) = \arg\min_{c_0, c_1} E(Z - c_0 - c_1^{\mathsf{T}} X)^2.$$

1. Show that
$$\beta_1 = \frac{E[\tilde{w}(X)\{\mu_1(X) - \mu_0(X)\}]}{E\{\tilde{w}(X)\}} + \frac{E[\{e(X) - \tilde{e}(X)\}\mu_0(X)]}{E\{\tilde{w}(X)\}}$$
where $\tilde{w}(X) = e(X)\{1 - \tilde{e}(X)\}$.
2. When X contains the dummy variables for a discrete covariate, show that
$$\beta_1 = \frac{E[w(X)\{\mu_1(X) - \mu_0(X)\}]}{E\{w(X)\}}$$
where $w(X) = e(X)\{1 - e(X)\}$ is the overlap weight introduced in Section 13.4.

Remark: Vansteelandt and Dukes (2022) gave the formula in part 1 without a detailed proof. The result in part 2 was derived many times in the literature (e.g., Angrist, 1998; Ding, 2021).

14.5 Analyzing a dataset from the Karolinska Institute

Revisit Problem 12.6. Estimate τ_{O} and τ based on the methods introduced in this chapter.

14.6 Recommended reading

Kang and Schafer (2007) gave a critical review of the doubly robust estimator, using simulation to compare it with many other estimators. Robins et al. (2007) gave a very insightful comment on Kang and Schafer (2007).

15

Matching in Observational Studies

Matching has a long history in empirical research (Greenwood, 1945; Chapin, 1947). W. Cochran and D. Rubin popularized it in statistical causal inference. Cochran and Rubin (1973) is an early review paper. Rubin (2006b) collects Rubin's contributions to this topic. This chapter also discusses modern contributions by Abadie and Imbens (2006, 2008, 2011) based on the asymptotic analysis of matching estimators.

15.1 A simple starting point: many more control units

Figure 15.1 illustrates the basic idea of matching in treatment-control observational studies. Consider a simple case with the number of control units n_0 being much larger than the number of treated units n_1. For unit $i = 1, \ldots, n_1$ in the treated group, we find a unit $m(i)$ in the control group such that $X_i = X_{m(i)}$. In the ideal case, we have exact matches. Therefore, the units within a matched pair have the same propensity score $e(X_i) = e(X_{m(i)})$. Consequently, conditional on the event that one unit receives the treatment and the other receives the control, the probability of unit i receiving the treatment and unit $m(i)$ receiving the control is

$$\text{pr}(Z_i = 1, Z_{m(i)} = 0 \mid Z_i + Z_{m(i)} = 1, X_i, X_{m(i)}) = 1/2$$

by a symmetry argument.[1] That is, the treatment assignment is identical to the MPE conditional on the covariates and the event that each pair has a treated unit and a control unit. So we can analyze the exactly matched observational study as if it is an MPE, using either the FRT or the Neymanian approach in Chapter 7. This gives us inference on the causal effect on the treated units.

We can also find multiple control units for each treated unit. In general, we can find M_i matched control units for the treated unit i. When the M_i's vary, it is called the *variable-ratio matching* (Ming and Rosenbaum, 2000, 2001; Pimentel et al., 2015). With perfect matching, the treatment assignment mechanism is identical to the general matched experiment discussed in Section 7.7. We can use the analytic results in that section to analyze the matched observational study.

[1]The rigorous argument is

$$\begin{aligned}
& \text{pr}(Z_i = 1, Z_{m(i)} = 0 \mid Z_i + Z_{m(i)} = 1, X_i, X_{m(i)}) \\
= & \frac{\text{pr}(Z_i = 1, Z_{m(i)} = 0 \mid X_i, X_{m(i)})}{\text{pr}(Z_i = 1, Z_{m(i)} = 0 \mid X_i, X_{m(i)}) + \text{pr}(Z_i = 0, Z_{m(i)} = 1 \mid X_i, X_{m(i)})} \\
= & \frac{e(X_i)\{1 - e(X_{m(i)})\}}{e(X_i)\{1 - e(X_{m(i)})\} + \{1 - e(X_i)\}e(X_{m(i)})} \\
= & \frac{1}{2}.
\end{aligned}$$

This type of calculation will appear in Chapter 19 when the symmetry breaks down.

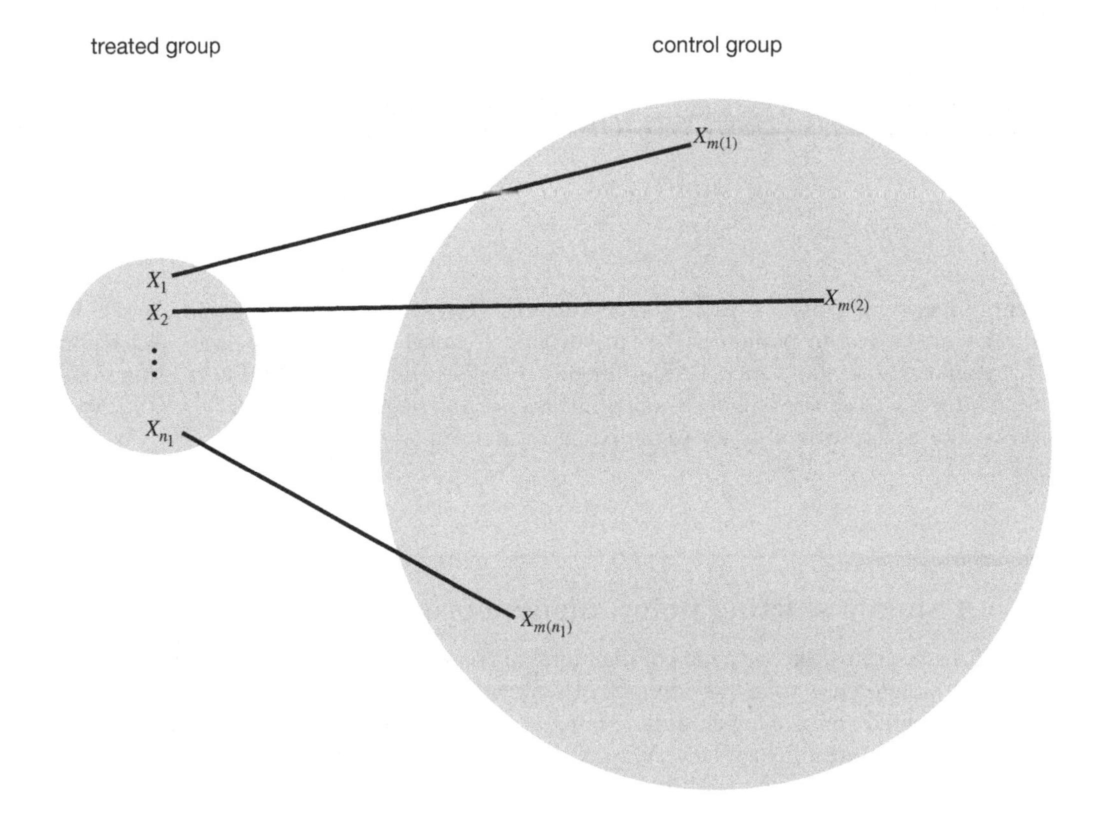

FIGURE 15.1: Illustration of matching in observational studies

Rosenbaum (2002b) advocated the above analysis strategy. In most observational studies, however, $X_i = X_{m(i)}$ does not hold for all units. The above reasoning for FRT does not hold. Recently, Guo and Rothenhäusler (2023) reported negative results on the consequence of inexact matching in FRT. This is a warning of this strategy.

15.2 A more complicated but realistic scenario

Even if the control group is large, we often do not have exact matches. What we can achieve is that $X_i \approx X_{m(i)}$ or $X_i - X_{m(i)}$ is small under some distance metric. So we have only approximate matches. For example, we define

$$m(i) = \arg \min_{k:Z_k=0} d(X_i, X_k),$$

where $d(X_i, X_k)$ measures the distance between X_i and X_k. Some canonical choices of the distance are the Euclidean distance

$$d(X_i, X_k) = (X_i - X_k)^{\mathsf{T}}(X_i - X_k),$$

and the Mahalanobis distance

$$d(X_i, X_k) = (X_i - X_k)^{\mathsf{T}}\Omega^{-1}(X_i - X_k)$$

with Ω being the sample covariance matrix of the X_i's from the whole population or only the control group.

I review some subtle issues about matching below. See Stuart (2010) for a review paper.

1. (one-to-one or one-to-M matching) The above discussion focused on one-to-one matching. We can also extend the discussion to one-to-M matching.
2. I focus on matching with replacement but some practitioners prefer matching without replacement. If the pool of control units is large, these two methods will not differ too much for the final result. Matching with replacement is computationally more convenient, but matching without replacement involves computationally intensive discrete optimization. Matching with replacement usually gives matches of higher quality but it introduces dependence by using the same units multiple times. In contrast, the advantage of matching without replacement is the independence of matched units and the simplicity in the subsequent data analysis.
3. Because of the residual covariate imbalance within matched pairs, it is crucial to use covariate adjustment when analyzing the data. In this case, covariate adjustment is not only for efficiency gain but also for bias correction.
4. If X is "high dimensional", it is likely that $d(X_i, X_k)$ is too large for some unit i in the treated group and for all choices of the units in the control group. If this happens, we may have to drop some units that are hard to find matches. By doing this, we effectively change the study population of interest.
5. It is hard to avoid the above problem. For example, if $X_i \sim \mathrm{N}(0, I_p), X_k \sim \mathrm{N}(0, I_p)$, and $X_i \perp\!\!\!\perp X_k$, then

$$(X_i - X_k)^{\mathsf{T}}(X_i - X_k) \sim 2\chi_p^2$$

which has mean $2p$ and variance $8p$ (see Section A.1.3). Theory shows that with large p, imperfect matching causes a large bias in causal effect estimation. This suggests that if p is large, we must do some dimension reduction before matching. Rosenbaum and Rubin (1983b) proposed to use matching based on the propensity score. With the estimated propensity score, we find pairs of units $\{i, m(i)\}$ with small values of $|\hat{e}(X_i) - \hat{e}(X_{m(i)})|$ or $|\mathrm{logit}\{\hat{e}(X_i)\} - \mathrm{logit}\{\hat{e}(X_{m(i)})\}|$,[2] i.e., we have a one-dimensional matching problem.

15.3 Matching estimator for the average causal effect

In a sequence of papers, Abadie and Imbens (AI) rigorously characterized the asymptotic properties of the matching estimator and proposed the corresponding large-sample confidence intervals for the average causal effect. They chose the standard setup for observational studies with $\{X_i, Z_i, Y_i(1), Y_i(0)\}_{i=1}^n \overset{\mathrm{IID}}{\sim} \{X, Z, Y(1), Y(0)\}$.

15.3.1 Point estimation and bias correction

AI focused on 1-M matching with replacement. For a treated unit i, we can simply impute the potential outcome under the treatment as $\hat{Y}_i(1) = Y_i$, and impute the potential outcome

[2]Define $\mathrm{logit}(w) = \log \frac{w}{1-w}$. The logit function is a map from $[0, 1]$ to $(-\infty, \infty)$.

under the control as

$$\hat{Y}_i(0) = M^{-1} \sum_{k \in J_i} Y_k,$$

where J_i is the set of matched units from the control group for unit i. For example, we can compute $d(X_i, X_k)$ for all k in the control group, and then define J_i as the indices of k with the M smallest values of $d(X_i, X_k)$.

For a control unit i, we simply impute the potential outcome under the control as $\hat{Y}_i(0) = Y_i$, and impute the potential outcome under the treatment as

$$\hat{Y}_i(1) = M^{-1} \sum_{k \in J_i} Y_k,$$

where J_i is the set of matched units from the treatment group for unit i.

The matching estimator is

$$\hat{\tau}^{\mathrm{m}} = n^{-1} \sum_{i=1}^{n} \{\hat{Y}_i(1) - \hat{Y}_i(0)\}.$$

AI showed that $\hat{\tau}^{\mathrm{m}}$ has non-negligible bias especially when X is multidimensional and the number of control units is comparable to the number of treated units. Through some technical derivations, they proposed the following estimator for the bias:

$$\hat{B} = n^{-1} \sum_{i=1}^{n} \hat{B}_i$$

where

$$\hat{B}_i = (2Z_i - 1)M^{-1} \sum_{k \in J_i} \{\hat{\mu}_{1-Z_i}(X_i) - \hat{\mu}_{1-Z_i}(X_k)\}$$

with $\{\hat{\mu}_1(X_i), \hat{\mu}_0(X_i)\}$ being the predicted outcomes by, for example, OLS fits. For a treated unit with $Z_i = 1$, the estimated bias is

$$\hat{B}_i = M^{-1} \sum_{k \in J_i} \{\hat{\mu}_0(X_i) - \hat{\mu}_0(X_k)\}$$

which corrects the discrepancy in predicted control potential outcomes due to the mismatch in covariates; for a control unit with $Z_i = 0$, the estimated bias is

$$\hat{B}_i = -M^{-1} \sum_{k \in J_i} \{\hat{\mu}_1(X_i) - \hat{\mu}_1(X_k)\}$$

which corrects the discrepancy in predicted treated potential outcomes due to the mismatch in covariates.

The final bias-corrected matching estimator is

$$\hat{\tau}^{\mathrm{mbc}} = \hat{\tau}^{\mathrm{m}} - \hat{B},$$

which has the following linear expansion.

Proposition 15.1 *We have*

$$\hat{\tau}^{\mathrm{mbc}} = n^{-1} \sum_{i=1}^{n} \hat{\psi}_i \tag{15.1}$$

where

$$\hat{\psi}_i = \hat{\mu}_1(X_i) - \hat{\mu}_0(X_i) + (2Z_i - 1)(1 + K_i/M)\{Y_i - \hat{\mu}_{Z_i}(X_i)\}$$

with K_i being the times that unit i is used as a match.

The linear expansion in Proposition 15.1 follows from simple but tedious algebra. I leave its proof as Problem 15.1. The linear expansion motivates a simple variance estimator

$$\hat{V}^{\text{mbc}} = \frac{1}{n^2}\sum_{i=1}^{n}(\hat{\psi}_i - \hat{\tau}^{\text{mbc}})^2,$$

by viewing $\hat{\tau}^{\text{mbc}}$ as sample averages of the $\hat{\psi}_i$'s. Abadie and Imbens (2008) first showed that the simple bootstrap by resampling the original data does not work for estimating the variance of the matching estimators, but their proposed variance estimation procedure is not easy to implement. Otsu and Rai (2017) proposed to bootstrap the $\hat{\psi}_i$'s in the linear expansion. However, the bootstrap proposed by Otsu and Rai (2017) essentially yields the variance estimator $\hat{V}^{\text{mbc}}$, which is simple to calculate.

15.3.2 Connection with the doubly robust estimators

The bias-corrected matching estimators and the doubly robust estimators are closely related. They both equal the outcome regression estimator with some modifications based on the residuals

$$\hat{R}_i = \begin{cases} Y_i - \hat{\mu}_1(X_i) & \text{if } Z_i = 1; \\ Y_i - \hat{\mu}_0(X_i) & \text{if } Z_i = 0. \end{cases}$$

For the average causal effect τ, recall the outcome regression estimator

$$\hat{\tau}^{\text{reg}} = n^{-1}\sum_{i=1}^{n}\{\hat{\mu}_1(X_i) - \hat{\mu}_0(X_i)\}$$

and the doubly robust estimator

$$\hat{\tau}^{\text{dr}} = \hat{\tau}^{\text{reg}} + n^{-1}\sum_{i=1}^{n}\left\{\frac{Z_i\hat{R}_i}{\hat{e}(X_i)} - \frac{(1-Z_i)\hat{R}_i}{1-\hat{e}(X_i)}\right\}.$$

Furthermore, we can verify that $\hat{\tau}^{\text{mbc}}$ has a form similar to $\hat{\tau}^{\text{dr}}$.

Proposition 15.2 *The bias-corrected matching estimator for τ equals*

$$\hat{\tau}^{\text{mbc}} = \hat{\tau}^{\text{reg}} + n^{-1}\sum_{i=1}^{n}\left\{\left(1+\frac{K_i}{M}\right)Z_i\hat{R}_i - \left(1+\frac{K_i}{M}\right)(1-Z_i)\hat{R}_i\right\}.$$

I leave the proof of Proposition 15.2 as Problem 15.2. From Proposition 15.2, we can view matching as a nonparametric method to estimate the propensity score, and the resulting bias-corrected matching estimator as a doubly robust estimator. [3] For instance, $1 + K_i/M$ should be close to $1/\hat{e}(X_i)$. When a treated unit has a small $e(X_i)$, the resulting weight based on the estimated propensity score $1/\hat{e}(X_i)$ will be large, and at the same time, it will be matched to many control units, resulting in large K_i and thus large $1+K_i/M$. However, this connection also raised an obvious question regarding matching. With a fixed M, the estimator $1 + K_i/M$ for $1/e(X_i)$ will be very noisy. Allowing M to grow with the sampling size is likely to improve the matching-based nonparametric estimator for the propensity score and thus improve the asymptotic properties of the matching and bias-corrected matching

[3] This might be obvious from the representation in Proposition 15.2. However, from the original definition of $\hat{\tau}^{\text{m}}$, it seems that the matching estimator focuses on predicting the outcomes rather than estimating the propensity score.

estimators. Lin et al. (2023) provided a formal theory that once we allow M to grow at a proper rate, the bias-corrected matching estimator $\hat{\tau}^{\text{mbc}}$ can achieve similar properties as the doubly robust estimator.

However, the theory of Lin et al. (2023) does not provide clear guidance on the choice of M in finite-sample data analysis. Without formal results on the optimal choice of M, I recommend reporting a sequence of point estimators and confidence intervals with varying M.

15.4 Matching estimator for the average causal effect on the treated

For the average causal effect on the treated

$$\tau_{\text{T}} = E(Y \mid Z=1) - E\{Y(0) \mid Z=1\},$$

we only need to impute the missing potential outcomes under control for all the treated units, resulting in the following estimator

$$\hat{\tau}_{\text{T}}^{\text{m}} = n_1^{-1} \sum_{i=1}^{n} Z_i\{Y_i - \hat{Y}_i(0)\}.$$

Again it is biased with multidimensional X. Otsu and Rai (2017) propose to estimate its bias by

$$\hat{B}_{\text{T}} = n_1^{-1} \sum_{i=1}^{n} Z_i \hat{B}_{\text{T},i}$$

where

$$\hat{B}_{\text{T},i} = M^{-1} \sum_{k \in J_i} \{\hat{\mu}_0(X_i) - \hat{\mu}_0(X_k)\}$$

corrects the bias due to the mismatch of covariates for a treated unit with $Z_i = 1$.

The final bias-corrected estimator is

$$\hat{\tau}_{\text{T}}^{\text{mbc}} = \hat{\tau}_{\text{T}}^{\text{m}} - \hat{B}_{\text{T}},$$

which has the following linear expansion.

Proposition 15.3 *We have*

$$\hat{\tau}_{\text{T}}^{\text{mbc}} = n_1^{-1} \sum_{i=1}^{n} \hat{\psi}_{\text{T},i}, \tag{15.2}$$

where

$$\hat{\psi}_{\text{T},i} = Z_i\{Y_i - \hat{\mu}_0(X_i)\} - (1 - Z_i)K_i/M\{Y_i - \hat{\mu}_0(X_i)\}.$$

I leave the proof to Problem 15.1. Motivated by Otsu and Rai (2017), we can view $\hat{\tau}_{\text{T}}^{\text{mbc}}$ as n/n_1 multiplied by the sample average of the $\hat{\psi}_{\text{T},i}$'s, so an intuitive variance estimator is

$$\begin{aligned} \hat{V}_{\text{T}}^{\text{mbc}} &= \left(\frac{n}{n_1}\right)^2 \frac{1}{n^2} \sum_{i=1}^{n} (\hat{\psi}_{\text{T},i} - \hat{\tau}_{\text{T}}^{\text{mbc}} n_1/n)^2 \\ &= \frac{1}{n_1^2} \sum_{i=1}^{n} (\hat{\psi}_{\text{T},i} - \hat{\tau}_{\text{T}}^{\text{mbc}} n_1/n)^2. \end{aligned}$$

Similar to the discussion in Section 15.3.2, we can compare the doubly robust and bias-corrected matching estimators with the outcome regression estimator. For the average causal effect on the treated units τ_{T}, recall the outcome regression estimator

$$\hat{\tau}_{\mathrm{T}}^{\mathrm{reg}} = n_1^{-1} \sum_{i=1}^{n} Z_i \{Y_i - \hat{\mu}_0(X_i)\},$$

and the doubly robust estimator

$$\hat{\tau}_{\mathrm{T}}^{\mathrm{dr}} = \hat{\tau}_{\mathrm{T}}^{\mathrm{reg}} - n_1^{-1} \sum_{i=1}^{n} \frac{\hat{e}(X_i)}{1 - \hat{e}(X_i)} (1 - Z_i) \hat{R}_i.$$

Furthermore, we can verify that $\hat{\tau}_{\mathrm{T}}^{\mathrm{mbc}}$ has a form similar to $\hat{\tau}_{\mathrm{T}}^{\mathrm{dr}}$.

Proposition 15.4 *The bias correction matching estimator for τ_{T} equals*

$$\hat{\tau}_{\mathrm{T}}^{\mathrm{mbc}} = \hat{\tau}_{\mathrm{T}}^{\mathrm{reg}} - n_1^{-1} \sum_{i=1}^{n} \frac{K_i}{M} (1 - Z_i) \hat{R}_i.$$

I leave the proof of Proposition 15.4 as Problem 15.3. Proposition 15.4 suggests that matching essentially uses K_i/M to estimate the odds of the treatment given covariates.

15.5 A case study

15.5.1 Experimental data

Now I revisit the LaLonde data using the `Matching` package by Sekhon (2011). We have used this package several times for the dataset `lalonde`, and now we will use its key function `Match`. The experimental part gives us the following results:

```
> library("car")
> library("Matching")
>
> ## Section 15.5.1
> ## experimental data
> data("lalonde")
> y = lalonde$re78
> z = lalonde$treat
> x = as.matrix(lalonde[, c("age", "educ", "black",
+                           "hisp", "married", "nodegr",
+                           "re74", "re75")])
>
> ## analysis the randomized experiment
> neymanols = lm(y ~ z)
> fisherols = lm(y ~ z + x)
> xc = scale(x)
> linols = lm(y ~ z*xc)
> resols = c(neymanols$coef[2],
+            fisherols$coef[2],
+            linols$coef[2],
+            sqrt(hccm(neymanols, type = "hc2")[2, 2]),
```

```
+              sqrt(hccm(fisherols, type = "hc2")[2, 2]),
+              sqrt(hccm(linols, type = "hc2")[2, 2]))
> resols = matrix(resols, 3, 2)
> rownames(resols) = c("neyman", "fisher", "lin")
> colnames(resols) = c("est", "se")
> resols
            est       se
neyman 1794.343 670.9967
fisher 1676.343 677.0493
lin    1621.584 694.7217
```

All regression estimators show positive significant results on the job training program. We can analyze the data as if it is an observational study based on 1-1 matching, yielding the following results:

```
> matchest.adj = Match(Y = y, Tr = z, X = x, BiasAdjust = TRUE)
> summary(matchest.adj)

Estimate...  2119.7
AI SE......  876.42
T-stat.....  2.4185
p.val......  0.015583

Original number of observations..............  445
Original number of treated obs...............  185
Matched number of observations...............  185
Matched number of observations  (unweighted).  268
```

Both the point estimator and standard error increase, but qualitatively, the conclusion remains the same.

15.5.2 Observational data

Then I revisit the observational counterpart of the data:

```
> dat <- read.table("cps1re74.csv", header = TRUE)
> dat$u74 <- as.numeric(dat$re74==0)
> dat$u75 <- as.numeric(dat$re75==0)
> y = dat$re78
> z = dat$treat
> x = as.matrix(dat[, c("age", "educ", "black",
+                       "hispan", "married", "nodegree",
+                       "re74", "re75", "u74", "u75")])
```

If we use simple OLS estimators, the results are far from the experimental benchmark and are sensitive to the specification of the regression:

```
> neymanols = lm(y ~ z)
> fisherols = lm(y ~ z + x)
> xc = scale(x)
> linols = lm(y ~ z*xc)
> resols = c(neymanols$coef[2],
+              fisherols$coef[2],
+              linols$coef[2],
+              sqrt(hccm(neymanols, type = "hc2")[2, 2]),
+              sqrt(hccm(fisherols, type = "hc2")[2, 2]),
+              sqrt(hccm(linols, type = "hc2")[2, 2]))
```

```
> resols = matrix(resols, 3, 2)
> rownames(resols) = c("neyman", "fisher", "lin")
> colnames(resols) = c("est", "se")
> resols
             est        se
neyman -8506.495  583.4426
fisher  1067.546  628.4389
lin    -4265.801 3211.7718
```

However, if we use 1-1 matching, the results almost recover those based on the experimental data:[4]

```
> matchest = Match(Y = y, Tr = z, X = x, BiasAdjust = TRUE)
> summary(matchest)

Estimate...  1747.8
AI SE......  916.59
T-stat.....  1.9068
p.val......  0.056543

Original number of observations..............  16177
Original number of treated obs...............  185
Matched number of observations...............  185
Matched number of observations  (unweighted).  248
```

Ignoring the ties in the matched data, we can also use the matched-pairs analysis, which again yields results similar to those based on the experimental data:

```
> diff = y[matchest$index.treated] -
+             y[matchest$index.control]
> round(summary(lm(diff ~ 1))$coef[1, ], 2)
  Estimate Std. Error    t value   Pr(>|t|)
   1581.44     558.55       2.83       0.01
>
> diff.x = x[matchest$index.treated, ] -
+               x[matchest$index.control, ]
> round(summary(lm(diff ~ diff.x))$coef[1, ], 2)
  Estimate Std. Error    t value   Pr(>|t|)
   1842.06     578.37       3.18       0.00
```

15.5.3 Covariate balance checks

Moreover, we can use simple OLS to check covariate balance. Before matching, the covariates are highly imbalanced, signified by many stars associated with the coefficients.

```
> lm.before = lm(z ~ x)
> summary(lm.before)

Residuals:
     Min        1Q    Median        3Q       Max
-0.18508 -0.01057  0.00303  0.01018  1.01355
```

[4]The default estimand of the **Match** function is τ_T, which is a natural choice in this example since the treated group is from the experimental data **lalonde** and the control group is from another large observational source. The default M equals 1 and the default distance metric is the Euclidean distance based on the standardized covariates with unit marginal variances.

```
Coefficients:
              Estimate Std. Error t value Pr(>|t|)
(Intercept)  1.404e-03  6.326e-03   0.222   0.8243
xage        -4.043e-04  8.512e-05  -4.750 2.05e-06 ***
xeduc        3.220e-04  4.073e-04   0.790   0.4293
xblack       1.070e-01  2.902e-03  36.871  < 2e-16 ***
xhispan      6.377e-03  3.103e-03   2.055   0.0399 *
xmarried    -1.525e-02  2.023e-03  -7.537 5.06e-14 ***
xnodegree    1.345e-02  2.523e-03   5.331 9.89e-08 ***
xre74        7.601e-07  1.806e-07   4.208 2.59e-05 ***
xre75       -1.231e-07  1.829e-07  -0.673   0.5011
xu74         4.224e-02  3.271e-03  12.914  < 2e-16 ***
xu75         2.424e-02  3.399e-03   7.133 1.02e-12 ***
```

However, after matching, the covariates are well-balanced, signified by the absence of stars for all coefficients.

```
> lm.after = lm(z ~ x,
+              subset = c(matchest$index.treated,
+                         matchest$index.control))
> summary(lm.after)

Residuals:
     Min       1Q   Median       3Q      Max
-0.66864 -0.49161 -0.03679  0.50378  0.65122

Coefficients:
              Estimate Std. Error t value Pr(>|t|)
(Intercept)  6.003e-01  2.427e-01   2.474   0.0137 *
xage         3.199e-03  3.427e-03   0.933   0.3511
xeduc       -1.501e-02  1.634e-02  -0.918   0.3590
xblack       6.141e-05  7.408e-02   0.001   0.9993
xhispan      1.391e-02  1.208e-01   0.115   0.9084
xmarried    -1.328e-02  6.729e-02  -0.197   0.8437
xnodegree   -3.023e-02  7.144e-02  -0.423   0.6723
xre74        6.754e-06  9.864e-06   0.685   0.4939
xre75       -9.848e-06  1.279e-05  -0.770   0.4417
xu74         2.179e-02  1.027e-01   0.212   0.8321
xu75        -2.642e-02  8.327e-02  -0.317   0.7512
```

15.6 Discussion

With many covariates, matching based on the original covariates may suffer from the curse of dimensionality. Rosenbaum and Rubin (1983b) suggested to use matching based on the estimated propensity score. Abadie and Imbens (2016) provided a form theory for this strategy.

15.7 Homework problems

15.1 Linear expansions of the bias-corrected estimators

Prove Propositions 15.1 and 15.3.

15.2 Doubly robust form of the bias-corrected matching estimator for τ

Prove Proposition 15.2.

15.3 Doubly robust form of the bias-corrected matching estimator for τ_{T}

Prove Proposition 15.4.

15.4 Revisit Example 10.3

Analyze the dataset in Example 10.3 using the matching estimator. Compare the results with previous results. You should check the covariate balance before and after matching. You can also choose a different number of matches for the matching estimator. Moreover, you can even apply various estimators to the matched data. Are your results sensitive to your choices?

15.5 Revisit Section 15.5

Section 15.5 analyzed the LaLonde observational study using matching. Matching performs well because it gives an estimator that is close to the experimental gold standard. Reanalyze the data using the outcome regression, propensity score stratification, two IPW, and the doubly robust estimators. Compare the results to the matching estimator and to the estimator from the experimental gold standard.

Note that you have many choices. For example, the number of strata for stratification and the threshold to trim to data based on the estimated propensity scores. You may consider fitting different propensity score and outcome models, e.g., including some quadratic terms of the basic covariates. You can even apply these estimators to the matched data.

This is a classic dataset and hundreds of papers have used it. You can read some references (Dehejia and Wahba, 1999; Hainmueller, 2012) and you can also be creative in your data analysis.

15.6 Data reanalyses

Ho et al. (2007) is an influential paper in political science, based on which the authors have developed an `R` package `MatchIt` (Ho et al., 2011). Ho et al. (2007) analyzed two datasets, both of which are available from the Harvard Dataverse.

Reanalyze these two datasets using the methods discussed so far. You can also try other methods as long as you can justify them.

15.7 Recommended reading

The literature on matching estimators is massive, and three excellent review papers are Sekhon (2009), Stuart (2010), and Imbens (2015).

Part IV

Difficulties and challenges of observational studies

16

Difficulties of Unconfoundedness in Observational Studies for Causal Effects

Part III of this book discusses causal inference with observational studies under two assumptions: unconfoundedness and overlap. Both are strong assumptions and are likely to be violated in practice. This chapter will discuss the difficulties of the unconfoundedness assumption. Chapters 17–19 will discuss various strategies for sensitivity analysis in observational studies with unmeasured confounding. Chapter 20 will discuss the difficulties of the overlap assumption.

16.1 Some basics of the causal diagram

Pearl (1995) introduced the causal diagram as a powerful tool for causal inference in empirical research. Pearl (2000) is a textbook on the causal diagram. Here I introduce the causal diagram as an intuitive tool for illustrating the causal relationships among variables.

For example, if we have the causal diagram

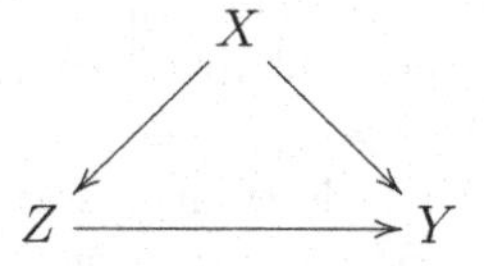

and focus on the causal effect of Z on Y, we can read it as the following data-generating process:[1]

$$\begin{cases} X \sim F_X(x), \\ Z = g_Z(X, \varepsilon_Z), \\ Y(z) = g_Y(X, z, \varepsilon_Y(z)), \end{cases}$$

where the g's are general functions and the random error terms satisfy $\varepsilon_Z \perp\!\!\!\perp \varepsilon_Y(z)$ for both $z = 0, 1$. In the above, covariates X are generated from a distribution $F_X(x)$, the treatment assignment is a function of X with a random error term ε_Z, and the potential outcome $Y(z)$ is a function of X, z, and a random error term $\varepsilon_Y(z)$. We can easily read from the equations that $Z \perp\!\!\!\perp Y(z) \mid X$, i.e., the unconfoundedness assumption holds.

If we have a causal diagram

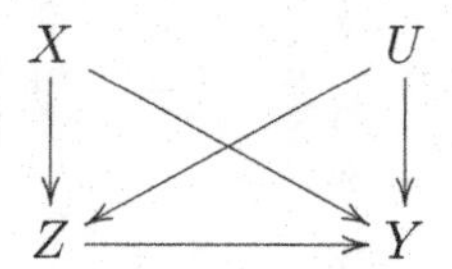

[1]In the literature, it is more common to see the last equation written as $Y = g_Y(X, Z, \varepsilon_Y)$ which is an equation for generating the observed Y. I choose $Y(z) = g_Y(X, z, \varepsilon_Y(z))$ to emphasize the data-generating process of the potential outcome, which allows for the dependence of ε_Y on z.

we can read it as the following data-generating process:

$$\begin{cases} X \sim F_X(x), \\ U \sim F_U(u), \\ Z = g_Z(X, U, \varepsilon_Z), \\ Y(z) = g_Y(X, U, z, \varepsilon_Y(z)), \end{cases}$$

where the g's are general functions and the random error terms satisfy $\varepsilon_Z \perp\!\!\!\perp \varepsilon_Y(z)$ for both $z = 0, 1$. We can easily read from the equations that $Z \perp\!\!\!\perp Y(z) \mid (X, U)$ but $Z \not\!\perp\!\!\!\perp Y(z) \mid X$, i.e., the unconfoundedness assumption holds conditional on (X, U) but does not hold conditional on X only. In this case, U is an unmeasured confounder. In this diagram, U is called an unmeasured confounder.

16.2 Assessing the unconfoundedness assumption

The unconfoundedness assumption

$$Z \perp\!\!\!\perp Y(1) \mid X, \quad Z \perp\!\!\!\perp Y(0) \mid X$$

implies that

$$\begin{aligned} \text{pr}\{Y(1) \mid Z = 1, X\} &= \text{pr}\{Y(1) \mid Z = 0, X\}, \\ \text{pr}\{Y(0) \mid Z = 1, X\} &= \text{pr}\{Y(0) \mid Z = 0, X\}. \end{aligned}$$

So the unconfoundedness assumption basically requires that the counterfactual distribution $\text{pr}\{Y(1) \mid Z = 0, X\}$ equals the observed distribution $\text{pr}\{Y(1) \mid Z = 1, X\}$, and the counterfactual distribution $\text{pr}\{Y(0) \mid Z = 1, X\}$ equals the observed distribution $\text{pr}\{Y(0) \mid Z = 0, X\}$. Because the counterfactual distributions are not directly identifiable from the data, the unconfoundedness assumption is fundamentally untestable without additional assumptions. I will discuss two strategies to assess the unconfoundedness assumption. Here, "assess" is a weaker notion than "test". The former is referred to as supplementary analysis that supports or undermines the initial analysis, but the latter is referred to as formal statistical testing.

16.2.1 Using negative outcomes

Assume that Y^{n} is an outcome similar to Y and ideally, shares the same confounding structure as Y. If we believe $Z \perp\!\!\!\perp Y(z) \mid X$, then we also tend to believe $Z \perp\!\!\!\perp Y^{\text{n}}(z) \mid X$. Moreover, we know, a priori, the effect of Z on Y^{n}:

$$\tau(Z \to Y^{\text{n}}) = E\{Y^{\text{n}}(1) - Y^{\text{n}}(0)\}.$$

An important example is that $\tau(Z \to Y^{\text{n}}) = 0$. A causal diagram satisfying these requirements is below:

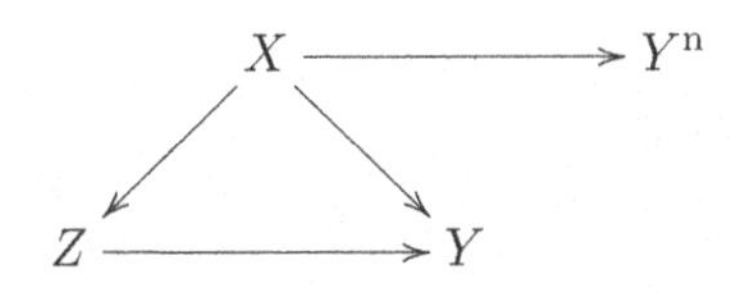

Example 16.1 *Cornfield et al. (1959) studied the causal role of cigarette smoking on lung cancer based on observational studies. They controlled for many important background variables but it is still possible to have some unmeasured confounders biasing the observed effects. To strengthen the evidence for causation, they also reported the effect of cigarette smoking on car accidents, which was close to zero, the anticipated effect based on biology. So even if they could not rule out unmeasured confounding in the analysis, this supplementary analysis based on a negative outcome makes the evidence of the causal effect of cigarette smoking on lung cancer stronger.*

Example 16.2 *Imbens and Rubin (2015) suggested using the lagged outcome as a negative outcome. In most cases, it is reasonable to believe that the lagged outcome and the outcome have a similar confounding structure. Since the lagged outcome happens before the treatment, the average causal effect on it must be zero. However, their suggestion should be used with caution since in most studies we simply treat lagged outcomes as an observed confounder.*

In some sense, the covariate balance check in Chapter 11 is a special case of using negative controls. Similar to the problem of using lagged outcomes as negative controls, those covariates are usually a part of the unconfoundedness assumption. Therefore, the failure of the covariate balance check does not really falsify the unconfoundedness assumption but rather the model specification of the propensity score.

Example 16.3 *Observational studies of elderly persons have shown that vaccination against influenza remarkably reduces one's risk of pneumonia/influenza hospitalization and all-cause mortality in the following season, after adjustment for measured covariates. Jackson et al. (2006) were skeptical about the large magnitude and thus conducted supplementary analysis on negative outcomes. Vaccination often begins in the fall, but influenza transmission is often minimal until the winter. Based on biology, the effect of vaccination should be most prominent during influenza season. But Jackson et al. (2006) found a greater effect before the influenza season, suggesting that the observed effect is due to unmeasured confounding.*

Jackson et al. (2006) seems the most convincing one since the influenza-related outcomes before and during the influenza season should have similar confounding patterns. The additional evidence from Cornfield et al. (1959) seems weaker since car accidents and lung cancer have very different causal mechanisms with respect to cigarette smoking. In fact, the critique from Fisher (1957) was that the relationship between cigarette smoking on lung cancer may be due to an unobserved genetic factor (see Chapter 17). Such a genetic factor might affect cigarette smoking and lung cancer simultaneously, but it seems unlikely that it also affects car accidents.

Lipsitch et al. (2010) is a recent article on negative outcomes. Rosenbaum (1989) discussed the role of known effects in causal inference.

16.2.2 Using negative exposures

Negative exposures are duals of negative outcomes. Assume Z^{n} is a treatment variable similar to Z and shares the same confounding structure as Z. If we believe $Z \perp\!\!\!\perp Y(z) \mid X$, then we tend to believe $Z^{\mathrm{n}} \perp\!\!\!\perp Y(z) \mid X$. Moreover, we know, a priori, the effect of Z^{n} on Y

$$\tau(Z^{\mathrm{n}} \to Y) = E\{Y(1^{\mathrm{n}}) - Y(0^{\mathrm{n}})\}.$$

An important example is that $\tau(Z^{\mathrm{n}} \to Y) = 0$. A causal diagram satisfying these requirements is below:

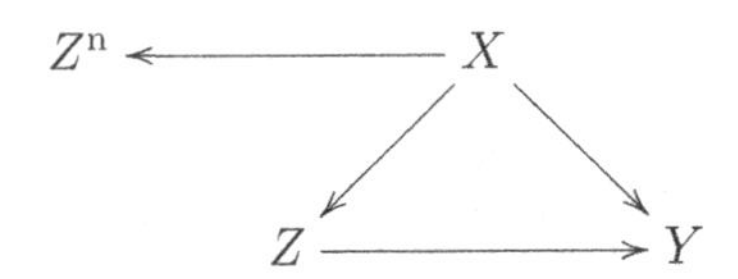

Example 16.4 *Sanderson et al. (2017) give many examples of negative exposures in determining the effect of intrauterine exposure on later outcomes by comparing the association of maternal exposure during pregnancy with the outcome of interest, with the association of paternal exposure with the same outcome. They review studies on the effect of maternal and paternal smoking on offspring outcomes, and studies on the effect of maternal and paternal BMI on later offspring BMI and autism spectrum disorder. In these examples, we expect the association of maternal exposure with the outcome to be larger than that of paternal exposure with the outcome.*

16.2.3 Summary

The unconfoundedness assumption is fundamentally untestable without additional assumptions. Although negative outcomes and negative controls in observational studies cannot prove or disprove unconfoundedness, using them in supplementary analyses can strengthen the evidence for causation. However, it is often non-trivial to conduct this type of supplementary analysis because it involves more data and more importantly, a deeper understanding of the causal problems to find convincing negative outcomes and negative controls.

16.3 Problems of over-adjustment

We have discussed many methods for estimating causal effects under the unconfoundedness assumption:

$$Z \perp\!\!\!\perp \{Y(1), Y(0)\} \mid X.$$

This is an assumption conditioning on X. It is crucial to select the right set of X that ensures the conditional independence. Rosenbaum (2002b) wrote that "there is no reason to avoid adjustment for a variable describing subjects before treatment." Similarly, Rubin (2007) wrote that "typically, the more conditional an assumption, the more acceptable it is." Both argued that we should control for all observed pretreatment covariates. VanderWeele and Shpitser (2011) called it the *pretreatment criterion*. Pearl disagreed with this recommendation and gave two counterexamples below.

16.3.1 M-bias

M-bias appears in the following causal diagram with an M-structure:

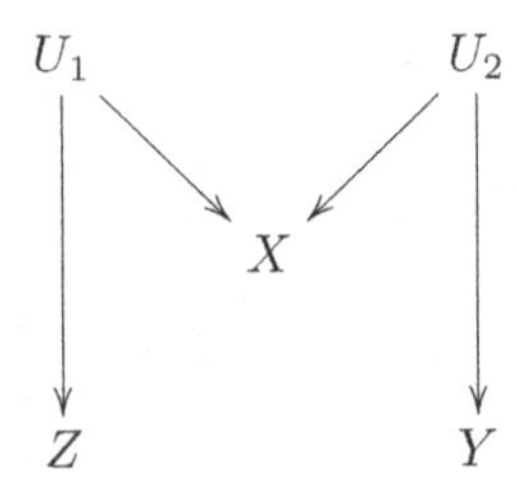

We can read from the diagram the data-generating process:

$$\begin{cases} U_1 \perp\!\!\!\perp U_2, \\ X = g_X(U_1, U_2, \varepsilon_X), \\ Z = g_Z(U_1, \varepsilon_Z), \\ Y = Y(z) = g_Y(U_2, \varepsilon_Y), \end{cases}$$

where the g's are general functions and $(\varepsilon_X, \varepsilon_Z, \varepsilon_Y)$ are independent random error terms. In the above causal diagram, X is observed, but U_1 and U_2 are unobserved. If we change the value of Z, the value of Y will not change at all. So the true causal effect of Z on Y must be 0. From the data-generating equations, we can read that $Z \perp\!\!\!\perp Y$, so the association between Z and Y is 0, and, in particular,

$$\begin{aligned} \tau_{\text{PF}} &= E(Y \mid Z = 1) - E(Y \mid Z = 0) \\ &= 0. \end{aligned}$$

This means that without adjusting for the covariate X, the simple estimator is unbiased for the true parameter.

However, if we condition on X, then $U_1 \not\!\perp\!\!\!\perp U_2 \mid X$, and consequently, $Z \not\!\perp\!\!\!\perp Y \mid X$ and

$$\int \{E(Y \mid Z = 1, X = x) - E(Y \mid Z = 0, X = x)\} f(x) \mathrm{d}x \neq 0$$

in general. To gain intuition, we consider the case with Normal linear models:[2]

$$\begin{cases} X = aU_1 + bU_2 + \varepsilon_X, \\ Z = cU_1 + \varepsilon_Z, \\ Y = Y(z) = dU_2 + \varepsilon_Y, \end{cases}$$

where $(U_1, U_2, \varepsilon_X, \varepsilon_Z, \varepsilon_Y) \overset{\text{IID}}{\sim} \mathrm{N}(0, 1)$. We have

$$\begin{aligned} \text{cov}(Z, Y) &= \text{cov}(cU_1 + \varepsilon_Z, dU_2 + \varepsilon_Y) \\ &= 0, \end{aligned}$$

but by the result in Problem 1.3, the partial correlation coefficient between Z and Y given X is[3]

$$\begin{aligned} \rho_{ZY|X} &= \frac{\rho_{ZY} - \rho_{ZX}\rho_{YX}}{\sqrt{1 - \rho_{ZX}^2}\sqrt{1 - \rho_{YX}^2}} \\ &\propto -\rho_{ZX}\rho_{YX} \\ &\propto -\text{cov}(Z, X)\text{cov}(Y, X) \\ &= -abcd, \end{aligned}$$

the product of the coefficients on the path from Z to Y. So the unadjusted estimator is unbiased but the adjusted estimator has a bias proportional to $abcd$.

The following simple example illustrates M-bias.

[2] It is not ideal for our discussion of binary Z, but it simplifies the derivations. Ding and Miratrix (2015) gave a detailed discussion with more natural models for binary Z.

[3] The notation $\propto$ reads as "proportional to" and allows us to drop some unimportant constants.

```
> ## M bias with large sample size
> n  = 10^6
> U1 = rnorm(n)
> U2 = rnorm(n)
> X  = U1 + U2 + rnorm(n)
> Y  = U2 + rnorm(n)
> ## with a continuous treatment Z
> Z  = U1 + rnorm(n)
> round(summary(lm(Y ~ Z))$coef[2, 1], 3)
[1] 0
> round(summary(lm(Y ~ Z + X))$coef[2, 1], 3)
[1] -0.2
>
> ## with a binary treatment Z
> Z  = (Z >= 0)
> round(summary(lm(Y ~ Z))$coef[2, 1], 3)
[1] 0.002
> round(summary(lm(Y ~ Z + X))$coef[2, 1], 3)
[1] -0.42
```

16.3.2 Z-bias

Consider the following causal diagram:

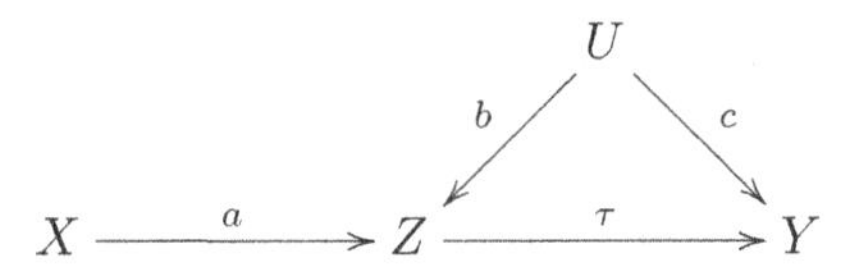

with the data generating process[4]

$$\begin{cases} Z = aX + bU + \varepsilon_Z, \\ Y(z) = \tau z + cU + \varepsilon_Y, \end{cases}$$

where $(U, X, \varepsilon_Z, \varepsilon_Y)$ are IID $\mathrm{N}(0,1)$. In this data-generating process, we have $X \perp\!\!\!\perp U$, $X \not\!\perp\!\!\!\perp Z$, and X affects Y only through Z.

The unadjusted estimator is

$$\begin{aligned} \tau_{\text{unadj}} &= \frac{\text{cov}(Z, Y)}{\text{var}(Z)} \\ &= \frac{\text{cov}(Z, \tau Z + cU)}{\text{var}(Z)} \\ &= \tau + \frac{c\text{cov}(aX + bU, U)}{\text{var}(Z)} \\ &= \tau + \frac{cb}{a^2 + b^2 + 1}, \end{aligned}$$

which has bias $bc/(a^2+b^2+1)$. The adjusted estimator from the OLS of Y on (Z, X) satisfies

$$\begin{cases} E\{Z(Y - \tau_{\text{adj}} Z - \alpha X)\} = 0, \\ E\{X(Y - \tau_{\text{adj}} Z - \alpha X)\} = 0. \end{cases}$$

[4]Again, we generate continuous Z from a linear model to simplify the derivations. Ding et al. (2017b) extended the theory to more general causal models, especially for binary Z.

Solve the above linear system of $(\tau_{\text{adj}}, \alpha)$ to obtain the formula of τ_{adj}:

$$\tau_{\text{adj}} = \tau + \frac{bc}{b^2+1}, \tag{16.1}$$

which has bias $bc/(b^2+1)$. I relegate the details for solving the linear system of $(\tau_{\text{adj}}, \alpha)$ as Problem 16.1.

So the unadjusted estimator has a smaller bias than the adjusted estimator. More interestingly, the stronger the association between X and Z is (measured by a), the larger the bias of the adjusted estimator is.

The mathematical derivation is not extremely hard. But this type of bias seems rather mysterious. Here is the intuition. The treatment is a function of X, U, and other random errors. If we condition on X, it is merely a function of U and other random errors. Therefore, conditioning on X makes Z less random, and more critically, makes the unmeasured confounder U play a more important role in Z. Consequently, the confounding bias due to U is amplified by conditioning on X. This idealized example illustrates the danger of over-adjusting for some covariates.

Heckman and Navarro-Lozano (2004) observed this phenomenon in simulation studies. Wooldridge (2016, technical report in 2006) verified it in linear models. Pearl (2010a, 2011) explained it using causal diagrams. Ding et al. (2017b) provided a more general theory as well as some intuition for this phenomenon. This type of bias is called Z-bias because, in Pearl's original papers, he used the symbol Z for our variable X. Throughout the book, however, Z is used for the treatment variable. In Part V of this book, we will call Z an instrumental variable if it satisfies the causal diagram presented in this subsection. This justifies the *instrumental variable bias* as another name for this type of bias.

The following simple example illustrates Z-bias.

```
> ## Z bias with large sample size
> n  = 10^6
> X = rnorm(n)
> U = rnorm(n)
> Z = X + U + rnorm(n)
> Y = U + rnorm(n)
>
> round(summary(lm(Y ~ Z))$coef[2, 1], 3)
[1] 0.333
> round(summary(lm(Y ~ Z + X))$coef[2, 1], 3)
[1] 0.501
>
> ## stronger association between X and Z
> Z = 2*X + U + rnorm(n)
> round(summary(lm(Y ~ Z))$coef[2, 1], 3)
[1] 0.167
> round(summary(lm(Y ~ Z + X))$coef[2, 1], 3)
[1] 0.501
>
> ## even stronger association between X and Z
> Z = 10*X + U + rnorm(n)
> round(summary(lm(Y ~ Z))$coef[2, 1], 3)
[1] 0.01
> round(summary(lm(Y ~ Z + X))$coef[2, 1], 3)
[1] 0.5
```

16.3.3 What covariates should we adjust for in observational studies?

We never know the true underlying data-generating process which can be quite complicated. However, the following causal diagram helps to clarify many ideas. It already rules out the possibility of M-bias discussed in Section 16.3.1.

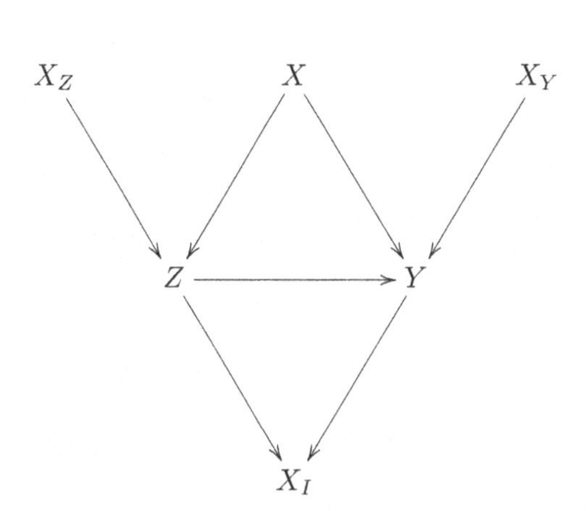

The covariates above have different features:

1. X affects both the treatment and the outcome. Conditioning on X ensures the unconfoundedness assumption, so we should control for X.
2. X_R is pure random noise not affecting either the treatment or the outcome. Including it in the analysis does not bias the estimate but introduces unnecessary variability in finite samples.
3. X_Z is an instrumental variable that affects the outcome only through the treatment. In the diagram above, including it in the analysis does not bias the estimate but increases the variability of the estimate. However, with unmeasured confounding, including it in the analysis amplifies the bias as shown in Section 16.3.1.
4. X_Y affects the outcome only but not the treatment. Without conditioning on it, the unconfoundedness assumption still holds. Since they are predictive of the outcome, including them in analysis often improves precision.
5. X_I is affected by the treatment and outcome. It is a post-treatment variable, not a pretreatment covariate. We should not include it if the goal is to infer the effect of the treatment on the outcome. We will discuss issues with post-treatment variables in causal inference in Part VI of this book.

If we believe the above causal diagram, we should adjust for at least X to remove bias and more ideally, further adjust for X_Y to reduce variance.

16.4 Homework problems

16.1 More details for the formula of Z-bias

Verify (16.1).

16.2 Cochran's formula or the omitted-variable bias formula

Sir David Cox calls the following result *Cochran's formula* (Cochran, 1938; Cox, 2007) and econometricians call it the *omitted-variable bias formula* (Angrist and Pischke, 2008). A special case appeared in Fisher (1925). It is also a counterpart of the Frisch–Waugh–Lovell theorem in Section B.3.

The formula has two versions. All vectors below are column vectors.

1. (Population version) Assume $(y_i, x_{i1}, x_{i2})_{i=1}^n$ are IID, where y_i is a scalar, x_{i1} has dimension K, and x_{i2} has dimension L.

   We have the following OLS decompositions of random variables

$$y_i = \beta_1^{\mathsf{T}} x_{i1} + \beta_2^{\mathsf{T}} x_{i2} + \varepsilon_i, \quad (16.2)$$

$$y_i = \gamma^{\mathsf{T}} x_{i1} + e_i, \quad (16.3)$$

$$x_{i2} = \delta^{\mathsf{T}} x_{i1} + v_i. \quad (16.4)$$

   Equation (16.2) is called the *long regression*, and Equation (16.3) is called the *short regression*. In Equation (16.4), δ is a matrix because it is a regression of a vector on a vector. You can view (16.4) as regression of each component of x_{i2} on x_{i1}.

   Show that $\gamma = \beta_1 + \delta\beta_2$.

2. (Sample version) We have an $n \times 1$ vector Y, an $n \times K$ matrix X_1, and an $n \times L$ matrix X_2. We do not assume any randomness. All results below are purely linear algebra.

   We can obtain the following OLS fits:

$$Y = X_1\hat{\beta}_1 + X_2\hat{\beta}_2 + \hat{\varepsilon},$$

$$Y = X_1\hat{\gamma} + \hat{e},$$

$$X_2 = X_1\hat{\delta} + \hat{v},$$

   where $\hat{\varepsilon}, \hat{e}, \hat{v}$ are the residuals. Again, the last OLS fit means the OLS fit of each column of X_2 on X_1, and therefore the residual $\hat{v}$ is an $n \times L$ matrix.

   Show that $\hat{\gamma} = \hat{\beta}_1 + \hat{\delta}\hat{\beta}_2$.

Remark: The product terms $\delta\beta_2$ and $\hat{\delta}\hat{\beta}_2$ are often referred to as the omitted-variable bias at the population level and sample level, respectively, because they arise due to omitting x_{i2} from the regression of y_i on x_{i1}.

16.3 Recommended reading

Imbens (2020) reviews and compares the roles of potential outcomes and causal diagrams for causal inference.

17

E-Value: Evidence for Causation in Observational Studies with Unmeasured Confounding

All the methods discussed in Part III rely crucially on the ignorability assumption. They require controlling for all confounding between the treatment and outcome. However, we cannot use the data to validate the ignorability assumption. Observational studies are often criticized due to the possibility of unmeasured confounding. The famous Yule–Simpson Paradox reviewed in Chapter 1 demonstrates that an unmeasured binary confounder can completely overturn an observed association between the treatment and outcome. However, to overturn a larger observed association, this unmeasured confounder must have a stronger association with the treatment and the outcome. In other words, not all observational studies are created equal. Some provide stronger evidence for causation than others.

The following three chapters will discuss various sensitivity analysis techniques that can quantify the evidence of causation based on observational studies in the presence of unmeasured confounding. This chapter starts with the E-value, introduced by VanderWeele and Ding (2017) based on the theory in Ding and VanderWeele (2016a). It is more useful for observational studies using logistic regressions to estimate the conditional risk ratio of a treatment on a binary outcome. Chapter 18 discusses sensitivity analysis for the average causal effect based on outcome regression, IPW, and doubly robust estimation. Chapter 19 discusses Rosenbaum's framework for sensitivity analysis for matched observational studies.

17.1 Cornfield-type sensitivity analysis

Although we do not assume ignorability given X:

$$Z \not\perp\!\!\!\perp \{Y(1), Y(0)\} \mid X,$$

we still assume latent ignorability given X and an unmeasured confounder U:

$$Z \perp\!\!\!\perp \{Y(1), Y(0)\} \mid (X, U).$$

The technique in this chapter works the best for a binary outcome Y although it can be extended to other non-negative outcomes (Ding and VanderWeele, 2016a). Focus on binary Y now. The true conditional causal effect on the risk ratio scale is defined as

$$\text{RR}_{ZY|x}^{\text{true}} = \frac{\text{pr}\{Y(1) = 1 \mid X = x\}}{\text{pr}\{Y(0) = 1 \mid X = x\}},$$

and the observed conditional risk ratio equals

$$\text{RR}_{ZY|x}^{\text{obs}} = \frac{\text{pr}(Y = 1 \mid Z = 1, X = x)}{\text{pr}(Y = 1 \mid Z = 0, X = x)}.$$

In general, with an unmeasured confounder U, they are different:

$$\mathrm{RR}_{ZY|x}^{\mathrm{true}} \neq \mathrm{RR}_{ZY|x}^{\mathrm{obs}}$$

because

$$\mathrm{RR}_{ZY|x}^{\mathrm{true}} = \frac{\int \mathrm{pr}(Y=1 \mid Z=1, X=x, U=u) f(u \mid X=x)\mathrm{d}u}{\int \mathrm{pr}(Y=1 \mid Z=0, X=x, U=u) f(u \mid X=x)\mathrm{d}u}$$

and

$$\mathrm{RR}_{ZY|x}^{\mathrm{obs}} = \frac{\int \mathrm{pr}(Y=1 \mid Z=1, X=x, U=u) f(u \mid Z=1, X=x)\mathrm{d}u}{\int \mathrm{pr}(Y=1 \mid Z=0, X=x, U=u) f(u \mid Z=0, X=x)\mathrm{d}u}$$

are averaged over different distributions of U.

Historically, Doll and Hill (1950) found that the risk ratio of cigarette smoking on lung cancer was 9 even after adjusting for many observed covariates X.[1] Fisher (1957) criticized their result to be noncausal because it is possible that a hidden gene simultaneously causes cigarette smoking and lung cancer although the true causal effect of cigarette smoking on lung cancer is absent. This is the *common cause* hypothesis, also discussed by Reichenbach (1957). Cornfield et al. (1959) took a more constructive perspective and asked: how strong this unmeasured confounder must be to explain away the observed association between cigarette smoking and lung cancer? Below we will use the general formulation of the problem due to Ding and VanderWeele (2016a).

Consider the following causal diagram:

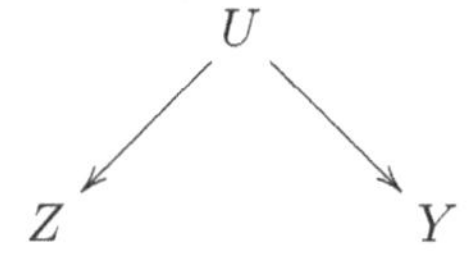

which conditions on X. So $Z \perp\!\!\!\perp Y \mid (X, U)$. Conditioning on X and U, we observe no association between Z and Y; but conditioning on only X, we observe an association between Z and Y. Although we can allow U to be general as Ding and VanderWeele (2016a), we assume that U is binary to simplify the presentation.

Define two sensitivity parameters:

$$\mathrm{RR}_{ZU|x} = \frac{\mathrm{pr}(U=1 \mid Z=1, X=x)}{\mathrm{pr}(U=1 \mid Z=0, X=x)}$$

measures the treatment-confounder association, and

$$\mathrm{RR}_{UY|x} = \frac{\mathrm{pr}(Y=1 \mid U=1, X=x)}{\mathrm{pr}(Y=1 \mid U=0, X=x)},$$

measures the confounder-outcome association, conditional on covariates $X=x$. We can show the main result below.

Theorem 17.1 *Under $Z \perp\!\!\!\perp Y \mid (X, U)$, assume*

$$\mathrm{RR}_{ZY|x}^{\mathrm{obs}} > 1, \quad \mathrm{RR}_{ZU|x} > 1, \quad \mathrm{RR}_{UY|x} > 1. \tag{17.1}$$

We have

$$\mathrm{RR}_{ZY|x}^{\mathrm{obs}} \leq \frac{\mathrm{RR}_{ZU|x}\mathrm{RR}_{UY|x}}{\mathrm{RR}_{ZU|x} + \mathrm{RR}_{UY|x} - 1}.$$

[1] Their original analysis was based on a case-control study and estimated the odds ratio of cigarette smoking on lung cancer. But the risk ratio is close to the odds ratio since lung cancer is a rare outcome; see Proposition 1.1.

In Theorem 17.1, we assume (17.1) without loss of generality. If $\text{RR}^{\text{obs}}_{ZY|x} < 1$, we can relabel the treatment and control levels to ensure $\text{RR}^{\text{obs}}_{ZY|x} > 1$. If $\text{RR}_{ZU|x} < 1$, we can redefine the unmeasured confounder U as $1-U$ to ensure $\text{RR}_{ZU|x} > 1$. If $\text{RR}^{\text{obs}}_{ZY|x} > 1$ and $\text{RR}_{ZU|x} > 1$, then $\text{RR}_{UY|x} > 1$ holds automatically. I relegate this subtle technical detail to Problem 17.2.

Theorem 17.1 shows the upper bound of the observed risk ratio of the treatment on the outcome if the conditional independence $Z \perp\!\!\!\perp Y \mid (X, U)$ holds. Under this conditional independence assumption, the association between the treatment and the outcome is purely due to the association between the treatment and the confounder $\text{RR}_{ZU|x}$, and the association between the confounder and the outcome, $\text{RR}_{UY|x}$. The upper bound equals $\text{RR}_{ZU|x} \times \text{RR}_{UY|x}/(\text{RR}_{ZU|x} + \text{RR}_{UY|x} - 1)$. A similar inequality appeared in Lee (2011). It is also related to Cochran's formula or the omitted-variable bias formula for linear models, which was reviewed in Problem 16.2.

Reversely, to generate a certain value of the observed risk ratio $\text{RR}^{\text{obs}}_{ZY|x}$, the two confounding measures $\text{RR}_{ZU|x}$ and $\text{RR}_{UY|x}$ cannot be arbitrary. Their function $\text{RR}_{ZU|x} \times \text{RR}_{UY|x}/(\text{RR}_{ZU|x} + \text{RR}_{UY|x} - 1)$ must be at least at large as $\text{RR}^{\text{obs}}_{ZY|x}$.

I will give the proof of Theorem 17.1 below.

Proof of Theorem 17.1: Define

$$f_{1,x} = \text{pr}(U = 1 \mid Z = 1, X = x), \quad f_{0,x} = \text{pr}(U = 1 \mid Z = 0, X = x).$$

We can decompose $\text{RR}^{\text{obs}}_{ZY|x}$ as

$$\begin{aligned}
&\text{RR}^{\text{obs}}_{ZY|x} \\
=\ & \frac{\text{pr}(Y = 1 \mid Z = 1, X = x)}{\text{pr}(Y = 1 \mid Z = 0, X = x)} \\
=\ & \frac{\left[\begin{array}{l}\text{pr}(U = 1 \mid Z = 1, X = x)\text{pr}(Y = 1 \mid Z = 1, U = 1, X = x) \\ \quad + \text{pr}(U = 0 \mid Z = 1, X = x)\text{pr}(Y = 1 \mid Z = 1, U = 0, X = x)\end{array}\right]}{\left[\begin{array}{l}\text{pr}(U = 1 \mid Z = 0, X = x)\text{pr}(Y = 1 \mid Z = 0, U = 1, X = x) \\ \quad + \text{pr}(U = 0 \mid Z = 0, X = x)\text{pr}(Y = 1 \mid Z = 0, U = 0, X = x)\end{array}\right]} \\
=\ & \frac{\left[\begin{array}{l}\text{pr}(U = 1 \mid Z = 1, X = x)\text{pr}(Y = 1 \mid U = 1, X = x) \\ \quad + \text{pr}(U = 0 \mid Z = 1, X = x)\text{pr}(Y = 1 \mid U = 0, X = x)\end{array}\right]}{\left[\begin{array}{l}\text{pr}(U = 1 \mid Z = 0, X = x)\text{pr}(Y = 1 \mid U = 1, X = x) \\ \quad + \text{pr}(U = 0 \mid Z = 0, X = x)\text{pr}(Y = 1 \mid U = 0, X = x)\end{array}\right]} \\
=\ & \frac{f_{1,x}\text{RR}_{UY|x} + 1 - f_{1,x}}{f_{0,x}\text{RR}_{UY|x} + 1 - f_{0,x}} \\
=\ & \frac{(\text{RR}_{UY|x} - 1)f_{1,x} + 1}{\frac{\text{RR}_{UY|x} - 1}{\text{RR}_{ZU|x}} f_{1,x} + 1}.
\end{aligned}$$

We can verify that $\text{RR}^{\text{obs}}_{ZY|x}$ is increasing in $f_{1,x}$ using the result in Problem 17.1. So letting $f_{1,x} = 1$, we have

$$\text{RR}^{\text{obs}}_{ZY|x} \le \frac{(\text{RR}_{UY|x} - 1) + 1}{\frac{\text{RR}_{UY|x} - 1}{\text{RR}_{ZU|x}} + 1} = \frac{\text{RR}_{ZU|x}\text{RR}_{UY|x}}{\text{RR}_{ZU|x} + \text{RR}_{UY|x} - 1}.$$

□

In the proof of Theorem 17.1, we have obtained an identity

$$\text{RR}^{\text{obs}}_{ZY|x} = \frac{(\text{RR}_{UY|x} - 1)f_{1,x} + 1}{\frac{\text{RR}_{UY|x} - 1}{\text{RR}_{ZU|x}} f_{1,x} + 1}. \tag{17.2}$$

But this identity involves three parameters

$$\{f_{1,x}, \text{RR}_{ZU|x}, \text{RR}_{UY|x}\};$$

see Problem 17.4 for a related formula. In contrast, the upper bound in Theorem 17.1 involves only two parameters

$$\{\text{RR}_{ZU|x}, \text{RR}_{UY|x}\}$$

which measure the strength of the confounder. Mathematically, the identity 17.2 is stronger than the inequality in Theorem 17.1. However, 17.2 involves more sensitivity parameters compared with Theorem 17.1. Therefore, we face a trade-off of accuracy and convenience in sensitivity analysis.

17.2 E-value

Lemma 17.1 below is useful for deriving interesting corollaries of Theorem 17.1. I relegate its proof to Problem 17.3.

Lemma 17.1 *Define* $\beta(w_1, w_2) = w_1 w_2/(w_1 + w_2 - 1)$ *for* $w_1 > 1$ *and* $w_2 > 1$.

1. $\beta(w_1, w_2)$ *is symmetric in* w_1 *and* w_2*;*
2. $\beta(w_1, w_2)$ *is increasing in both* w_1 *and* w_2*;*
3. $\beta(w_1, w_2) \leq w_1$ *and* $\beta(w_1, w_2) \leq w_2$*;*
4. $\beta(w_1, w_2) \leq w^2/(2w - 1)$, *where* $w = \max(w_1, w_2)$.

Using Theorem 17.1 and Lemma 17.1(3), we have

$$\text{RR}_{ZU|x} \geq \text{RR}^{\text{obs}}_{ZY|x}, \quad \text{RR}_{UY|x} \geq \text{RR}^{\text{obs}}_{ZY|x},$$

or, equivalently,

$$\min(\text{RR}_{ZU|x}, \text{RR}_{UY|x}) \geq \text{RR}^{\text{obs}}_{ZY|x}.$$

Therefore, to explain away the observed relative risk, both confounding measures $\text{RR}_{ZU|x}$ and $\text{RR}_{UY|x}$ must be at least as large as $\text{RR}^{\text{obs}}_{ZY|x}$. Cornfield et al. (1959) first derived the inequality $\text{RR}_{ZU|x} \geq \text{RR}^{\text{obs}}_{ZY|x}$, also called the *Cornfield inequality* (Gastwirth et al., 1998). Schlesselman (1978) derived the inequality $\text{RR}_{UY|x} \geq \text{RR}^{\text{obs}}_{ZY|x}$. These are related to to the *data processing inequality* in information theory.[2]

If we define $w = \max(\text{RR}_{ZU|x}, \text{RR}_{UY|x})$, then we can use Theorem 17.1 and Lemma 17.1(4) to obtain

$$w^2/(2w - 1) \geq \beta(\text{RR}_{ZU|x}, \text{RR}_{UY|x}) \geq \text{RR}^{\text{obs}}_{ZY|x},$$

[2]In information theory, the *mutual information*

$$I(A, B) = \iint f(a, b) \log_2 \frac{f(a, b)}{f(a)f(b)} \text{d}a\text{d}b$$

measures the dependence between two random variables A and B, where the f's denote the joint or marginal density of (A, B). The *data processing inequality* is a famous result: if $Z \perp\!\!\!\perp Y \mid U$, then $I(Z, Y) \leq I(Z, U)$ and $I(Z, Y) \leq I(U, Y)$. Lihua Lei and Bin Yu both pointed out to me the connection between Cornfield's inequality and the data processing inequality.

which implies a quadratic inequality

$$w^2 - 2\text{RR}^{\text{obs}}_{ZY|x} w + \text{RR}^{\text{obs}}_{ZY|x} \geq 0.$$

One root $\text{RR}^{\text{obs}}_{ZY|x} - \sqrt{\text{RR}^{\text{obs}}_{ZY|x}(\text{RR}^{\text{obs}}_{ZY|x} - 1)}$ is always smaller than or equal to 1, so we have

$$w = \max(\text{RR}_{ZU|x}, \text{RR}_{UY|x}) \geq \text{RR}^{\text{obs}}_{ZY|x} + \sqrt{\text{RR}^{\text{obs}}_{ZY|x}(\text{RR}^{\text{obs}}_{ZY|x} - 1)}.$$

Therefore, to explain away the observed relative risk, the maximum of the confounding measures $\text{RR}_{ZU|x}$ and $\text{RR}_{UY|x}$ must be at least as large as $\text{RR}^{\text{obs}}_{ZY|x} + \sqrt{\text{RR}^{\text{obs}}_{ZY|x}(\text{RR}^{\text{obs}}_{ZY|x} - 1)}$. Based on this result, VanderWeele and Ding (2017) introduced the following notion of E-value for measuring the *evidence* of causation with observational studies.

Definition 17.1 (E-Value) *With the observed conditional risk ratio* $\text{RR}^{\text{obs}}_{ZY|x}$, *define the E-Value as*

$$\text{RR}^{\text{obs}}_{ZY|x} + \sqrt{\text{RR}^{\text{obs}}_{ZY|x}(\text{RR}^{\text{obs}}_{ZY|x} - 1)}.$$

The E-value is defined for the parameter $\text{RR}^{\text{obs}}_{ZY|x}$. In practice, $\text{RR}^{\text{obs}}_{ZY|x}$ is estimated with sampling error. We can calculate the E-value based on the estimated $\text{RR}^{\text{obs}}_{ZY|x}$, as well as the corresponding E-values for the lower and upper confidence limits of $\text{RR}^{\text{obs}}_{ZY|x}$.

Fisher's p-value measures the evidence for causal effects in randomized experiments. We have discussed the p-value based on the FRT in Part II of this book. However, in observational studies with large sample sizes, p-values can be a poor measure of evidence for causal effects. Even if the true causal effects are 0, a tiny amount of unmeasured confounding can bias the estimate, which can result in extremely small p-values given the small sampling uncertainty. The sampling uncertainty is usually secondary in observational studies with large sample sizes, but the uncertainty due to unmeasured confounding is often the first-order problem that does not diminish with increased sample sizes. VanderWeele and Ding (2017) argued that the E-value is a better measure of the evidence for causal effects in observational studies.

17.3 A classic example

I revisit a classic example below.

Example 17.1 *Hammond and Horn (1958) used the U.S. population to study the cigarette smoking and lung cancer relationship. Ignoring covariates, their data can be represented by a two-by-two table below:*

	lung cancer	*no lung cancer*
smoker	*397*	*78557*
non-smoker	*51*	*108778*

Based on the data, they obtained an estimate of the risk ratio 10.73 *with a 95% confidence interval* [8.02, 14.36] *(see Section A.3.2 for the formulas). To explain away the point estimate, the E-value is*

$$10.73 + \sqrt{10.73 \times (10.73 - 1)} = 20.95;$$

to explain away the lower confidence limit, the E-value is

$$8.02 + \sqrt{8.02 \times (8.02 - 1)} = 15.52.$$

Figure 17.1 shows the joint values of the two confounding measures to explain away the point estimate and lower confidence limit of the risk ratio. In particular, to explain away the point estimate, they must lie in the area above the solid curve; to explain away the lower confidence limit, they must lie in the area above the dashed curve.[3]

The simple R code below computes the numbers in Example 17.1:

```
> ## e-value based on RR
> evalue = function(rr)
+ {
+   rr + sqrt(rr*(rr - 1))
+ }
>
> p1    = 397/(397+78557)
> p0    = 51/(51+108778)
> rr    = p1/p0
> logrr = log(p1/p0)
> se    = sqrt(1/397+1/51-1/(397+78557)-1/(51+108778))
> upper = exp(logrr+1.96*se)
> lower = exp(logrr-1.96*se)
>
> ## point estimate
> rr
[1] 10.72978
> ## lower CI
> lower
[1] 8.017414
> ## e-value based on rr
> evalue(rr)
[1] 20.94733
> ## e-value based on lower CI
> evalue(lower)
[1] 15.51818
```

17.4 Extensions

17.4.1 E-value and Bradford Hill's criteria for causation

The E-value provides evidence for causation. However, evidence is not a proof. With a larger E-value, we need a stronger unmeasured confounder to explain away the observed risk ratio; the evidence for causation is stronger. With a smaller E-value, we need a weaker unmeasured confounder to explain away the observed risk ratio; the evidence for causation is weaker. Coupled with the discussion in Section 17.5.1, a larger observed risk ratio has stronger evidence for causation. This is closely related to Sir Bradford Hill's first criterion for causation: *strength* of the association (Bradford Hill, 1965). Theorem 17.1 provides a mathematical quantification of his heuristic argument.

[3]Based on some simple algebra, the solid curve and dashed curve are both hyperbolas.

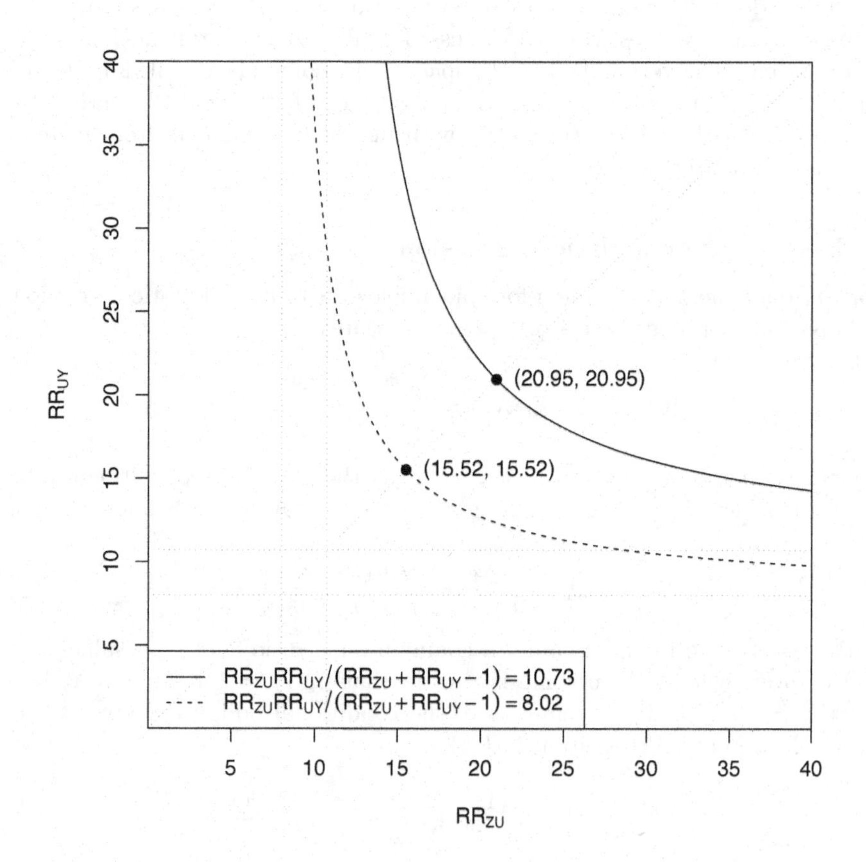

FIGURE 17.1: Magnitude of confounding to explain away the observed risk ratio based on the data from Hammond and Horn (1958)

In a famous paper, Bradford Hill (1965) proposed a set of nine criteria to provide evidence for causation between a presumed cause and outcome.

Definition 17.2 *Bradford Hill gave nine criteria for causation:*

1. *strength;*
2. *consistency;*
3. *specificity;*
4. *temporality;*
5. *biological gradient;*
6. *plausibility;*
7. *coherence;*
8. *experiment;*
9. *analogy.*

The E-value is a way to justify his first criterion. That is, stronger association often provides stronger evidence for causation because to explain away stronger association, we need stronger confounding measures. We have discussed randomized experiments in Part II, which corroborates his eighth criterion. Due to the space limit, I omit the detailed discussion of his other criteria and encourage the readers to read Bradford Hill (1965). Recently, this paper has been reprinted as Bradford Hill (2020) with insightful comments from many leading researchers in causal inference.

17.4.2 E-value after logistic regression

With a binary outcome, it is common for epidemiologists to use a logistic regression of the outcome Y_i on the treatment indicator Z_i and covariates X_i:

$$\mathrm{pr}(Y_i = 1 \mid Z_i, X_i) = \frac{e^{\beta_0+\beta_1 Z_i+\beta_2^\top X_i}}{1+e^{\beta_0+\beta_1 Z_i+\beta_2^\top X_i}}.$$

In the logistic model above, the coefficient of Z_i is the log of the conditional odds ratio between the treatment and the outcome given the covariates (see Section B.6 for more details):

$$\beta_1 = \log \frac{\mathrm{pr}(Y_i = 1 \mid Z_i = 1, X_i = x)/\mathrm{pr}(Y_i = 0 \mid Z_i = 1, X_i = x)}{\mathrm{pr}(Y_i = 1 \mid Z_i = 0, X_i = x)/\mathrm{pr}(Y_i = 0 \mid Z_i = 0, X_i = x)}.$$

Importantly, the logistic model assumes a common odds ratio across all values of the covariates. Moreover, when the outcome is rare in that $\mathrm{pr}(Y_i = 1 \mid Z_i = 1, X_i = x)$ and $\mathrm{pr}(Y_i = 1 \mid Z_i = 0, X_i = x)$ are close to 0, the conditional odds ratio approximates the conditional risk ratio (see Proposition 1.1(3)):

$$\begin{aligned}\beta_1 &\approx \log \frac{\mathrm{pr}(Y_i = 1 \mid Z_i = 1, X_i = x)}{\mathrm{pr}(Y_i = 1 \mid Z_i = 0, X_i = x)} \\ &= \log \mathrm{RR}^{\mathrm{obs}}_{ZY\mid x}.\end{aligned}$$

Therefore, based on the estimated logistic regression coefficient and the corresponding confidence limits, we can calculate the E-value immediately. This is the leading application of the E-value.

Example 17.2 *The `NCHS2003.txt` contains the National Center for Health Statistics birth certificate data, with the following binary indicator variables useful for us:*

`PTbirth`	*pre-term birth*
`preeclampsia`	*pre-eclampsia*
`ageabove35`	*an older mother with age* ≥ 35 *(the treatment)*
`somecollege`	*college education*
`mar`	*marital status*
`smoking`	*smoking status*
`drinking`	*drinking status*
`hispanic`	*mother's ethnicity*
`black`	*mother's ethnicity*
`nativeamerican`	*mother's ethnicity*
`asian`	*mother's ethnicity*

This version of the data is from Valeri and Vanderweele (2014). This example focuses on the outcome `PTbirth` and Problem 17.5 focuses on the outcome `pre-eclampsia`, a multisystem hypertensive disorder of pregnancy. The following `R` code computes the E-values after fitting a logistic regression. Based on the E-values, we conclude that to explain away the point *estimate, the maximum confounding measure must be larger than 1.94, and to explain away the lower confidence limit, the maximum confounding measure must be larger than 1.91. Although these confounding measures are not as strong as those in Section 17.3, they appear to be fairly large in epidemiologic studies.*

The simple R code below computes the numbers in Example 17.2:

```
> NCHS2003 = read.table("NCHS2003.txt", header = TRUE, sep = "\t")
> ## outcome: PTbirth
> y_logit = glm(PTbirth ~ ageabove35 +
+                 mar + smoking + drinking + somecollege +
+                 hispanic + black + nativeamerican + asian,
+               data = NCHS2003,
+               family = binomial)
> log_or   = summary(y_logit)$coef[2, 1:2]
> est      = exp(log_or[1])
> lower.ci = exp(log_or[1] - 1.96*log_or[2])
> est
Estimate
1.305982
> evalue(est)
Estimate
1.938127
> lower.ci
Estimate
1.294619
> evalue(lower.ci)
Estimate
1.912211
```

17.4.3 Non-zero true causal effect

Theorem 17.1 assumes no true causal effect of the treatment on the outcome. Ding and VanderWeele (2016a) proved a general theorem allowing for a non-zero true causal effect.

Theorem 17.2 *Modify the definition of* $\text{RR}_{UY|x}$ *as*

$$\text{RR}_{UY|x} = \max_{z=0,1} \frac{\text{pr}(Y=1 \mid Z=z, U=1, X=x)}{\text{pr}(Y=1 \mid Z=z, U=0, X=x)}.$$

Assume (17.1). *We have*

$$\text{RR}^{\text{true}}_{ZY|x} \geq \text{RR}^{\text{obs}}_{ZY|x} \Big/ \frac{\text{RR}_{ZU|x}\text{RR}_{UY|x}}{\text{RR}_{ZU|x} + \text{RR}_{UY|x} - 1}.$$

Theorem 17.1 is a special case of Theorem 17.2 with $\text{RR}^{\text{true}}_{ZY|x} = 1$. See the original paper of Ding and VanderWeele (2016a) for the proof of Theorem 17.2. Without assuming any additional assumptions, Theorem 17.2 states a lower bound of the true risk ratio $\text{RR}^{\text{true}}_{ZY|x}$ given the observed risk ratio $\text{RR}^{\text{obs}}_{ZY|x}$ and the two sensitivity parameters $\text{RR}_{ZU|x}$ and $\text{RR}_{UY|x}$.

When the treatment is apparently preventive to the outcome, the observed risk ratio is smaller than 1. In this case, Theorems 17.1 and 17.2 are not directly useful, and we must relabel the treatment levels and calculate the E-value based on $1/\text{RR}^{\text{obs}}_{ZY|x}$.

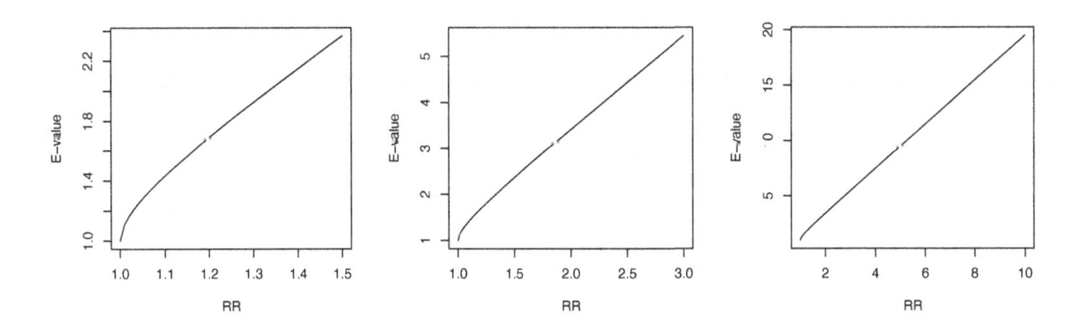

FIGURE 17.2: E-value as a monotone transformation of the risk ratio: three figures have different ranges of the risk ratio.

17.5 Critiques and responses

Since the original paper was published, the E-value has become a standard number reported in many epidemiologic studies. Nevertheless, it also attracted critiques (Ioannidis et al., 2019; Sjölander and Greenland, 2022). I will review some limitations of E-values below.

17.5.1 E-value is just a monotone transformation of the risk ratio

From Figure 17.2, we can see that if the risk ratio is large, then the E-value $\text{RR}^{\text{obs}}_{ZY|x} + \sqrt{\text{RR}^{\text{obs}}_{ZY|x}(\text{RR}^{\text{obs}}_{ZY|x} - 1)}$ is nearly $2\text{RR}^{\text{obs}}_{ZY|x}$ which is linear in the risk ratio. For a small risk ratio, the E-value is more nonlinear in $\text{RR}^{\text{obs}}_{ZY|x}$. Critics often say that the E-value is merely a monotone transformation of the point estimator or the confidence limits of the risk ratio. So it does not provide any additional information.

This is partially true. Indeed, the E-value is entirely based on the point estimator or the confidence limits of the risk ratio. However, it has a meaningful interpretation based on Theorem 17.1: to explain away the observed risk ratio, the maximum of the confounding measures must be at least as large as the E-value.

17.5.2 Calibration of the E-value

The E-value equals the maximum value of the association between the confounder and the treatment and that between the confounder and the outcome to completely explain away an observed association. An obvious problem is that this confounder is fundamentally latent. So it is not trivial to decide whether a certain E-value is large or small. Another related problem is that the E-value depends on how many observed covariates X we have controlled for since it quantifies the strength of the residual confounding given X. Therefore, E-values across studies are not directly comparable. The E-value provides evidence for causation but this evidence should be assessed carefully based on background knowledge of the problem of interest.

The following leave-one-covariate-out approach is an intuitive approach to calibrating the E-value. With $X = (X_1, \dots, X_p)$, we can pretend that the component X_j were not observed and compute the Z-X_j and X_j-Y risk ratios given other observed covariates ($j = 1, \dots, p$).

These risk ratios provide the range for the confounding measures due to U if we believe that the unmeasured U is not as strong as some of the observed covariates. However, I am not aware of any formal justification for this approach.

17.5.3 It works the best for a binary outcome and the risk ratio

Theorem 17.1 works well for a binary outcome and the risk ratio. Ding and VanderWeele (2016a) also proposed sensitivity analysis methods for other causal parameters, including the risk difference for binary outcomes, the mean ratio for non-negative outcomes, and the hazard ratio for survival outcomes. However, they are not as elegant as the E-value for binary outcomes based on the risk ratio. The next chapter will propose a simple sensitivity analysis method for the average causal effect that can include the outcome regression, IPW, and doubly robust estimators in Part III as special cases.

17.6 Homework problems

17.1 A technical lemma for the proof of Theorem 17.1

Show that

$$f(x) = \frac{ax+1}{bx+1}$$

is increasing in x if $a > b$ and decreasing in x is $a < b$.

17.2 Technical assumption for Theorem 17.1

Revisit the proof of Theorem 17.1. Assume $Z \perp\!\!\!\perp Y \mid (X, U)$. Show that if $\text{RR}_{ZU|x} > 1$ but $\text{RR}_{UY|x} < 1$, then $\text{RR}^{\text{obs}}_{ZY|x} < 1$.

Remark: This result is intuitive. Condition on X. It says that if the Z-U relationship is positive and the U-Y relationship is negative, then the Z-Y relationship is negative given the conditional independence of Z and Y given U. Based on this result, if we assume $\text{RR}^{\text{obs}}_{ZY|x} > 1$ and $\text{RR}_{ZU|x} > 1$, then $\text{RR}_{UY|x} > 1$ must be true. Therefore, the third condition in (17.1) is in fact redundant.

17.3 Lemma 17.1

Prove Lemma 17.1.

17.4 Schlesselman's formula

For simplicity, we condition on X implicitly in this problem. Consider a binary treatment Z, outcome Y, and unmeasured confounder U. Assume a common risk ratio of the treatment on the outcome within both $U = 0$ and $U = 1$:

$$\text{RR}_{ZY|U=0} = \text{RR}_{ZY|U=1},$$

and also a common risk ratio of the confounder on the outcome within both $Z = 0$ and $Z = 1$:

$$\text{RR}_{UY|Z=0} = \text{RR}_{UY|Z=1}, \text{ denoted by } \gamma.$$

Show that

$$\frac{\text{RR}^{\text{obs}}_{ZY}}{\text{RR}^{\text{true}}_{ZY}} = \frac{1 + (\gamma - 1)\text{pr}(U = 1 \mid Z = 1)}{1 + (\gamma - 1)\text{pr}(U = 1 \mid Z = 0)}.$$

Remark: This formula is due to Schlesselman (1978). To prove it, you should first verify that if $\text{RR}_{ZY|U=0} = \text{RR}_{ZY|U=1}$ then

$$\text{RR}_{ZY}^{\text{true}} = \text{RR}_{ZY|U=0} = \text{RR}_{ZY|U=1}.$$

This identity shows the *collapsibility* of the risk ratio. In epidemiology, the risk ratio is a *collapsible* measure of association.

Schlesselman's formula does not assume conditional independence $Z \perp\!\!\!\perp Y \mid U$, but assumes homogeneity of the Z-Y and U-Y risk ratios. It is a classic formula for sensitivity analysis. It is an identity that is simple to implement with pre-specified

$$\{\gamma, \text{pr}(U = 1 \mid Z = 1), \text{pr}(U = 1 \mid Z = 0)\}.$$

However, it involves more sensitivity parameters than Theorem 17.1. Even though Theorem 17.1 only gives an inequality, it is not a loose inequality compared to Schlesselman's formula under stronger assumptions. With Theorem 17.1, Schlesselman's formula is only of historical interest.

17.5 E-value after logistic regression: data analysis

This problem uses the same dataset as Example 17.2.

Report the E-value for the outcome `preeclampsia`.

17.6 Cornfield-type inequalities for the risk difference

Consider binary Z, Y, U, and condition on X implicitly. Assume latent ignorability given U. Show that under $Z \perp\!\!\!\perp Y \mid U$, we have

$$\text{RD}_{ZY}^{\text{obs}} = \text{RD}_{ZU} \times \text{RD}_{UY} \tag{17.3}$$

where

$$\text{RD}_{ZY}^{\text{obs}} = \text{pr}(Y = 1 \mid Z = 1) - \text{pr}(Y = 1 \mid Z = 0)$$

is the observed risk difference of Z on Y, and

$$\text{RD}_{ZU} = \text{pr}(U = 1 \mid Z = 1) - \text{pr}(U = 1 \mid Z = 0)$$

and

$$\text{RD}_{UY} = \text{pr}(Y = 1 \mid U = 1) - \text{pr}(Y = 1 \mid U = 0)$$

are the treatment-confounder and confounder-outcome risk differences, respectively (recall the definition of the risk difference in Section 1.2.2).

Remark: Without loss of generality, assume that $\text{RD}_{ZY}^{\text{obs}}, \text{RD}_{ZU}, \text{RD}_{UY}$ are all positive. Then (17.3) implies that

$$\min(\text{RD}_{ZU}, \text{RD}_{UY}) \geq \text{RD}_{ZY}^{\text{obs}}$$

and

$$\max(\text{RD}_{ZU}, \text{RD}_{UY}) \geq \sqrt{\text{RD}_{ZY}^{\text{obs}}}.$$

These are the Cornfield inequalities for the risk difference with a binary confounder. They show that for an unmeasured confounder to explain away an observed risk difference $\text{RD}_{ZY}^{\text{obs}}$, the treatment-confounder and confounder-outcome risk differences must both be larger than $\text{RD}_{ZY}^{\text{obs}}$, and the maximum of them must be larger than the square root of $\text{RD}_{ZY}^{\text{obs}}$.

Cornfield et al. (1959) obtained, but did not appreciate the significance of (17.3). Gastwirth et al. (1998) and Poole (2010) discussed the first Cornfield condition for the risk difference, and Ding and VanderWeele (2014) discussed the second.

Ding and VanderWeele (2014) also derived more general results without assuming a binary U. Unfortunately, the results for a general U are weaker than those above for a binary U, that is, the inequalities become looser with more levels of U. This motivated Ding and VanderWeele (2016a) to focus on the Cornfield inequalities for the risk ratio, which do not deteriorate with more levels of U.

17.7 Recommended reading

Ding and VanderWeele (2016a) extended and unified the Cornfield-type sensitivity analysis, which is the theoretical basis for the notion of E-value.

18

Sensitivity Analysis for the Average Causal Effect with Unmeasured Confounding

Cornfield-type sensitivity analysis works best for binary outcomes on the risk ratio scale, conditional on the observed covariates. Although Ding and VanderWeele (2016a) also proposed Cornfield-type sensitivity analysis methods for the average causal effect, they are not general enough and are not convenient to apply. Below I give a more direct approach to sensitivity analysis based on the conditional expectations of the potential outcomes. The advantage of this approach is that it can deal with commonly used estimators for the average causal effect under the sensitivity analysis framework. The idea appeared in the early work of Robins (1999) and Scharfstein et al. (1999). This chapter is based on the formulation of Lu and Ding (2023).

The approach is closely related to the idea of deriving worst-case bounds on the average potential outcomes. I will first review the simpler idea of bounds, and then extend the approach to sensitivity analysis.

18.1 Introduction

Recall the canonical setup of an observational study with $\{Z_i, X_i, Y_i(1), Y_i(0)\}_{i=1}^n \stackrel{\text{IID}}{\sim} \{Z, X, Y(1), Y(0)\}$ and focus on the average causal effect

$$\tau = E\{Y(1) - Y(0)\}.$$

It decomposes to

$$\begin{aligned}\tau &= [E(Y \mid Z=1)\text{pr}(Z=1) + E\{Y(1) \mid Z=0\}\text{pr}(Z=0)] \\ &\quad - [E\{Y(0) \mid Z=1\}\text{pr}(Z=1) + E(Y \mid Z=0)\text{pr}(Z=0)].\end{aligned}$$

So the fundamental difficulty is to estimate the counterfactual means

$$E\{Y(1) \mid Z=0\}$$

and

$$E\{Y(0) \mid Z=1\}.$$

There are in general two extreme strategies to estimate them.

We have discussed the first strategy in Part III, which relies on ignorability. Assuming

$$\begin{aligned}E\{Y(1) \mid Z=1, X\} &= E\{Y(1) \mid Z=0, X\}, \\ E\{Y(0) \mid Z=1, X\} &= E\{Y(0) \mid Z=0, X\},\end{aligned}$$

we can identify the counterfactual means by the observables:

$$E\{Y(1) \mid Z=0\} = E\{E(Y \mid Z=1, X) \mid Z=0\}$$

TABLE 18.1: Science Table with bounded outcome $[\underline{y}, \overline{y}]$, where $\underline{y}$ and $\overline{y}$ are two constants

Z	$Y(1)$	$Y(0)$	Lower $Y(1)$	Upper $Y(1)$	Lower $Y(0)$	Upper $Y(0)$
1	$Y_1(1)$	?	$Y_1(1)$	$Y_1(1)$	$\underline{y}$	$\overline{y}$
$\vdots$	$\vdots$	$\vdots$	$\vdots$	$\vdots$	$\vdots$	$\vdots$
1	$Y_{n_1}(1)$	?	$Y_{n_1}(1)$	$Y_{n_1}(1)$	$\underline{y}$	$\overline{y}$
0	?	$Y_{n_1+1}(0)$	$\underline{y}$	$\overline{y}$	$Y_{n_1+1}(0)$	$Y_{n_1+1}(0)$
$\vdots$	$\vdots$	$\vdots$	$\vdots$	$\vdots$	$\vdots$	$\vdots$
0	?	$Y_n(0)$	$\underline{y}$	$\overline{y}$	$Y_n(0)$	$Y_n(0)$

and, similarly,

$$E\{Y(0) \mid Z=1\} = E\{E(Y \mid Z=0, X) \mid Z=1\}.$$

The second strategy in the next section assumes nothing except that the outcomes are bounded between $\underline{y}$ and $\overline{y}$. This is natural for binary outcomes with $\underline{y}=0$ and $\overline{y}=1$. With this assumption, the two counterfactual means are also bounded between $\underline{y}$ and $\overline{y}$, which implies the worst-case bounds on τ. Table 18.1 illustrates the basic idea and Section 18.2 below reviews this strategy in more detail.

18.2 Manski-type worst-case bounds on the average causal effect without assumptions

Assume that the outcome is bounded between $\underline{y}$ and $\overline{y}$. From the decomposition

$$E\{Y(1)\} = E\{Y(1) \mid Z=1\}\text{pr}(Z=1) + E\{Y(1) \mid Z=0\}\text{pr}(Z=0),$$

we can derive that $E\{Y(1)\}$ has lower bound

$$E\{Y \mid Z=1\}\text{pr}(Z=1) + \underline{y}\text{pr}(Z=0)$$

and upper bound

$$E\{Y \mid Z=1\}\text{pr}(Z=1) + \overline{y}\text{pr}(Z=0).$$

Similarly, from the decomposition

$$E\{Y(0)\} = E\{Y(0) \mid Z=1\}\text{pr}(Z=1) + E\{Y(0) \mid Z=0\}\text{pr}(Z=0),$$

we can derive that $E\{Y(0)\}$ has lower bound

$$\underline{y}\text{pr}(Z=1) + E\{Y \mid Z=0\}\text{pr}(Z=0)$$

and upper bound

$$\overline{y}\text{pr}(Z=1) + E\{Y \mid Z=0\}\text{pr}(Z=0).$$

Combining these bounds, we can derive that the average causal effect $\tau = E\{Y(1)\} - E\{Y(0)\}$ has the lower bound

$$E\{Y \mid Z=1\}\text{pr}(Z=1) + \underline{y}\text{pr}(Z=0) - \overline{y}\text{pr}(Z=1) - E\{Y \mid Z=0\}\text{pr}(Z=0)$$

and the upper bound

$$E\{Y \mid Z=1\}\text{pr}(Z=1) + \overline{y}\text{pr}(Z=0) - \underline{y}\text{pr}(Z=1) - E\{Y \mid Z=0\}\text{pr}(Z=0).$$

The length of the bounds is $\overline{y} - \underline{y}$. The bounds are not informative but are better than the a priori bounds $[\underline{y} - \overline{y}, \overline{y} - \underline{y}]$ with length $2(\overline{y} - \underline{y})$. Without further assumptions, the observed data distribution does not uniquely determine τ. In this case, we say that τ is *partially identified*, with the formal definition below.

Definition 18.1 (partial identification) *A parameter θ is partially identified if the observed data distribution is compatible with multiple values of θ.*

Compare Definitions 10.1 and 18.1. If the parameter θ is uniquely determined by the observed data distribution, then it is identifiable; otherwise, it is only partially identifiable. Therefore, τ is identifiable with the ignorability assumption, but only partially identifiable without the ignorability assumption.

Cochran (1953) used the idea of worst-case bounds in surveys with missing data but abandoned the idea because it often gives very conservative results. Similarly, the above worst-case bounds on τ are often uninteresting from a practical perspective because they often cover 0. Moreover, this strategy does not apply to the settings with unbounded outcomes.

Manski applied the idea to causal inference (Manski, 1990) and many other econometric models (Manski, 2003). This idea of bounding causal parameters with minimal assumptions is powerful when coupled with other qualitative assumptions. Manski (2003) surveyed many strategies. For instance, we may believe that the treatment does not harm any units, so the monotonicity assumption holds: $Y(1) \geq Y(0)$. Then the lower bound on τ is zero but the upper bound is unchanged. Another type of assumption is $Z = I\{Y(1) \geq Y(0)\}$, that is, the treatment selection is based on the difference between the latent potential outcomes. This assumption can also improve the bounds on τ. A more detailed discussion of this approach is beyond the scope of this book.

18.3 Sensitivity analysis for the average causal effect

The first strategy is optimistic and assumes that the potential outcomes do not differ across treatment and control groups, conditional on the observed covariates. The second strategy is pessimistic and does not infer the counterfactual means based on the observed data at all. The following strategy is in-between.

18.3.1 Identification formulas

Define

$$\frac{E\{Y(1) \mid Z=1, X\}}{E\{Y(1) \mid Z=0, X\}} = \varepsilon_1(X),$$
$$\frac{E\{Y(0) \mid Z=1, X\}}{E\{Y(0) \mid Z=0, X\}} = \varepsilon_0(X),$$

which are the sensitivity parameters. For simplicity, we can further assume that they are constant independent of X. In practice, we need to fix them or vary them in a pre-specified range.

Recall that $\mu_1(X) = E(Y \mid Z = 1, X)$ and $\mu_0(X) = E(Y \mid Z = 0, X)$ are the conditional mean functions of the observed outcomes under treatment and control, respectively. We can identify the two counterfactual means and the average causal effect as follows.

Theorem 18.1 *With known $\varepsilon_1(X)$ and $\varepsilon_0(X)$, we have*

$$\begin{aligned} E\{Y(1) \mid Z = 0\} &= E\left\{\mu_1(X)/\varepsilon_1(X) \mid Z = 0\right\}, \\ E\{Y(0) \mid Z = 1\} &= E\left\{\mu_0(X)\varepsilon_0(X) \mid Z = 1\right\} \end{aligned}$$

and therefore

$$\begin{aligned} \tau &= E\{ZY + (1-Z)\mu_1(X)/\varepsilon_1(X)\} \\ &\quad -E\{Z\mu_0(X)\varepsilon_0(X) + (1-Z)Y\} \qquad (18.1) \\ &= E\{Z\mu_1(X) + (1-Z)\mu_1(X)/\varepsilon_1(X)\} \\ &\quad -E\{Z\mu_0(X)\varepsilon_0(X) + (1-Z)\mu_0(X)\}. \qquad (18.2) \end{aligned}$$

I leave the proof of Theorem 18.1 to Problem 18.1. With the fitted outcome model, (18.1) and (18.2) motivate the following predictive and projective estimators for τ:

$$\begin{aligned} \hat{\tau}^{\text{pred}} &= \left\{ n^{-1}\sum_{i=1}^{n} Z_i Y_i + n^{-1}\sum_{i=1}^{n}(1-Z_i)\hat{\mu}_1(X_i)/\varepsilon_1(X_i) \right\} \\ &\quad - \left\{ n^{-1}\sum_{i=1}^{n} Z_i \hat{\mu}_0(X_i)\varepsilon_0(X_i) + n^{-1}\sum_{i=1}^{n}(1-Z_i)Y_i \right\}, \end{aligned}$$

and

$$\begin{aligned} \hat{\tau}^{\text{proj}} &= \left\{ n^{-1}\sum_{i=1}^{n} Z_i \hat{\mu}_1(X_i) + n^{-1}\sum_{i=1}^{n}(1-Z_i)\hat{\mu}_1(X_i)/\varepsilon_1(X_i) \right\} \\ &\quad - \left\{ n^{-1}\sum_{i=1}^{n} Z_i \hat{\mu}_0(X_i)\varepsilon_0(X_i) + n^{-1}\sum_{i=1}^{n}(1-Z_i)\hat{\mu}_0(X_i) \right\}. \end{aligned}$$

The terminology "predictive" and "projective" is from the survey sampling literature (Firth and Bennett, 1998; Ding and Li, 2018); see also Section 6.2.2.2. The estimators $\hat{\tau}^{\text{pred}}$ and $\hat{\tau}^{\text{proj}}$ differ slightly: the former uses the observed outcomes when available, whereas the latter replaces the observed outcomes with the fitted values.

More interestingly, we can also identify τ by an inverse probability weighting formula. Recall that $e(X) = \text{pr}(Z = 1 \mid X)$ is the propensity score.

Theorem 18.2 *With known $\varepsilon_1(X)$ and $\varepsilon_0(X)$, we have*

$$\begin{aligned} E\{Y(1)\} &= E\left\{ w_1(X)\frac{Z}{e(X)}Y \right\}, \\ E\{Y(0)\} &= E\left\{ w_0(X)\frac{1-Z}{1-e(X)}Y \right\}, \end{aligned}$$

where

$$\begin{aligned} w_1(X) &= e(X) + \{1-e(X)\}/\varepsilon_1(X), \\ w_0(X) &= e(X)\varepsilon_0(X) + 1 - e(X). \end{aligned}$$

I leave the proof of Theorem 18.2 to Problem 18.2. Theorem 18.2 modifies the classic IPW formulas with two extra factors $w_1(X)$ and $w_0(X)$, which depend on both the propensity score and the sensitivity parameters. With the fitted propensity scores, Theorem 18.2 motivates the following estimators for τ:

$$\begin{aligned}\hat{\tau}^{\text{ht}} &= n^{-1}\sum_{i=1}^{n}\frac{\{\hat{e}(X_i)\varepsilon_1(X_i)+1-\hat{e}(X_i)\}Z_iY_i}{\varepsilon_1(X_i)\hat{e}(X_i)} \\ &\quad -n^{-1}\sum_{i=1}^{n}\frac{\{\hat{e}(X_i)\varepsilon_0(X_i)+1-\hat{e}(X_i)\}(1-Z_i)Y_i}{1-\hat{e}(X_i)}\end{aligned}$$

and

$$\begin{aligned}\hat{\tau}^{\text{haj}} &= \sum_{i=1}^{n}\frac{\{\hat{e}(X_i)\varepsilon_1(X_i)+1-\hat{e}(X_i)\}Z_iY_i}{\varepsilon_1(X_i)\hat{e}(X_i)}\Big/\sum_{i=1}^{n}\frac{Z_i}{\hat{e}(X_i)} \\ &\quad -\sum_{i=1}^{n}\frac{\{\hat{e}(X_i)\varepsilon_0(X_i)+1-\hat{e}(X_i)\}(1-Z_i)Y_i}{1-\hat{e}(X_i)}\Big/\sum_{i=1}^{n}\frac{1-Z_i}{1-\hat{e}(X_i)}.\end{aligned}$$

More interestingly, with fitted propensity score and outcome models, the following estimator for τ is doubly robust:

$$\hat{\tau}^{\text{dr}} = \hat{\tau}^{\text{ht}} - n^{-1}\sum_{i=1}^{n}\{Z_i-\hat{e}(X_i)\}\left\{\frac{\hat{\mu}_1(X_i)}{\hat{e}(X_i)\varepsilon_1(X_i)}+\frac{\hat{\mu}_0(X_i)\varepsilon_0(X_i)}{1-\hat{e}(X_i)}\right\}.$$

That is, with known $\varepsilon_1(X_i)$ and $\varepsilon_0(X_i)$, the estimator $\hat{\tau}^{\text{dr}}$ is consistent for τ if either the propensity score model or the outcome model is correctly specified. We can use the bootstrap to approximate the variance of the above estimators. See Lu and Ding (2023) for the proof of the double robustness property.

When $\varepsilon_1(X_i) = \varepsilon_0(X_i) = 1$, the above estimators reduce to the predictive estimator, IPW estimator, and the doubly robust estimators introduced in Part III.

18.3.2 Example

18.3.2.1 R functions for sensitivity analysis

The following R function can compute the point estimates for sensitivity analysis.

```
OS_est_sa = function(z, y, x, out.family = gaussian,
                     truncps = c(0, 1), e1 = 1, e0 = 1)
{
     ## fitted propensity score
     pscore    = glm(z ~ x, family = binomial)$fitted.values
     pscore    = pmax(truncps[1], pmin(truncps[2], pscore))

     ## fitted potential outcomes
     outcome1  = glm(y ~ x, weights = z,
                     family = out.family)$fitted.values
     outcome0  = glm(y ~ x, weights = (1 - z),
                     family = out.family)$fitted.values

     ## outcome regression estimator
     ace.reg   = mean(z*y) + mean((1-z)*outcome1/e1) -
                 mean(z*outcome0*e0) - mean((1-z)*y)
```

```
    ## IPW estimators
    w1 = pscore + (1-pscore)/e1
    w0 = pscore*e0 + (1-pscore)
    y.treat     = mean(z*y*w1/pscore)
    y.control   = mean((1 - z)*y*w0/(1 - pscore))
    one.treat   = mean(z/pscore)
    one.control = mean((1 - z)/(1 - pscore))
    ace.ipw0    = y.treat - y.control
    ace.ipw     = y.treat/one.treat - y.control/one.control

    ## doubly robust estimator
    aug = outcome1/pscore/e1 + outcome0*e0/(1-pscore)
    ace.dr   = ace.ipw0 - mean((z-pscore)*aug)

    return(c(ace.reg, ace.ipw0, ace.ipw, ace.dr))
}
```

I relegate the calculation of the standard errors to Problem 18.3.

18.3.2.2 Revisit Example 10.3

With

$$\varepsilon_1(X) = \varepsilon_0(X) \in \{1/2, 1/1.7, 1/1.5, 1/1.3, 1, 1.3, 1.5, 1.7, 2\},$$

we obtain an array of doubly robust estimates of τ based on the following R code:

```
nhanes_bmi = read.csv("nhanes_bmi.csv")[, -1]
z = nhanes_bmi$School_meal
y = nhanes_bmi$BMI
x = as.matrix(nhanes_bmi[, -c(1, 2)])
x = scale(x)


E1 = c(1/2, 1/1.7, 1/1.5, 1/1.3, 1, 1.3, 1.5, 1.7, 2)
E0 = c(1/2, 1/1.7, 1/1.5, 1/1.3, 1, 1.3, 1.5, 1.7, 2)
EST = outer(E1, E0)
ll1 = length(E1)
ll0 = length(E0)
for(i in 1:ll1)
  for(j in 1:ll0)
    EST[i, j] = OS_est_sa(z, y, x, e1 = E1[i], e0 = E0[j])[4]
```

TABLE 18.2: Sensitivity analysis for the average causal effect

	1/2	1/1.7	1/1.5	1/1.3	1	1.3	1.5	1.7	2
1/2	14.62	13.62	12.73	11.57	8.96	5.57	3.30	1.04	-2.36
1/1.7	11.93	10.93	10.04	8.88	6.27	2.87	0.61	-1.66	-5.05
1/1.5	10.13	9.13	8.24	7.08	4.47	1.08	-1.19	-3.45	-6.85
1/1.3	8.33	7.33	6.45	5.29	2.67	-0.72	-2.98	-5.25	-8.64
1	5.64	4.64	3.75	2.59	-0.02	-3.41	-5.68	-7.94	-11.34
1.3	3.57	2.57	1.68	0.52	-2.09	-5.49	-7.75	-10.01	-13.41
1.5	2.65	1.65	0.76	-0.40	-3.01	-6.41	-8.67	-10.93	-14.33
1.7	1.94	0.94	0.06	-1.11	-3.72	-7.11	-9.38	-11.64	-15.03
2	1.15	0.15	-0.74	-1.90	-4.51	-7.90	-10.17	-12.43	-15.83

Table 18.2 presents the point estimates. The signs of the estimates are not sensitive to sensitivity parameters larger than 1, but they are quite sensitive to sensitivity parameters smaller than 1. When the participants of the meal plan tend to have higher BMI (that is, $\varepsilon_1(X) > 1$ and $\varepsilon_0(X) > 1$), the average causal effect of the meal plan on BMI is negative. However, this conclusion can be quite sensitive if the participants of the meal plan tend to have lower BMI.

18.3.3 Practical implementation of the sensitivity analysis method

To implement the sensitivity analysis estimators proposed in Section 18.3, we need to specify the sensitivity parameters $\varepsilon_1(X)$ and $\varepsilon_0(X)$. Below I will comment on two issues. First, how do we specify the range of them? Second, do we allow them to vary with X?

18.3.3.1 Calibrating the range of the sensitivity parameters

It is fundamentally challenging to specify the ranges of the sensitivity parameters $\varepsilon_1(X)$ and $\varepsilon_0(X)$ because the observed data do not contain any information about them. An initial motivation of the parametrization of $\varepsilon_1(X)$ and $\varepsilon_0(X)$ is to characterize the violation of ignorability without imposing any restrictions on the observed data. They compare the conditional means of the observed outcomes with those of the counterfactual outcomes given the observed covariates.

Lu and Ding (2023) recommend a leave-one-covariate-out approach to calibrate the sensitivity parameters; recall Section 17.5.2. For each observed covariate X_j in the p-dimensional $X = (X_1, \ldots, X_p)$, $j = 1, \ldots, p$, we first drop it as if it is an unobserved confounder, then estimate the value of sensitivity parameters with X_j unobserved. Under ignorability $Z \perp\!\!\!\perp Y(z) \mid X$, we have

$$\begin{aligned} \varepsilon_z(X_{-j}) &= \frac{E\{Y(z) \mid Z=1, X_{-j}\}}{E\{Y(z) \mid Z=0, X_{-j}\}} \\ &= \frac{E\{\mu_z(X) \mid Z=1, X_{-j}\}}{E\{\mu_z(X) \mid Z=0, X_{-j}\}}, \quad (z=0,1) \end{aligned}$$

where X_{-j} denotes the $p-1$ dimensional observed covariates after dropping covariate X_j. The $\varepsilon_z(X_{-j})$ measures how much the observed data deviates from ignorability when we delete an observed confounder X_j, and can be estimated using observed data. Thus, we can summarize the distribution of $\varepsilon_z(X_{-j})$ to characterize the strength of each covariate X_j as a confounder. To summarize $\varepsilon_z(X_{-j})$ from a function of X_{-j} to a scalar, we can further marginalize over the distribution of X_{-j} to compute the mean of sensitivity parameter ignoring X_j, or use other summary statistics such as maximum, minimum, upper, or lower quantiles of the estimated distribution to guide us on the interpretation of the magnitude of the sensitivity parameter $\varepsilon_z(X)$.

18.3.3.2 Constant or covariate-specific sensitivity parameters?

Our theory allows $\varepsilon_1(X)$ and $\varepsilon_0(X)$ to vary with X. However, for the simplicity of implementation, we often assume that they are constant $\varepsilon_1(X) = \varepsilon_1$ and $\varepsilon_0(X) = \varepsilon_0$ to generate tables or figures of the estimates with respect to $(\varepsilon_1, \varepsilon_0)$. Advanced researchers may want to assume that $\varepsilon_z(X) = \exp(\alpha_z + \beta_z^{\mathrm{T}} X)$ and treat (α_z, β_z) as the sensitivity parameters $(z = 0, 1)$. I omit the discussion of this strategy. Although this strategy is more flexible than assuming constant $(\varepsilon_1, \varepsilon_0)$, it challenges the presentation of the sensitivity analysis results due to multi-dimensional (α_z, β_z).

Alternatively, when the outcomes are non-negative (e.g., binary), we can assume that $(\varepsilon_1(X), \varepsilon_0(X))$ are constants independent of X but interpret the estimates as the worst-case results under the assumptions that the maximum or minimum values of $(\varepsilon_1(X), \varepsilon_0(X))$ are $(\varepsilon_1, \varepsilon_0)$, respectively. Lu and Ding (2023) provide a justification of this strategy; see Problem 18.4.

18.4 Homework problems

18.1 Proof of Theorem 18.1

Prove Theorem 18.1.

18.2 Proof of Theorem 18.2

Prove Theorem 18.2.

18.3 Standard errors in sensitivity analysis

Section 18.3.2 only presents the point estimates. Report the corresponding bootstrap standard errors.

18.4 Bounds on τ

Prove Theorem 18.3 below.

Theorem 18.3 *Let $\varepsilon_{z,\mathrm{L}}$ and $\varepsilon_{z,\mathrm{U}}$ denote the lower and upper bound of $\varepsilon_z(X)$, i.e., $\varepsilon_z(X) \in [\varepsilon_{z,\mathrm{L}}, \varepsilon_{z,\mathrm{U}}]$ for $z = 0, 1$. In the special case when the potential outcomes are non-negative (e.g., binary), we have $\tau \in [\tau_{\mathrm{L}}, \tau_{\mathrm{U}}]$, where*

$$
\begin{aligned}
\tau_{\mathrm{L}} &= E\left\{Z\mu_1(X) + \frac{(1-Z)\mu_1(X)}{\varepsilon_{1,\mathrm{U}}}\right\} - E\left\{Z\mu_0(X)\varepsilon_{0,\mathrm{U}} + (1-Z)\mu_0(X)\right\} \\
&= E\left[\left\{e(X) + \frac{1-e(X)}{\varepsilon_{1,\mathrm{U}}}\right\}\frac{ZY}{e(X)}\right] - E\left[\left\{e(X)\varepsilon_{0,\mathrm{U}} + 1 - e(X)\right\}\frac{(1-Z)Y}{1-e(X)}\right] \\
&= E\left[\left\{e(X) + \frac{1-e(X)}{\varepsilon_{1,\mathrm{U}}}\right\}\frac{ZY}{e(X)} - \frac{\{Z-e(X)\}\mu_1(X)}{e(X)\varepsilon_{1,\mathrm{U}}}\right] \\
&\quad -E\left[\left\{e(X)\varepsilon_{0,\mathrm{U}} + 1 - e(X)\right\}\frac{(1-Z)Y}{1-e(X)} - \frac{\{e(X)-Z\}\mu_0(X)\varepsilon_{0,\mathrm{U}}}{1-e(X)}\right],
\end{aligned}
$$

and τ_{U} is computed using the same formulas as τ_{L}, with the replacement of $\varepsilon_{1,\mathrm{U}}$ and $\varepsilon_{0,\mathrm{U}}$ by $\varepsilon_{1,\mathrm{L}}$ and $\varepsilon_{0,\mathrm{L}}$, respectively.

Remark: Theorem 18.3 gives bounds based on the three identification formulas which allow us to construct the corresponding estimators.

18.5 Sensitivity analysis for the average causal effect on the treated units τ_{T}

This problem extends Chapter 13 to allow for unmeasured confounding for estimating

$$
\begin{aligned}
\tau_{\mathrm{T}} &= E\{Y(1) - Y(0) \mid Z = 1\} \\
&= E(Y \mid Z = 1) - E\{Y(0) \mid Z = 1\}.
\end{aligned}
$$

We can easily estimate $E(Y \mid Z=1)$ by the sample moment, $\hat{\mu}_{\mathrm{T}1} = \sum_{i=1}^n Z_i Y_i / \sum_{i=1}^n Z_i$. The only counterfactual term is $E\{Y(0) \mid Z=1\}$. Therefore, we only need the sensitivity parameter $\varepsilon_0(X)$. We have the following two identification formulas with a known $\varepsilon_0(X)$.

Theorem 18.4 *With known $\varepsilon_0(X)$, we have*

$$\begin{aligned} E\{Y(0) \mid Z=1\} &= E\{Z\mu_0(X)\varepsilon_0(X)\}/e \\ &= E\left\{e(X)\varepsilon_0(X)\frac{1-Z}{1-e(X)}Y\right\}/e, \end{aligned}$$

where $e = \mathrm{pr}(Z=1)$.

Prove Theorem 18.4.

Remark: Theorem 18.4 motivates using $\hat{\tau}_{\mathrm{T}}^* = \hat{\mu}_{\mathrm{T}1} - \hat{\mu}_{\mathrm{T}0}^*$ to estimate τ_{T}, where

$$\begin{aligned} \hat{\mu}_{\mathrm{T}0}^{\mathrm{reg}} &= n_1^{-1}\sum_{i=1}^n Z_i\varepsilon_0(X_i)\hat{\mu}_0(X_i), \\ \hat{\mu}_{\mathrm{T}0}^{\mathrm{ht}} &= n_1^{-1}\sum_{i=1}^n \varepsilon_0(X_i)\hat{o}(X_i)(1-Z_i)Y_i, \\ \hat{\mu}_{\mathrm{T}0}^{\mathrm{haj}} &= \sum_{i=1}^n \varepsilon_0(X_i)\hat{o}(X_i)(1-Z_i)Y_i \Big/ \sum_{i=1}^n \hat{o}(X_i)(1-Z_i), \end{aligned}$$

with $\hat{o}(X_i) = \hat{e}(X_i)/\{1-\hat{e}(X_i)\}$ being the estimated conditional odds of the treatment given the observed covariates. Moreover, we can construct the doubly robust estimator $\hat{\tau}_{\mathrm{T}}^{\mathrm{dr}} = \hat{\mu}_{\mathrm{T}1} - \hat{\mu}_{\mathrm{T}0}^{\mathrm{dr}}$ for τ_{T}, where

$$\hat{\mu}_{\mathrm{T}0}^{\mathrm{dr}} = \hat{\mu}_{\mathrm{T}0}^{\mathrm{ht}} - n_1^{-1}\sum_{i=1}^n \varepsilon_0(X_i)\frac{\hat{e}(X_i)-Z_i}{1-\hat{e}(X_i)}\hat{\mu}_0(X_i).$$

Lu and Ding (2023) provide more details.

18.6 R code for τ_{T}

Implement the estimators in Problem 18.5. Analyze the data used in Section 18.3.2.

18.7 Bounds on τ_{T}

Prove Theorem 18.5 below.

Theorem 18.5 *Let $\varepsilon_{0,\mathrm{L}}$ and $\varepsilon_{0,\mathrm{U}}$ denote the lower and upper bound of $\varepsilon_0(X)$, i.e., $\varepsilon_0(X) \in [\varepsilon_{0,\mathrm{L}}, \varepsilon_{0,\mathrm{U}}]$. In the special case when the potential outcomes are non-negative (e.g., binary), we have $\tau_{\mathrm{T}} \in [\tau_{\mathrm{T,L}}, \tau_{\mathrm{T,U}}]$, where*

$$\begin{aligned} \tau_{\mathrm{T,L}} &= E(Y \mid Z=1) - \varepsilon_{0,\mathrm{U}} E\{Z\mu_0(X)\}/e \\ &= E(Y \mid Z=1) - \varepsilon_{0,\mathrm{U}} E\left[\frac{e(X)(1-Z)Y}{e\{1-e(X)\}}\right] \\ &= E(Y \mid Z=1) - \varepsilon_{0,\mathrm{U}} E\left[Z\mu_0(X) + e(X)\frac{(1-Z)\{Y-\mu_0(X)\}}{1-e(X)}\right]/e, \end{aligned}$$

and $\tau_{\mathrm{T,U}}$ is computed using the same formula as $\tau_{\mathrm{T,L}}$, with the replacement of $\varepsilon_{0,\mathrm{U}}$ by $\varepsilon_{0,\mathrm{L}}$.

Remark: Theorem 18.5 gives bounds based on the three identification formulas which allow us to construct the corresponding estimators.

18.8 Recommended reading

Rosenbaum and Rubin (1983a) and Imbens (2003) are two classic papers on sensitivity analysis which, however, involve more complicated procedures.

19

Rosenbaum-Style p-Values for Matched Observational Studies with Unmeasured Confounding

Rosenbaum (1987b) introduced a sensitivity analysis technique for matched observational studies. Although it works for general matched studies (Rosenbaum, 2002b), the theory is most elegant for one-to-one matching. Different from Chapters 17 and 18, Rosenbaum-type sensitivity analysis works best for matched observational studies for testing the sharp null hypothesis of no individual treatment effect.

19.1 A model for sensitivity analysis with matched data

Consider a matched observational study, with (i, j) indexing unit j in pair i $(i = 1, \dots, n; j = 1, 2)$. With exactly matched pairs, unit $(i, 1)$ and unit $(i, 2)$ have the same covariates X_i. Assume IID sampling, and extend Definition 11.1 to define the propensity score as

$$e_{ij} = \text{pr}\{Z_{ij} = 1 \mid X_i, Y_{ij}(1), Y_{ij}(0)\}.$$

Let $\mathbb{S}_i = \{Y_{i1}(1), Y_{i1}(0), Y_{i2}(1), Y_{i2}(0)\}$ denote the set of all potential outcomes within pair i. Conditioning on the event that $Z_{i1} + Z_{i2} = 1$, we have

$$\begin{aligned}
\pi_{i1} &= \text{pr}\{Z_{i1} = 1 \mid X_i, \mathbb{S}_i, Z_{i1} + Z_{i2} = 1\} \\
&= \frac{\text{pr}\{Z_{i1} = 1, Z_{i2} = 0 \mid X_i, \mathbb{S}_i\}}{\text{pr}\{Z_{i1} + Z_{i2} = 1 \mid X_i, \mathbb{S}_i\}} \\
&= \frac{\text{pr}\{Z_{i1} = 1, Z_{i2} = 0 \mid X_i, \mathbb{S}_i\}}{\text{pr}\{Z_{i1} = 1, Z_{i2} = 0 \mid X_i, \mathbb{S}_i\} + \text{pr}\{Z_{i1} = 0, Z_{i2} = 1 \mid X_i, \mathbb{S}_i\}} \\
&= \frac{e_{i1}(1 - e_{i2})}{e_{i1}(1 - e_{i2}) + (1 - e_{i1})e_{i2}}.
\end{aligned}$$

Define $o_{ij} = e_{ij}/(1 - e_{ij})$ as the odds of the treatment for unit (i, j), and we have

$$\pi_{i1} = \frac{o_{i1}}{o_{i1} + o_{i2}}.$$

Under ignorability, e_{ij} is only a function of X_i, and therefore, $e_{i1} = e_{i2}$ and $\pi_{i1} = 1/2$. Thus the treatment assignment mechanism conditional on covariates and potential outcomes is equivalent to that from an MPE with equal treatment and control probabilities. This is a strategy to analyze matched observational studies we discussed in Section 15.1.

In general, e_{ij} is also a function of the unobserved potential outcomes, and it can range from 0 to 1. Rosenbaum's model for sensitivity analysis imposes bounds on the odds ratio o_{i1}/o_{i2} (Rosenbaum, 1987b).

Assumption 19.1 (Rosenbaum's sensitivity analysis model) *The odds ratios are bounded by*

$$o_{i1}/o_{i2} \leq \Gamma, \quad o_{i2}/o_{i1} \leq \Gamma, \quad (i = 1, \ldots, n)$$

for some pre-specified $\Gamma \geq 1$. *Equivalently,*

$$\frac{1}{1+\Gamma} \leq \pi_{i1} \leq \frac{\Gamma}{1+\Gamma} \quad (i = 1, \ldots, n)$$

for some pre-specified $\Gamma \geq 1$.

Under Assumption 19.1, we have a biased MPE with unequal and varying treatment and control probabilities across pairs. When $\Gamma = 1$, we have $\pi_{i1} = 1/2$ and thus a standard MPE. Therefore, $\Gamma > 1$ measures the deviation from the ideal MPE due to the omitted variables in matching.

19.2 Worst-case p-values under Rosenbaum's sensitivity analysis model

Consider testing the sharp null hypothesis

$$H_{0\text{F}} : Y_{ij}(1) = Y_{ij}(0) \text{ for } i = 1, \ldots, n \text{ and } j = 1, 2$$

based on the within-pair differences $\hat{\tau}_i = (2Z_{i1} - 1)(Y_{i1} - Y_{i2})$ $(i = 1, \ldots, n)$. Under $H_{0\text{F}}$, $|\hat{\tau}_i|$ is fixed but $S_i = I(\hat{\tau}_i > 0)$ is random if $\hat{\tau}_i \neq 0$. Consider the following class of test statistics:

$$T = \sum_{i=1}^{n} S_i q_i,$$

where $q_i \geq 0$ is a function of $(|\hat{\tau}_1|, \ldots, |\hat{\tau}_n|)$. Special cases include the sign statistic, the pair t statistic (up to some constant shift), and the Wilcoxon signed-rank statistic:

$$T = \sum_{i=1}^{n} S_i,$$

$$T = \sum_{i=1}^{n} S_i |\hat{\tau}_i|,$$

$$T = \sum_{i=1}^{n} S_i R_i,$$

where $(R_1, \ldots, R_n)$ are the ranks of $(|\hat{\tau}_1|, \ldots, |\hat{\tau}_n|)$.

What is the null distribution of the test statistic T under Assumption 19.1 with a general Γ? It can be quite complicated because Assumption 19.1 does not fully specify the exact values of the π_{i1}'s. Fortunately, we do not need to know the exact distribution of T but rather the worst-case distribution of T that yields the largest p-value under Assumption 19.1. This worst-case distribution corresponds to

$$S_i \overset{\text{IID}}{\sim} \text{Bernoulli}\left(\frac{\Gamma}{1+\Gamma}\right).$$

This can be intuitive if we interpret the "worst-case distribution of T" as the "largest possible value" of T. Of course, T is a random variable, so we need to make the intuition of the "largest possible value" of T rigorous; see Problem 19.2. The corresponding distribution of T has mean

$$E_\Gamma(T) = \frac{\Gamma}{1+\Gamma}\sum_{i=1}^n q_i,$$

and variance

$$\mathrm{var}_\Gamma(T) = \frac{\Gamma}{(1+\Gamma)^2}\sum_{i=1}^n q_i^2,$$

where E and var are both indexed by the parameter Γ. We can further prove a CLT for T, yielding the Normal approximation

$$\frac{T - \frac{\Gamma}{1+\Gamma}\sum_{i=1}^n q_i}{\sqrt{\frac{\Gamma}{(1+\Gamma)^2}\sum_{i=1}^n q_i^2}} \to \mathrm{N}(0,1)$$

in distribution. In practice, we can report a sequence of p-values as a function of Γ.

19.3 Examples

19.3.1 Revisiting the LaLonde data

We conduct Rosenbaum-style sensitivity analysis in the matched LaLonde data. Using the `Matching` package, we can construct the matched dataset.

```
library("Matching")
library("sensitivitymv")
library("sensitivitymw")
dat <- read.table("cps1re74.csv",header=T)
dat$u74 <- as.numeric(dat$re74==0)
dat$u75 <- as.numeric(dat$re75==0)
y = dat$re78
z = dat$treat
x = as.matrix(dat[, c("age", "educ", "black",
                      "hispan", "married", "nodegree",
                      "re74", "re75", "u74", "u75")])
matchest = Match(Y = y, Tr = z, X = x)
ytreated = y[matchest$index.treated]
ycontrol = y[matchest$index.control]
datamatched = cbind(ytreated, ycontrol)
```

We consider using the test statistic $T = \sum_{i=1}^n S_i|\hat{\tau}_i|$. Under the ideal MPE with $\Gamma = 1$, we can simulate the distribution of T and obtain the p-value 0.002, as shown in the first subfigure in Figure 19.1. With a slightly larger $\Gamma = 1.1$, the worst-case distribution of T shifts to the right, and the p-value increases to 0.011. If we further increase Γ to 1.3, then the worst-case distribution of T shifts further and the p-value exceeds 0.05. Figure 19.2 shows the histogram of the $\hat{\tau}_i$'s and the p-value as a function of Γ; $\Gamma = 1.233$ measures the maximum confounding that we can still reject the null hypothesis at level 0.05.

We can also use the `senmw` function in the `sensitivitymw` package to obtain a sequence of p-values against Γ.

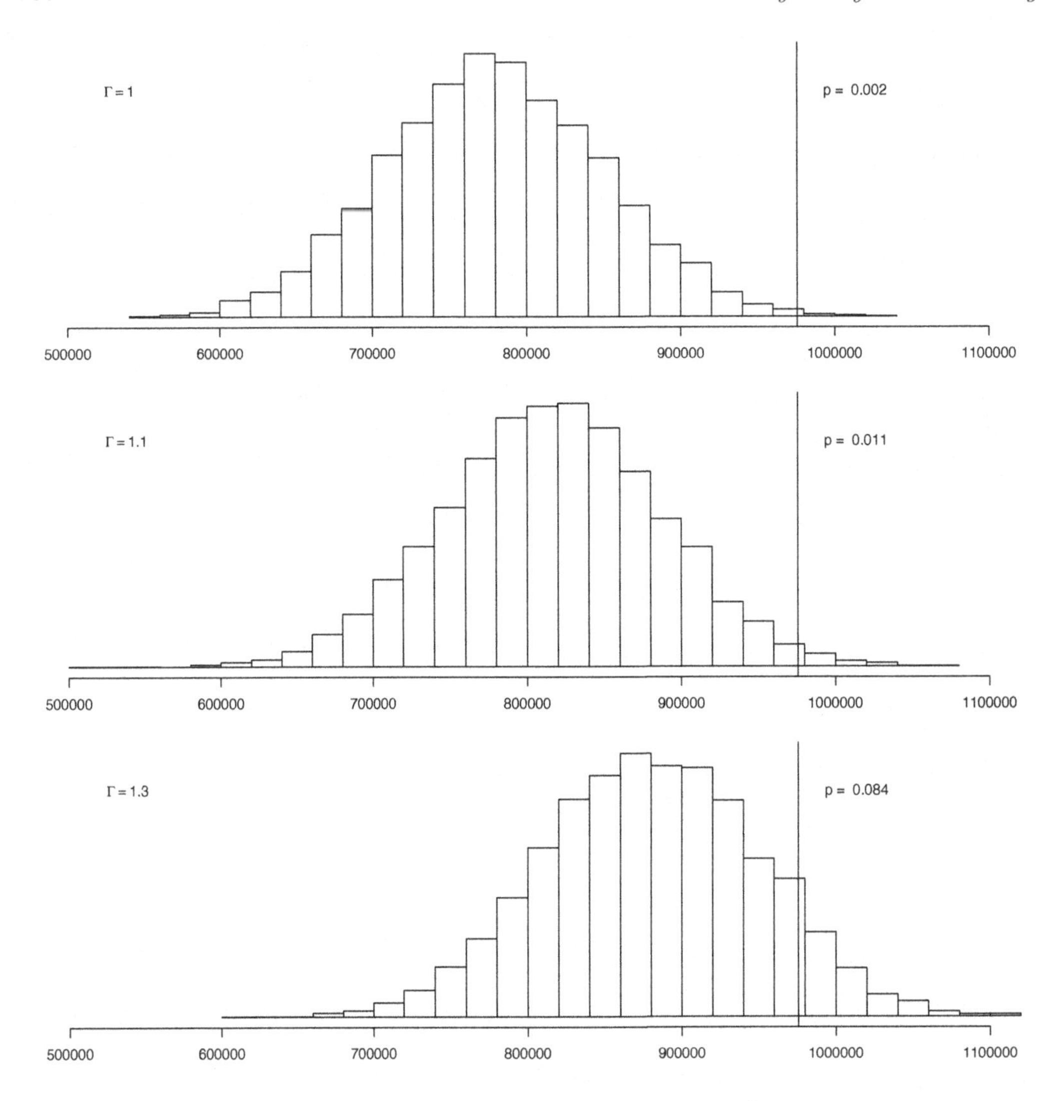

FIGURE 19.1: The worst-case distributions of $T = \sum_{i=1}^{n} S_i|\hat{\tau}_i|$ with S_i's IID Bernoulli$(\Gamma/(1+\Gamma))$, based on the matched LaLonde data

```
Gamma  = seq(1, 1.4, 0.001)
Pvalue = Gamma
for(i in 1:length(Gamma))
{
  Pvalue[i] = senmw(datamatched, gamma = Gamma[i],
                    method = "t")$pval
}
```

Figure 19.2 shows the plot of the p-value against Γ.

19.3.2 Two examples from Rosenbaum's packages

The `erpcp` dataset is from the `R` package `sensitivitymw`. It contains $n = 39$ matched pairs of a welder and a control, based on observed covariates age and smoking. The outcome is

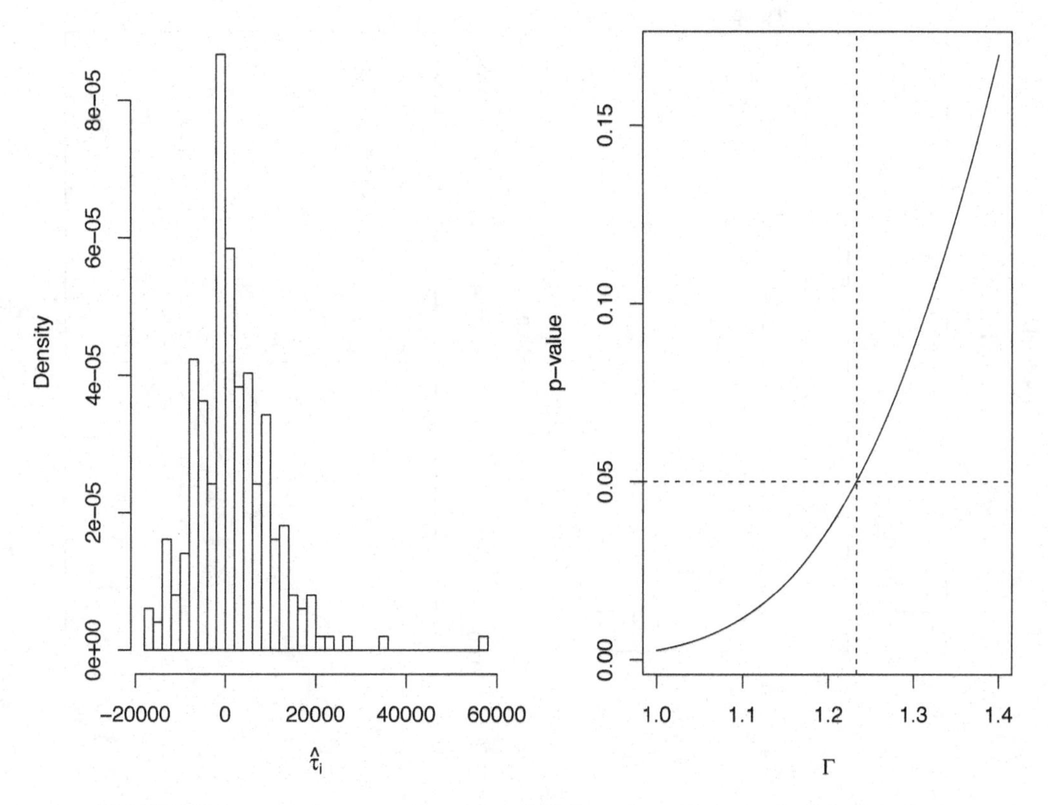

FIGURE 19.2: p-value as a function of Γ, based on the matched LaLonde data

the DNA elution rate. Figure 19.3a shows the histogram of the within-pair differences of the outcomes and the p-value against Γ based on the pair t-statistic. The following R code generates Figure 19.3a.

```
par(mfrow = c(1, 2), mai = c(0.8, 0.8, 0.3, 0.3))
data(erpcp)
hist(erpcp[, 1] - erpcp[, 2], main = "erpcp",
     xlab = expression(hat(tau)[i]),
     freq = FALSE)

Gamma  = seq(1, 5, 0.005)
Pvalue = Gamma
for(i in 1:length(Gamma))
{
  Pvalue[i] = senmw(erpcp, gamma = Gamma[i], method = "t")$pval
}
gammastar = Gamma[which(Pvalue >= 0.05)[1]]
gammastar

plot(Pvalue ~ Gamma, type = "l",
     xlab = expression(Gamma),
     ylab = "p-value")
abline(h = 0.05, lty = 2)
abline(v = gammastar, lty = 2)
```

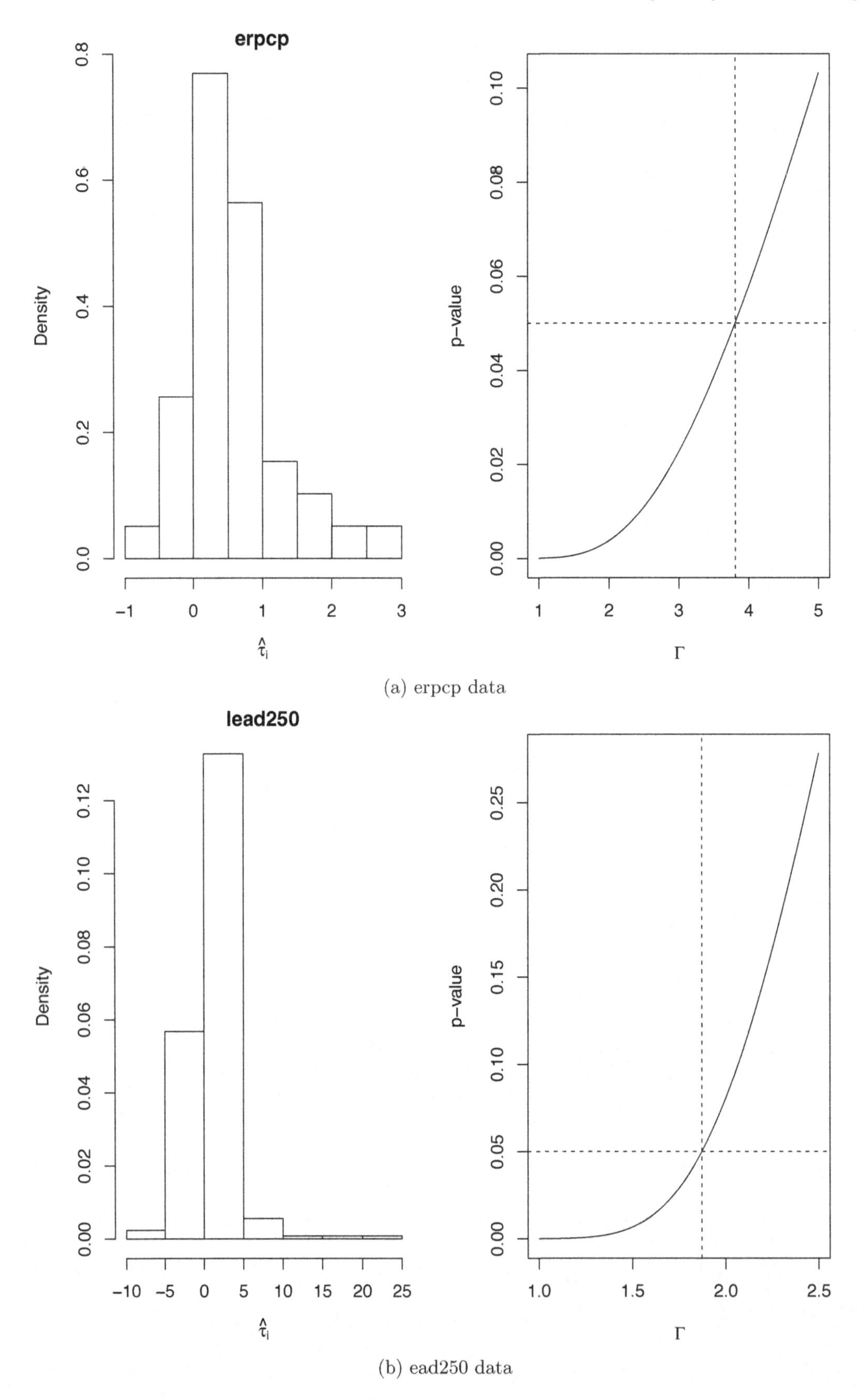

(a) erpcp data

(b) ead250 data

FIGURE 19.3: Two examples: histogram of the $\hat{\tau}_i$'s and p-value as a function of Γ

The `lead250` dataset is from the `R` package `sensitivitymw`. It contains $n = 250$ matched pairs of a daily smoker and a control of nonsmoker, based on observed covariates gender, age, race education level, and household income from NHANES. The outcome is the blood lead level in $\mu g/l$. Figure 19.3b shows the histogram of the within-pair differences of the outcomes and the p-value against Γ based on the pair t-statistic. The following `R` code generates Figure 19.3b.

```
par(mfrow = c(1, 2), mai = c(0.8, 0.8, 0.3, 0.3))
data(lead250)
hist(lead250[, 1] - lead250[, 2],
     main = "lead250",
     xlab = expression(hat(tau)[i]),
     freq = FALSE)

Gamma  = seq(1, 2.5, 0.001)
Pvalue = Gamma
for(i in 1:length(Gamma))
{
  Pvalue[i] = senmw(lead250, gamma = Gamma[i], method = "t")$pval
}
gammastar = Gamma[which(Pvalue >= 0.05)[1]]
gammastar

plot(Pvalue ~ Gamma, type = "l",
     xlab = expression(Gamma), ylab = "p-value")
abline(h = 0.05, lty = 2)
abline(v = gammastar, lty = 2)
```

19.4 Homework problems

19.1 A model for Assumption 19.1

Assumption 19.1 has the following equivalent form.

Assumption 19.2 *The propensity score satisfies the following model*

$$\log \frac{\pi_{ij}}{1-\pi_{ij}} = g(X_i) + \gamma U_{ij}, \quad (i = 1, \ldots, n; j = 1, 2)$$

where $g(\cdot)$ is an unknown function and $U_{ij} \in [0, 1]$ is a bounded unobserved covariate for unit (i, j).

Show that if Assumption 19.2 holds, then Assumption 19.1 must hold with $\Gamma = e^{\gamma}$.

Remark: Rosenbaum (2002b, page 108) also shows that if Assumption 19.1 holds, then Assumption 19.2 must hold for some U_{ij}'s. Assumption 19.2 may be easier to interpret because γ measures the log of the conditional odds ratio of U on the treatment; see Section B.6 for the interpretation of the coefficients in the logistic regression.

19.2 Worst-case distribution

Assume $q_i \geq 0$ for all $i = 1, \dots, n$. Assume that the S_i's are Bernoulli(π_i) and the $\tilde{S}_i$'s are Bernoulli(π), with $\pi_i \leq \pi$ for all $i = 1, \dots, n$. Then for any fixed t, we have

$$\text{pr}\left(\sum_{i=1}^n S_i q_i \geq t\right) \leq \text{pr}\left(\sum_{i=1}^n \tilde{S}_i q_i \geq t\right).$$

Remark: You can use a "coupling" argument by generating S_i and $\tilde{S}_i$ from a common uniform random variable.

19.3 Application of Rosenbaum's approach

Reanalyze Example 10.3 using Rosenbaum's approach based on matching.

19.4 Recommended reading

Rosenbaum (2015) provides a tutorial for his two R packages for sensitivity analysis with matched observational studies.

20

Overlap in Observational Studies: Difficulties and Opportunities

In Part III of this book, causal inference with observational studies relies on two critical assumptions: ignorability

$$Z \perp\!\!\!\perp \{Y(1), Y(0)\} \mid X$$

and overlap

$$0 < e(X) < 1.$$

Both are strong assumptions. Chapters 16–19 in Part IV discuss the difficulties of the ignorability assumption. This chapter has two sections: Section 20.1 will discuss the difficulties of the overlap assumption, and Section 20.1.2 will discuss *regression discontinuity*, an extreme case without overlap.

Regression discontinuity is a popular tool for empirical research in many fields, but it is usually not discussed together with the overlap assumption. I expect many readers might find the logic flow of this chapter strange. If so, please treat Section 20.1 and Section 20.1.2 as two independent chapters.

20.1 Implications of overlap

D'Amour et al. (2021) pointed out the tension between the ignorability and overlap assumptions: typically, more covariates make the ignorability assumption more plausible,[1] but more covariates make the overlap assumption less plausible because the treatment becomes more predictable given more covariates.

If some units have $e(X) = 0$ or $e(X) = 1$, then we have philosophical difficulty in thinking about the counterfactual potential outcomes (King and Zeng, 2006). In particular, if a unit deterministically receives the treatment, then it may not be meaningful to conceive its potential outcome under the control; if a unit deterministically receives the control, then it may not be meaningful to conceive its potential outcome under the treatment. Even if the true propensity score is not exactly 0 or 1, the estimated propensity score can be very close to 0 or 1 in finite samples, which makes the estimators based on IPW numerically unstable. I have discussed this issue in Chapter 11.

Many statistical analyses in fact require a strict version of overlap:

Assumption 20.1 (strict overlap) *$\eta \leq e(X) \leq 1 - \eta$ for some $\eta \in (0, 1/2)$.*

However, D'Amour et al. (2021, Corollary 1) showed that Assumption 20.1 has very strong implications when the number of covariates is large. For simplicity, I present only

[1] Mathematically, this statement is incorrect, recalling the M-bias discussed in Section 16.3.1. However, it is correct if we control for more common causes of the treatment and outcome.

one of their results. Let X_k $(k = 1, \ldots, p)$ be the kth component of the covariate $X = (X_1, \ldots, X_p)$, and

$$e = \mathrm{pr}(Z = 1)$$

be the proportion of the treated units.

Theorem 20.1 *Assumption 20.1 implies that* $\eta \le e \le 1 - \eta$ *and*

$$\begin{aligned} & p^{-1} \sum_{k=1}^{p} \left| E(X_k \mid Z = 1) - E(X_k \mid Z = 0) \right| \\ \le \; & p^{-1/2} C^{1/2} \left\{ e\lambda_1^{1/2} + (1-e)\lambda_0^{1/2} \right\}, \end{aligned} \tag{20.1}$$

where

$$C = \frac{(e - \eta)(1 - e - \eta)}{e^2 (1 - e)^2 \eta (1 - \eta)}$$

is a positive constant depending only on (e, η), *and* λ_1 *and* λ_0 *are the maximum eigenvalues of the covariance matrices* $\mathrm{cov}(X \mid Z = 1)$ *and* $\mathrm{cov}(X \mid Z = 0)$, *respectively.*

What is the order of the maximum eigenvalues in Theorem 20.1? D'Amour et al. (2021) showed that it is usually smaller than $O(p)$ unless the components of X are highly correlated. If the components of X are highly correlated, then some components are redundant after including other components. If the components of X are not highly correlated, then the right-hand side of (20.1) converges to zero. So the average difference in means of the covariates is close to zero, that is, the treatment and control groups are nearly balanced in means averaged over all dimensions of the covariates. Mathematically, the left-hand side of (20.1) converging to zero rules out the possibility that all dimensions of X have non-vanishing differences in means across the treatment and control groups. It is a strong requirement in observational studies with many covariates.

20.1.1 Trimming in the presence of limited overlap

When Assumption 20.1 does not hold, it is common to trim the units based on the estimated propensity scores (Crump et al., 2009; Yang and Ding, 2018b). Trimming drops units within regions of little overlap, which changes the population and estimand. The restrictive implications of overlap in Assumption 20.1 suggest that trimming must be employed more often and one may need to trim a large proportion of units to achieve desirable overlap in high dimensions.

Another strategy is to use the overlap weight introduced in Section 13.4. That is, we can focus on

$$\tau_O = \frac{E\left[e(X)\{1 - e(X)\}\tau(X)\right]}{E\left[e(X)\{1 - e(X)\}\right]}$$

which downweights the units with extreme propensity scores. Similar to trimming, it changes the estimand to a weighted population.

20.1.2 Outcome modeling in the presence of limited overlap

The somewhat negative results in D'Amour et al. (2021) also highlight the limitation of focusing only on the propensity score in the presence of limited overlap. This in some sense challenges Rubin's view that "for objective causal inference, design trumps analysis" (Rubin, 2008). Rubin (2008) argued strongly for the role of the "design" stage in observational studies. The "design" stage focuses on the propensity score which may not satisfy the overlap condition with many covariates.

With high dimensional covariates, outcome modeling becomes more important. In particular, if the outcome means only depend on a function of the original covariates in that

$$E\{Y(z) \mid X\} = f_z(r(X)), \quad (z = 0, 1)$$

then it suffices to control for $r(X)$, a lower-dimensional summary of the original covariates. See Problem 20.1 for more details. Due to the dimension reduction, the strict overlap condition on $r(X)$ can be much weaker than the strict overlap condition on X. This is conceptually straightforward, but the corresponding theory and methods are missing.

20.2 Causal inference with no overlap: regression discontinuity

Let us start from the simple case with a univariate X. An extreme treatment assignment mechanism is a deterministic one:

$$Z = I(X \geq x_0),$$

where x_0 is a predetermined threshold. An interesting consequence of this assignment is that the ignorability assumption holds automatically:

$$Z \perp\!\!\!\perp \{Y(1), Y(0)\} \mid X$$

because Z is a deterministic function of X and a constant is independent of any random variables. However, the overlap assumption is violated by definition:

$$\begin{aligned} e(X) &= \text{pr}(Z = 1 \mid X) \\ &= I(X \geq x_0) \\ &= \begin{cases} 1 & \text{if } X \geq x_0, \\ 0 & \text{if } X < x_0. \end{cases} \end{aligned}$$

So our analytic strategies discussed in Part IV are no longer applicable here. We must change our perspective.

The discussion above seems contrived, with a deterministic treatment assignment. Interestingly, it has many applications in practice and is called *regression discontinuity*. Below, I first review some canonical examples and then give a mathematical formulation of this type of study.

20.2.1 Examples and graphical diagnostics

Example 20.1 *Thistlethwaite and Campbell (1960) first proposed the idea of regression-discontinuity analysis. Their motivating example was to study the effect of students' winning the Certificate of Merit on later career plans, where the Certificate of Merit was determined by whether the Scholarship Qualifying Test score was above a certain threshold. Their initial analysis was mainly graphical. Figure 20.1 shows one of their graphs.*

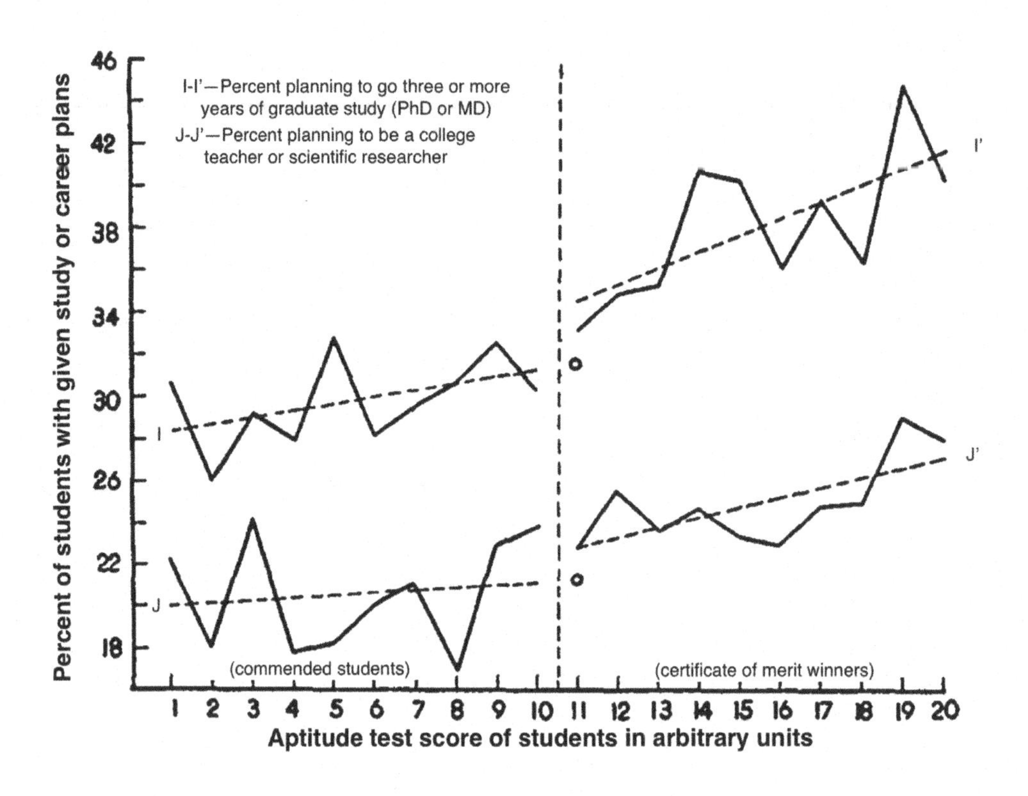

FIGURE 20.1: A graph from Thistlethwaite and Campbell (1960) with minor modifications of the unclear text in the original paper

Example 20.2 *Bor et al. (2014) used regression discontinuity to study the effect of when to start HIV patients with antiretroviral on their mortality, where the treatment is determined by whether the patients' CD4 counts were below 200 cells/μL (note that CD4 cells are white blood cells that fight infection.)*

Example 20.3 *Carpenter and Dobkin (2009) studied the effect of alcohol consumption on mortality, which leverages the minimum legal drinking age as a discontinuity for alcohol consumption. They derived mortality data from the National Center for Health Statistics, including the decedent's date of birth and date of death. They computed the age profile of deaths per 100,000 person-years with outcomes measured by the following nine variables:*

`all`	*all deaths, the sum of* `internal` *and* `external`
`internal`	*deaths due to internal causes*
`external`	*deaths due to external causes, the sum of the rest*
`homicide`	*homicides*
`suicide`	*suicides*
`mva`	*motor vehicle accidents*
`alcohol`	*deaths with a mention of alcohol*
`drugs`	*deaths with a mention of drug use*
`externalother`	*deaths due to other external causes*

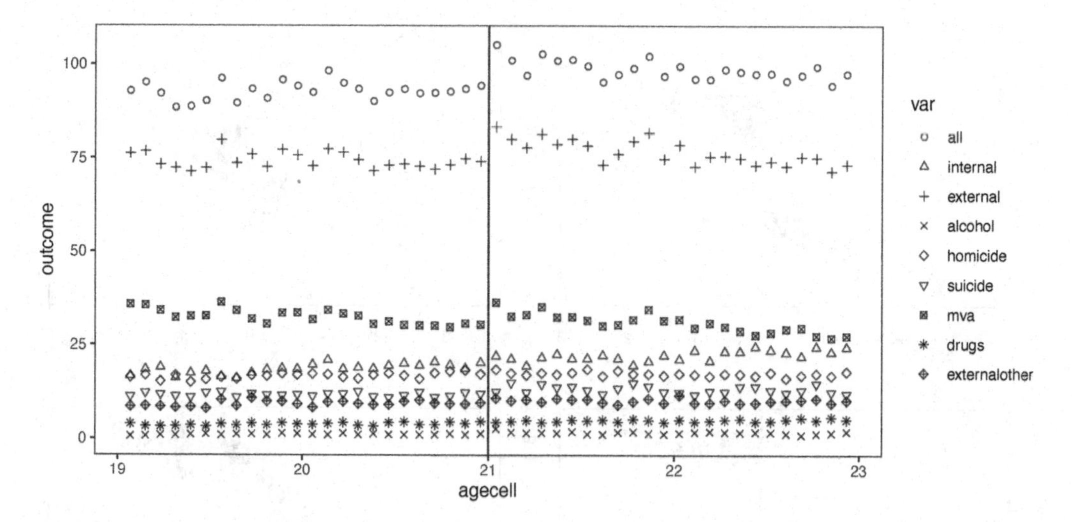

FIGURE 20.2: Minimum legal drinking age example from Carpenter and Dobkin (2009)

Figure 20.2 plots the number of deaths per 100,000 person-years for nine measures based on the data used by Angrist and Pischke (2014). From the jumps at age 21, it seems obvious that there is an increase in mortality at age 21, primarily due to motor vehicle accidents. I leave the formal analysis as Problem 20.4.

20.2.2 A mathematical formulation of regression discontinuity

The technical term for the variable X that determines the treatment is the *running variable.* Intuitively, regression discontinuity can identify a *local average causal effect* at the cutoff point x_0:

$$\tau(x_0) = E\{Y(1) - Y(0) \mid X = x_0\}.$$

In particular, for the potential outcome under treatment, we have

$$\begin{aligned} E\{Y(1) \mid X = x_0\} &= \lim_{\varepsilon \to 0+} E\{Y(1) \mid X = x_0 + \varepsilon\} && (20.2) \\ &= \lim_{\varepsilon \to 0+} E\{Y(1) \mid Z = 1, X = x_0 + \varepsilon\} && (20.3) \\ &= \lim_{\varepsilon \to 0+} E(Y \mid Z = 1, X = x_0 + \varepsilon), && (20.4) \end{aligned}$$

where (20.2) holds if $E\{Y(1) \mid X = x\}$ is continuous from the right at x_0 and (20.3) follows by the definition of Z. Similarly, for the potential outcome under control, we have

$$E\{Y(0) \mid X = x_0\} = \lim_{\varepsilon \to 0+} E(Y \mid Z = 0, X = x_0 - \varepsilon)$$

if $E\{Y(0) \mid X = x\}$ is continuous from the left at x_0. So the local average causal effect at x_0 can be identified by the difference of the two limits. I summarize the key identification result below.

Theorem 20.2 *Assume that the treatment is determined by $Z = I(X \geq x_0)$ where x_0 is a predetermined threshold. Assume that $E\{Y(1) \mid X = x\}$ is continuous from the right at x_0 and $E\{Y(0) \mid X = x\}$ is continuous from the left at x_0. Then the local average treatment effect at $X = x_0$ is identified by*

$$\tau(x_0) = \lim_{\varepsilon \to 0+} E(Y \mid Z = 1, X = x_0 + \varepsilon) - \lim_{\varepsilon \to 0+} E(Y \mid Z = 0, X = x_0 - \varepsilon).$$

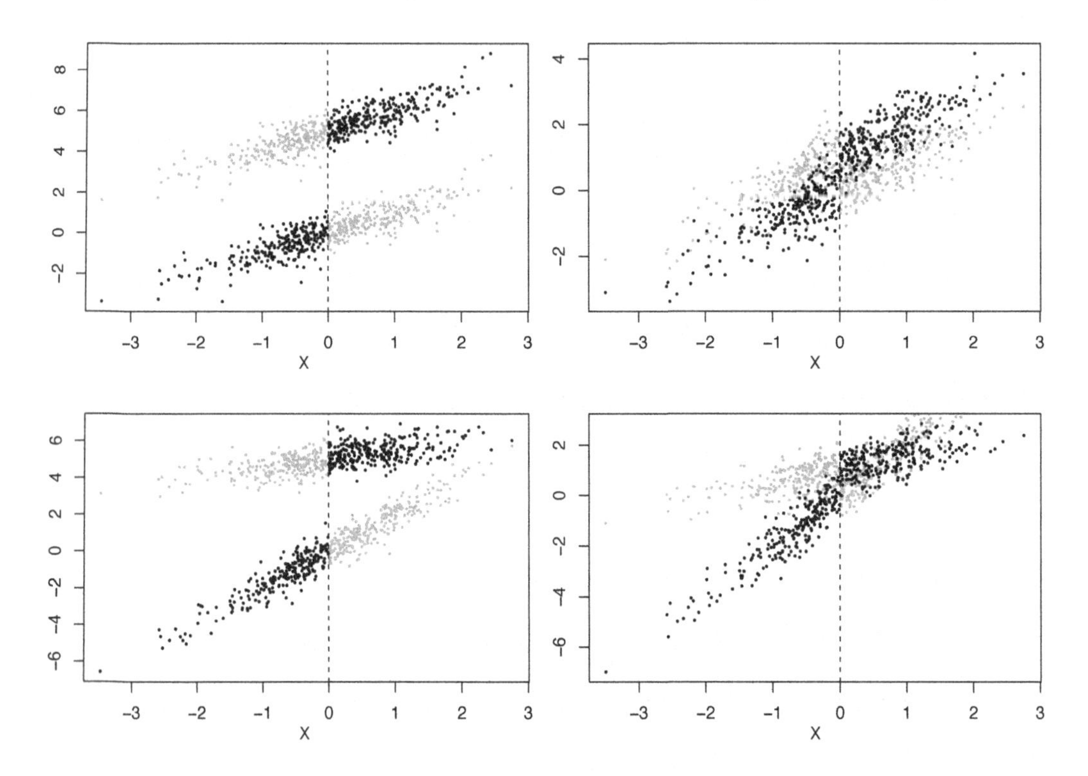

FIGURE 20.3: Examples of regression discontinuity. The black points are the observed outcomes whereas the grey points are the unobserved counterfactual outcomes. In the first column, the data-generating processes result in visible jumps at the cutoff points; in the second column, the jumps are not so visible. In the first row, the data generating processes have constant $\tau(x)$; in the second row, $\tau(x)$ varies with x.

Since the right-hand side of the above equation only involves observables, the parameter $\tau(x_0)$ is nonparametrically identified. However, the form of the identification formula is totally different from what we derived before. In particular, the identification formula involves limits of two conditional expectation functions.

20.2.3 Regressions near the boundary

If we are lucky, graphical diagnostics sometimes can clearly show the causal effect at the cutoff point. However, many outcomes are noisy so graphical diagnostics are not enough in finite samples. Figure 20.3 shows two simulated examples with obvious jumps at the cutoff point and two examples without obvious jumps, although the underlying data-generating processes all have discontinuities.

Assume that

$$\begin{aligned} E(Y \mid Z=1, X=x) &= \gamma_1 + \beta_1 x, \\ E(Y \mid Z=0, X=x) &= \gamma_0 + \beta_0 x \end{aligned}$$

are both linear in x. We can run OLS based on the treated and control data to obtain the fitted lines $\hat{\gamma}_1 + \hat{\beta}_1 x$ and $\hat{\gamma}_0 + \hat{\beta}_0 x$, respectively. We can then estimate the average causal effect at the point $X = x_0$ as

$$\hat{\tau}(x_0) = (\hat{\gamma}_1 - \hat{\gamma}_0) + (\hat{\beta}_1 - \hat{\beta}_0)x_0.$$

Numerically, $\hat{\tau}(x_0)$ is identical to the coefficient of Z_i in the OLS

$$Y_i \sim \{1, Z_i, X_i - x_0, Z_i(X_i - x_0)\}, \tag{20.5}$$

and it is also identical to the coefficient of Z_i in the OLS

$$Y_i \sim \{1, Z_i, R_i, L_i\}, \tag{20.6}$$

where

$$R_i = \max(X_i - x_0, 0), \qquad L_i = \min(X_i - x_0, 0)$$

indicate the right and left parts of $X_i - x_0$, respectively. I leave the algebraic details to Problem 20.2.

However, this approach may be sensitive to the violation of the linear model assumption. Theory suggests running regression using only the local observations near the cutoff point.[2] However, the rules for choosing the "local points" are quite involved. Fortunately, the `rdrobust` function in the `rdrobust` package in `R` implements various choices of "local points." Since choosing the "local points" is the key to regression discontinuity, it seems more sensible to report estimates and confidence intervals based on various choices of the "local points." This can be viewed as a sensitivity analysis for regression discontinuity.

20.2.4 An example

Lee (2008) gave a famous example of using regression discontinuity to study the incumbency advantage in the U.S. House. He wrote that "incumbents are, by definition, those politicians who were successful in the previous election. If what makes them successful is somewhat persistent over time, they should be expected to be somewhat more successful when running for re-election." Therefore, this is a fundamentally challenging causal inference problem. The regression discontinuity is a clever study design to study this problem.

The running variable is the lagged vote in the previous election centered at 0, and the outcome is the vote in the current election, with units being the congressional districts. The treatment is the binary indicator for being the current incumbent party in a district, determined by the lagged vote. The following `R` code generates Figure 20.4 that shows the raw data.

```
house = read.csv("house.csv")[, -1]
plot(y ~ x, data = house, pch = 19, cex = 0.1)
abline(v = 0, col = "grey")
```

The `rdrobust` function gives three sets of the point estimate and confidence intervals. They all suggest a positive incumbency advantage.

```
> library(rdrobust)
> RDDest = rdrobust(house$y, house$x)
Warning message:
In rdrobust(house$y, house$x) :
  Mass points detected in the running variable.
> cbind(RDDest$coef, RDDest$ci)
```

[2] This is called *local linear regression* in nonparametric statistics, which belongs to a broader class of *local polynomial regression* (Fan and Gijbels, 1996).

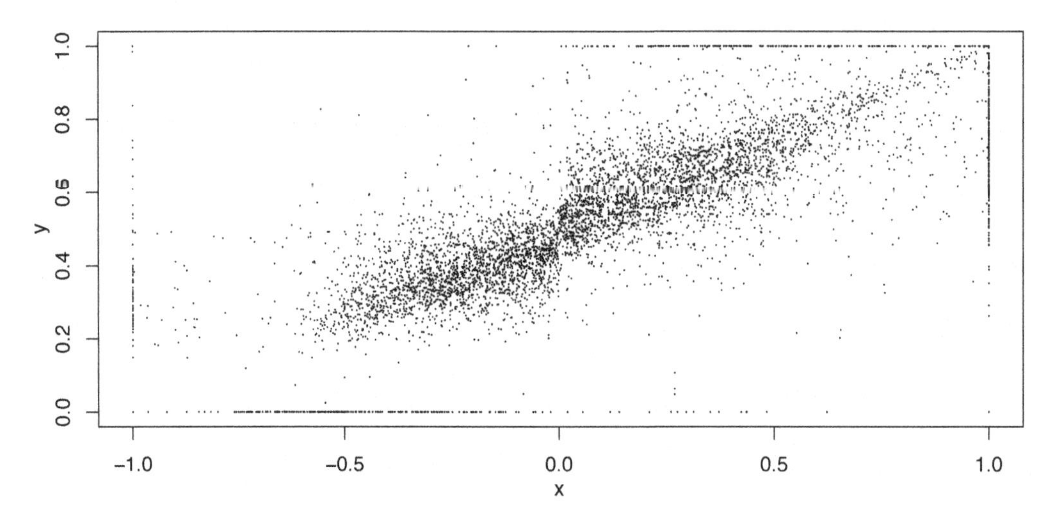

FIGURE 20.4: Raw data of Lee (2008)

```
                    Coeff      CI Lower    CI Upper
Conventional    0.06372533 0.04224798 0.08520269
Bias-Corrected 0.05937028 0.03789292 0.08084763
Robust          0.05937028 0.03481238 0.08392818
```

We can also obtain the point estimates and the confidence intervals based on OLS with different choices of the local points defined by $|X| < h$ based on the following R code.

```
house$z = (house$x >= 0)
hh = seq(0.05, 1, 0.01)
local.lm = sapply(hh, function(h){
  Greg = lm(y ~ z + x + z*x, data = house,
            subset = (abs(x)<=h))
  cbind(coef(Greg)[2], confint(Greg, 'zTRUE'))
})
plot(local.lm[1, ] ~ hh, type = "p",
     pch = 19, cex = 0.3,
     ylim = range(local.lm),
     xlab = "h",
     ylab = "point and interval estimates",
     main = "subset linear regression: |X|<h")
lines(local.lm[2, ] ~ hh, type = "p",
      pch = 19, cex = 0.1)
lines(local.lm[3, ] ~ hh, type = "p",
      pch = 19, cex = 0.1)
```

Figure 20.5 shows the point estimates and the confidence intervals as a function of h. While the point estimates and the confidence intervals are sensitive to the choice of h, the qualitative result remains the same as above.

20.2.5 Problems of regression discontinuity

What can go wrong with the regression discontinuity analysis? The technical challenge is to specify the neighborhood near the cutoff point. We have discussed this issue above.

In addition, Theorem 20.2 holds under a continuity condition. It may be violated in practice. For instance, if the mortality rate jumps at the age of 21, then we may not be

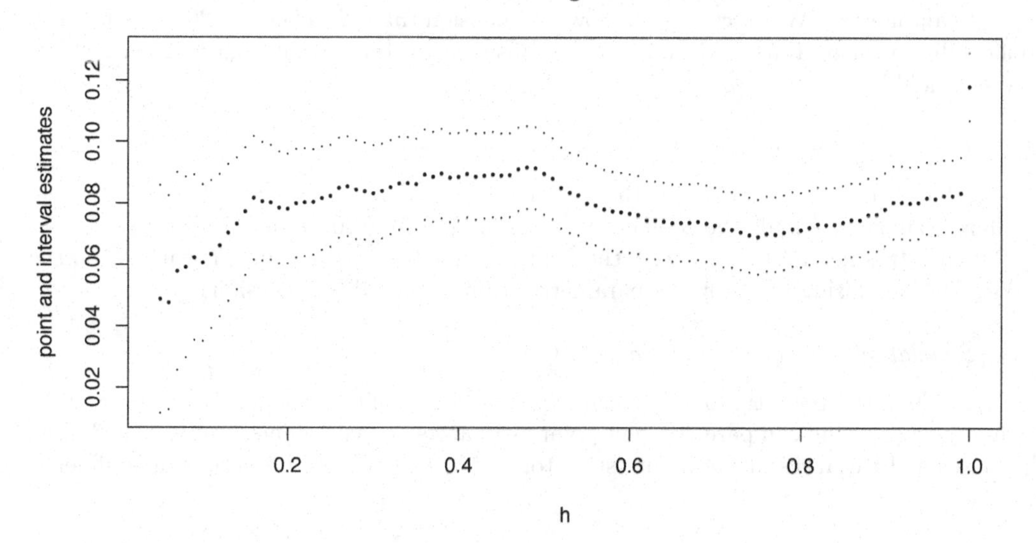

FIGURE 20.5: Estimates and confidence intervals based on local linear regressions

able to attribute the jumps in Figure 20.2 to the change in drinking behavior due to the legal drinking age. However, it is hard to check the violation of the continuity condition empirically.

McCrary (2008) proposed an indirect test for the validity of the regression discontinuity. He suggested checking the density of the running variable at the cutoff point. The discontinuity in the density of the running variable at the cutoff point may suggest that some units were able to manipulate their treatment status perfectly. The `DCdensity` function in the `R` package `rdd` implements this test. I omit the details.

20.3 Homework problems

20.1 A theorem for the role of outcome modeling in the presence of limited overlap

This problem extends the discussion in Section 20.1.2. I state a formal theorem below.

Theorem 20.3 *Assume*

$$Z \perp\!\!\!\perp Y(z) \mid X, \quad (z = 0, 1)$$

and

$$E\{Y(z) \mid X\} = f_z(r(X)), \quad (z = 0, 1)$$

for some function $r(X)$. The average causal effect $\tau = E\{Y(1) - Y(0)\}$ can be identified by

$$\tau = E\{E(Y \mid Z = 1, r(X)) - E(Y \mid Z = 0, r(X))\}$$

or

$$\tau = E\left\{\frac{ZY}{e(r(X))}\right\} - E\left\{\frac{(1 - Z)Y}{1 - e(r(X))}\right\}$$

where $e(r(X)) = \text{pr}\{Z = 1 \mid r(X)\}$.

Prove Theorem 20.3.

Remark: When $r(X) = X$, Theorem 20.3 reduces to the standard IPW formula for the average causal effect. When $r(X)$ has a lower dimension than X, Theorem 20.3 has broader applicability if the overlap condition on $e(X)$ fails whereas the overlap condition on $e(r(X))$ may still hold.

20.2 *Linear potential outcome models*

This problem gives more details for the numerical equivalence in Section 20.2.3.

Show that $\hat{\tau}(x_0)$ equals the coefficients of Z_i in OLS fits (20.5) and (20.6).

Remark: It is helpful to start with the figures of $Z_i(X_i - x_0)$, L_i, and R_i with X_i on the x-axis. The conclusion holds by a reparametrization of the OLS regressions.

20.3 *Simulation for regression discontinuity*

In Figure 20.3, the potential outcomes are simulated from linear models. Change them to nonlinear models, and compare different point estimators and confidence intervals, including the biases and variances of the point estimators, and the coverage properties of confidence intervals.

20.4 *Reanalyzing the data on the minimum legal drinking age*

Figure 20.2 shows the jumps at the cutoff point. Analyze the data `mlda.csv` of Carpenter and Dobkin (2009).

20.5 *Reanalyzing the data from Lee (2008)*

Figure 20.5 plots the confidence intervals based on the standard errors assuming homoskedasticity. Generate a figure with confidence intervals based on the EHW standard errors in OLS.

20.6 *Recommended reading*

D'Amour et al. (2021) discussed the implications of overlap with high dimensional covariates.

The original paper by Thistlethwaite and Campbell (1960) on regression discontinuity was reprinted as Thistlewaite and Campbell (2016) with many insightful comments. Coincidentally, Thistlethwaite and Campbell (1960) and Rubin (1974) were both published in the *Journal of Educational Psychology*.

Part V

Instrumental variables

21

An Experimental Perspective of the Instrumental Variable

The *instrumental variable method* has been a powerful tool in econometrics. It identifies causal effects in studies without unconfoundedness between the treatment and the outcome. It relies on an additional variable, called the *instrumental variable* (IV), that satisfies certain conditions. These conditions may not be easy to digest when you learn them for the first time. In some sense, the IV method is like magic. This chapter presents a not-so-magic perspective based on the encouragement design. This again echos the suggestion from Dorn (1953) that the planner of an observational study should always ask himself the following question:[1]

> How would the study be conducted if it were possible to do it by controlled experimentation?

The experimental analog of the IV method is *noncompliance* in the *encouragement design* (Zelen, 1979; Powers and Swinton, 1984; Holland, 1986).

21.1 Encouragement design and noncompliance

Consider an experiment with units indexed by $i = 1, \dots, n$. Let Z_i be the treatment assigned, with 1 for the treatment and 0 for the control. Let D_i be the treatment received, with 1 for the treatment and 0 for the control. When $Z_i \neq D_i$ for some unit i, the noncompliance problem arises. Noncompliance is a very common problem, especially in encouragement designs involving human beings as experimental units. In those cases, the experimenters can not force the units to take the treatment but rather only encourage them to do so. Let Y_i be the outcome of interest.

Consider complete randomization of Z and ignore covariates X for now. We have the potential values for the treatment received $\{D_i(1), D_i(0)\}$ and the potential values for the outcome $\{Y_i(1), Y_i(0)\}$, all with respect to the treatment assignment levels 1 and 0. Their observed values are $D_i = Z_i D_i(1) + (1 - Z_i) D_i(0)$ and $Y_i = Z_i Y_i(1) + (1 - Z_i) Y_i(0)$, respectively. For notational simplicity, we assume $\{Z_i, D_i(1), D_i(0), Y_i(1), Y_i(0)\}_{i=1}^n \stackrel{\text{IID}}{\sim} \{Z, D(1), D(0), Y(1), Y(0)\}$ and sometimes drop the subscript i when it should not cause confusions.

We start with the CRE.

Assumption 21.1 (randomization) $Z \perp\!\!\!\perp \{D(1), D(0), Y(1), Y(0)\}$.

[1] This quote also appeared in Chapter 10 before.

Randomization allows for the identification of the average causal effects on D and Y by

$$\begin{aligned}\tau_D &= E\{D(1) - D(0)\} \\ &= E(D \mid Z = 1) - E(D \mid Z = 0)\end{aligned}$$

and

$$\begin{aligned}\tau_Y &= E\{Y(1) - Y(0)\} \\ &= E(Y \mid Z = 1) - E(Y \mid Z = 0).\end{aligned}$$

We can use simple difference-in-means estimators $\hat{\tau}_D$ and $\hat{\tau}_Y$ to estimate τ_D and τ_Y, respectively.

Reporting the estimate $\hat{\tau}_Y$ with the associated standard error is called the intention-to-treat (ITT) analysis. It estimates the effect of the treatment assignment on the outcome, and the CRE in Assumption 21.1 justifies this analysis. However, it may not answer the scientific question, that is, the causal effect of the treatment received on the outcome.

21.2 Latent compliance status and effects

21.2.1 Nonparametric identification

Following Imbens and Angrist (1994) and Angrist et al. (1996), we stratify the population based on the joint potential values of $U_i = \{D_i(1), D_i(0)\}$. Because D is binary, we have four possible combinations:

$$U_i = \begin{cases} \text{a}, & \text{if } D_i(1) = 1 \text{ and } D_i(0) = 1; \\ \text{c}, & \text{if } D_i(1) = 1 \text{ and } D_i(0) = 0; \\ \text{d}, & \text{if } D_i(1) = 0 \text{ and } D_i(0) = 1; \\ \text{n}, & \text{if } D_i(1) = 0 \text{ and } D_i(0) = 0, \end{cases}$$

where "a" is for "always taker," "c" is for "complier," "d" is for "defier," and "n" is for "never taker." Because we cannot observe $D_i(1)$ and $D_i(0)$ simultaneously, U_i is a latent variable for the compliance behavior of unit i.

Based on U, we can use the law of total probability to decompose the average causal effect on Y into four terms:

$$\begin{aligned}\tau_Y = \; & E\{Y(1) - Y(0) \mid U = \text{a}\}\text{pr}(U = \text{a}) \\ & + E\{Y(1) - Y(0) \mid U = \text{c}\}\text{pr}(U = \text{c}) \\ & + E\{Y(1) - Y(0) \mid U = \text{d}\}\text{pr}(U = \text{d}) \\ & + E\{Y(1) - Y(0) \mid U = \text{n}\}\text{pr}(U = \text{n}). \qquad (21.1)\end{aligned}$$

Therefore, τ_Y is a weighted average of four latent subgroup effects. We will look into more details of the latent groups below.

Assumption 21.2 below restricts the third term in (21.1) to be zero.

Assumption 21.2 (monotonicity) $\text{pr}(U_i = \text{d}) = 0$ *or* $D_i(1) \geq D_i(0)$ *for all* i*, that is, there are no defiers.*

Assumption 21.2 holds automatically with *one-sided noncompliance* when the units assigned to the control arm have no access to the treatment, i.e., $D_i(0) = 0$ for all units; see Problem 22.10. Under randomization, Assumption 21.2 has a testable implication that

$$\text{pr}(D = 1 \mid Z = 1) \geq \text{pr}(D = 1 \mid Z = 0). \qquad (21.2)$$

Assumption 21.2 is much stronger than the inequality in equation (21.2). The former restricts $D_i(1)$ and $D_i(0)$ at the individual level whereas the latter restricts them only on average. Nevertheless, when the testable implication in equation (21.2) holds, we cannot use the observed data to refute Assumption 21.2.

Assumption 21.3 below restricts the first and last terms in (21.1) to be zero based on the mechanism of the treatment assignment on the outcome through only the treatment received.

Assumption 21.3 (exclusion restriction) $Y_i(1) = Y_i(0)$ *for always takers with* $U_i = \mathrm{a}$ *and never takers with* $U_i = \mathrm{n}$.

An equivalent statement of Assumption 21.3 is that $D_i(1) = D_i(0)$ must imply $Y_i(1) = Y_i(0)$ for all i. It requires that the treatment assignment affects the outcome only if it affects the treatment received. In double-blind RCTs,[2] this assumption is plausible because the outcome only depends on the actual treatment received. That is, if the treatment assignment does not change the treatment received, it does not change the outcome either. It can be violated if the treatment assignment has *direct effects*[3] on the outcome, not through the treatment received. Some RCTs are not double-blinded, and the treatment assignment can have some unknown pathways to the outcome. For instance, doctors may give additional health instructions to the patients in addition to the drugs.

Under Assumptions 21.2 and 21.3, the decomposition (21.1) only has the second term

$$\tau_Y = E\{Y(1) - Y(0) \mid U = \mathrm{c}\}\mathrm{pr}(U = \mathrm{c}). \tag{21.3}$$

Similarly, we can decompose the average causal effect on D into four terms:

$$\begin{aligned}
\tau_D &= E\{D(1) - D(0) \mid U = \mathrm{a}\}\mathrm{pr}(U = \mathrm{a}) \\
&\quad + E\{D(1) - D(0) \mid U = \mathrm{c}\}\mathrm{pr}(U = \mathrm{c}) \\
&\quad + E\{D(1) - D(0) \mid U = \mathrm{d}\}\mathrm{pr}(U = \mathrm{d}) \\
&\quad + E\{D(1) - D(0) \mid U = \mathrm{n}\}\mathrm{pr}(U = \mathrm{n}) \\
&= 0 \times \mathrm{pr}(U = \mathrm{a}) \\
&\quad + 1 \times \mathrm{pr}(U = \mathrm{c}) \\
&\quad + (-1) \times \mathrm{pr}(U = \mathrm{d}) \\
&\quad + 0 \times \mathrm{pr}(U = \mathrm{n}),
\end{aligned}$$

which, under Assumption 21.2, reduces to

$$\tau_D = \mathrm{pr}(U = \mathrm{c}). \tag{21.4}$$

By (21.4), the proportion of the compliers $\pi_\mathrm{c} = \mathrm{pr}(U = \mathrm{c})$ equals the average causal effect of the treatment assigned on D, an identifiable quantity under the CRE. Although we do not know all the compliers based on the observed data, we can identify their proportion in the whole population based on (21.4). Combining (21.3) and (21.4), we have the following result.

[2]In general, it is better to blind the experiment to avoid various biases arising from placebo effects, patients' expectations, etc. In double-blind RCTs, both doctors and patients do not know the treatment; in single-blind RCTs, the patients do not know the treatment but the doctors know. Sometimes, it is impossible to conduct double or even single-blind trials; those trials are called open trials.

[3]I use "direct effects" informally here. See more detailed discussion of the concept in Chapters 27 and 28 later.

Theorem 21.1 *Under Assumptions 21.2–21.3, we have*

$$E\{Y(1) - Y(0) \mid U = \mathrm{c}\} = \frac{\tau_Y}{\tau_D}$$

if $\tau_D \neq 0$.

Following Imbens and Angrist (1994) and Angrist et al. (1996), we define a new causal effect below.

Definition 21.1 (CACE or LATE) *Define*

$$\tau_{\mathrm{c}} = E\{Y(1) - Y(0) \mid U = \mathrm{c}\}$$

as the "complier average causal effect (CACE)" or the "local average treatment effect (LATE)". It has alternative forms:

$$\begin{aligned} \tau_{\mathrm{c}} &= E\{Y(1) - Y(0) \mid D(1) = 1, D(0) = 0\} \\ &= E\{Y(1) - Y(0) \mid D(1) > D(0)\}. \end{aligned}$$

The CACE is the average causal effect of Z on Y among compliers with $\{D(1) = 1, D(0) = 0\}$, or, equivalently, units with $D(1) > D(0)$ under the monotonicity. Based on Definition 21.1, we can rewrite Theorem 21.1 as

$$\tau_{\mathrm{c}} = \frac{\tau_Y}{\tau_D},$$

that is, the CACE or LATE equals the ratio of the average causal effects on Y over that on D. Under Assumption 21.1, we further identify the CACE below.

Corollary 21.1 *Under Assumptions 21.1–21.3, we have*

$$\tau_{\mathrm{c}} = \frac{E(Y \mid Z = 1) - E(Y \mid Z = 0)}{E(D \mid Z = 1) - E(D \mid Z = 0)}$$

if $\tau_D \neq 0$.

Therefore, under CRE, monotonicity, and exclusion restriction, we can nonparametrically identify the CACE as the ratio of the difference in means of the outcome over the difference in means of the treatment received.

21.2.2 Estimation

Based on Corollary 21.1, we can estimate τ_{c} by a simple ratio

$$\hat{\tau}_{\mathrm{c}} = \frac{\hat{\tau}_Y}{\hat{\tau}_D},$$

which is called the Wald estimator (Wald, 1940) or the IV estimator. In the above discussion, Z acts as the IV for D.

We can obtain the variance estimator based on the following heuristics (see Example A.3):

$$\begin{aligned} \hat{\tau}_{\mathrm{c}} - \tau_{\mathrm{c}} &= (\hat{\tau}_Y - \tau_{\mathrm{c}}\hat{\tau}_D)/\hat{\tau}_D \\ &\approx (\hat{\tau}_Y - \tau_{\mathrm{c}}\hat{\tau}_D)/\tau_D \\ &= \hat{\tau}_A/\tau_D, \end{aligned}$$

where $\hat{\tau}_A$ is the difference in means of the adjusted outcome $A_i = Y_i - \tau_{\mathrm{c}} D_i$. So the asymptotic variance of $\hat{\tau}_{\mathrm{c}}$ is close to the variance of $\hat{\tau}_A$ divided by τ_D^2. The variance estimation proceeds in the following steps:

1. obtain the adjusted outcomes $\hat{A}_i = Y_i - \hat{\tau}_c D_i$ $(i = 1, \ldots, n)$;
2. obtain the Neyman-type variance estimate based on the adjusted outcomes:

$$\hat{V}_{\hat{A}} = \frac{\hat{S}^2_{\hat{A}}(1)}{n_1} + \frac{\hat{S}^2_{\hat{A}}(0)}{n_0},$$

   where $\hat{S}^2_{\hat{A}}(1)$ and $\hat{S}^2_{\hat{A}}(0)$ are the sample variances of the $\hat{A}_i$'s under the treatment and control, respectively;
3. obtain the final variance estimator $\hat{V}_{\hat{A}}/\hat{\tau}_D^2$.

See Problem 21.2 for the justification of the above variance estimator. Alternatively, we can also use the bootstrap to approximate the variance of $\hat{\tau}_c$. The following functions can calculate the point estimator $\hat{\tau}_c$ and the standard error with the formula $\hat{V}_{\hat{A}}$ and the bootstrap.

```
## IV point estimator
IV_Wald = function(Z, D, Y)
{
      tau_D = mean(D[Z==1]) - mean(D[Z==0])
      tau_Y = mean(Y[Z==1]) - mean(Y[Z==0])
      CACE  = tau_Y/tau_D

      c(tau_D, tau_Y, CACE)
}

## IV se via the delta method
IV_Wald_delta = function(Z, D, Y)
{
      est          = IV_Wald(Z, D, Y)
      AdjustedY    = Y - D*est[3]
      VarAdj       = var(AdjustedY[Z==1])/sum(Z) +
                          var(AdjustedY[Z==0])/sum(1 - Z)

      c(est[3], sqrt(VarAdj)/abs(est[1]))
}

##IV se via the bootstrap
IV_Wald_bootstrap = function(Z, D, Y, n.boot = 200)
{
      est       = IV_Wald(Z, D, Y)

      CACEboot   = replicate(n.boot, {
        id.boot  = sample(1:length(Z), replace = TRUE)
        IV_Wald(Z[id.boot], D[id.boot], Y[id.boot])[3]
      })

      c(est[3], sd(CACEboot))
}
```

Under the null hypothesis that $\tau_c = 0$, we can approximate the variance by $\hat{V}_Y/\hat{\tau}_D^2$, where $\hat{V}_Y$ is the Neyman-type variance estimate for the difference in means of Y. This variance estimator is inconsistent if the true τ_c is not zero. Therefore, it works for testing but not for estimation. Nevertheless, it gives insights for the ITT estimator and the Wald estimator. The ITT estimator $\hat{\tau}_Y$ has estimated standard error $\sqrt{\hat{V}_Y}$. The Wald estimator

$\hat{\tau}_Y/\hat{\tau}_D$ essentially equals the ITT estimator multiplied by $1/\hat{\tau}_D > 1$, which is larger in magnitude but at the same time, its estimated standard error increases by the same factor. Based on the variance estimators $\hat{V}_Y$ and $\hat{V}_Y/\hat{\tau}_D^2$, the confidence intervals for τ_Y and τ_c are

$$\hat{\tau}_Y \pm z_{1-\alpha/2}\sqrt{\hat{V}_Y}$$

and

$$\hat{\tau}_Y/\hat{\tau}_D \pm z_{1-\alpha/2}\sqrt{\hat{V}_Y}/\hat{\tau}_D = \left(\hat{\tau}_Y \pm z_{1-\alpha/2}\sqrt{\hat{V}_Y}\right)/\hat{\tau}_D,$$

respectively, where $z_{1-\alpha/2}$ is the $1-\alpha/2$ upper quantile of the standard Normal random variable. These confidence intervals give the same qualitative conclusions about the causal effects since they will both cover zero or not. In some sense, the IV analysis provides the same qualitative information as the ITT analysis of Y although it involves more complicated procedures.

21.3 Covariates

21.3.1 Covariate adjustment in the CRE

We now consider completely randomized experiments with covariates and assume

$$Z \perp\!\!\!\perp \{D(1), D(0), Y(1), Y(0), X\}.$$

With covariates X, we can obtain Lin's (2013) estimators $\hat{\tau}_{D,\mathrm{L}}$ and $\hat{\tau}_{Y,\mathrm{L}}$ for both D and Y, resulting in $\hat{\tau}_{\mathrm{c,L}} = \hat{\tau}_{Y,\mathrm{L}}/\hat{\tau}_{D,\mathrm{L}}$. We can approximate the asymptotic variance of $\hat{\tau}_{\mathrm{c,L}}$ using the bootstrap. The following functions can calculate the point estimator $\hat{\tau}_{\mathrm{c,L}}$ and the standard error based on the bootstrap.

```
## covariate adjustment in IV analysis
IV_Lin = function(Z, D, Y, X)
{
  X     = scale(as.matrix(X))
  tau_D = lm(D ~ Z + X + Z*X)$coef[2]
  tau_Y = lm(Y ~ Z + X + Z*X)$coef[2]
  CACE  = tau_Y/tau_D

  c(tau_D, tau_Y, CACE)
}


##IV_adj se via the bootstrap
IV_Lin_bootstrap = function(Z, D, Y, X, n.boot = 200)
{
  X        = scale(as.matrix(X))
  est      = IV_Lin(Z, D, Y, X)
  CACEboot = replicate(n.boot, {
    id.boot = sample(1:length(Z), replace = TRUE)
    IV_Lin(Z[id.boot], D[id.boot], Y[id.boot], X[id.boot, ])[3]
  })

  c(est[3], sd(CACEboot))
}
```

21.3.2 Covariates in conditional randomization or unconfounded observational studies

If randomization holds conditionally, i.e.,

$$Z \perp\!\!\!\perp \{D(1), D(0), Y(1), Y(0)\} \mid X,$$

then we must adjust for covariates to avoid bias. The analysis is also straightforward since we already have discussed many estimators in Part III for estimating the effects of Z on D and Y, respectively. We can just use them in the ratio formula $\hat{\tau}_c = \hat{\tau}_Y / \hat{\tau}_D$ and use the bootstrap to approximate the asymptotic variance. I do not implement the corresponding estimator and variance estimator but relegate it as Problem 21.8.

21.4 Weak IV

21.4.1 Some simulation

Even if $\tau_D > 0$, there is a positive probability that $\hat{\tau}_D$ is zero, so the variance of $\hat{\tau}_c$ is infinity (see Problem 21.1). The variance from the Normal approximation discussed before is not the variance of $\hat{\tau}_c$ but rather the variance of its asymptotic distribution. This is a subtle technical point.

When τ_D is close to 0, which is referred to as the weak IV case, the ratio estimator $\hat{\tau}_c = \hat{\tau}_Y / \hat{\tau}_D$ has poor finite-sample properties. In this scenario, the sampling distribution of $\hat{\tau}_c$ is centered away from τ_c and deviates far from the Normal distribution. Consequently, the corresponding Wald-type confidence intervals have poor coverage properties.[4] In the simple case with a binary outcome Y, we know that τ_Y must be bounded between -1 and 1, but there is no guarantee that $\hat{\tau}_c$ is bounded between -1 and 1.

Figures 21.1a and 21.1b show the histograms of $\hat{\tau}_c$ and $\hat{\tau}_{c,L}$ over simulation with different π_c. I leave the detailed data-generating processes to the `R` code. From the figures, the distributions of the estimators $\hat{\tau}_c$ and $\hat{\tau}_{c,L}$ are far from Normal when π_c equals 0.2 and 0.1. Statistical inference based on asymptotic Normality is unlikely to be reliable.

21.4.2 A procedure robust to weak IV

How do we deal with a weak IV? From a testing perspective, there is an easy solution. Because $\tau_c = \tau_Y / \tau_D$, the null hypothesis

$$H_0 : \tau_c = 0$$

is equivalent to the null hypothesis

$$H_0' : \tau_Y = 0$$

if $\tau_D > 0$. Therefore, we simply test H_0', i.e., the average causal effect of Z on Y is zero. This echoes our discussion in Section 21.2.2 about the relationship between the ITT analysis and the IV analysis.

[4]The theory often assumes that τ_D has the order $n^{-1/2}$. Under this regime, the proportion of compliers goes to 0 at the rate of $n^{-1/2}$ as n goes to infinity. The IV method can only identify a subgroup average causal effect with the proportion shrinking to 0. This is a contrived regime for theoretical analysis. It is hard to justify this assumption in practice. The following discussion does not assume it.

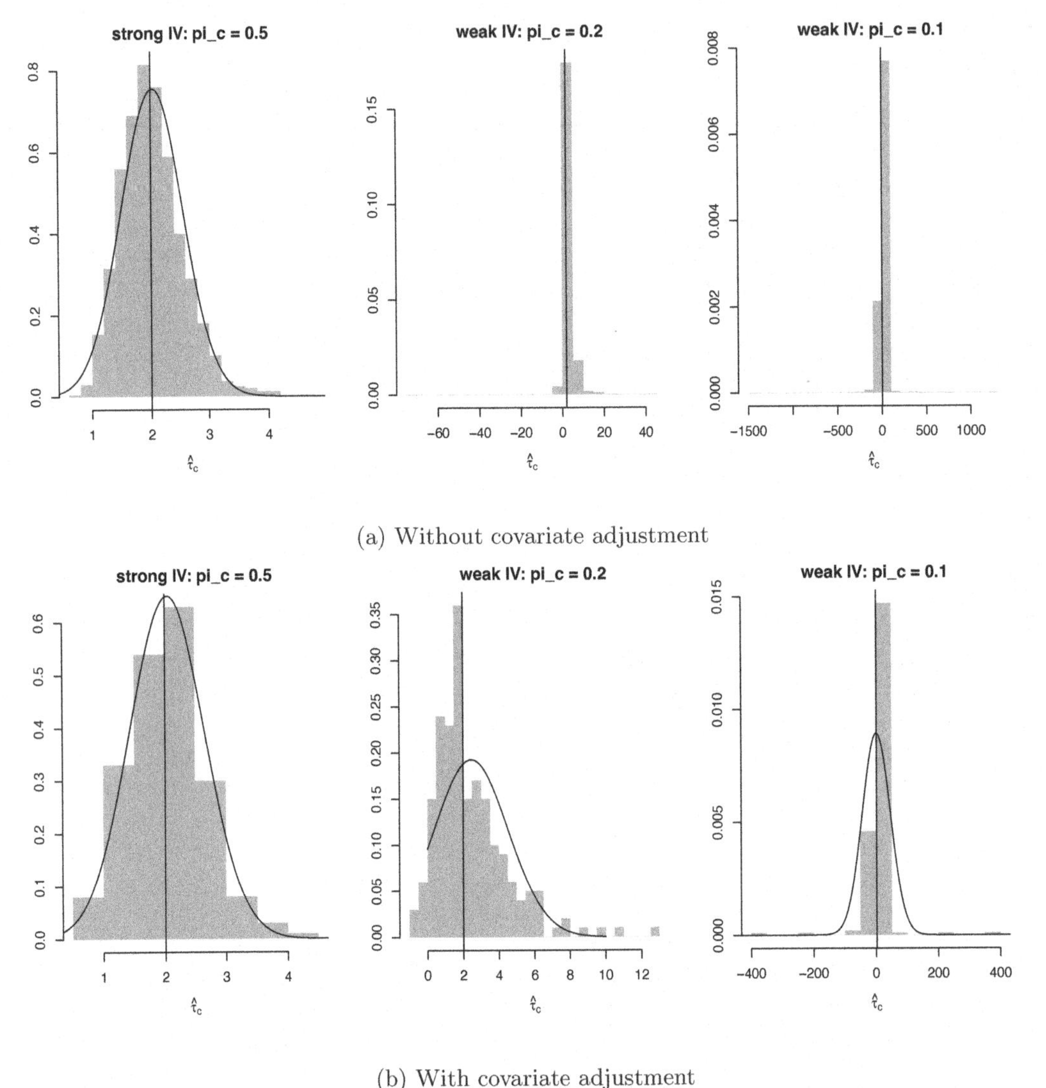

(a) Without covariate adjustment

(b) With covariate adjustment

FIGURE 21.1: Simulation with strong and weak IVs

From an estimation perspective, we can focus on the confidence interval although the point estimator has poor finite-sample properties. Because $\tau_c = \tau_Y/\tau_D$, this is similar to the classical Fieller–Creasy problem in statistics. Below we discuss a strategy for constructing confidence intervals for τ_c motivated by Fieller (1954); see Section A.4.2. By the duality between confidence intervals and hypothesis tests (see Section A.2.5), we can construct a confidence set for τ_c by inverting a sequence of null hypotheses

$$H_0(b) : \tau_c = b.$$

Given the true value $\tau_c = b$, we have

$$\tau_Y - b\tau_D = 0.$$

The null hypothesis $H_0(b)$ is equivalent to the null hypothesis of zero average causal effect on the outcome $A_i(b) = Y_i - bD_i$:

$$H_0(b) : \tau_{A(b)} = 0.$$

Let $\hat{\tau}_A(b)$ be a generic point estimator for $\tau_{A(b)}$ with the associated variance estimator $\hat{V}_A(b)$. The point and variance estimators depend on the settings. In the CRE without covariates, $\hat{\tau}_A(b)$ is the difference in means of the outcome $A_i(b)$, and $\hat{V}_A(b)$ is the Neyman-type variance estimator. In the CRE with covariates, $\hat{\tau}_A(b)$ is Lin's (2013) estimator for the outcome $A_i(b)$, and $\hat{V}_A(b)$ is the EHW variance estimator in the associated OLS fit of $Y_i - bD_i$ on (Z_i, X_i, Z_iX_i) with the correction term[5] discussed in Section 9.1. In unconfounded observational studies, we can obtain the estimator for the average causal effect on $A_i(b)$ and the associated variance estimator based on many existing strategies in Part III of this book.

Based on $\hat{\tau}_A(b)$ and $\hat{V}_A(b)$, we can construct a Wald-type test for $H_0(b)$. Inverting tests, we can construct the following confidence set for τ_c:

$$\left\{ b : \frac{\hat{\tau}_A^2(b)}{\hat{V}_A(b)} \leq z_{1-\alpha/2}^2 \right\}$$

where $z_{1-\alpha/2}$ is the $1-\alpha/2$ upper quantile of the standard Normal random variable. This is close to the Anderson–Rubin-type confidence interval in econometrics (Anderson and Rubin, 1950). Due to its connection to Fieller (1954), I will call it the Fieller–Anderson–Rubin (FAR) confidence interval. These weak-IV confidence intervals reduce to the asymptotic confidence intervals when the IV is strong. But they have additional guarantees when the IV is weak. I recommend using them in practice.

Example 21.1 *To gain intuition about the FAR confidence interval, we look into the simple case of the CRE without covariates. The quadratic inequality in the confidence interval reduces to*

$$\begin{aligned} & (\hat{\tau}_Y - b\hat{\tau}_D)^2 \\ \leq\ & z_{1-\alpha/2} \Big[n_1^{-1} \{ \hat{S}_Y^2(1) + b^2 \hat{S}_D^2(1) - 2b\hat{S}_{YD}(1) \} \\ & + n_0^{-1} \{ \hat{S}_Y^2(0) + b^2 \hat{S}_D^2(0) - 2b\hat{S}_{YD}(0) \} \Big], \end{aligned}$$

where $\{\hat{S}_Y^2(1), \hat{S}_D^2(1), \hat{S}_{YD}(1)\}$ and $\{\hat{S}_Y^2(0), \hat{S}_D^2(0), \hat{S}_{YD}(0)\}$ are the sample variances and covariances of Y and D under the treatment and control, respectively. The confidence set can be a finite bounded interval, two disconnected intervals, an empty set, or the whole real line, which is similar to the Fieller–Creasy problem in Section A.4.2. I relegate the detailed discussion to Problem 21.4.

21.4.3 Implementation and simulation

The following functions can compute a sequence of p-values as a function of b. The function `FARci` does not use covariates, and the function `FARciX` uses covariates which relies on the function `linestimator` defined in Section 9.2.

```
FARci = function(Z, D, Y, Lower, Upper, grid)
{
```

[5] If we adopt the finite-population perspective, then we do not need this correction.

```
  CIrange = seq(Lower, Upper, grid)
  Pvalue  = sapply(CIrange, function(t){
    Y_t         = Y - t*D
    Tauadj      = mean(Y_t[Z==1]) - mean(Y_t[Z==0])
    VarAdj      = var(Y_t[Z==1])/sum(Z) +
      var(Y_t[Z==0])/sum(1 - Z)
    Tstat       = Tauadj/sqrt(VarAdj)
    (1 - pnorm(abs(Tstat)))*2
  })

  return(list(CIrange = CIrange, Pvalue  = Pvalue))
}

FARciX = function(Z, D, Y, X, Lower, Upper, grid)
{
  CIrange = seq(Lower, Upper, grid)
  X       = scale(X)
  Pvalue  = sapply(CIrange, function(t){
    Y_t         = Y - t*D
    linest      = linestimator(Z, Y_t, X)
    Tstat       = linest[1]/linest[3]
    (1 - pnorm(abs(Tstat)))*2
  })

  return(list(CIrange = CIrange, Pvalue  = Pvalue))
}
```

Figure 21.2 displays the p-value as a function of b in simulated data with different π_c. Figures 21.2a and 21.2b are based on two realizations of the same data-generating process, with details relegated to the `R` code. In Figure 21.2a, the FAR confidence sets are all closed intervals, whereas, in Figure 21.2b, the confidence sets are not closed intervals with $\pi_c = 0.2$ and $\pi_c = 0.1$. The shapes of the confidence sets remain stable across two realizations with $\pi_c = 0.5$.

21.5 Application

The `mediation` package contains a dataset `jobs` from the Job Search Intervention Study (JOBS II), which was a randomized field experiment that investigated the efficacy of a job training intervention on unemployed workers. The variable `treat` is the indicator for whether a participant was randomly selected for the JOBS II training program, and the variable `comply` is the indicator for whether a participant actually participated in the JOBS II program. An outcome of interest is `job_seek` for measuring the level of job-search self-efficacy with values from 1 to 5. Covariates include `sex`, `age`, `marital`, `nonwhite`, `educ`, and `income`.

```
> jobsdata = read.csv("jobsdata.csv")
> Z = jobsdata$treat
> D = jobsdata$comply
> Y = jobsdata$job_seek
> getX      = lm(treat ~ sex + age + marital
+                 + nonwhite + educ + income,
+                 data = jobsdata)
> X = model.matrix(getX)[, -1]
```

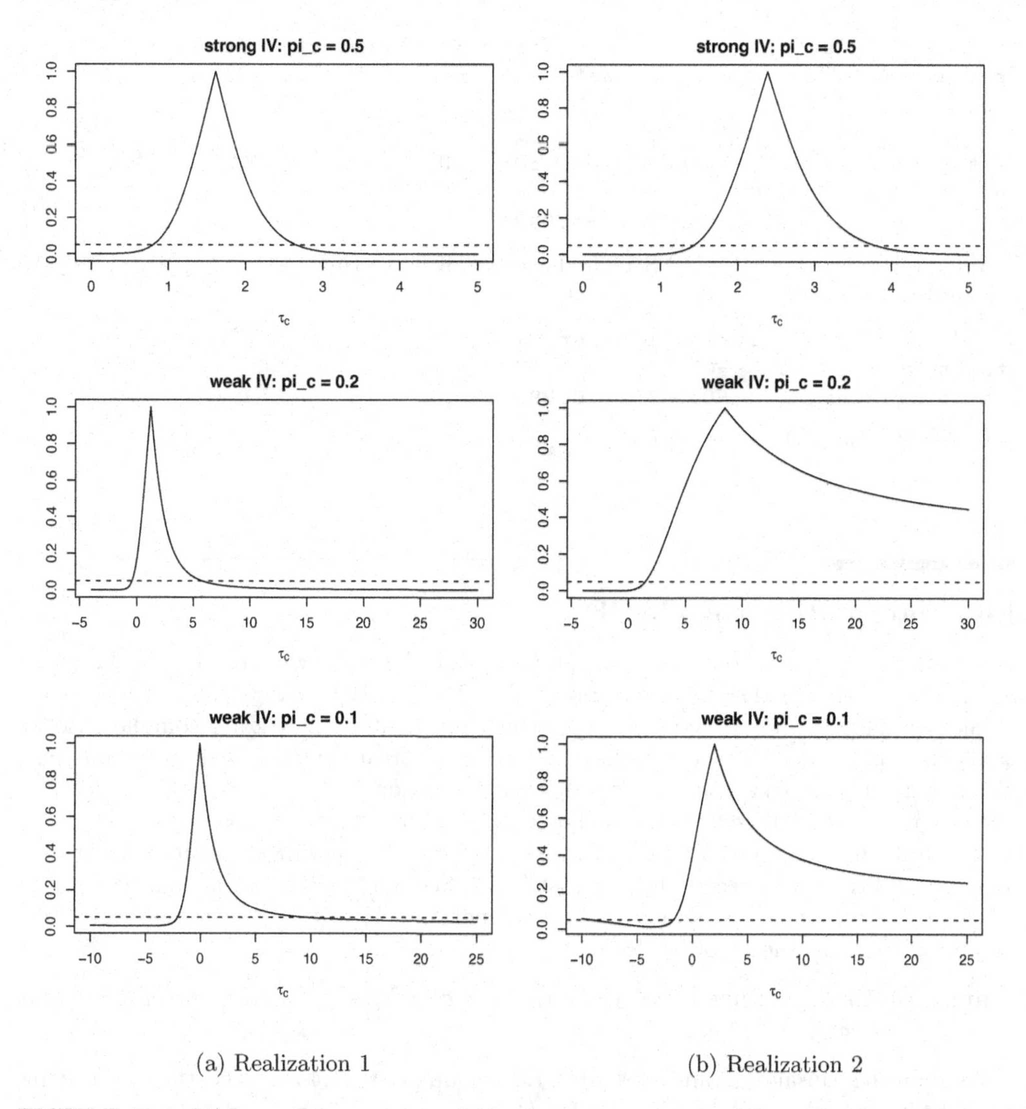

(a) Realization 1

(b) Realization 2

FIGURE 21.2: FAR confidence intervals by inverting tests based on simulated data with different π_c: two realizations. The x-axis denotes the possible values for τ_c, and the y-axis denotes the p-values from the corresponding hypothesis testing.

We can estimate τ_c by $\hat{\tau}_c$ and obtain the standard error based on $\hat{V}_{\hat{A}}$ and the bootstrap. We can further conduct covariate adjustment to obtain $\hat{\tau}_{c,L}$ and obtain the standard error based on the bootstrap. The results are below. The point estimators and the standard errors are stable across methods.

```
> ## without covariates
> res = rbind(IV_Wald_delta(Z, D, Y),
+             IV_Wald_bootstrap(Z, D, Y, n.boot = 10^3))
> ## with covariates
> res = rbind(res,
+             IV_Lin_bootstrap(Z, D, Y, X, n.boot = 10^3))
> res = cbind(res, res[, 1] - 1.96*res[, 2],
+             res[, 1] + 1.96*res[, 2])
```

```
> row.names(res) = c("delta", "bootstrap", "with covariates")
> colnames(res)  = c("est", "se", "lower CI", "upper CI")
> round(res, 3)
                  est    se lower CI upper CI
delta           0.109 0.081   -0.050    0.268
bootstrap       0.109 0.083   -0.054    0.271
with covariates 0.118 0.082   -0.042    0.278
```

We can also construct the FAR confidence sets by inverting tests. They are similar to the confidence intervals above.

```
                   lower CI upper CI
without covariates   -0.050    0.267
with covariates      -0.047    0.282
```

Figure 21.3 plots the p-values for a sequence of tests.

21.6 Interpreting the CACE

The notation of potential outcomes $\{D(1), D(0), Y(1), Y(0)\}$ is with respect to the hypothetical intervention of the treatment assigned Z. So τ_c is the average causal effect of the treatment assigned on the outcome for compliers. Fortunately, $D = Z$ for compliers, so we can also interpret τ_c as the average causal effect of the treatment received on the outcome for compliers. This partially answers the scientific question.

Some papers use different notation. For instance, Angrist et al. (1996) use $Y_i(z, d)$ for the potential outcome of unit i under a two-by-two factorial experiment[6] with the treatment assigned z and treatment received d. Angrist (2022, Section 3.1)[7] comments on the intellectual history of this choice of notation. With the notation $Y_i(z, d)$, the exclusion restriction assumption then has the following form.

Assumption 21.4 (exclusion restriction) $Y_i(z, d) = Y_i(d)$ *for all* i*, that is, the potential outcome is only a function of* d*.*

Based on the causal diagram in Figure 21.4, Assumption 21.4 rules out the direct arrow from Z to Y. In such a case, Z is an IV for D.

Under Assumption 21.4, the augmented notation $Y_i(z, d)$ reduces to $Y_i(d)$, which justifies the name of "exclusion restriction." Therefore, $Y_i(1, d) = Y_i(0, d)$ for $d = 0, 1$, which, coupled with Assumption 21.2, implies that

$$
\begin{aligned}
Y_i(z=1) - Y_i(z=0) &= Y_i(1, D_i(1)) - Y_i(0, D_i(0)) \\
&= \begin{cases} 0, & \text{if } U_i = \text{a}, \\ 0, & \text{if } U_i = \text{n}, \\ Y_i(d=1) - Y_i(d=0), & \text{if } U_i = \text{c}. \end{cases}
\end{aligned}
$$

[6] The name "factorial experiment" is from the experimental design literature (Dasgupta et al., 2015). The experimenter randomizes multiple factors to each unit. Equivalently, the treatment in the factorial experiment has multiple levels. In particular, the notation $Y_i(z, d)$ corresponds to a two-by-two factorial design, so each unit i can take one of the four treatment values.

[7] This paper is based on Angrist's Nobel Memorial Lecture. David Card received the 2021 Nobel Prize in Economics "for his empirical contributions to labour economics," and shared the prize with Joshua Angrist and Guido Imbens "for their methodological contributions to the analysis of causal relationships." This chapter reviews some fundamental contributions of Joshua Angrist and Guido Imbens, as well as their collaborator Donald Rubin. Section 23.7 will revisit a canonical study of Card (1993) for an application of IV.

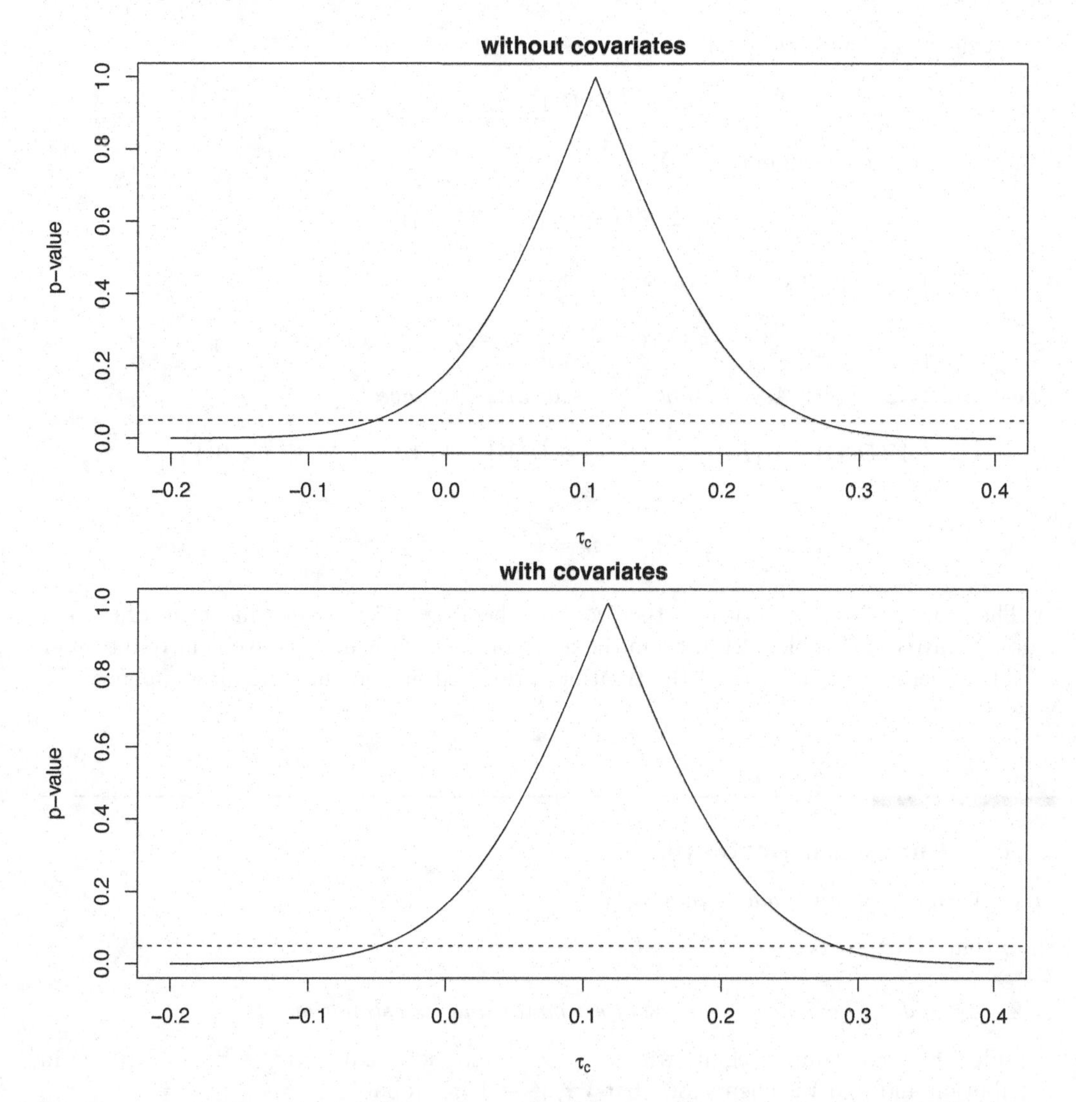

FIGURE 21.3: The FAR confidence intervals of τ_c by inverting tests: upper panel without covariate adjustment, and lower panel with covariate adjustment

In the above, I emphasize that the potential outcomes are with respect to z, d, or both, to avoid confusion. The previous decomposition of τ_Y holds and we have the following result from Imbens and Angrist (1994) and Angrist et al. (1996).

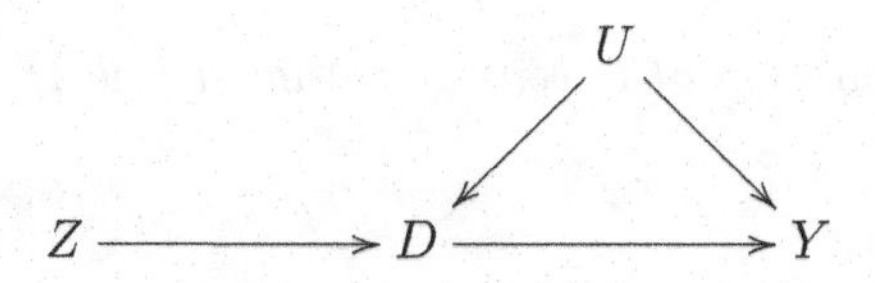

FIGURE 21.4: Causal diagram for IV. The node U denotes the unmeasured confounder between D and Y. Based on the discussion in this chapter, we can simply define $U = \{D(1), D(0)\}$. Then D is a deterministic function of Z and U.

Recall the average causal effect on D,

$$\tau_D = E\{D(1) - D(0)\},$$

define the average causal effect on Y as

$$\tau_Y = E\{Y(D(1)) - Y(D(0))\},$$

and define the complier average causal effect as

$$\tau_{\mathrm{c}} = E\{Y(d=1) - Y(d=0) \mid U = \mathrm{c}\}.$$

Theorem 21.2 *Under Assumptions 21.2 and 21.4, we have*

$$Y(D(1)) - Y(D(0)) = \{D(1) - D(0)\} \times \{Y(d=1) - Y(d=0)\}$$

and

$$\tau_{\mathrm{c}} = \frac{\tau_Y}{\tau_D}.$$

The proof is almost identical to the proof of Theorem 21.1 with modifications of the notation. I leave it as Problem 21.3. From the notation $Y_i(d)$, it is more convenient to interpret τ_{c} as the average causal effect of the treatment received on the outcome for compliers.

21.7 Homework problems

21.1 Variance of the Wald estimator

Show that $\mathrm{var}(\hat{\tau}_{\mathrm{c}}) = \infty$.

21.2 Asymptotic variance of the Wald estimator and its estimation

Consider the large-sample regime with $n \to \infty$. First show that $\sqrt{n}(\hat{\tau}_{\mathrm{c}} - \tau_{\mathrm{c}}) \to \mathrm{N}(0, V)$ in distribution and find V. Then show that $\hat{V}_{\hat{A}}/V \to 1$ in probability.

Remark: You may find Example A.1 useful for your proof. The asymptotic distribution of $\hat{\tau}_{\mathrm{c}}$ holds under some regularity conditions. I omit those conditions, but you need to specify them in your proof.

Problem 21.1 states that the variance of $\hat{\tau}_{\mathrm{c}}$ is infinite, whereas Problem 21.2 states that the asymptotic variance of $\hat{\tau}_{\mathrm{c}}$ can be finite under some regularity conditions. These two problems are not in contradiction with each other because the asymptotic variance of $\hat{\tau}_{\mathrm{c}}$ equals the variance of its asymptotic distribution, not the limiting value of its variance. This is a subtle issue in asymptotic theory of mathematical statistics.

21.3 Proof of the main theorem of Imbens and Angrist (1994) and Angrist et al. (1996)

Prove Theorem 21.2.

21.4 More on the FAR confidence set

The confidence set in Example 21.1 can be a close interval, two disconnected intervals, an empty set, or the whole real line. Find the precise condition for each case.

21.5 More simulation for the FAR confidence set

Figure 21.2 shows the FAR confidence sets without using covariates. Conduct parallel simulation for the FAR confidence sets adjusting for covariates in CRE.

21.6 Binary IV and ordinal treatment received

Angrist and Imbens (1995) discussed a more general setting with a binary IV Z, an ordinal treatment received $D \in \{0, 1, \ldots, J\}$, and an outcome Y. The ordinal treatment received has potential outcomes $D(1)$ and $D(0)$ with respect to the binary IV, and the outcome has potential outcomes $Y(z, d)$ with respect to both the binary IV and the ordinal treatment received. Extend the discussion in Section 21.6 and the corresponding IV assumptions as below.

Assumption 21.5 *We have (1) randomization that* $Z \perp\!\!\!\perp \{D(z), Y(z, d) : z = 0, 1; d = 0, 1, \ldots, J\}$*; (2) monotonicity that* $D(1) \geq D(0)$*; and (3) exclusion restriction that* $Y(z, d) = Y(d)$ *for all* $z = 0, 1$ *and* $d = 0, 1, \ldots, J$.

They proved Theorem 21.3 below.

Theorem 21.3 *Under Assumption 21.5, we have*

$$\frac{E(Y \mid Z=1) - E(Y \mid Z=0)}{E(D \mid Z=1) - E(D \mid Z=0)} = \sum_{j=1}^{J} w_j E\{Y(j) - Y(j-1) \mid D(1) \geq j > D(0)\}$$

where

$$w_j = \frac{\text{pr}\{D(1) \geq j > D(0)\}}{\sum_{j'=1}^{J} \text{pr}\{D(1) \geq j' > D(0)\}}.$$

Prove Theorem 21.3.

Remark: When $J = 1$, Theorem 21.3 reduces to Theorem 21.2. With a general J, it states that the standard IV formula identifies a weighted average of some latent subgroup effects. The weights are proportional to the probability of the latent groups defined by $D(1) \geq j > D(0)$, and the latent subgroup effect $E\{Y(j) - Y(j-1) \mid D(1) \geq j > D(0)\}$ compares the adjacent levels of the treatment received. However, this weighted average may not be easy to interpret because the latent groups overlap.

The proof can be tedious. A trick is to write the treatment received and outcome under treatment assignment z as

$$D(z) = \sum_{j=0}^{J} j 1\{D(z) = j\}, \quad Y(D(z)) = \sum_{j=0}^{J} Y(j) 1\{D(z) = j\}$$

to obtain

$$D(1) - D(0) = \sum_{j=0}^{J} j[1\{D(1) = j\} - 1\{D(0) = j\}]$$

and

$$Y(D(1)) - Y(D(0)) = \sum_{j=0}^{J} Y(j)[1\{D(1) = j\} - 1\{D(0) = j\}].$$

Then use the following *Abel's lemma*, also called *summation by parts*:

$$\sum_{j=0}^{J} f_j(g_{j+1} - g_j) = f_J g_{J+1} - f_0 g_0 - \sum_{j=1}^{J} g_j(f_j - f_{j-1})$$

for appropriately specified sequences (f_j) and (g_j).

21.7 Data analysis: a flu shot encouragement design (McDonald et al., 1992)

The dataset in `fludata.txt` is from a randomized encouragement design of McDonald et al. (1992), which was also reanalyzed by Hirano et al. (2000). It contains the following variables:

`assign`	binary encouragement to receive the flu shot Z
`receive`	binary indicator for receiving the flu shot D
`outcome`	binary outcome for flu-related hospitalization Y
`age`	age of the patient
`sex`	sex of the patient
`race`	race of the patient
`copd`	chronic obstructive pulmonary disease
`dm`	diabetes
`heartd`	heart disease
`renal`	renal disease
`liverd`	liver disease

Analyze the data with and without adjusting for the covariates.

21.8 IV estimation conditional on covariates

Implement the estimators and corresponding variance estimators mentioned in Section 21.3.2 in `R`.

Remark: The problem is useful for Problem 21.9.

21.9 Data analysis: the Karolinska data

Revisit Problem 12.6. Rubin (2008) used the Karolinska data as an example for the IV method. In `karolinska.txt`, whether a patient was diagnosed at a large volume hospital can be viewed as an IV for whether a patient was treated at a large volume hospital. This is plausible conditional on other observed covariates. See Rubin (2008) for more details.

Reanalyze the data assuming that the IV is randomly assigned conditional on observed covariates.

21.10 Data analysis: a job training program

The file `jobtraining.rtf` contains the description of the data files `X.csv` and `Y.csv`.

The dataset `X.csv` contains the pretreatment covariates. You can view the sampling weight variable `wgt` as a covariate too. Many previous analyses made this simplification although this is always a controversial issue in statistical analysis of survey data with weights. Conduct analyses with and without covariates.

The dataset `Y.csv` contains the sampling weight, treatment assigned, treatment received, and many post-treatment variables. Therefore, this dataset contains many outcomes depending on your questions of interest. The data also have many complications. First, some outcomes are missing. Second, unemployed individuals do not have wages. Third, the outcomes are repeatedly observed over time. When you analyze the data, please give details about your choice of the questions of interest and estimators.

Remark: Schochet et al. (2008) analyzed the original data. Frumento et al. (2012) provided a more sophisticated analysis based on the framework in Chapter 26 later.

21.11 Recommended reading

Angrist et al. (1996) bridged the econometric IV perspective and statistical causal inference based on potential outcomes and demonstrated its usefulness with an application.

Some other early references on IV are Permutt and Hebel (1989), Sommer and Zeger (1991), Baker and Lindeman (1994), and Cuzick et al. (1997).

22

Disentangle Mixture Distributions and Instrumental Variable Inequalities

The IV model in Chapter 21 imposes Assumptions 21.1–21.3:

1. $Z \perp\!\!\!\perp \{D(1), D(0), Y(1), Y(0)\}$;
2. $\text{pr}(U = \text{d}) = 0$;
3. $Y(1) = Y(0)$ for $U = \text{a}$ or n.

Table 22.1 summarizes the observed groups and the corresponding latent groups under the monotonicity assumption. The observed group with $(Z = 1, D = 1)$ is a mixture distribution with two latent group types c and a, and the observed group with $(Z = 0, D = 0)$ is a mixture distribution with two latent group types c and n.

TABLE 22.1: Observed groups and latent groups under Assumption 21.2

Z	D	$D(1)$	$D(0)$	latent groups
$Z = 1$	$D = 1$	$D(1) = 1$	$D(0) = ?$	$U = \text{c or a}$
$Z = 1$	$D = 0$	$D(1) = 0$	$D(0) = 0$	$U = \text{n}$
$Z = 0$	$D = 1$	$D(1) = 1$	$D(0) = 1$	$U = \text{a}$
$Z = 0$	$D = 0$	$D(1) = ?$	$D(0) = 0$	$U = \text{c or n}$

Interestingly, Assumptions 21.1–21.3 together have some testable implications. Balke and Pearl (1997) called them the *instrumental variable inequalities*. This chapter will give an intuitive derivation of a special case of these inequalities. The proof is a direct consequence of identifying the means of the potential outcomes for all latent groups in Table 22.1 defined by U. The key challenge in the proof is to disentangle mixture distributions in the observed groups with $(Z = 1, D = 1)$ and $(Z = 0, D = 0)$, that is, to identify the outcome distributions for all the latent group types.

22.1 Disentangle mixture distributions

I summarize the main results in Theorem 22.1 below. Define

$$\pi_u = \text{pr}(U = u) \qquad (u = \text{a}, \text{n}, \text{c})$$

as the proportion of type $U = u$, and

$$\mu_{zu} = E\{Y(z) \mid U = u\} \qquad (z = 0, 1; u = \text{a}, \text{n}, \text{c})$$

as the mean of the potential outcome $Y(z)$ for type $U = u$. Exclusion restriction implies that $\mu_{1\text{n}} = \mu_{0\text{n}}$ and $\mu_{1\text{a}} = \mu_{0\text{a}}$. Let μ_n and μ_a denote them, respectively.

Theorem 22.1 *Under Assumptions 21.1–21.3, we can identify the proportions of the latent types by*

$$
\begin{aligned}
\pi_{\mathrm{n}} &= \mathrm{pr}(D=0 \mid Z=1), \\
\pi_{\mathrm{a}} &= \mathrm{pr}(D=1 \mid Z=0), \\
\pi_{\mathrm{c}} &= E(D \mid Z=1) - E(D \mid Z=0),
\end{aligned}
$$

and the type-specific means of the potential outcomes by

$$
\begin{aligned}
\mu_{\mathrm{n}} &= E(Y \mid Z=1, D=0), \\
\mu_{\mathrm{a}} &= E(Y \mid Z=0, D=1), \\
\mu_{1\mathrm{c}} &= \pi_{\mathrm{c}}^{-1}\{E(DY \mid Z=1) - E(DY \mid Z=0)\}, \\
\mu_{0\mathrm{c}} &= \pi_{\mathrm{c}}^{-1}\big[E\{(1-D)Y \mid Z=0\} - E\{(1-D)Y \mid Z=1\}\big].
\end{aligned}
$$

Proof of Theorem 22.1: Part I: We first identify the proportions of the latent compliance types. We can identify the proportion of the never takers by

$$
\begin{aligned}
\mathrm{pr}(D=0 \mid Z=1) &= \mathrm{pr}(U=\mathrm{n} \mid Z=1) \\
&= \mathrm{pr}(U=\mathrm{n}) \\
&= \pi_{\mathrm{n}},
\end{aligned}
$$

and the proportion of the always takers by

$$
\begin{aligned}
\mathrm{pr}(D=1 \mid Z=0) &= \mathrm{pr}(U=\mathrm{a} \mid Z=0) \\
&= \mathrm{pr}(U=\mathrm{a}) \\
&= \pi_{\mathrm{a}}.
\end{aligned}
$$

Therefore, the proportion of compliers is

$$
\begin{aligned}
\pi_{\mathrm{c}} &= \mathrm{pr}(U=\mathrm{c}) \\
&= 1 - \pi_{\mathrm{n}} - \pi_{\mathrm{a}} \\
&= 1 - \mathrm{pr}(D=0 \mid Z=1) - \mathrm{pr}(D=1 \mid Z=0) \\
&= E(D \mid Z=1) - E(D \mid Z=0) \\
&= \tau_D,
\end{aligned}
$$

which is coherent with our previous discussion. Although we do not know individual latent compliance types for all units, we can identify the proportions of never takers, always takers, and compliers.

Part II: We then identify the means of the potential outcomes within latent compliance types. The observed group $(Z=1, D=0)$ only has never takers, so

$$
\begin{aligned}
E(Y \mid Z=1, D=0) &= E\{Y(1) \mid Z=1, U=\mathrm{n}\} \\
&= E\{Y(1) \mid U=\mathrm{n}\} \\
&= \mu_{\mathrm{n}}.
\end{aligned}
$$

The observed group $(Z=0, D=1)$ only has always takers, so

$$
\begin{aligned}
E(Y \mid Z=0, D=1) &= E\{Y(0) \mid Z=0, U=\mathrm{a}\} \\
&= E\{Y(0) \mid U=\mathrm{a}\} \\
&= \mu_{\mathrm{a}}.
\end{aligned}
$$

The observed group $(Z = 1, D = 1)$ has both always takers and compliers, so

$$\begin{aligned}
& E(Y \mid Z = 1, D = 1) \\
= \; & E\{Y(1) \mid Z = 1, D(1) = 1\} \\
= \; & E\{Y(1) \mid D(1) = 1\} \\
= \; & \text{pr}\{D(0) = 1 \mid D(1) = 1\} E\{Y(1) \mid D(1) = 1, D(0) = 1\} \\
& + \text{pr}\{D(0) = 0 \mid D(1) = 1\} E\{Y(1) \mid D(1) = 1, D(0) = 0\} \\
= \; & \frac{\pi_{\text{a}}}{\pi_{\text{c}} + \pi_{\text{a}}} \mu_{\text{a}} + \frac{\pi_{\text{c}}}{\pi_{\text{c}} + \pi_{\text{a}}} \mu_{1\text{c}}.
\end{aligned}$$

Solve the linear equation above to obtain

$$\begin{aligned}
\mu_{1\text{c}} &= \pi_{\text{c}}^{-1} \{(\pi_{\text{c}} + \pi_{\text{a}}) E(Y \mid Z = 1, D = 1) - \pi_{\text{a}} E(Y \mid Z = 0, D = 1)\} \\
&= \pi_{\text{c}}^{-1} \{\text{pr}(D = 1 \mid Z = 1) E(Y \mid Z = 1, D = 1) \\
&\qquad - \text{pr}(D = 1 \mid Z = 0) E(Y \mid Z = 0, D = 1)\} \\
&= \pi_{\text{c}}^{-1} \{E(DY \mid Z = 1) - E(DY \mid Z = 0)\}.
\end{aligned}$$

The observed group $(Z = 0, D = 0)$ has both compliers and never takers, so

$$\begin{aligned}
& E(Y \mid Z = 0, D = 0) \\
= \; & E\{Y(0) \mid Z = 0, D(0) = 0\} \\
= \; & E\{Y(0) \mid D(0) = 0\} \\
= \; & \text{pr}\{D(1) = 1 \mid D(0) = 0\} E\{Y(0) \mid D(1) = 1, D(0) = 0\} \\
& + \text{pr}\{D(1) = 0 \mid D(0) = 0\} E\{Y(0) \mid D(1) = 0, D(0) = 0\} \\
= \; & \frac{\pi_{\text{c}}}{\pi_{\text{c}} + \pi_{\text{n}}} \mu_{0\text{c}} + \frac{\pi_{\text{n}}}{\pi_{\text{c}} + \pi_{\text{n}}} \mu_{\text{n}}.
\end{aligned}$$

Solve the linear equation above to obtain

$$\begin{aligned}
\mu_{0\text{c}} &= \pi_{\text{c}}^{-1} \{(\pi_{\text{c}} + \pi_{\text{n}}) E(Y \mid Z = 0, D = 0) - \pi_{\text{n}} E(Y \mid Z = 1, D = 0)\} \\
&= \pi_{\text{c}}^{-1} \{\text{pr}(D = 0 \mid Z = 0) E(Y \mid Z = 0, D = 0) \\
&\qquad - \text{pr}(D = 0 \mid Z = 1) E(Y \mid Z = 1, D = 0)\} \\
&= \pi_{\text{c}}^{-1} \left[E\{(1 - D)Y \mid Z = 0\} - E\{(1 - D)Y \mid Z = 1\}\right].
\end{aligned}$$

□

Based on the formulas of $\mu_{1\text{c}}$ and $\mu_{0\text{c}}$ in Theorem 22.1, we can simplify $\tau_{\text{c}} = \mu_{1\text{c}} - \mu_{0\text{c}}$ as

$$\tau_{\text{c}} = \{E(Y \mid Z = 1) - E(Y \mid Z = 0)\} / \pi_{\text{c}},$$

which is the same as the formula in Theorem 21.1 before.

Theorem 22.1 focuses on identifying the means of the potential outcomes, μ_{zu}. Imbens and Rubin (1997) derived more general identification formulas for the distribution of the potential outcomes; I leave the details to Problem 22.3.

22.2 Testable implications: instrumental variable inequalities

Is there any additional value of the detour to derive the formula of τ_c through Theorem 22.1? The answer is yes. For binary outcome, the following inequalities must be true:

$$\begin{aligned} 0 \leq \ & \mu_{1c} \ \leq 1, \\ 0 \leq \ & \mu_{0c} \ \leq 1, \end{aligned}$$

which implies four inequalities

$$\begin{aligned} E(DY \mid Z=1) - E(DY \mid Z=0) &\geq 0, \\ E(DY \mid Z=1) - E(DY \mid Z=0) &\leq E(D \mid Z=1) - E(D \mid Z=0), \\ E\{(1-D)Y \mid Z=0\} - E\{(1-D)Y \mid Z=1\} &\geq 0, \\ E\{(1-D)Y \mid Z=0\} - E\{(1-D)Y \mid Z=1\} &\leq E(D \mid Z=1) - E(D \mid Z=0). \end{aligned}$$

Rearranging terms, we obtain the following unified inequalities.

Theorem 22.2 (Instrumental Variable Inequalities) *With a binary outcome Y, Assumptions 21.1–21.3 imply*

$$E(Q \mid Z=1) - E(Q \mid Z=0) \geq 0, \tag{22.1}$$

where $Q = DY, D(1-Y), (D-1)Y$, or $D+Y-DY$.

Under the IV assumptions 21.1–21.3, the difference in means for $Q = DY, D(1-Y), (D-1)Y$ and $D+Y-DY$ must all be non-negative. Importantly, these implications only involve the distribution of the observed variables. They are testable based on the observed data. Rejection of the IV inequalities leads to rejection of the IV assumptions.

Balke and Pearl (1997) derived more general IV inequalities with and without assuming monotonicity. The proving strategy above is due to Jiang and Ding (2020) for a slightly more complex setting. Theorem 22.2 states the testable implications only for a binary outcome. Problem 22.4 gives an equivalent form, and Problem 22.5 gives the result for a general outcome.

22.3 Examples

For a binary outcome, we can estimate all the parameters by the method of moments below.

```
## function for binary data (Z, D, Y)
## n_{zdy}'s are the counts from 2X2X2 table
IVbinary = function(n111, n110, n101, n100,
                    n011, n010, n001, n000){

  n_tr = n111 + n110 + n101 + n100
  n_co = n011 + n010 + n001 + n000
  n    = n_tr + n_co

  ## proportions of the latent strata
  pi_n = (n101 + n100)/n_tr
```

```
  pi_a = (n011 + n010)/n_co
  pi_c = 1 - pi_n - pi_a

  ## four observed means of the outcomes (Z=z,D=d)
  mean_y_11 = n111/(n111 + n110)
  mean_y_10 = n101/(n101 + n100)
  mean_y_01 = n011/(n011 + n010)
  mean_y_00 = n001/(n001 + n000)

  ## means of the outcomes of two strata
  mu_n1 = mean_y_10
  mu_a0 = mean_y_01
  ## ER implies the following two means
  mu_n0 = mu_n1
  mu_a1 = mu_a0
  ## stratum (Z=1,D=1) is a mixture of c and a
  mu_c1 = ((pi_c + pi_a)*mean_y_11 - pi_a*mu_a1)/pi_c
  ## stratum (Z=0,D=0) is a mixture of c and n
  mu_c0 = ((pi_c + pi_n)*mean_y_00 - pi_n*mu_n0)/pi_c

  ## identifiable quantities from the observed data
  list(pi_c = pi_c,
       pi_n = pi_n,
       pi_a = pi_a,
       mu_c1= mu_c1,
       mu_c0= mu_c0,
       mu_n1= mu_n1,
       mu_n0= mu_n0,
       mu_a1= mu_a1,
       mu_a0= mu_a0,
       tau_c= mu_c1 - mu_c0)
}
```

We then revisit two canonical examples with binary data.

Example 22.1 *Investigators et al. (2014) assess the effectiveness of the emergency endovascular versus the open surgical repair strategies for patients with a clinical diagnosis of ruptured aortic aneurism. Patients are randomized to either the emergency endovascular or the open repair strategy. The primary outcome is the survival status after 30 days. Let Z be the treatment assigned, with $Z = 1$ for the endovascular strategy and $Z = 0$ for the open repair. Let D be the treatment received. Let Y be the survival status, with $Y = 1$ for dead, and $Y = 0$ for alive. Table 22.2a summarizes the observed data. Using the* `IVbinary` *function above, we can obtain the following estimates:*

```
> investigators_analysis = IVbinary(n111 = 107,
+                                   n110 = 42,
+                                   n101 = 68,
+                                   n100 = 42,
+                                   n011 = 24,
+                                   n010 = 8,
+                                   n001 = 131,
+                                   n000 = 79)
>
> investigators_analysis
$pi_c
[1] 0.4430582
```

```
$pi_n
[1] 0.4247104

$pi_a
[1] 0.1322314

$mu_c1
[1] 0.7086064

$mu_c0
[1] 0.6292042

$mu_n1
[1] 0.6181818

$mu_n0
[1] 0.6181818

$mu_a1
[1] 0.75

$mu_a0
[1] 0.75

$tau_c
[1] 0.07940223
```

There is no evidence of violating the IV assumptions since all $\hat{\mu}_{zu}$'s are positive.

Example 22.2 *In Hirano et al. (2000), physicians are randomly selected to receive a letter encouraging them to inoculate patients at risk for flu. The treatment is the actual flu shot, and the outcome is an indicator of flu-related hospital visits. However, some patients do not comply with their assignments. Let Z_i be the indicator of encouragement to receive the flu shot, with $Z = 1$ if the physician receives the encouragement letter, and $Z = 0$ otherwise. Let D be the treatment received. Let Y be the outcome, with $Y = 0$ if for a flu-related hospitalization during the winter, and $Y = 1$ otherwise. See Problem 21.7 for more details of the data. Table 22.2b summarizes the observed data. Using the* `IVbinary` *function above, we can obtain the following estimates:*

```
> flu_analysis = IVbinary(n111 = 31,
+                         n110 = 422,
+                         n101 = 84,
+                         n100 = 935,
+                         n011 = 30,
+                         n010 = 233,
+                         n001 = 99,
+                         n000 = 1027)
> flu_analysis
$pi_c
[1] 0.1183997
```

TABLE 22.2: Binary data and IV inequalities

(a) Data from Investigators et al. (2014)

	Z = 1		Z = 0	
	D = 1	D = 0	D = 1	D = 0
Y = 1	107	68	24	131
Y = 0	42	42	8	79

(b) Data from Hirano et al. (2000)

	Z = 1		Z = 0	
	D = 1	D = 0	D = 1	D = 0
Y = 1	31	85	30	99
Y = 0	424	944	237	1041

```
$pi_n
[1] 0.6922554

$pi_a
[1] 0.1893449

$mu_c1
[1] -0.004548064

$mu_c0
[1] 0.1200094

$mu_n1
[1] 0.08243376

$mu_n0
[1] 0.08243376

$mu_a1
[1] 0.1140684

$mu_a0
[1] 0.1140684

$tau_c
[1] -0.1245575
```

Since $\hat{\mu}_{1c} < 0$, there is evidence of violating the IV assumptions.

22.4 Homework problems

22.1 More detailed data analysis

Examples 22.1 and 22.2 ignored the uncertainty in the estimates. Calculate the confidence intervals for the true parameters.

22.2 Risk ratio for compliers

With a binary outcome, we can define the risk ratio for compliers as

$$\mathrm{RR_c} = \frac{\mathrm{pr}\{Y(1) = 1 \mid U = \mathrm{c}\}}{\mathrm{pr}\{Y(0) = 1 \mid U = \mathrm{c}\}}.$$

Show that under Assumptions 21.1–21.3, we can identify it by

$$\mathrm{RR_c} = \frac{E(DY \mid Z = 1) - E(DY \mid Z = 0)}{E\{(D-1)Y \mid Z = 1\} - E\{(D-1)Y \mid Z = 0\}}.$$

Remark: Using Theorem 22.1, we can identify any comparisons between $E\{Y(1) \mid U = \mathrm{c}\}$ and $E\{Y(0) \mid U = \mathrm{c}\}$.

22.3 Disentangle the mixtures: distributional results

This problem extends Theorem 22.1. Define

$$f_{zu}(y) = \mathrm{pr}\{Y(z) = y \mid U = u\}, \quad (z = 0, 1; u = \mathrm{a}, \mathrm{n}, \mathrm{c})$$

as the density of $Y(z)$ for latent stratum $U = u$, and define

$$g_{zd}(y) = \mathrm{pr}(Y = y \mid Z = z, D = d)$$

as the density of the outcome within the observed group $(Z = z, D = d)$. Exclusion restriction implies that $f_{1\mathrm{n}}(y) = f_{0\mathrm{n}}(y)$ and $f_{1\mathrm{a}}(y) = f_{0\mathrm{a}}(y)$. Let $f_{\mathrm{n}}(y)$ and $f_{\mathrm{a}}(y)$ denote them, respectively.

Prove Theorem 22.3 below.

Theorem 22.3 *Under Assumptions 21.1–21.3, we can identify the type-specific densities of the potential outcomes by*

$$\begin{aligned}
f_{\mathrm{n}}(y) &= g_{10}(y), \\
f_{\mathrm{a}}(y) &= g_{01}(y), \\
f_{1\mathrm{c}}(y) &= \pi_{\mathrm{c}}^{-1}\{\mathrm{pr}(D = 1 \mid Z = 1) g_{11}(y) - \mathrm{pr}(D = 1 \mid Z = 0) g_{01}(y)\}, \\
f_{0\mathrm{c}}(y) &= \pi_{\mathrm{c}}^{-1}\{\mathrm{pr}(D = 0 \mid Z = 0) g_{00}(y) - \mathrm{pr}(D = 0 \mid Z = 1) g_{10}(y)\}.
\end{aligned}$$

22.4 Alternative form of Theorem 22.2

The inequalities in (22.1) can be rewritten as

$$\begin{aligned}
\mathrm{pr}(D = 1, Y = y \mid Z = 1) &\geq \mathrm{pr}(D = 1, Y = y \mid Z = 0), \\
\mathrm{pr}(D = 0, Y = y \mid Z = 0) &\geq \mathrm{pr}(D = 0, Y = y \mid Z = 1)
\end{aligned}$$

for both $y = 0, 1$.

22.5 IV inequalities for a general outcome

For a general outcome Y, show that Assumptions 21.1–21.3 imply

$$\begin{aligned}
\mathrm{pr}(D = 1, Y \geq y \mid Z = 1) &\geq \mathrm{pr}(D = 1, Y \geq y \mid Z = 0), \\
\mathrm{pr}(D = 1, Y < y \mid Z = 1) &\geq \mathrm{pr}(D = 1, Y < y \mid Z = 0), \\
\mathrm{pr}(D = 0, Y \geq y \mid Z = 0) &\geq \mathrm{pr}(D = 0, Y \geq y \mid Z = 1), \\
\mathrm{pr}(D = 0, Y < y \mid Z = 0) &\geq \mathrm{pr}(D = 0, Y < y \mid Z = 1)
\end{aligned}$$

for all y.

Remark: Imbens and Rubin (1997) and Kitagawa (2015) discussed similar results. We can test the first inequality based on an analog of the Kolmogorov–Smirnov statistic:

$$\mathrm{KS}_1 = \max_y \left| \frac{\sum_{i=1}^n Z_i D_i 1(Y_i \leq y)}{\sum_{i=1}^n Z_i} - \frac{\sum_{i=1}^n (1-Z_i) D_i 1(Y_i \leq y)}{\sum_{i=1}^n (1-Z_i)} \right|.$$

22.6 Example for the IV inequalities

Give an example in which all the IV inequalities hold and another example in which not all the IV inequalities hold. You need to specify the joint distribution of (Z, D, Y) with binary variables.

22.7 Violations of the key assumptions

Theorem 21.1 relies on randomization, monotonicity, and exclusion restriction. The latter two are not testable even in randomized experiments. When they are violated, the IV estimator no longer identifies the CACE. This problem gives two cases below, which are restatements of Propositions 2 and 3 in Angrist et al. (1996). Recall $\pi_u = \mathrm{pr}(U = u)$ and $\tau_u = E\{Y(1) - Y(0) \mid U = u\}$ for $u = \mathrm{a, n, c, d}$.

Theorem 22.4 *(a) Under Assumptions 21.1 and 21.2 without the exclusion restriction, we have*

$$\frac{E(Y \mid Z=1) - E(Y \mid Z=0)}{E(D \mid Z=1) - E(D \mid Z=0)} - \tau_{\mathrm{c}} = \frac{\pi_{\mathrm{a}}\tau_{\mathrm{a}} + \pi_{\mathrm{n}}\tau_{\mathrm{n}}}{\pi_{\mathrm{c}}}.$$

(b) Under Assumptions 21.1 and 21.3 without the monotonicity, we have

$$\frac{E(Y \mid Z=1) - E(Y \mid Z=0)}{E(D \mid Z=1) - E(D \mid Z=0)} - \tau_{\mathrm{c}} = \frac{\pi_{\mathrm{d}}(\tau_{\mathrm{c}} + \tau_{\mathrm{d}})}{\pi_{\mathrm{c}} - \pi_{\mathrm{d}}}.$$

Prove Theorem 22.4.

22.8 Problems of other analyses

In the process of deriving the IV inequalities in Section 22.1, we disentangled the mixture distributions by identifying the proportions of the latent strata as well as the conditional means of their potential outcomes. These results help to understand the drawbacks of other seemingly reasonable analyses. I review three estimators below and suppose Assumptions 21.1–21.3 hold.

1. The *as-treated analysis* compares the means of the outcomes among units receiving the treatment and control, yielding

$$\tau_{\mathrm{AT}} = E(Y \mid D=1) - E(Y \mid D=0).$$

Show that

$$\tau_{\mathrm{AT}} = \frac{\pi_{\mathrm{a}}\mu_{\mathrm{a}} + \mathrm{pr}(Z=1)\pi_{\mathrm{c}}\mu_{1\mathrm{c}}}{\mathrm{pr}(D=1)} - \frac{\pi_{\mathrm{n}}\mu_{\mathrm{n}} + \mathrm{pr}(Z=0)\pi_{\mathrm{c}}\mu_{0\mathrm{c}}}{\mathrm{pr}(D=0)}.$$

2. The *per-protocol analysis* compares the units that comply with the treatment assigned in treatment and control groups, yielding

$$\tau_{\mathrm{PP}} = E(Y \mid Z=1, D=1) - E(Y \mid Z=0, D=0).$$

Show that

$$\tau_{\mathrm{PP}} = \frac{\pi_{\mathrm{a}}\mu_{\mathrm{a}} + \pi_{\mathrm{c}}\mu_{1\mathrm{c}}}{\pi_{\mathrm{a}} + \pi_{\mathrm{c}}} - \frac{\pi_{\mathrm{n}}\mu_{\mathrm{n}} + \pi_{\mathrm{c}}\mu_{0\mathrm{c}}}{\pi_{\mathrm{n}} + \pi_{\mathrm{c}}}.$$

3. We may also want to compare the outcomes among units receiving the treatment and control, conditioning on their treatment assignment, yielding

$$\begin{aligned} \tau_{Z=1} &= E(Y \mid Z=1, D=1) - E(Y \mid Z=1, D=0), \\ \tau_{Z=0} &= E(Y \mid Z=0, D=1) - E(Y \mid Z=0, D=0). \end{aligned}$$

Show that they reduce to

$$\tau_{Z=1} = \frac{\pi_{\mathrm{a}}\mu_{\mathrm{a}} + \pi_{\mathrm{c}}\mu_{1\mathrm{c}}}{\pi_{\mathrm{a}} + \pi_{\mathrm{c}}} - \mu_{\mathrm{n}}, \quad \tau_{Z=0} = \mu_{\mathrm{a}} - \frac{\pi_{\mathrm{n}}\mu_{\mathrm{n}} + \pi_{\mathrm{c}}\mu_{0\mathrm{c}}}{\pi_{\mathrm{n}} + \pi_{\mathrm{c}}}.$$

22.9 Bounds on the average causal effect on the whole population

Extend the discussion in Section 22.1 based on the notation in Section 21.6. With the potential outcome $Y(d)$, define the average causal effect of the treatment received on the outcome as

$$\delta = E\{Y(d=1) - Y(d=0)\},$$

and modify the definition of μ_{zu} as

$$m_{du} = E\{Y(d) \mid U = u\}, \quad (d = 0, 1; u = \mathrm{a}, \mathrm{n}, \mathrm{c})$$

due to the change of the notation. They satisfy

$$\delta = \sum_{u=\mathrm{a},\mathrm{n},\mathrm{c}} \pi_u (m_{1u} - m_{0u}).$$

Section 22.1 identifies π_{a}, π_{n}, π_{c}, $m_{1\mathrm{a}} = \mu_{1\mathrm{a}}$, $m_{0\mathrm{n}} = \mu_{0\mathrm{n}}$, $m_{1\mathrm{c}} = \mu_{1\mathrm{c}}$, and $m_{0\mathrm{c}} = \mu_{0\mathrm{c}}$. But the data do not contain any information about $m_{0\mathrm{a}}$ and $m_{1\mathrm{n}}$. Therefore, we cannot identify δ. With a bounded outcome, we can bound δ.

Theorem 22.5 *Under Assumptions 21.2–21.4 with a bounded outcome in* $[\underline{y}, \overline{y}]$, *we have* $\underline{\delta} \leq \delta \leq \overline{\delta}$, *where*

$$\underline{\delta} = \delta' - \overline{y}\mathrm{pr}(D = 1 \mid Z = 0) + \underline{y}\mathrm{pr}(D = 0 \mid Z = 1)$$

and

$$\overline{\delta} = \delta' - \underline{y}\mathrm{pr}(D = 1 \mid Z = 0) + \overline{y}\mathrm{pr}(D = 0 \mid Z = 1)$$

with $\delta' = E(DY \mid Z = 1) - E(Y - DY \mid Z = 0)$.

Prove Theorem 22.5.

Remark: In the special case with a binary outcome, the bounds simplify to

$$\underline{\delta} = E(DY \mid Z = 1) - E(D + Y - DY \mid Z = 0)$$

and

$$\overline{\delta} = E(DY + 1 - D \mid Z = 1) - E(Y - DY \mid Z = 0).$$

22.10 One-sided noncompliance and statistical inference

Consider a randomized encouragement design where the units assigned to the control have no access to the treatment. For unit i, let Z_i be the binary treatment assigned, D_i be the binary treatment received, and Y_i be the outcome of interest. One-sided noncompliance happens when $Z_i = 0$ implies $D_i = 0$ for all units $i = 1, \ldots, n$. Suppose that Assumption 21.1 holds.

1. Does monotonicity Assumption 21.2 hold in this case? How many latent strata defined by $\{D_i(1), D_i(0)\}$ are there in this problem? How do we identify their proportions by the observed data distribution?
2. State the assumption of exclusion restriction. Under exclusion restriction, show that $E\{Y(z) \mid U = u\}$ can be identified by the observed data distributions. Give the formulas for all possible values of z and u. How do we identify the CACE in this case?
3. If we observe pretreatment covariates X_i for all units i, how do we use the covariate information to improve the estimation efficiency of the CACE?
4. Under Assumption 21.1, the exclusion restriction Assumption 21.3 has testable implications, which are the IV inequalities for one-sided noncompliance. State the IV inequalities.
5. Sommer and Zeger (1991) provided the following dataset:

	$Z=1$		$Z=0$	
	$D=1$	$D=0$	$D=1$	$D=0$
$Y=1$	9663	2385	0	11514
$Y=0$	12	34	0	74

The treatment assigned Z is whether or not the child was assigned to the vitamin A supplement, the treatment received indicator D is whether or not the child received the vitamin A supplement, and the binary outcome Y is the survival indicator. The original RCT was conducted in Indonesia. Reanalyze the data.

Remark: Bloom (1984) first discussed one-sided noncompliance and proposed the estimator $\hat{\tau}_c = \hat{\tau}_Y / \hat{\tau}_D$, which is sometimes called the Bloom estimator. The notation in Bloom (1984) is different from the notation in this chapter.

22.11 One-sided noncompliance with partial adherence

Sanders and Karim (2021, Table 3) reported the following data from an RCT aiming to estimate the efficacy of smoking cessation interventions among individuals with psychotic disorders.

group assigned	treatment received	group size	# positive outcomes
Control	None	151	25
Treatment	None	35	7
Treatment	Partial	42	17
Treatment	Full	70	40

Three tiers of treatment received are defined as follows: "full" treatment corresponds to attending all 8 treatment sessions, "partial" corresponds to attending 5 to 7 sessions, and "none" corresponds to < 5 sessions. The outcome is defined as the binary indicator of smoking reduction of 50% or greater relative to baseline, measured at three months.

In this problem, the treatment assignment Z is binary but the treatment received D takes three values $0, 0.5, 1$ for "none," "partial," and "full." The three-leveled D causes complications, but it can only be 0 under the control assignment. How many latent strata $U = \{D(1), D(0)\}$ do we have in this problem? Can we identify their proportions?

How do we extend the exclusion restriction to this problem? What can be the causal effects of interest? Can we identify them?

Analyze the data based on the questions above.

22.12 Recommended reading

Balke and Pearl (1997) derived more general IV inequalities.

23

An Econometric Perspective of the Instrumental Variable

Chapters 21 and 22 discuss the IV method from the experimental perspective. Figure 23.1 illustrates the intuition behind the discussion.

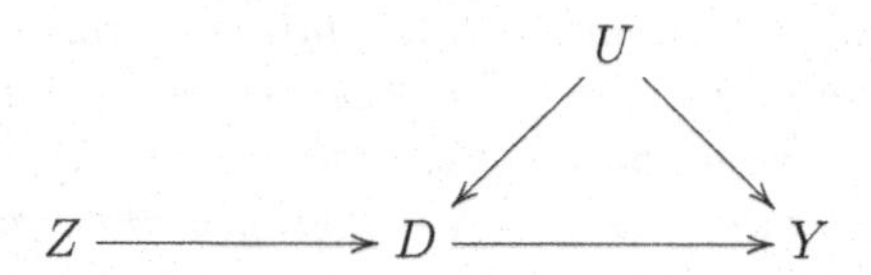

FIGURE 23.1: Causal diagram for IV. Z denotes the IV, D denotes the treatment, Y denotes the outcome, and U denotes the unmeasured confounder between the treatment and outcome.

In an encouragement design with noncompliance, Z is randomized, so it is independent of the confounder U between the treatment received D and the outcome Y. Importantly, the treatment assignment Z does not have any direct effect on the outcome Y. It acts as an IV for the treatment received D in the sense that it affects the outcome Y only through the treatment received D. This IV is generated by the experimenter.

In many applications, randomization is infeasible. Then how can we draw causal inference in the presence of unmeasured confounding between the treatment D and outcome Y? A clever idea from econometrics is to find *natural experiments* to mimic the setting of encouragement designs. To identify the causal effect of D on Y with unmeasured confounding, we can find another variable Z that satisfies the assumptions of the diagram in Figure 23.1. The variable Z must satisfy the following assumptions:

1. it should be close to being randomized so that it is independent of the unmeasured confounding U;
2. it should change the distribution of D;
3. it should affect the outcome Y only indirectly through D but not directly.

If all these conditions hold, then Z is a valid IV for estimating the effect of D on Y.

This chapter provides the traditional econometric perspective on IV. It is based on linear regression. Imbens and Angrist (1994) and Angrist et al. (1996) made a fundamental contribution by clarifying the connection between this perspective and the experimental perspective in Chapters 21 and 22. I will start with examples and then give more algebraic details.

23.1 Examples of studies with IVs

Finding IV for causal inference is more an art than a science. The challenge lies not in the algebraic details but in finding IVs in empirical research. Below are some famous examples.

Example 23.1 *In an encouragement design, Z is the randomly assigned treatment, D is the final treatment received, and Y is the outcome. The IV assumptions encoded by Figure 23.1 are plausible in double-blind RCTs as discussed in Chapter 21. This is the ideal case for IV.*

Example 23.2 *Hearst et al. (1986) reported that men with low lottery numbers in the Vietnam Era draft lottery had higher mortality rates afterward. They attributed this to the negative effect of military service. Angrist (1990) further reported that men with low lottery numbers in the Vietnam Era draft lottery had lower subsequent earnings. He attributed this to the negative effect of military service. These explanations are plausible because the lottery numbers were randomly generated, men with low lottery numbers were more likely to have military service, and the lottery numbers were unlikely to affect the subsequent mortality or earnings directly. That is, Figure 23.1 is plausible. Angrist et al. (1996) reanalyzed the data using the IV framework. Here, the lottery number is the IV, military service is the treatment, and mortality or earnings is the outcome.*

Example 23.3 *Angrist and Krueger (1991) studied how schooling in years impacted earnings, using the quarter of birth as an IV. This IV is plausible because of the pseudo-randomization of the quarter of birth. It affected the years of schooling because (1) most states in the U.S. required the students to enter school in the calendar year in which they turned six, and (2) compulsory schooling laws typically required students to remain in school before their sixteenth birthday. More importantly, it is plausible that the quarter of birth did not affect earnings directly.*

Example 23.4 *Angrist and Evans (1998) studied the effect of family size on mothers' employment and work, using the sibling sex composition as an IV. This IV is plausible because of the pseudo-randomization of the sibling sex composition. Moreover, parents in the U.S. with two children of the same sex are more likely to have a third child than those parents with two children of different sex. It is also plausible that the sibling sex composition does not affect the mother's employment and work directly.*

Example 23.5 *Card (1993) studied the effect of schooling on wages, using the geographic variation in college proximity as an IV. In particular, Z contains dummy variables for whether a subject grew up near a two-year college or a four-year college. Although this study is classic, it might be a poor example for IV because parents' choices of where to live might not be random, and moreover, where a subject grew up might affect the subsequent wage through channels other than college education.*

Example 23.6 *Voight et al. (2012) studied the causal effect of plasma high-density lipoprotein (HDL) cholesterol on the risk of heart attack based on Mendelian randomization. They used some single-nucleotide polymorphisms (SNPs) as genetic IV for HDL, which are random with respect to the unmeasured confounders between HDL and heart attack by Mendel's second law, and affect heart attack only through HDL. I will give more details of Mendelian randomization in Chapter 25.*[1]

[1]This is not an example from econometrics. However, the literature of IV in econometrics has heavily

23.2 Brief review of the ordinary least squares

Before discussing the econometric view of IV, I will first review the OLS (see Appendix B). This is a standard topic in statistics. However, it has different formulations, and the choice of formulation matters for the interpretation.

The first view is based on projection. Given a random variable Y and a random variable or vector D with finite second moments, define the population OLS coefficient as

$$\begin{aligned} \beta &= \arg\min_b E(Y - D^\top b)^2 \\ &= E(DD^\top)^{-1}E(DY), \end{aligned}$$

and then define the population residual as $\varepsilon = Y - D^\top \beta$. By definition, Y decomposes into

$$Y = D^\top \beta + \varepsilon, \tag{23.1}$$

which must satisfy

$$E(D\varepsilon) = 0.$$

Based on $(D_i, Y_i)_{i=1}^n \stackrel{\text{IID}}{\sim} (D, Y)$, the OLS estimator of β is the moment estimator

$$\hat{\beta} = \left(\sum_{i=1}^n D_i D_i^\top\right)^{-1} \sum_{i=1}^n D_i Y_i.$$

Because

$$\begin{aligned} \hat{\beta} &= \left(\sum_{i=1}^n D_i D_i^\top\right)^{-1} \sum_{i=1}^n D_i (D_i^\top \beta + \varepsilon_i) \\ &= \beta + \left(\sum_{i=1}^n D_i D_i^\top\right)^{-1} \sum_{i=1}^n D_i \varepsilon_i, \end{aligned}$$

we can show that $\hat{\beta}$ is consistent for β by the law of large numbers and the fact $E(D\varepsilon) = 0$. The classic EHW robust variance estimator for $\text{cov}(\hat{\beta})$ is

$$\hat{V}_{\text{EHW}} = \left(\sum_{i=1}^n D_i D_i^\top\right)^{-1} \left(\sum_{i=1}^n \hat{\varepsilon}_i^2 D_i D_i^\top\right) \left(\sum_{i=1}^n D_i D_i^\top\right)^{-1}$$

where $\hat{\varepsilon}_i = Y_i - D_i^\top \hat{\beta}$ is the residual.

The second view is to treat

$$Y = D^\top \beta + \varepsilon, \tag{23.2}$$

as a true model for the data-generating process. That is, given the random variables (D, ε), we generate Y based on the linear equation (23.2). Importantly, in the data-generating process, ε and D may be correlated which leads to $E(D\varepsilon) \neq 0$. In model (23.2), we can

influenced the literature of Mendelian randomization in genetics. More interestingly, the idea of IV has both genetic and econometric origins, due to the son Sewall Wright who was a geneticist, and the father Philip Wright who was an economist. See Stock and Trebbi (2003) for some historical comments on the origins of the idea of IV.

interpret the coefficient β as the effect of D on Y. Figure 23.2 gives an example for model (23.2). The fact that $E(D\varepsilon) \neq 0$ is the fundamental difference compared with the first view where $E(D\varepsilon) = 0$ holds by the definition of the population OLS. Consequently, the OLS estimator can be inconsistent:

$$\hat{\beta} \to \beta + E(DD^{\mathsf{T}})^{-1}E(D\varepsilon) \neq \beta$$

in probability, as the sample size n approaches infinity.

I end this section with definitions of *endogenous* and *exogenous* regressors based on (23.2), although their definitions are not unique in econometrics.

Definition 23.1 *When $E(D\varepsilon) \neq 0$, the regressor D is called endogenous; when $E(D\varepsilon) = 0$, the regressor D is called exogenous.*

The terminology in Definition 23.1 is standard in econometrics. When $E(D\varepsilon) \neq 0$, we also say that we have *endogeneity*; when $E(D\varepsilon) = 0$, we also say that we have *exogeneity*.

In the first view of OLS, the notions of endogeneity and exogeneity do not play any roles because $E(D\varepsilon) = 0$ by definition. Statisticians holding the first view usually find the notions of endogeneity and exogeneity strange, and consequently, find the idea of IV unnatural. To understand the econometric view of IV, we must switch to the second view of OLS.

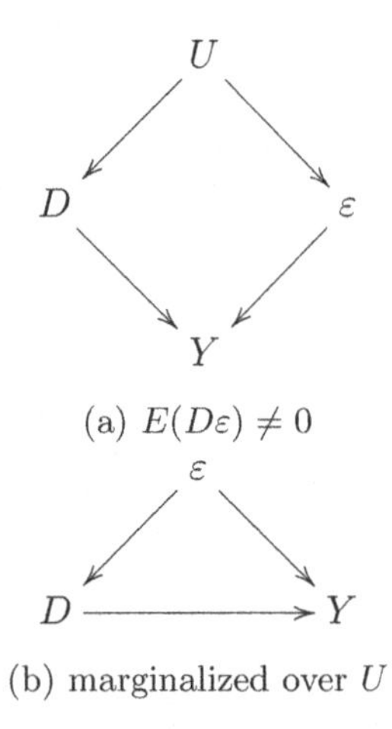

FIGURE 23.2: Different representations of the endogenous regressor D. In the upper panel, U represents the unmeasured common cause of D and ε. In the lower panel, ε represents the unmeasured common cause of D and Y, without showing U explicitly.

23.3 Linear instrumental variable model

When D is endogenous, the OLS estimator is inconsistent. We must use additional information to construct a consistent estimator for β. I will focus on the following linear IV model:

Definition 23.2 (linear IV model) *We have*

$$Y = D^{\mathsf{T}}\beta + \varepsilon,$$

with

$$E(Z\varepsilon) = 0. \tag{23.3}$$

The linear IV model in Definition 23.2 can be illustrated by the following causal graph:

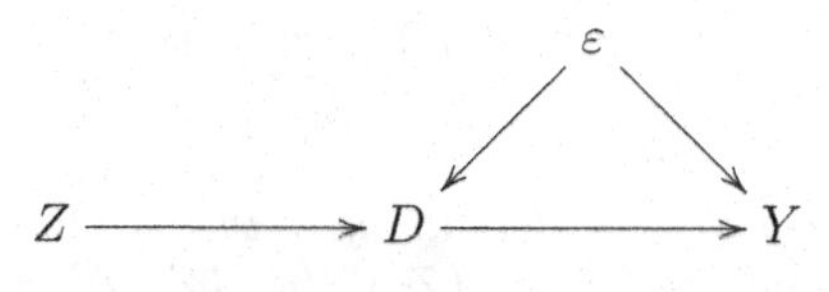

The above linear IV model allows that $E(D\varepsilon) \neq 0$ but requires an alternative moment condition $E(Z\varepsilon) = 0$ in (23.3). With $E(\varepsilon) = 0$ by incorporating the intercept, the new condition states that Z is uncorrelated with the error term ε. But any randomly generated noise is uncorrelated with ε, so an additional condition must hold to ensure that Z is useful for estimating β. Intuitively, the additional condition requires that Z is correlated to D, with more technical details stated below.

The mathematical requirement (23.3) seems simple. However, it is a key challenge in empirical research to find such a variable Z that satisfies (23.3). Since the condition (23.3) involves the unobservable ε, it is generally untestable.

23.4 The just-identified case

We first consider the case in which Z and D have the same dimension and $E(ZD^{\mathsf{T}})$ has full rank. The condition $E(Z\varepsilon) = 0$ implies that

$$E\{Z(Y - D^{\mathsf{T}}\beta)\} = 0,$$

which implies the linear equations

$$E(ZY) = E(ZD^{\mathsf{T}})\beta.$$

Solve the linear equations to obtain

$$\beta = E(ZD^{\mathsf{T}})^{-1}E(ZY) \tag{23.4}$$

if $E(ZD^{\mathsf{T}})$ is invertible. The OLS is a special case if $E(D\varepsilon) = 0$, i.e., D acts as an IV for itself. The resulting moment estimator is

$$\hat{\beta}_{\text{IV}} = \left(\sum_{i=1}^{n} Z_i D_i^{\mathsf{T}}\right)^{-1} \sum_{i=1}^{n} Z_i Y_i. \tag{23.5}$$

It can be insightful to work out the details for the case with a scalar D and Z. See Example 23.7 below.

Example 23.7 *In the simple case with an intercept and scalar D and Z, we have the model*

$$\begin{cases} Y = \alpha + \beta D + \varepsilon, \\ E(\varepsilon) = 0, \quad \text{cov}(\varepsilon, Z) = 0. \end{cases}$$

Under this model, we have

$$\text{cov}(Z, Y) = \beta\text{cov}(Z, D)$$

which implies

$$\beta = \frac{\text{cov}(Z, Y)}{\text{cov}(Z, D)}.$$

Standardize the numerator and denominator by $\text{var}(Z)$ *to obtain*

$$\beta = \frac{\text{cov}(Z, Y)/\text{var}(Z)}{\text{cov}(Z, D)/\text{var}(Z)},$$

which equals the ratio between the coefficients of Z *in the OLS fits of* Y *and* D *on* $(1, Z)$. *As a side note, the IV formula for* β, $\text{cov}(Z, Y)/\text{cov}(Z, D)$, *and the OLS formula for* β, $\text{cov}(D, Y)/\text{var}(D)$, *have different signs if and only if*

$$\text{cov}(Z, Y) \cdot \text{cov}(Z, D) \cdot \text{cov}(D, Y) < 0,$$

that is, the product of the off-diagonal terms in the covariance matrix of (Z, D, Y) *is negative.*

If Z *is binary, these coefficients are differences in means (see Problem B.2), and the IV formula for* β *reduces to*

$$\beta = \frac{E(Y \mid Z = 1) - E(Y \mid Z = 0)}{E(D \mid Z = 1) - E(D \mid Z = 0)}.$$

This is identical to the identification formula in Theorem 21.1. That is, with a binary IV Z *and a binary treatment* D, *the IV estimator recovers the CACE under the potential outcomes framework. This is a key result in Imbens and Angrist (1994) and Angrist et al. (1996).*

23.5 The over-identified case

The discussion in Section 23.4 focuses on the *just-identified* case because the IV model uniquely determines β by (23.4). It relies on the conditions that Z and D have the same dimensions and $E(ZD^{\mathsf{T}})$ is invertible.

When Z has a lower dimension than D, or, more generally, $E(ZD^{\mathsf{T}})$ does not have full column rank, the equation $E(ZY) = E(ZD^{\mathsf{T}})\beta$ has infinitely many solutions. This is the *under-identified* case in which the coefficient β cannot be uniquely determined even with Z. It is a challenging case beyond the scope of this book. To ensure identifiability, we need at least as many IVs as the endogenous regressors.

When Z has a higher dimension than D and $E(ZD^{\mathsf{T}})$ has full column rank, we have many ways to determine β from $E(ZY) = E(ZD^{\mathsf{T}})\beta$. This is the *over-identified* case. What is more, the sample analog

$$n^{-1}\sum_{i=1}^{n} Z_i Y_i = n^{-1}\sum_{i=1}^{n} Z_i D_i^{\mathsf{T}}\beta$$

may not have any solution because the number of equations is larger than the number of unknown parameters.

A computational trick for the over-identified case is the *two-stage least squares* (TSLS) estimator (Theil, 1953; Basmann, 1957). It is a clever computational trick, which has two steps.

Definition 23.3 (Two-stage least squares) *Define the TSLS estimator of the coefficient of D with Z being the IV as follows.*

1. *Run OLS of D on Z, and obtain the fitted value $\hat{D}_i$ ($i = 1, \ldots, n$). If D_i is a vector, then we need to run component-wise OLS to obtain $\hat{D}_i$. Put the fitted vectors in a matrix $\hat{D}$ with rows $\hat{D}_i^{\mathsf{T}}$;*
2. *Run OLS of Y on $\hat{D}$, and obtain the coefficient $\hat{\beta}_{\text{TSLS}}$.*

To see why TSLS works, we need more algebra. Write it more explicitly as

$$\begin{aligned}
\hat{\beta}_{\text{TSLS}} &= \left(\sum_{i=1}^n \hat{D}_i \hat{D}_i^{\mathsf{T}}\right)^{-1} \sum_{i=1}^n \hat{D}_i Y_i \quad (23.6)\\
&= \left(\sum_{i=1}^n \hat{D}_i \hat{D}_i^{\mathsf{T}}\right)^{-1} \sum_{i=1}^n \hat{D}_i (D_i^{\mathsf{T}} \beta + \varepsilon_i)\\
&= \left(\sum_{i=1}^n \hat{D}_i \hat{D}_i^{\mathsf{T}}\right)^{-1} \sum_{i=1}^n \hat{D}_i D_i^{\mathsf{T}} \beta + \left(\sum_{i=1}^n \hat{D}_i \hat{D}_i^{\mathsf{T}}\right)^{-1} \sum_{i=1}^n \hat{D}_i \varepsilon_i.
\end{aligned}$$

The first stage OLS fit ensures $D_i = \hat{D}_i + \check{D}_i$ with orthogonal fitted value $\hat{D}_i$'s and residual $\check{D}_i$'s, that is,

$$\sum_{i=1}^n \hat{D}_i \check{D}_i^{\mathsf{T}} = 0 \quad (23.7)$$

is a zero square matrix with the same dimension as D_i; see Problem 23.4. The orthogonality (23.7) implies

$$\sum_{i=1}^n \hat{D}_i D_i^{\mathsf{T}} = \sum_{i=1}^n \hat{D}_i \hat{D}_i^{\mathsf{T}},$$

which further implies that

$$\hat{\beta}_{\text{TSLS}} = \beta + \left(\sum_{i=1}^n \hat{D}_i \hat{D}_i^{\mathsf{T}}\right)^{-1} \sum_{i=1}^n \hat{D}_i \varepsilon_i. \quad (23.8)$$

The first stage OLS fit also ensures

$$\hat{D}_i = \hat{\Gamma}^{\mathsf{T}} Z_i \quad (23.9)$$

with OLS coefficient matrix $\hat{\Gamma}^{\mathsf{T}}$, which implies that

$$\hat{\beta}_{\text{TSLS}} = \beta + \left\{\hat{\Gamma}^{\mathsf{T}} \left(n^{-1} \sum_{i=1}^n Z_i Z_i^{\mathsf{T}}\right) \hat{\Gamma}\right\}^{-1} \hat{\Gamma}^{\mathsf{T}} \left(n^{-1} \sum_{i=1}^n Z_i \varepsilon_i\right). \quad (23.10)$$

Based on (23.10), we can see the consistency of the TSLS estimator by the law of large numbers and the fact that the term $n^{-1} \sum_{i=1}^n Z_i \varepsilon_i$ has probability limit $E(Z\varepsilon) = 0$. We can also use (23.10) to show that when Z and D have the same dimension, $\hat{\beta}_{\text{TSLS}}$ is numerically identical to $\hat{\beta}_{\text{IV}}$ defined in Section 23.4, which is left as Problem 23.2.

Based on (23.8), we can obtain the standard error as follows. We first obtain the residual $\hat{\varepsilon}_i = Y_i - \hat{\beta}_{\text{TSLS}}^{\mathsf{T}} D_i$, and then obtain the robust variance estimator as

$$\hat{V}_{\text{TSLS}} = \left(\sum_{i=1}^n \hat{D}_i \hat{D}_i^{\mathsf{T}}\right)^{-1} \left(\sum_{i=1}^n \hat{\varepsilon}_i^2 \hat{D}_i \hat{D}_i^{\mathsf{T}}\right) \left(\sum_{i=1}^n \hat{D}_i \hat{D}_i^{\mathsf{T}}\right)^{-1}.$$

Importantly, the $\hat{\varepsilon}_i$'s are not the residual from the second stage OLS $Y_i - \hat{\beta}_{\text{TSLS}}^{\mathsf{T}} \hat{D}_i$, so $\hat{V}_{\text{TSLS}}$ differs from the robust variance estimator from the second stage OLS.

23.6 A special case: a single IV for a single endogenous treatment

This section focuses on a simple case with a single IV and a single endogenous treatment. It has wide applications. Consider the following *structural equations*:

$$\begin{cases} Y_i = \beta_0 + \beta_1 D_i + \beta_2^{\mathsf{T}} X_i + \varepsilon_i, \\ D_i = \gamma_0 + \gamma_1 Z_i + \gamma_2^{\mathsf{T}} X_i + \varepsilon_{i2}, \end{cases} \tag{23.11}$$

where D_i is a scalar endogenous regressor representing the treatment variable of interest (i.e., $E(D_i\varepsilon_i) \neq 0$), Z_i is a scalar IV for D_i (i.e., $E(Z_i\varepsilon_i) = 0$), and X_i contains other exogenous regressors (i.e., $E(X_i\varepsilon_i) = 0$). This is a special case with D replaced by $(1, D, X)$ and Z replaced by $(1, Z, X)$.

23.6.1 Two-stage least squares

The TSLS estimator in Definition 23.3 simplifies to the following form.

Definition 23.4 (TSLS with a single endogenous regressor) *Based on* (23.11), *the TSLS estimator has the following two steps.*

1. *Run OLS of D on $(1, Z, X)$, obtain the fitted values $\hat{D}_i$ $(i = 1, \ldots, n)$, and vectorize them as $\hat{D}$;*
2. *Run OLS of Y on $(1, \hat{D}, X)$, and obtain the coefficient $\hat{\beta}_{\text{TSLS}}$, and in particular, $\hat{\beta}_{1,\text{TSLS}}$, the coefficient of $\hat{D}$.*

23.6.2 Indirect least squares

The structural equation (23.11) implies

$$\begin{aligned} Y_i &= \beta_0 + \beta_1(\gamma_0 + \gamma_1 Z_i + \gamma_2^{\mathsf{T}} X_i + \varepsilon_{i2}) + \beta_2^{\mathsf{T}} X_i + \varepsilon_i \\ &= (\beta_0 + \beta_1\gamma_0) + \beta_1\gamma_1 Z_i + (\beta_2 + \beta_1\gamma_2)^{\mathsf{T}} X_i + (\varepsilon_i + \beta_1\varepsilon_{i2}). \end{aligned}$$

Define $\Gamma_0 = \beta_0 + \beta_1\gamma_0, \Gamma_1 = \beta_1\gamma_1, \Gamma_2 = \beta_2 + \beta_1\gamma_2$, and $\varepsilon_{i1} = \varepsilon_i + \beta_1\varepsilon_{i2}$. We have the following equations

$$\begin{cases} Y_i = \Gamma_0 + \Gamma_1 Z_i + \Gamma_2^{\mathsf{T}} X_i + \varepsilon_{i1}, \\ D_i = \gamma_0 + \gamma_1 Z_i + \gamma_2^{\mathsf{T}} X_i + \varepsilon_{i2}, \end{cases} \tag{23.12}$$

which is called the *reduced form*, in contrast to the *structural form* in (23.11). The parameter of interest equals the ratio of two coefficients

$$\beta_1 = \Gamma_1/\gamma_1.$$

In the reduced form, the left-hand sides are dependent variables Y and D, and the right-hand sides are the exogenous variables Z and X satisfying

$$\begin{aligned} E(Z\varepsilon_{i1}) = E(Z\varepsilon_{i2}) &= 0, \\ E(X\varepsilon_{i1}) = E(X\varepsilon_{i2}) &= 0. \end{aligned}$$

More importantly, OLS gives consistent estimators for the coefficients in the reduced form (23.12).

The reduced form (23.12) suggests that the ratio of two OLS coefficients $\hat{\Gamma}_1$ and $\hat{\gamma}_1$ is a reasonable estimator for β_1. This is called the *indirect least squares* (ILS) estimator:

$$\hat{\beta}_{1,\text{ILS}} = \hat{\Gamma}_1/\hat{\gamma}_1.$$

Interestingly, it is numerically identical to the TSLS estimator under (23.11).

Theorem 23.1 *With a single endogenous treatment and a single IV in* (23.11), *we have*

$$\hat{\beta}_{1,\text{ILS}} = \hat{\beta}_{1,\text{TSLS}}.$$

Theorem 23.1 is an algebraic fact. Imbens (2014, Section A.3) pointed it out without giving a proof. I relegate its proof to Problem 23.5. The ratio formula makes it clear that the TSLS estimator has poor finite sample properties when the IV Z is weak, i.e., when γ_1 is close to zero.

23.6.3 Weak IV

The following inferential procedure is simpler, more transparent, and more robust to weak IV. It is more computationally intensive though.

The reduced form (23.12) also implies that

$$Y_i - bD_i = (\Gamma_0 - b\gamma_0) + (\Gamma_1 - b\gamma_1)Z_i + (\Gamma_2 - b\gamma_2)^{\mathrm{T}} X_i + (\varepsilon_{i1} - b\varepsilon_{i2}) \tag{23.13}$$

for any b. At the true value $b = \beta_1$, the coefficient of Z_i must be 0. This simple fact suggests a confidence interval for β_1 by inverting tests for $H_0(b) : \beta_1 = b$:

$$\left\{b : |t_Z(b)| \leq z_{1-\alpha/2}\right\},$$

where $t_Z(b)$ is the t-statistic for the coefficient of Z based on the OLS fit of (23.13) with the EHW standard error, and $z_{1-\alpha/2}$ is the $1 - \alpha/2$ upper quantile of the standard Normal random variable. This confidence interval is more robust than the Wald-type confidence interval based on the TSLS estimator. It is similar to the FAR confidence set discussed in Chapter 21. This procedure makes the TSLS estimator unnecessary. What is more, we only need to run the OLS fit of Y based on the reduced form if the goal is to test $\beta_1 = 0$ under (23.11).

23.7 Application

Revisit Example 23.5. Card (1993) used the National Longitudinal Survey of Young Men to estimate the causal effect of education on earnings. The data set contains 3010 men with ages between 14 and 24 in the year 1966, and Card (1993) leveraged the geographic variation in college proximity as an IV for education. Here, Z is the indicator of growing up near a four-year college, D measures the years of education, and the outcome Y is the log wage in the year 1976, ranging from 4.6 to 7.8. Additional covariates are years of labor force experience, the squared years of labor force experience, race, and variables representing the living areas.

```
> library("car")
> ## Card Data
```

```
> card.data = read.csv("card1995.csv")
> Y = card.data[, "lwage"]
> D = card.data[, "educ"]
> Z = card.data[, "nearc4"]
> X = card.data[, c("exper", "expersq", "black", "south",
+                    "smsa", "reg661", "reg662", "reg663",
+                    "reg664", "reg665", "reg666",
+                    "reg667", "reg668", "smsa66")]
> X = as.matrix(X)
```

Based on TSLS, we can obtain the following the point estimator and 95% confidence interval.

```
> Dhat     = lm(D ~ Z + X)$fitted.values
> tslsreg  = lm(Y ~ Dhat + X)
> tslsest  = coef(tslsreg)[2]
> ## correct se by changing the residuals
> res.correct         = Y - cbind(1, D, X)%*%coef(tslsreg)
> tslsreg$residuals = as.vector(res.correct)
> tslsse = sqrt(hccm(tslsreg, type = "hc0")[2, 2])
> res = c(tslsest, tslsest - 1.96*tslsse, tslsest + 1.96*tslsse)
> names(res) = c("TSLS", "lower CI", "upper CI")
> round(res, 3)
   TSLS lower CI upper CI
  0.132    0.026    0.237
```

Using the strategy of the FAR confidence set, we can compute p-value as a function of b.

```
> BetaAR    = seq(-0.1, 0.4, 0.001)
> PvalueAR = sapply(BetaAR, function(b){
+   Y_b    = Y - b*D
+   ARreg  = lm(Y_b ~ Z + X)
+   coefZ  = coef(ARreg)[2]
+   seZ    = sqrt(hccm(ARreg)[2, 2])
+   Tstat  = coefZ/seZ
+   (1 - pnorm(abs(Tstat)))*2
+ })
```

Figure 23.3 shows the p-values for a sequence of tests for the coefficient of D based on following R code:

```
> plot(PvalueAR ~ BetaAR, type = "l",
+      xlab = "coefficient of D",
+      ylab = "p-value",
+      main = "FAR interval based on Card's data")
> point.est = BetaAR[which.max(PvalueAR)]
> abline(h = 0.05, lty = 2, col = "grey")
> abline(v = point.est, lty = 2, col = "grey")
> ARCI = range(BetaAR[PvalueAR >= 0.05])
> abline(v = ARCI[1], lty = 2, col = "grey")
> abline(v = ARCI[2], lty = 2, col = "grey")
```

We report the point estimate as the value of b with the largest p-value as well as the confidence interval as the region of b with p-values larger than 0.05.

```
> FARres = c(point.est, ARCI)
> names(FARres) = c("FAR est", "lower CI", "upper CI")
> round(FARres, 3)
```

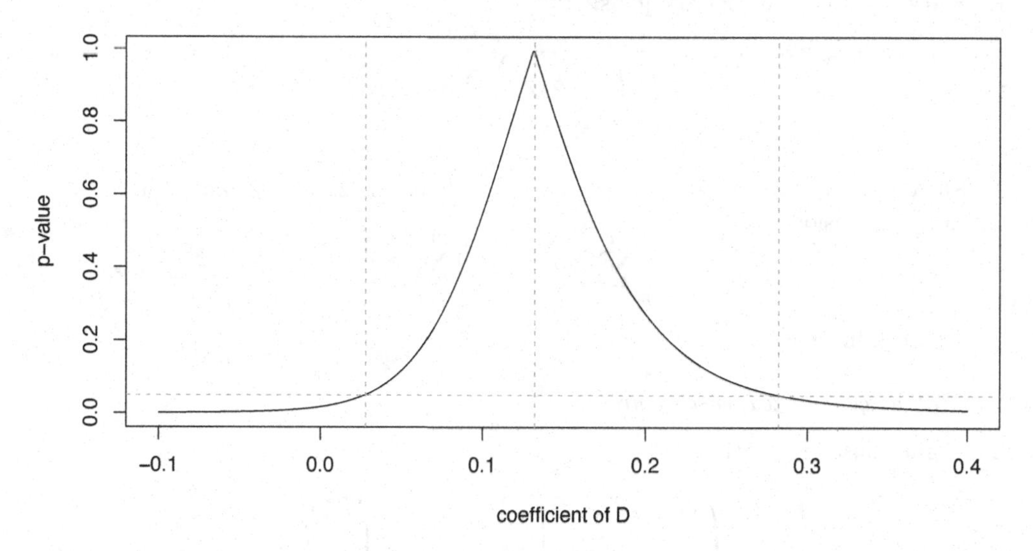

FIGURE 23.3: Reanalyzing the data from Card (1993) by inverting tests

```
FAR est lower CI upper CI
  0.132    0.028    0.282
```

Comparing the TSLS and FAR methods, the lower confidence limits are very close but the upper confidence limits are slightly different due to the possibly heavy right tail of the distribution of the TSLS estimator. Overall, the TSLS and FAR methods give similar results in this example because the IV is not weak.

23.8 Homework problems

23.1 Finite-sample analog of Example 23.7

Revisit Example 23.7 with a binary D and a binary Z. Show that the second coordinate of the general IV formula for β,

$$\hat{\beta}_{\text{IV}} = \left(\sum_{i=1}^{n} Z_i D_i^{\top}\right)^{-1} \sum_{i=1}^{n} Z_i Y_i$$

reduces to

$$\hat{\tau}_{\text{c}} = \frac{\hat{\tau}_Y}{\hat{\tau}_D}$$

which was introduced in Chapter 21 as the "Wald estimator."

Remark: Be careful that $\hat{\beta}_{\text{IV}}$ is the abstract formula for the IV estimator based on the linear IV model assumption. To apply the formula of $\hat{\beta}_{\text{IV}}$, you need to specify Z_i as $(1, Z_i)^{\top}$ and D_i as $(1, D_i)^{\top}$.

23.2 More algebra for TSLS in Section 23.5

1. Show that the $\hat{\Gamma}$ in (23.9) equals

$$\hat{\Gamma} = \left(\sum_{i=1}^{n} Z_i Z_i^{\mathsf{T}} \right)^{-1} \sum_{i=1}^{n} Z_i D_i^{\mathsf{T}}.$$

2. Show $\hat{\beta}_{\text{TSLS}}$ defined in (23.6) reduces to $\hat{\beta}_{\text{IV}}$ defined in (23.5) if Z and D have the same dimension and

$$n^{-1} \sum_{i=1}^{n} Z_i Z_i^{\mathsf{T}}, \quad n^{-1} \sum_{i=1}^{n} Z_i D_i^{\mathsf{T}}$$

are both invertible.

23.3 TSLS as generalized least squares

With a slight abuse of notation, let

$$Y = \begin{pmatrix} Y_1 \\ \vdots \\ Y_n \end{pmatrix}, \quad D = \begin{pmatrix} D_1^{\mathsf{T}} \\ \vdots \\ D_n^{\mathsf{T}} \end{pmatrix}, \quad Z = \begin{pmatrix} Z_1^{\mathsf{T}} \\ \vdots \\ Z_n^{\mathsf{T}} \end{pmatrix}$$

denote the vector of the Y_i's, the matrix of the D_i's, and the matrix of the Z_i's, respectively. Let $H_Z = Z(Z^{\mathsf{T}} Z)^{-1} Z^{\mathsf{T}}$ denote the hat matrix formed by Z (recall the definition of the hat matrix in OLS in Section B.2).

Show that $\hat{\beta}_{\text{TSLS}}$ solves the following minimization problem:

$$\min_b (Y - Db)^{\mathsf{T}} H_Z (Y - Db).$$

Remark: The OLS can be written as $\min_b (Y - Db)^{\mathsf{T}} (Y - Db)$, which is inconsistent for estimating β when D is endogenous. The modification $\min_b (Y - Db)^{\mathsf{T}} H_Z (Y - Db)$ is called generalized least squares, with the hat matrix H_Z formed by the instrumental variable Z. The result above show that TSLS is numerically identical to generalized least squares.

23.4 Properties of OLS

Prove (23.7).

23.5 Equivalence between TSLS and ILS

Prove Theorem 23.1.

Remark: Use the FWL theorem in Appendix B.

23.6 Control function estimator in the linear instrumental variable model

Definition 23.5 below parallels Definition 23.3 above.

Definition 23.5 (control function estimator) *Define the control function estimator* $\hat{\beta}_{\text{CF}}$ *as follows.*

1. *Run OLS of* D *on* Z*, and obtain the residuals* $\check{D}_i$ $(i = 1, \ldots, n)$*. If* D_i *is a vector, then we need to run component-wise OLS to obtain* $\check{D}_i$*. Put the residual vectors in a matrix* $\check{D}$ *with rows* $\check{D}_i^{\mathrm{T}}$*;*

2. *Run OLS of* Y *on* D *and* $\check{D}$*, and obtain the coefficient of* D*,* $\hat{\beta}_{\text{CF}}$*.*

Show that $\hat{\beta}_{\text{CF}} = \hat{\beta}_{\text{TSLS}}$.

Remark: To prove the result, you can use the results in Problems B.4 and B.5. In Definition 23.5, $\check{D}$ from Step 1 is called the control function for Step 2. Hausman (1978) pointed out this result. Wooldridge (2015) provided a more general discussion of the control function methods in more complex models.

23.7 Data analysis: Efron and Feldman (1991)

Efron and Feldman (1991) was one of the early studies dealing with noncompliance under the potential outcomes framework. The original randomized experiment, the Lipid Research Clinics Coronary Primary Prevention Trial (LRC-CPPT), was designed to evaluate the effect of the drug cholestyramine on cholesterol levels.

In the dataset `EF.csv`, the first column contains the binary indicators for treatment and control, the second column contains the proportions of the nominal cholestyramine dose actually taken, and the last three columns are cholesterol levels. Note that the individuals did not know whether they were assigned to cholestyramine or to the placebo, but differences in adverse side effects could induce differences in compliance behavior by treatment status. All individuals were assigned the same nominal dose of the drug or placebo, for the same time period. Column 3, C_3, was taken before communication about the benefits of a low-cholesterol diet, Column 4, C_4, was taken after this suggestion, but before the random assignment to cholestyramine or placebo, and Column 5, C_5, an average of post-randomization cholesterol readings, averaged over two-month readings for a period of time averaging 7.3 years for all the individuals in the study. Efron and Feldman (1991) used the change in cholesterol level as the final outcome of interest, defined as $C_5 - 0.25C_3 - 0.75C_4$. Their original paper contains more detailed descriptions.

The data structure here is more complicated than the noncompliance problem discussed in Chapters 21 and 22. Jin and Rubin (2008) reanalyzed the data based on the idea in Chapter 26 later. You can analyze the data based on your understanding of the problem, but you need to justify your choice of the method. There is no gold-standard solution for this problem.

23.8 Recommended reading

Imbens (2014) gave an econometrician's perspective of IV.

24

Application of the Instrumental Variable Method: Fuzzy Regression Discontinuity

The regression discontinuity method introduced in Chapter 20 and the IV method introduced in Chapters 21–23 are two important examples of *natural experiments*. The study designs are not as ideal as randomized experiments in Part II, but they have features similar to randomized experiments. That's why they are called natural experiments.

Compounding the regression discontinuity method with the IV method yields the *fuzzy regression discontinuity* method, another important natural experiment. I will start with examples and then provide a mathematical formulation.

24.1 Motivating examples

Chapter 20 introduces the regression discontinuity method. The following three examples are slightly different because the treatments received are not deterministic functions of the running variables. Rather, the running variables discontinuously change the probabilities of the treatments received at the cutoff points.

Example 24.1 *In 2000, the Government of India launched the Prime Minister's Village Road Program, and by 2015, this program had funded the construction of all-weather roads to nearly 200,000 villages. Based on village-level data, Asher and Novosad (2020) use the regression discontinuity method to estimate the effect of new feeder roads on various economic variables. The national program guidelines prioritized larger villages according to some arbitrary thresholds based on the 2001 Population Census. The treatment variable equals 1 if the village received a new road before the year in which the outcomes were measured. The difference between the population size of a village and the threshold did not determine the treatment variable but affected its probability discontinuously at the cutoff point 0.*

Example 24.2 *Li et al. (2015) used the data on the first-year students enrolled in 2004 to 2006 from two Italian universities to evaluate the causal effect of a university grant on the dropout rate. The students were eligible for this grant if their standardized family income was below 15,000 euros. For simplicity, we use the running variable defined as 15,000 minus the standardized family income. To receive this grant, the students must apply first. Therefore, the eligibility and the application status jointly determined the final treatment status. The running variable alone did not determine the treatment status although it changed the treatment probability at the cutoff point 0.*

Example 24.3 *Amarante et al. (2016) estimated the impact of in-utero exposure to a social assistance program on children's birth outcomes. They used a regression discontinuity induced by the Uruguayan Plan de Atención Nacional a la Emergencia Social. It was a temporary social assistance program targeted to the poorest 10 percent of households, implemented between April 2005 and December 2007. Households with a predicted low-income*

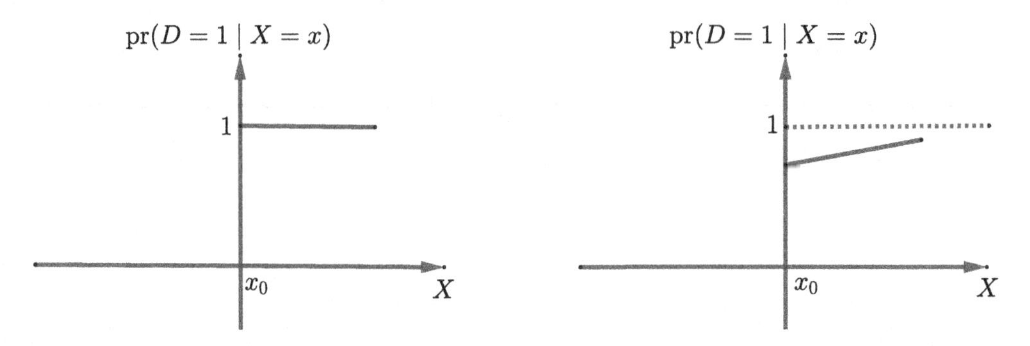

FIGURE 24.1: The treatment assignments of sharp regression discontinuity (left) and fuzzy regression discontinuity (right)

score below a predetermined threshold were assigned to the program. The predicted income score did not determine whether the mother received at least one program transfer during the pregnancy but it changed the probability of the final treatment received. The birth outcomes included birth weight, weeks of gestation, etc.

The above examples are called fuzzy regression discontinuity in contrast to the (sharp) regression discontinuity in Chapter 20. I will analyze the data in Examples 24.1 and 24.2 in Section 24.3 below.

24.2 Mathematical formulation

Let X_i denote the running variable which determines

$$Z_i = I(X_i \geq x_0)$$

with the cutoff point x_0. The treatment received D_i may not equal Z_i, but $\text{pr}(D_i = 1 \mid X_i = x)$ has a jump at x_0. Figure 24.1 compares the treatment received probabilities of the sharp regression discontinuity and fuzzy regression discontinuity. It shows a special case of fuzzy regression discontinuity with $\text{pr}(D = 1 \mid X < x_0) = 0$, which is coherent to Example 24.2.

Let Y_i denote the outcome of interest. Viewing Z_i as the treatment assigned, we can define potential outcomes $\{D_i(1), D_i(0), Y_i(1), Y_i(0)\}$. The sharp regression discontinuity of Z allows for the identification of

$$\begin{aligned} \tau_D(x_0) &= E\{D(1) - D(0) \mid X = x_0\} \\ &= \lim_{\varepsilon \to 0+} E(D \mid Z = 1, X = x_0 + \varepsilon) - \lim_{\varepsilon \to 0+} E(D \mid Z = 0, X = x_0 - \varepsilon) \end{aligned}$$

and

$$\begin{aligned} \tau_Y(x_0) &= E\{Y(1) - Y(0) \mid X = x_0\} \\ &= \lim_{\varepsilon \to 0+} E(Y \mid Z = 1, X = x_0 + \varepsilon) - \lim_{\varepsilon \to 0+} E(Y \mid Z = 0, X = x_0 - \varepsilon) \end{aligned}$$

based on Theorem 20.2. Using Z as an IV for D and imposing the IV assumptions at $X = x_0$, we can identify the local complier average causal effect by applying Theorem 21.1.

Theorem 24.1 *Assume that the treatment is determined by $Z = I(X \geq x_0)$ where x_0 is a predetermined threshold. Assume monotonicity*

$$D_i(1) \geq D_i(0)$$

and exclusion restriction

$$D_i(1) = D_i(0) \Longrightarrow Y_i(1) = Y_i(0)$$

in the infinitesimal neighborhood of x_0. The local complier average causal effect, defined as

$$\tau_c(x_0) = E\{Y(1) - Y(0) \mid D(1) > D(0), X = x_0\},$$

equals

$$\tau_c(x_0) = \frac{E\{Y(1) - Y(0) \mid X = x_0\}}{E\{D(1) - D(0) \mid X = x_0\}}.$$

Further assume that $E\{D(1) \mid X = x\}$ and $E\{Y(1) \mid X = x\}$ are continuous from the right at $x = x_0$, and $E\{D(0) \mid X = x\}$ and $E\{Y(0) \mid X = x\}$ are continuous from the left at $x = x_0$. The local complier average causal effect can be identified by

$$\tau_c(x_0) = \frac{\lim_{\varepsilon \to 0+} E(Y \mid Z = 1, X = x_0 + \varepsilon) - \lim_{\varepsilon \to 0+} E(Y \mid Z = 0, X = x_0 - \varepsilon)}{\lim_{\varepsilon \to 0+} E(D \mid Z = 1, X = x_0 + \varepsilon) - \lim_{\varepsilon \to 0+} E(D \mid Z = 0, X = x_0 - \varepsilon)}$$

if $E(D \mid X = x)$ has a non-zero jump at $x = x_0$.

Theorem 24.1 is a superposition of Theorems 20.2 and 21.1. I leave its proof as Problem 24.1.

In both sharp and fuzzy regression discontinuity, the key is to specify the neighborhood around the cutoff point. Practically, a smaller neighborhood leads to a smaller bias but a larger variance, while a larger neighborhood leads to a larger bias but a smaller variance. That is, we face a bias-variance trade-off. Some automatic procedures exist based on some statistical criteria, which rely on some strong conditions. It seems wiser to conduct sensitivity analysis over a range of the choice of the neighborhood.

Assume that we have specified the neighborhood of x_0 determined by a bandwidth h. Recall the definitions $R_i = \max(X_i - x_0, 0)$ and $L_i = \min(X_i - x_0, 0)$, and the regression in (20.6). For data with $X_i \in [x_0 - h, x_0 + h]$, we can estimate $\tau_D(x_0)$ by

$$\hat{\tau}_D(x_0) = \text{the coefficient of } Z_i \text{ in the OLS fit of } D_i \text{ on } \{1, Z_i, R_i, L_i\},$$

and estimate $\tau_Y(x_0)$ by

$$\hat{\tau}_Y(x_0) = \text{the coefficient of } Z_i \text{ in the OLS fit of } Y_i \text{ on } \{1, Z_i, R_i, L_i\}.$$

Then we can estimate the local complier average causal effect by

$$\hat{\tau}_c(x_0) = \hat{\tau}_Y(x_0) / \hat{\tau}_D(x_0).$$

This is an indirect least squares estimator. By Theorem 23.1, it is numerically identical to

$$\text{the coefficient of } D_i \text{ in the TSLS fit of } Y_i \text{ on } \{1, D_i, R_i, L_i\}$$

with D_i instrumented by Z_i. In sum, after specifying h, the estimation of $\tau_c(x_0)$ reduces to a TSLS procedure with the local data around the cutoff point.

24.3 Application

24.3.1 Reanalyzing the data from Asher and Novosad (2020)

Revisit Example 24.1. We can compute the point estimates and standard errors for a sequence of h based on the outcome `occupation_index_andrsn`.

```
library("car")
road_dat = read.csv("indianroad.csv")
table(road_dat$t, road_dat$r2012)
road_dat$runv = road_dat$left + road_dat$right
## sensitivity analysis
seq.h  = seq(10, 80, 1)
frd_sa = lapply(seq.h, function(h){
  road_sub = subset(road_dat, abs(runv)<=h)
  road_sub$r2012hat = lm(r2012 ~ t + left + right,
                         data = road_sub)$fitted.values
  tslsreg = lm(occupation_index_andrsn ~ r2012hat + left + right,
               data = road_sub)
  res = with(road_sub,
             {
               occupation_index_andrsn -
                 cbind(1, r2012, left, right)%*%coef(tslsreg)
             })
  tslsreg$residuals = as.vector(res)

  c(coef(tslsreg)[2],
    sqrt(hccm(tslsreg, type = "hc2")[2, 2]),
    length(res))
})
frd_sa = do.call(rbind, frd_sa)
```

Figure 24.2 shows the results. The treatment effect is not significant unless h is large.

The package `rdrobust` selects the bandwidth automatically. The results suggest that receiving a new road did not affect the outcome significantly.

```
> library("rdrobust")
> frd_road = with(road_dat,
+                 {
+                   rdrobust(y = occupation_index_andrsn,
+                            x = runv,
+                            c = 0,
+                            fuzzy = r2012)
+                 })
> res = cbind(frd_road$coef, frd_road$se)
> round(res, 3)
                 Coeff Std. Err.
Conventional    -0.253     0.301
Bias-Corrected  -0.283     0.301
Robust          -0.283     0.359
```

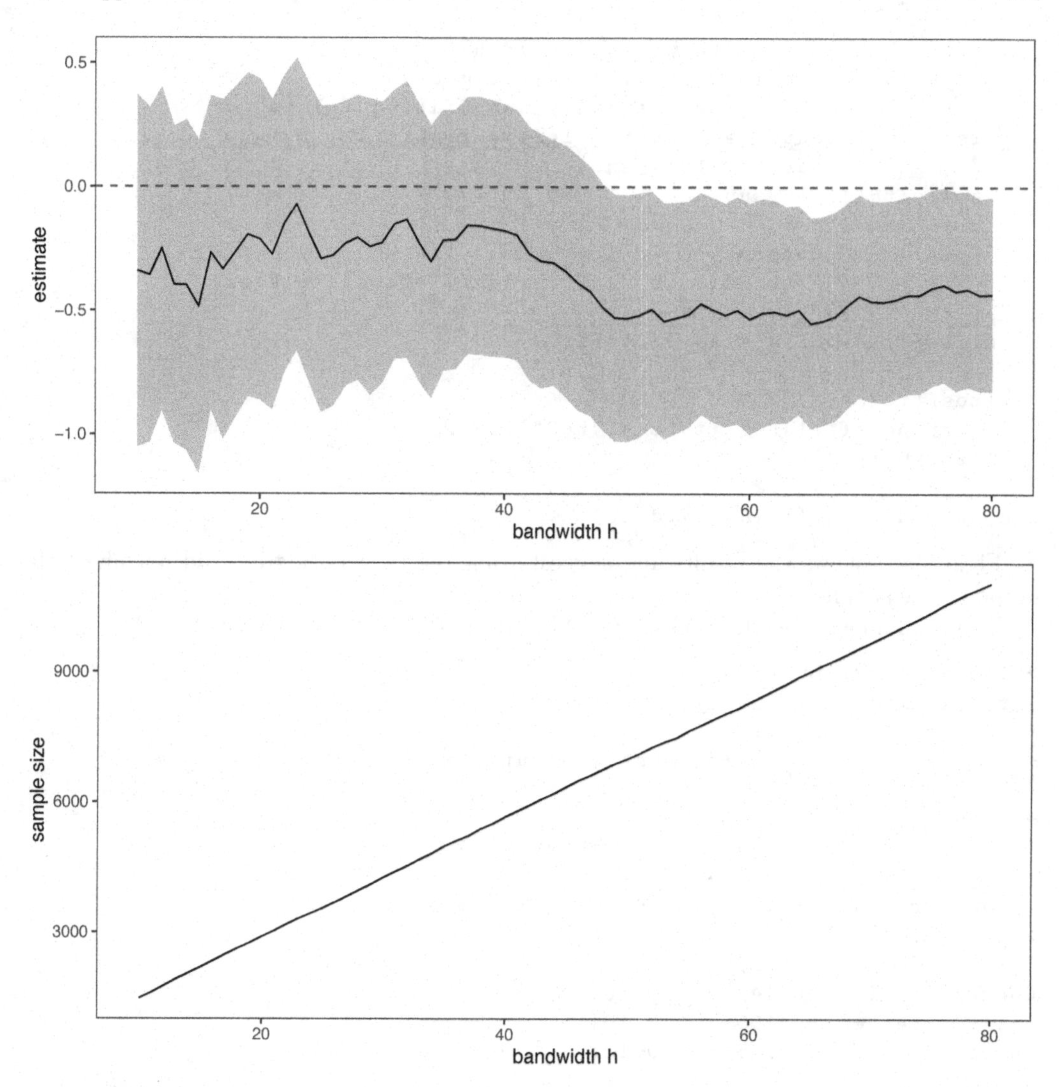

FIGURE 24.2: Reanalyzing the data from Asher and Novosad (2020), with point estimates and standard errors from TSLS

24.3.2 Reanalyzing the data from Li et al. (2015)

Revisit Example 24.2. Recall that the running variable is 15,000 minus the standardized income. In the analysis, I restrict the data to a subset with this running between $[-5000, 5000]$, and then divide the running variable by 5000 so that the running variable is bounded between $[-1, 1]$ at cutoff point zero. We can compute the point estimates and standard errors for a sequence of h.

```
library("car")
italy = read.csv("italy.csv")
italy$left  = pmin(italy$rv0, 0)
italy$right = pmax(italy$rv0, 0)
## sensitivity analysis
seq.h  = seq(0.1, 1, 0.01)
frd_sa = lapply(seq.h, function(h){
```

```
  italy_sub = subset(italy, abs(rv0)<=h)
  italy_sub$Dhat = lm(D ~ Z + left + right,
                      data = italy_sub)$fitted.values
  tslsreg = lm(outcome ~ Dhat + left + right,
               data = italy_sub)
  res = with(italy_sub,
             {
               outcome -
                 cbind(1, D, left, right)%*%coef(tslsreg)
             })
  tslsreg$residuals = as.vector(res)

  c(coef(tslsreg)[2],
    sqrt(hccm(tslsreg, type = "hc2")[2, 2]),
    length(res))
})
frd_sa = do.call(rbind, frd_sa)
```

Figure 24.3 shows the results and suggests that the university grant did not affect the dropout rate significantly.

The results based on the package `rdrobust` reach the same conclusion.

```
> library("rdrobust")
> frd_italy = with(italy,
+                  {
+                    rdrobust(y = outcome,
+                             x = rv0,
+                             c = 0,
+                             fuzzy = D)
+                  })
> res = cbind(frd_italy$coef, frd_italy$se)
> round(res, 3)
                Coeff Std. Err.
Conventional   -0.149     0.101
Bias-Corrected -0.155     0.101
Robust         -0.155     0.121
```

24.4 Discussion

Both Chapter 20 and this chapter formulate regression discontinuity based on the continuity of the conditional expectations of the potential outcomes given the running variables. This perspective is mathematically simpler but it only identifies the local effects precisely at the cutoff point of the running variable. Hahn et al. (2001) started this line of literature.

An alternative perspective is based on *local randomization* (Cattaneo et al., 2015; Li et al., 2015). If we view the running variable as a noisy measure of some underlying truth and the cutoff point is somewhat arbitrarily chosen, the units near the cutoff point do not differ systematically. This suggests that in a small neighborhood of the cutoff point, the units receive the treatment and the control in a random fashion just as in a randomized experiment. Similar to the issue of choosing h in the first perspective, it is crucial to decide how local the randomized experiment should be under the regression discontinuity. It is

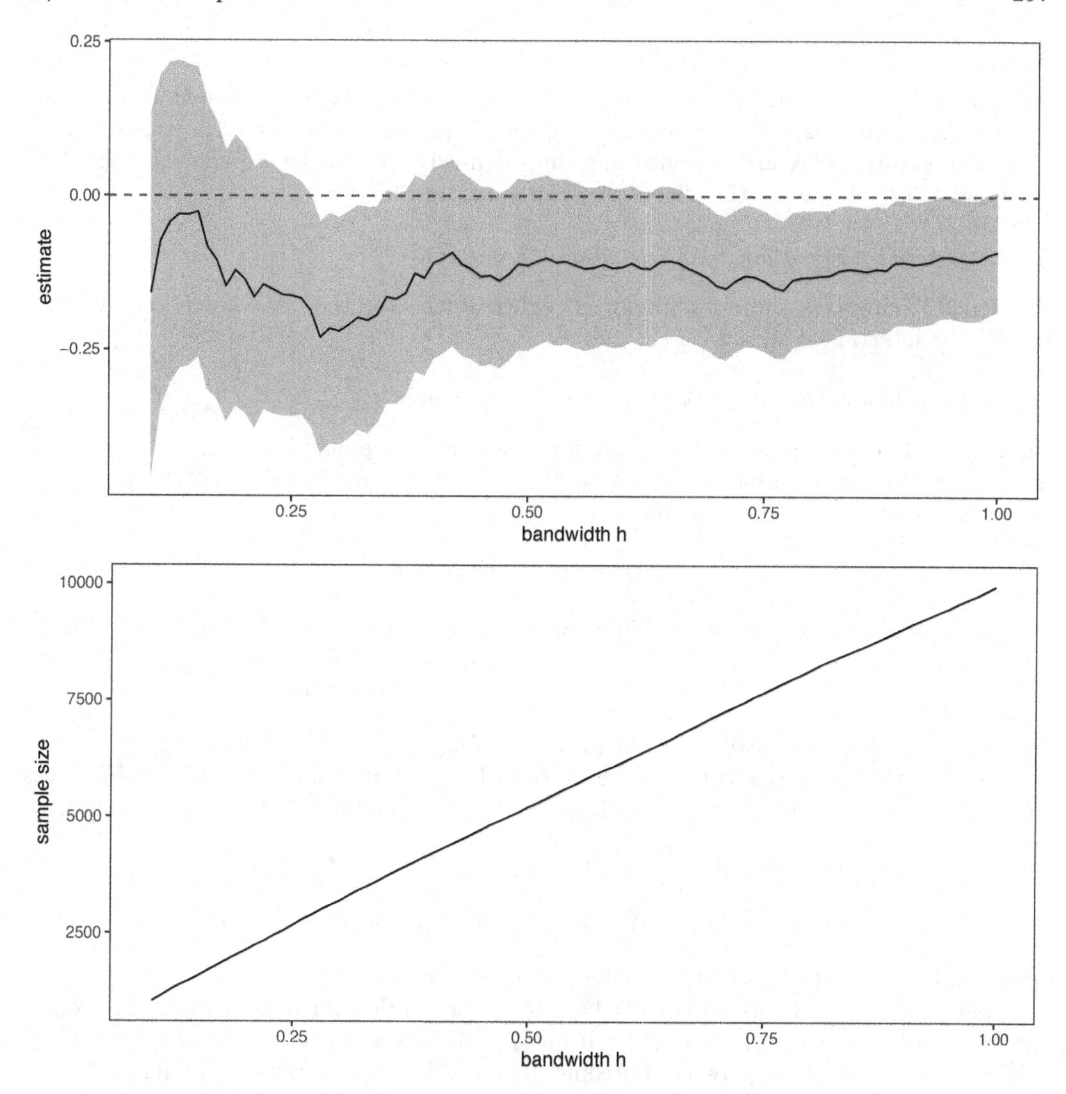

FIGURE 24.3: Reanalyzing the data from Li et al. (2015), with point estimates and standard errors from TSLS

not easy to quantify the intuition mathematically, and again conducting sensitivity analysis with a range of h seems a reasonable approach in the second perspective as well.

See Sekhon and Titiunik (2017) for a more conceptual discussion of regression discontinuity.

24.5 Homework problems

24.1 Proof of Theorem 24.1

Prove Theorem 24.1.

24.2 Data analysis

Section 24.3.1 reports the estimate of the effect on `occupation_index_andrsn`. Four other outcome variables are `transport_index_andrsn`, `firms_index_andrsn`, `consumption_index_andrsn`, and `agriculture_index_andrsn`, with meanings defined in the original paper of Asher and Novosad (2020). Estimate the effects on these outcomes.

24.3 Weak IV method for fuzzy regression discontinuity

Section 24.3 analyzes the data based on TSLS. Reanalyze the data based on the FAR confidence interval for weak IVs.

24.4 Reflection on the analysis of the data from Li et al. (2015)

In Li et al. (2015), a key variable determining the treatment status is the binary application status A, which has potential outcomes $A(1)$ and $A(0)$ corresponding to the treatment $Z = 1$ and control $Z = 0$. By definition,

$$D(1) = A(1), \quad D(0) = 0,$$

so the compliers $\{D(1), D(0)\} = (1, 0)$ is equivalent to $A(1) = 1$. So

$$\tau_{\mathrm{c}}(x_0) = E\{Y(1) - Y(0) \mid A(1) = 1, X = x_0\}.$$

Section 24.3.2 used the whole data set to estimate $\tau_{\mathrm{c}}(x_0)$.

An alternative analysis is to use the units with $A = 1$ only. Then the treatment status is determined by X. However, this analysis can be problematic because

$$\begin{aligned} & \lim_{\varepsilon \to 0+} E\{Y \mid A = 1, X = x_0 + \varepsilon\} - \lim_{\varepsilon \to 0+} E\{Y \mid A = 1, X = x_0 - \varepsilon\} \\ = \; & E\{Y(1) \mid A(1) = 1, X = x_0\} - E\{Y(0) \mid A(0) = 1, X = x_0\}. \end{aligned} \tag{24.1}$$

Prove (24.1) and explain why this analysis can be problematic.

Remark: The left-hand side of (24.1) is the identification formula of the local average treatment effect at $X = x_0$, conditional on $A = 1$. The right-hand side of (24.1) is the difference in means of the potential outcomes for the subgroups of units with $(A(1) = 1, X = x_0)$ and $(A(0) = 1, X = x_0)$, respectively. See Chapter 26 later for related discussions.

24.5 Recommended reading

Imbens and Lemieux (2008) gave practical guidance to regression discontinuity based on the potential outcomes framework. Lee and Lemieux (2010) reviewed regression discontinuity and its applications in economics.

25

Application of the Instrumental Variable Method: Mendelian Randomization

Katan (1986) was concerned with the observational studies suggesting that low serum cholesterol levels were associated with the risk of cancer. As we have discussed, however, observational studies suffer from unmeasured confounding. Consequently, it is difficult to interpret the observed association as causality. In the particular problem studied by Katan (1986), it is even possible that early stages of cancer reversely cause low serum cholesterol levels. Disentangling the causal effect of the serum cholesterol level on cancer seems a hard problem using standard epidemiologic studies. Katan (1986) argued that Apolipoprotein E genes are associated with serum cholesterol levels but do not directly affect the cancer status. So if low serum cholesterol levels cause cancer, we should observe differences in cancer risks among people with and without the genotype that leads to different serum cholesterol levels. Using our language for causal inference, Katan (1986) proposed to use Apolipoprotein E genes as IVs.

Katan (1986) did not conduct any data analysis but just proposed a conceptual design that could address not only *unmeasured confounding* but also *reverse causality*. Since then, more complicated and sophisticated studies have been conducted thanks to the modern genome-wide association studies. These studies used genetic information as IVs for exposures in epidemiologic studies to estimate the causal effects of exposures on outcomes. They were all motivated by *Mendel's second law*, the law of *random assortment*, which suggests that the inheritance of one trait is independent of the inheritance of other traits. Therefore, the method of using genetic information as IV is called *Mendelian Randomization* (MR).

25.1 Background and motivation

Graphically, Figure 25.1 shows the causal diagram on the treatment D, outcome Y, unmeasured confounder U, as well as the genetic IVs $G_1, \ldots, G_p$. In many MR studies, the genetic IVs are single nucleotide polymorphisms (SNPs). Because of pleiotropy,[1] it is possible that the genetic IVs have a direct effect on the outcome of interest, so Figure 25.1 also allows for the violation of the exclusion restriction assumption.

The standard linear IV model assumes away the direct effect of the IVs on the outcome. Definition 25.1 below gives both the structural and reduced forms.

Definition 25.1 (linear IV model) *The standard linear IV model*

$$Y = \beta_0 + \beta D + \beta_u U + \varepsilon_Y, \tag{25.1}$$

$$D = \gamma_0 + \gamma_1 G_1 + \cdots + \gamma_p G_p + \gamma_u U + \varepsilon_D, \tag{25.2}$$

[1] Pleiotropy occurs when one gene influences two or more seemingly unrelated phenotypic traits.

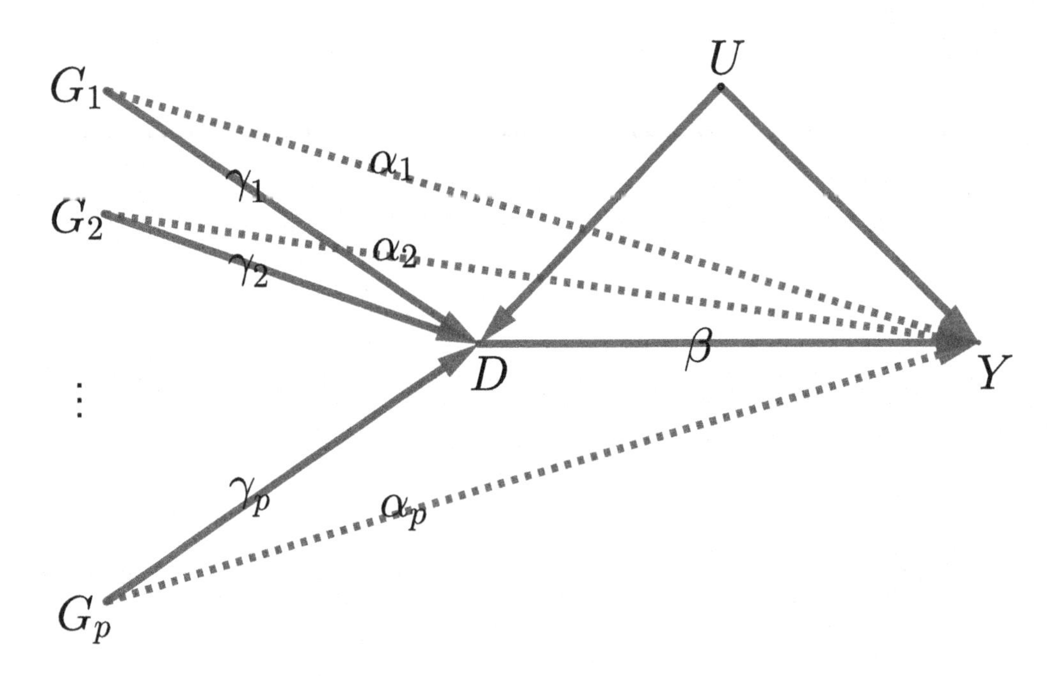

FIGURE 25.1: Causal diagram for Mendelian Randomization, with β representing the effect of D on Y

has reduced form

$$Y = \beta_0 + \beta\gamma_0 + \beta\gamma_1 G_1 + \cdots + \beta\gamma_p G_p + (\beta_u + \beta_0\gamma_u)U + \varepsilon_Y, \quad (25.3)$$
$$D = \gamma_0 + \gamma_1 G_1 + \cdots + \gamma_p G_p + \gamma_u U + \varepsilon_D. \quad (25.4)$$

Definition 25.2 below allows for the violation of exclusion restriction. Then, $G_1, \ldots, G_p$ are not valid IVs.

Definition 25.2 (linear model with possibly invalid IVs) *The linear model*

$$Y = \beta_0 + \beta D + \alpha_1 G_1 + \cdots + \alpha_p G_p + \beta_u U + \varepsilon_Y, \quad (25.5)$$
$$D = \gamma_0 + \gamma_1 G_1 + \cdots + \gamma_p G_p + \gamma_u U + \varepsilon_D, \quad (25.6)$$

has reduced form

$$Y = (\beta_0 + \beta\gamma_0) + (\alpha_1 + \beta\gamma_1)G_1 + \cdots + (\alpha_p + \beta\gamma_p)G_p + (\beta_u + \beta\gamma_u)U + \varepsilon_Y, \quad (25.7)$$
$$D = \gamma_0 + \gamma_1 G_1 + \cdots + \gamma_p G_p + \gamma_u U + \varepsilon_D. \quad (25.8)$$

Definitions 25.1 and 25.2 are slightly different from the linear IV model in Chapter 23. They include the confounder U explicitly in the models. However, this slight difference does not change the discussion fundamentally.

Under Definition 25.1 with exclusion restriction, we further define

$$\Gamma_j = \beta\gamma_j, \quad (j = 1, \ldots, p)$$

as the coefficients of the G's in the model of Y. Under Definition 25.2 without exclusion restriction, we further define

$$\Gamma_j = \alpha_j + \beta\gamma_j, \quad (j = 1, \ldots, p)$$

as the coefficients of the G's in the model of Y.

If we have individual data, we can apply the classic TSLS estimator to estimate β under the linear IV model in Definition 25.1. However, most MR studies do not have individual data but rather summary statistics from multiple genome-wide association studies. A canonical MR study with summary statistics has the following data structure.

Assumption 25.1 (MR study with summary statistics) *(a) We have the regression coefficients of the treatment on the genetic IVs* $\hat{\gamma}_1, \ldots, \hat{\gamma}_p$ *as well as the standard errors* $\text{se}_{D1}, \ldots, \text{se}_{Dp}$. *Assume*

$$\hat{\gamma}_1 \to \gamma_1, \ldots, \hat{\gamma}_p \to \gamma_p \tag{25.9}$$

in probability, and ignore the uncertainty in the standard errors.

(b) We have the regression coefficients of the outcome on the genetic IVs $\hat{\Gamma}_1, \ldots, \hat{\Gamma}_p$ *as well as with standard errors* $\text{se}_{Y1}, \ldots, \text{se}_{Yp}$. *Assume*

$$\hat{\Gamma}_1 \to \Gamma_1, \ldots, \hat{\Gamma}_p \to \Gamma_p \tag{25.10}$$

in probability, and ignore the uncertainty in the standard errors.

(c) The estimated coefficients $\hat{\gamma}_1, \ldots, \hat{\gamma}_p, \hat{\Gamma}_1, \ldots, \hat{\Gamma}_p$ *are jointly Normal and independent.*

The asymptotic Normality of the regression coefficients can be justified by the CLT. The standard errors are accurate estimates of the true standard errors in large samples. Therefore, the only subtle assumption is the joint independence of the regression coefficients. The independence of the $\hat{\gamma}_j$'s and the $\hat{\Gamma}_j$'s are reasonable because they are often calculated based on different samples.

The independence among the $\hat{\gamma}_j$'s can be reasonable if the G_j's are independent and the true linear model for D holds with homoskedastic error terms. See Section B.4. However, this assumption can be tricky if the error term of the linear model is heteroskedastic. Without the independence of the G_j's, it is hard to justify the independence. If the regression coefficients are from nonlinear models, then it is even harder to justify the independence. The independence among the $\hat{\Gamma}_j$'s follows from a similar argument.

This chapter focuses on the statistical inference of β based on the above summary statistics in Assumption 25.1.

25.2 MR based on summary statistics

25.2.1 Fixed-effect estimator

Under Definition 25.1, $\alpha_j = 0$ which implies that

$$\beta = \Gamma_j / \gamma_j \text{ for all } j.$$

A simple approach is based on the so-called *meta-analysis* (Bowden et al., 2018), that is, combining multiple estimates

$$\hat{\beta}_j = \hat{\Gamma}_j / \hat{\gamma}_j$$

for the common parameter β, also called the fixed effect. Using the delta method (see Example A.3), the ratio estimator $\hat{\beta}_j$ has approximate squared standard error

$$\text{se}_j^2 = (\text{se}_{Yj}^2 + \hat{\beta}_j^2 \text{se}_{Dj}^2) / \hat{\gamma}_j^2.$$

Therefore, the best linear combination to estimate β is the Fisher weighting based on the inverses of the variances (see Problem A.6):

$$\hat{\beta}_{\text{fisher0}} = \frac{\sum_{j=1}^{p} \hat{\beta}_j / \text{se}_j^2}{\sum_{j=1}^{p} 1/\text{se}_j^2}$$

which has variance $(\sum_{j=1}^{p} 1/\text{se}_j^2)^{-1}$. Ignoring the uncertainty due to $\hat{\gamma}_j$ quantified by se_{Dj}, the estimator reduces to

$$\begin{aligned} \hat{\beta}_{\text{fisher1}} &= \frac{\sum_{j=1}^{p} \hat{\beta}_j \hat{\gamma}_j^2 / \text{se}_{Yj}^2}{\sum_{j=1}^{p} \hat{\gamma}_j^2 / \text{se}_{Yj}^2} \\ &= \frac{\sum_{j=1}^{p} \hat{\Gamma}_j \hat{\gamma}_j / \text{se}_{Yj}^2}{\sum_{j=1}^{p} \hat{\gamma}_j^2 / \text{se}_{Yj}^2}, \end{aligned}$$

which has variance $(\sum_{j=1}^{p} \hat{\gamma}_j^2 / \text{se}_{Yj}^2)^{-1}$. Inference based on $\hat{\beta}_{\text{fisher1}}$ is suboptimal although it is more widely used in practice (Bowden et al., 2018). Both $\hat{\beta}_{\text{fisher0}}$ and $\hat{\beta}_{\text{fisher1}}$ are called the *fixed-effect estimators*.

Focus on the suboptimal yet simpler estimator $\hat{\beta}_{\text{fisher1}}$. Under Definition 25.2, we can show that

$$\begin{aligned} \hat{\beta}_{\text{fisher1}} &\to \frac{\sum_{j=1}^{p} \Gamma_j \gamma_j / \text{se}_{Yj}^2}{\sum_{j=1}^{p} \gamma_j^2 / \text{se}_{Yj}^2} \\ &= \beta + \frac{\sum_{j=1}^{p} \alpha_j \gamma_j / \text{se}_{Yj}^2}{\sum_{j=1}^{p} \gamma_j^2 / \text{se}_{Yj}^2} \end{aligned}$$

in probability. If $\alpha_j = 0$ for all j, $\hat{\beta}_{\text{fisher1}}$ is consistent. Even if this does not hold, it is still possible that $\hat{\beta}_{\text{fisher1}}$ is consistent as long as the inner product between α_j and γ_j weighted by $1/\text{se}_{Yj}^2$ is zero. This holds if we have many genetic instruments and violation of the exclusion restriction, captured by α_j, is an independent random draw from a distribution with mean zero.

25.2.2 Egger regression

Start with Definition 25.1. With the true parameters, we have

$$\Gamma_j = \beta \gamma_j \quad (j = 1, \ldots, p);$$

with the estimates, the above identity holds only approximately

$$\hat{\Gamma}_j \approx \beta \hat{\gamma}_j \quad (j = 1, \ldots, p).$$

This seems a classic least-squares problem of $\{\hat{\Gamma}_j\}_{j=1}^{p}$ on $\{\hat{\gamma}_j\}_{j=1}^{p}$. We can fit a WLS of $\hat{\Gamma}_j$ on $\hat{\gamma}_j$, with or without an intercept, possibly weighted by w_j, to estimate β. The following results hold thanks to the algebraic properties of the WLS reviewed in Section B.5.

Without an intercept, the coefficient of $\hat{\gamma}_j$ is

$$\hat{\beta}_{\text{egger1}} = \frac{\sum_{j=1}^{p} \hat{\gamma}_j \hat{\Gamma}_j w_j}{\sum_{j=1}^{p} \hat{\gamma}_j^2 w_j},$$

which reduces to $\hat{\beta}_{\text{fisher1}}$ if $w_j = 1/\text{se}^2_{Y_j}$. The WLS is called the *Egger regression*. It is more general than the fixed-effect estimators in Section 25.2.1. With an intercept, the coefficient of $\hat{\gamma}_j$ is (see the formula in Problem B.3)

$$\hat{\beta}_{\text{egger0}} = \frac{\sum_{j=1}^p (\hat{\gamma}_j - \hat{\gamma}_w)(\hat{\Gamma}_j - \hat{\Gamma}_w) w_j}{\sum_{j=1}^p (\hat{\gamma}_j - \hat{\gamma}_w)^2 w_j}$$

where $\hat{\gamma}_w = \sum_{j=1}^p \hat{\gamma}_j w_j / \sum_{j=1}^p w_j$ and $\hat{\Gamma}_w = \sum_{j=1}^p \hat{\Gamma}_j w_j / \sum_{j=1}^p w_j$ are the weighted averages of the $\hat{\gamma}_j$'s and $\hat{\Gamma}_j$'s, respectively.

Even without assuming that all γ_j's are zero under Definition 25.2, we have

$$\begin{aligned}\hat{\beta}_{\text{egger0}} &\rightarrow \frac{\sum_{j=1}^p (\gamma_j - \gamma_w)(\Gamma_j - \Gamma_w) w_j}{\sum_{j=1}^p (\gamma_j - \gamma_w)^2 w_j} \\ &= \beta + \frac{\sum_{j=1}^p (\gamma_j - \gamma_w)(\alpha_j - \alpha_w) w_j}{\sum_{j=1}^p (\gamma_j - \gamma_w)^2 w_j}\end{aligned}$$

in probability, where γ_w, Γ_w and α_w are the corresponding weighted averages of the true parameters. So $\hat{\beta}_{\text{egger0}}$ is consistent for β as long as the WLS coefficient of α_j on γ_j is zero. This is weaker than $\alpha_j = 0$ for all j. This weaker assumption holds if γ_j and α_j are realizations of independent random variables, which is called the Instrument Strength Independent of Direct Effect (InSIDE) assumption (Bowden et al., 2015). More interestingly, the intercept from the Egger regression is

$$\hat{\alpha}_{\text{egger0}} = \hat{\Gamma}_w - \hat{\beta}_{\text{egger0}} \hat{\gamma}_w,$$

which, under the InSIDE assumption, converges to

$$\Gamma_w - \beta \gamma_w = \alpha_w$$

in probability. So the intercept estimates the weighted average of the direct effects.

25.3 An example

First, the following function can implement Fisher weighting.

```
fisher.weight = function(est, se)
{
  n = sum(est/se^2)
  d = sum(1/se^2)
  res = c(n/d, sqrt(1/d))
  names(res) = c("est", "se")
  res
}
```

I use the `bmi.sbp` data in the `mr.raps` package (Zhao et al., 2020) to illustrate the Fisher weighting based on different variance estimates. The results based on $\hat{\beta}_{\text{fisher0}}$ and $\hat{\beta}_{\text{fisher1}}$ are similar.

```
> bmisbp = read.csv("mr_bmisbp.csv")
> bmisbp$iv        = with(bmisbp, beta.outcome/beta.exposure)
```

```
> bmisbp$se.iv    = with(bmisbp, se.outcome/beta.exposure)
> bmisbp$se.iv1   = with(bmisbp,
+                        sqrt(se.outcome^2 + iv^2*se.exposure^2)/
                         beta.exposure)
> fisher.weight(bmisbp$iv, bmisbp$se.iv)
       est          se
0.31727680 0.05388827
> fisher.weight(bmisbp$iv, bmisbp$se.iv1)
       est          se
0.31576007 0.05893783
```

The Egger regressions with or without the intercept also give very similar results. However, the standard errors are quite different from those based on the Fisher weighting.

```
> mr.egger = lm(beta.outcome ~ 0 + beta.exposure,
+               data = bmisbp,
+               weights = 1/se.outcome^2)
> summary(mr.egger)$coef
               Estimate Std. Error  t value    Pr(>|t|)
beta.exposure 0.3172768  0.1105994 2.868704 0.004681659
>
> mr.egger.w = lm(beta.outcome ~ beta.exposure,
+                 data = bmisbp,
+                 weights = 1/se.outcome^2)
> summary(mr.egger.w)$coef
                  Estimate  Std. Error    t value    Pr(>|t|)
(Intercept)   0.0001133328 0.002079418 0.05450217 0.956603931
beta.exposure 0.3172989306 0.110948506 2.85987564 0.004811017
```

Figure 25.2 shows the raw data as well as the fitted Egger regression line with the intercept.

25.4 Critiques of the analysis based on Mendelian Randomization

Mendelian Randomization (MR) is an application of the idea of the instrumental variable (IV). It relies on strong assumptions. I provide three sets of critiques from the conceptual, biological, and technical perspectives.

Conceptually, the treatments in most studies based on MR are not well defined from the potential outcomes perspective. For instance, the treatments are often defined as the cholesterol level or the body mass index (BMI). They are composite variables and can correspond to complex, non-unique definitions of the hypothetical experiments. The stable unit treatment value assumption (SUTVA) often does not hold for these treatments; recall the discussion in Chapter 2.

Biologically, the fundamental assumptions for the IV analysis may not hold. Mendel's second law ensures that the inheritances of different traits are independent. However, it does not ensure that the candidate IVs are independent of the hidden confounders between the treatment and the outcome. It is possible that these IVs have direct effects on the confounders. It is also possible that some unmeasured genes affect both the IVs and the confounders. Mendel's second law does not ensure the exclusion restriction assumption either. It is possible that the IVs have other causal pathways to the outcome, beyond the pathway through the treatment of interest.

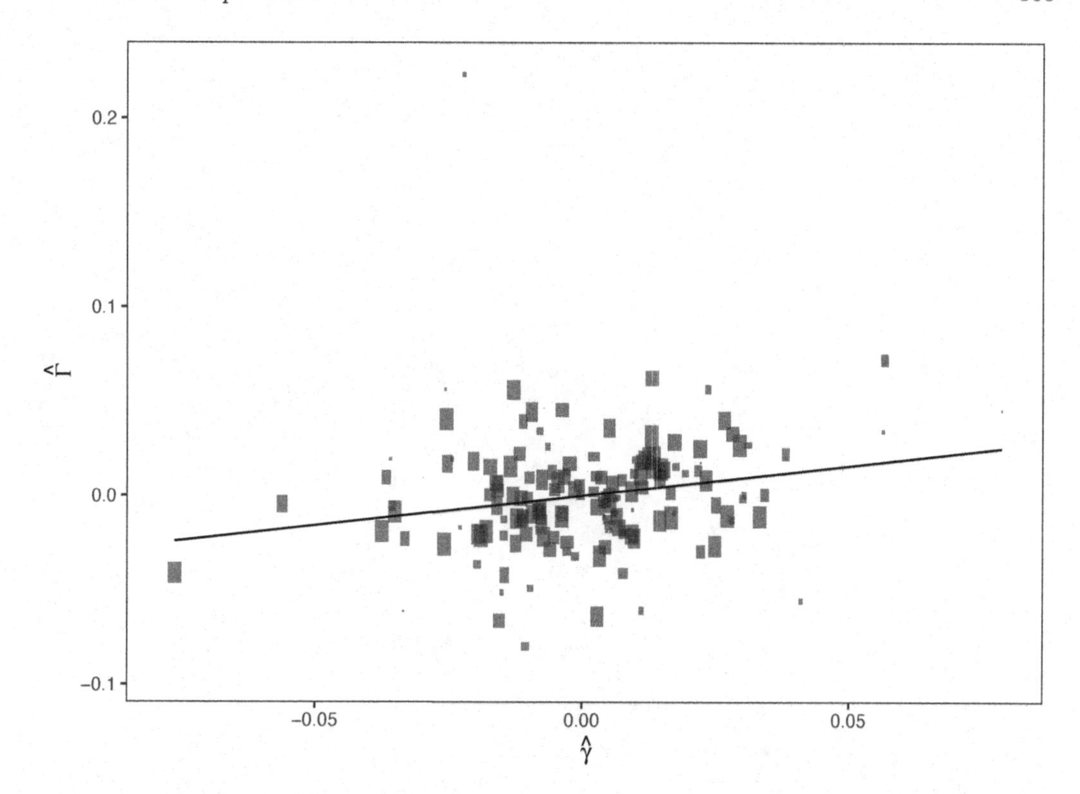

FIGURE 25.2: Scatter plot proportional to the inverse of the variance, with the Egger regression line including the intercept

Technically, the statistical assumptions for MR are quite strong. Clearly, the linear IV model is a strong modeling assumption. The independence of the $\hat{\gamma}_j$'s and the $\hat{\Gamma}_j$'s is also quite strong. Other issues in the data-collecting process can further complicate the interpretation of the IV assumptions. For instance, the treatments and outcomes are often measured with errors, and the genome-wide associate studies are often based on the case-control design with the samples conditional on the outcomes (see Section B.6.3).

VanderWeele et al. (2014) is an excellent review paper that discusses the methodological challenges in MR.

25.5 Homework problems

25.1 Data analysis

Analyze the `bmi.bmi` data in the `R` package `mr.raps`. See the package and Zhao et al. (2020, Section 7.2) for more details.

25.2 Recommended reading

Davey Smith and Ebrahim (2003) reviewed the potentials and limitations of MR.

Part VI

Causal mechanisms with post-treatment variables

26

Principal Stratification

Parts II–V of this book focus on the causal effects of a treatment on an outcome, possibly adjusting for some observed pretreatment covariates. Many applications also have some post-treatment variable M which happens after the treatment and is related to the outcome. An important question is how to use the post-treatment variable M appropriately. I will start with several motivating examples and then introduce the formulation by Frangakis and Rubin (2002) based on potential outcomes.

26.1 Motivating examples

Example 26.1 (noncompliance) *In randomized experiments with noncompliance, we can use M to represent the treatment received, which is affected by the treatment assignment Z and affects the outcome Y. In this example, M has the same meaning as D in Chapter 21.*

Example 26.2 (truncation by death) *In randomized experiments on patients with severe diseases, some patients may die before the measurement of the outcome Y, e.g., the quality of life. The post-treatment variable M in this example is the binary indicator of the survival status.*

Example 26.3 (unemployment) *In job training programs, units are randomly assigned to treatment and control groups, and report their employment status M and wage Y. Then the post-treatment variable is the binary indicator of the employment status M.*

Example 26.4 (surrogate endpoint) *In clinical trials, the outcomes of interest (e.g., 30 years of survival) require a long or costly follow-up. Practitioners instead collect data on some other variables early in the follow-up that are easy to measure. These variables are called the "surrogate endpoints." A concrete example is from clinical trials on HIV patients, where the candidate surrogate endpoint M is the CD4 cell count (recall that CD4 cells are white blood cells that fight infection).*

Examples 26.1–26.4 above have the similarity that a variable M occurs after the treatment and is related to the outcome. It is possible that M is on the causal pathway from Z to Y. Figure 26.1(a) illustrates this mechanism. Example 26.1 corresponds to Figure 26.1(a). It is also possible that M is not on the causal pathway from Z to Y. Figure 26.1(b) illustrates this mechanism. It is even possible that Y is on the causal pathway from Z to M. Figure 26.1(c) illustrates this mechanism. It is not entirely clear which causal diagrams Examples 26.2, 26.3, and 26.4 correspond to, respectively.

In practice, the underlying causal diagrams can be much more complex than those in Figure 26.1. This chapter follows the formulation by Frangakis and Rubin (2002), which does not assume the underlying causal diagrams. Nevertheless, causal diagrams can sometimes be helpful for thinking about post-treatment variables.

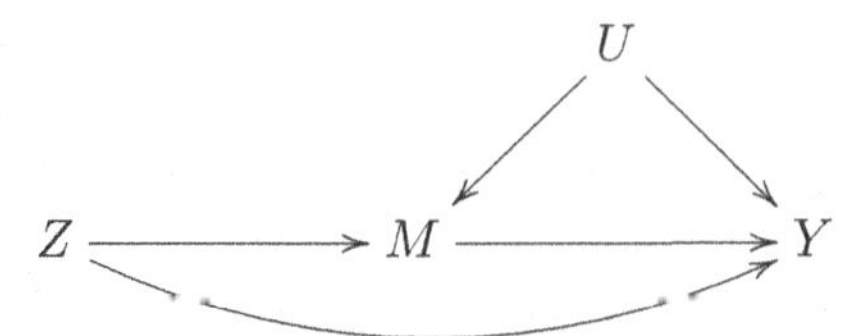

(a) M is on the causal pathway from Z to Y, with Z randomized and U representing unmeasured confounding

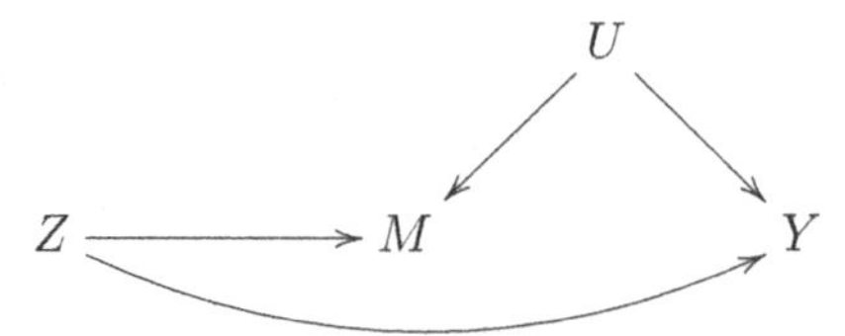

(b) M is not on the causal pathway from Z to Y, with Z randomized and U representing unmeasured confounding

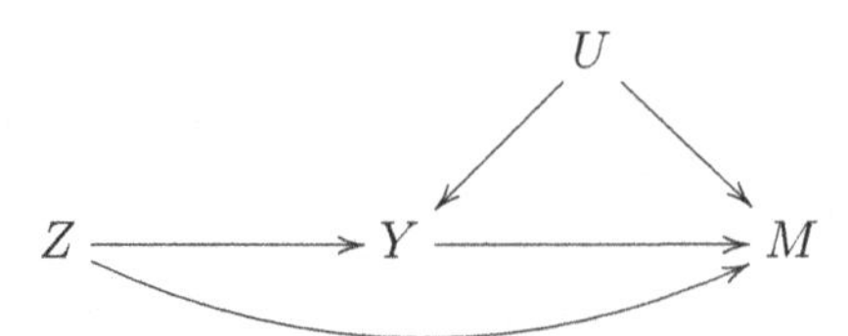

(c) Y is on the causal pathway from Z to M, with Z randomized and U representing unmeasured confounding

FIGURE 26.1: Causal diagrams with a post-treatment variable M

26.2 The problem of conditioning on the post-treatment variable

A naive method to deal with the post-treatment variable M is to condition on its observed value as if it were a pretreatment covariate. However, M is fundamentally different from X, because M is affected by the treatment in general whereas X is not. It is also a "rule of thumb" that data analyzers should not condition on any post-treatment variables in evaluating the average causal effect of the treatment on the outcome (Cochran, 1957; Rosenbaum, 1984). Based on potential outcomes, Frangakis and Rubin (2002) gave the following insightful explanation.

For simplicity, we focus on the CRE in this chapter.

Assumption 26.1 (CRE with an intermediate variable) *We have*

$$Z \perp\!\!\!\perp \{M(1), M(0), Y(1), Y(0), X\}.$$

Conditioning on $M = m$, we compare

$$\text{pr}(Y \mid Z = 1, M = m) \tag{26.1}$$

and

$$\text{pr}(Y \mid Z = 0, M = m). \tag{26.2}$$

This comparison seems intuitive because it measures the difference in the outcome distributions in the treated and control groups given the same value of the post-treatment variable. When M is a pretreatment covariate, this comparison yields a reasonable subgroup effect. However, when M is a post-treatment variable, the interpretation of this comparison is problematic. Under Assumption 26.1, we can rewrite the probabilities in (26.1) and (26.2) as

$$\begin{aligned} \text{pr}(Y \mid Z=1, M=m) &= \text{pr}\{Y(1) \mid Z=1, M(1)=m\} \\ &= \text{pr}\{Y(1) \mid M(1)=m\} \end{aligned}$$

and

$$\begin{aligned} \text{pr}(Y \mid Z=0, M=m) &= \text{pr}\{Y(0) \mid Z=0, M(0)=m\} \\ &= \text{pr}\{Y(0) \mid M(0)=m\}. \end{aligned}$$

Therefore, under CRE, comparing (26.1) and (26.2) is equivalent to comparing the distributions of $Y(1)$ and $Y(0)$ for different subsets of units because the units with $M(1) = m$ are different from the units with $M(0) = m$ if the Z affects M. Consequently, the comparison conditioning on $M = m$ does not have a causal interpretation in general unless $M(1) = M(0)$.[1]

Revisit Example 26.1. Comparing $\text{pr}(Y \mid Z = 1, M = 1)$ and $\text{pr}(Y \mid Z = 0, M = 1)$ is equivalent to comparing the treated potential outcomes for compliers and always-takers with the control potential outcomes for always-takers, under the monotonicity assumption that $M(1) \geq M(0)$. Part 3 of Problem 22.8 has pointed out the drawbacks of this analysis.

Revisit Example 26.2. If the treatment improves the survival status, the treatment can save more weak patients than the control. In this case, units with $M(1) = 1$ are weaker than units with $M(0) = 1$, so the naive comparison gives biased results that are in favor of the control.

26.3 Conditioning on the potential values of the post-treatment variable

Frangakis and Rubin (2002) proposed to condition on the joint potential value of the post-treatment variable $U = \{M(1), M(0)\}$ and compare

$$\text{pr}\{Y(1) \mid M(1) = m_1, M(0) = m_0\}$$

and

$$\text{pr}\{Y(0) \mid M(1) = m_1, M(0) = m_0\}$$

for some (m_1, m_0). This is a comparison of the potential outcomes under treatment and control for the same subset of units with $M(1) = m_1$ and $M(0) = m_0$. Frangakis and Rubin (2002) called this strategy *principal stratification*, viewing $\{M(1), M(0)\}$ as a pre-treatment covariate. Based on this idea, we can define

$$\tau(m_1, m_0) = E\{Y(1) - Y(0) \mid M(1) = m_1, M(0) = m_0\}$$

[1]Based on the causal diagrams, we can reach the same conclusion. In Figure 26.1, even though $Z \perp\!\!\!\perp U$ by randomization of Z, conditioning on M introduces the "collider bias" that causes $Z \not\perp\!\!\!\perp U$.

as the principal stratification average causal effect for the subgroup with $M(1) = m_1, M(0) = m_0$. For a binary M, we have four subgroups

$$\begin{cases} \tau(1,1) & = & E\{Y(1) - Y(0) \mid M(1) = 1, M(0) = 1\}, \\ \tau(1,0) & = & E\{Y(1) - Y(0) \mid M(1) = 1, M(0) = 0\}, \\ \tau(0,1) & = & E\{Y(1) - Y(0) \mid M(1) = 0, M(0) = 1\}, \\ \tau(0,0) & = & E\{Y(1) - Y(0) \mid M(1) = 0, M(0) = 0\}. \end{cases} \quad (26.3)$$

Since $\{M(1), M(0)\}$ is unaffected by the treatment, it is a covariate so $\tau(m_1, m_0)$ is a subgroup causal effect. For subgroups with $M(1) = M(0)$, the treatment does not change the intermediate variable, so $\tau(1,1)$ and $\tau(0,0)$ measure the *dissociative effects.* For other subgroups with $m_1 \neq m_0$, the principal stratification average causal effects $\tau(m_1, m_0)$ measure the *associative effects.* The terminology is from Frangakis and Rubin (2002), which does not assume that M is on the causal pathway from Z to Y. When we have Figure 26.1(a), we can interpret the dissociative effects as direct effects of Z on Y that act independent of M, although we cannot simply interpret the associative effects as direct or indirect effects of Z on Y.

Example 26.1 (noncompliance) *With noncompliance,* (26.3) *consists of the average causal effects for always takers, compliers, defiers, and never takers (Imbens and Angrist, 1994; Angrist et al., 1996).*

Example 26.2 (truncation by death) *Because the outcome is well-defined only if the patient survives, three subgroup causal effects in* (26.3) *are not meaningful, and the only well-defined subgroup effect is*

$$\tau(1,1) = E\{Y(1) - Y(0) \mid M(1) = 1, M(0) = 1\}. \quad (26.4)$$

It is called the survivor average causal effect (Rubin, 2006a). It is the average causal effect of the treatment on the outcome for those units who survive regardless of the treatment status.

Example 26.3 (unemployment) *The unemployment problem is isomorphic to the truncation-by-death problem because the wage is well-defined only if the unit is employed in the first place. Therefore, the only well-defined subgroup effect is* (26.4)*, the employed average causal effect. Previously, Heckman (1979) proposed a model, now called the Heckman Selection Model, to deal with unemployment in modeling the wage, viewing the wages of those unemployed as missing values.*[2] *However, Zhang and Rubin (2003) and Zhang et al. (2009) argued that* $\tau(1,1)$ *is a more meaningful quantity under the potential outcomes framework.*

Example 26.4 (surrogate endpoint) *Intuitively, we want to assess the effect of the treatment on the outcome via the effect of the treatment on the surrogate endpoint. Therefore, a good surrogate endpoint should satisfy two conditions: first, if the treatment does not affect the surrogate, then it does not affect the outcome either; second, if the treatment affects the surrogate, then it affects the outcome too. The first condition is called the "causal*

[2] Heckman won the Nobel prize of economics in 2000 "for his development of theory and methods for analyzing selective samples." His model contains two stages. First, the employment status is determined by a latent linear model

$$M_i = 1(X_i^{\top}\beta + u_i \geq 0).$$

Second, the latent log wage is determined by a linear model

$$Y_i^* = W_i^{\top}\gamma + v_i$$

and Y_i^* is observed as Y_i only if $M_i = 1$. In his two-stage model, the covariates X_i and W_i may differ, and the errors (u_i, v_i) are correlated bivariate Normal.

necessity" by Frangakis and Rubin (2002), and the second condition is called the "causal sufficiency" by Gilbert and Hudgens (2008). Based on (26.3) *for a binary surrogate endpoint, causal necessity requires that* $\tau(1,1)$ *and* $\tau(0,0)$ *are zero, and causal sufficiency requires that* $\tau(1,0)$ *and* $\tau(0,1)$ *are not zero.*

26.4 Statistical inference and its difficulty

In Example 26.1, if we have randomization, monotonicity, and exclusion restriction, then we can identify the complier average causal effect $\tau(1,0)$. This is the key result derived in Chapter 21.

However, in other examples, we cannot impose the exclusion restriction assumption. For instance, $\tau(1,1)$ is the main parameter of interest in Examples 26.2 and 26.3, and $\tau(1,1)$ and $\tau(0,0)$ are both of interest in Example 26.4. Without the exclusion restriction assumption, it is very challenging to identify the principal stratification average causal effect. Sometimes, we cannot even impose the monotonicity assumption, and thus cannot identify the proportions of the latent strata in the first place.

26.4.1 Special case: truncation by death with binary outcome

I use the simple setting with a binary treatment, binary survival status, and binary outcome to illustrate the idea and especially the difficulty of statistical inference based on principal stratification.

In addition to Assumption 26.1, we impose the monotonicity.

Assumption 26.2 (monotonicity) $M(1) \geq M(0)$.

Theorem 22.1 demonstrates that under Assumptions 26.1 and 26.2, we can identify the proportions of the three latent strata by

$$\begin{aligned}
\pi_{(1,1)} &= \text{pr}(M = 1 \mid Z = 0), \\
\pi_{(0,0)} &= \text{pr}(M = 0 \mid Z = 1), \\
\pi_{(1,0)} &= \text{pr}(M = 1 \mid Z = 1) - \text{pr}(M = 1 \mid Z = 0).
\end{aligned}$$

Our goal is to identify the survivor average causal effect $\tau(1,1)$. First, we can easily identify $E\{Y(0) \mid M(1) = 1, M(0) = 1\}$ because the observed group $(Z = 0, M = 1)$ consists of only survivors:

$$E\{Y(0) \mid M(1) = 1, M(0) = 1\} = E(Y \mid Z = 0, M = 1).$$

The key is then to identify $E\{Y(1) \mid M(1) = 1, M(0) = 1\}$. The observed group $(Z = 1, M = 1)$ is a mixture of two strata $(1,1)$ and $(1,0)$, therefore we have

$$\begin{aligned}
E(Y \mid Z = 1, M = 1) &= \frac{\pi_{(1,1)}}{\pi_{(1,1)} + \pi_{(1,0)}} E\{Y(1) \mid M(1) = 1, M(0) = 1\} \\
&\quad + \frac{\pi_{(1,0)}}{\pi_{(1,1)} + \pi_{(1,0)}} E\{Y(1) \mid M(1) = 1, M(0) = 0\}.
\end{aligned} \tag{26.5}$$

We have two unknown parameters but only one equation in (26.5). So we cannot uniquely determine $E\{Y(1) \mid M(1)=1, M(0)=1\}$ from the above equation (26.5). Nevertheless, (26.5) contains some information about the quantity of interest. That is, $E\{Y(1) \mid M(1)= 1, M(0)=1\}$ is partially identified by Definition 18.1.

For a binary outcome Y, we know that $E\{Y(1) \mid M(1)=1, M(0)=0\}$ is bounded between 0 and 1, and consequently, $E\{Y(1) \mid M(1)=1, M(0)=1\}$ is bounded between the solutions to the following two equations:

$$\begin{aligned} E(Y \mid Z=1, M=1) &= \frac{\pi_{(1,1)}}{\pi_{(1,1)}+\pi_{(1,0)}} E\{Y(1) \mid M(1)=1, M(0)=1\} \\ &\quad + \frac{\pi_{(1,0)}}{\pi_{(1,1)}+\pi_{(1,0)}} \end{aligned}$$

and

$$E(Y \mid Z=1, M=1) = \frac{\pi_{(1,1)}}{\pi_{(1,1)}+\pi_{(1,0)}} E\{Y(1) \mid M(1)=1, M(0)=1\}.$$

Therefore, $E\{Y(1) \mid M(1)=1, M(0)=1\}$ has lower bound

$$\frac{\{\pi_{(1,1)}+\pi_{(1,0)}\} E(Y \mid Z=1, M=1)-\pi_{(1,0)}}{\pi_{(1,1)}},$$

and upper bound

$$\frac{\{\pi_{(1,1)}+\pi_{(1,0)}\} E(Y \mid Z=1, M=1)}{\pi_{(1,1)}}.$$

We can then derive the bounds on $\tau(1,1)$, summarized in Theorem 26.1 below.

Theorem 26.1 *Under Assumptions 26.1 and 26.2 with a binary Y, we have*

$$\begin{aligned} & \frac{\{\pi_{(1,1)}+\pi_{(1,0)}\} E(Y \mid Z=1, M=1)-\pi_{(1,0)}}{\pi_{(1,1)}} - E(Y \mid Z=0, M=1) \\ \leq\ & \tau(1,1) \\ \leq\ & \frac{\{\pi_{(1,1)}+\pi_{(1,0)}\} E(Y \mid Z=1, M=1)}{\pi_{(1,1)}} - E(Y \mid Z=0, M=1). \end{aligned}$$

To construct the confidence interval for $\tau(1,1)$, we can use the approach proposed by Imbens and Manski (2004), which involves two steps:

1. we obtain the estimated lower and upper bounds $[\hat{l}, \hat{u}]$ with estimated standard errors $(\text{se}_l, \text{se}_u)$;
2. we construct the confidence interval as $[\hat{l} - z_{1-\alpha}\text{se}_l, \hat{u} + z_{1-\alpha}\text{se}_u]$, where $z_{1-\alpha}$ is the $1-\alpha$ quantile of the standard Normal distribution.

The validity of their confidence interval relies on some regularity conditions. In most truncation by death problems, those conditions hold because the lower and upper bounds are quite different, and they are bounded away from the extreme values -1 and 1. I omit the technical details here.

To summarize, this is a challenging problem since we cannot identify the parameter based on the observed data even with an infinite sample size. We can derive large-sample bounds for $\tau(1,1)$ but the statistical inference based on the bounds is not standard. If we do not have monotonicity, the large-sample bounds have even more complex forms; see Zhang and Rubin (2003) and Jiang et al. (2016, Appendix A).

26.4.2 Implementation of the bounds

Based on the data with binary $(Z_i, M_i, Y_i)_{i=1}^n$, we can implement the bounds on as well as the confidence intervals on $\tau(1,1)$ using the following R code.

```
## function for SACE with a binary outcome
SACE01.fit = function(Z, M, Y)
{
  ## summary statistics
  pM1 = mean(M[Z==1])
  pM0 = mean(M[Z==0])

  mu11 = mean(Y[Z==1&M==1])
  mu01 = mean(Y[Z==0&M==1])

  ## proporitions of the strata
  pi11 = pM0
  pi00 = 1 - pM1
  pi10 = 1 - pi11 - pi00

  ## bounds on the treatment potential outcomes
  lb = ((pi11 + pi10)*mu11 - pi10)/pi11
  ub = ((pi11 + pi10)*mu11)/pi11

  ## bounds on the SACE
  c(lb - mu01, ub - mu01)
}

SACE01 = function(Z, M, Y, n.boot = 1000, alpha = 0.05)
{
   bounds = SACE01.fit(Z, M, Y)
   index  = 1:length(Z)
   cv = qnorm(1 - alpha)

   ## bootstrap se of the upper and lower bounds
   b.bounds = sapply(1:n.boot, FUN = function(b){
     id.boot = sample(index, replace = TRUE)
     SACE01.fit(Z[id.boot], M[id.boot], Y[id.boot])
   })
   b.se = apply(b.bounds, 1, sd)

   ## Imbens and Manski confidence interval
   l.ci = bounds[1] - cv*b.se[1]
   u.ci = bounds[2] + cv*b.se[2]

   res = cbind(bounds, b.se, c(l.ci, u.ci))
   colnames(res) = c("est", "se", "confidence limit")
   row.names(res) = c("lower", "upper")

   return(res)
}
```

TABLE 26.1: Data truncated by death with * indicating the outcomes for dead patients

	Treatment $Z = 1$		
	$Y = 1$	$Y = 0$	total
$M = 1$	54	268	322
$M = 0$	*	*	109

	Control $Z = 0$		
	$Y = 1$	$Y = 0$	total
$M = 1$	59	218	277
$M = 0$	*	*	152

26.4.3 An application

I use the data in Yang and Small (2016) from the Acute Respiratory Distress Syndrome Network study which involves 861 patients with lung injury and acute respiratory distress syndrome. Patients were randomized to receive mechanical ventilation with either lower tidal volumes or traditional tidal volumes. The outcome is the binary indicator for whether patients could breathe without assistance by day 28. Table 26.1 summarizes the observed data.

We first obtain the point estimators of the latent strata:

$$\begin{aligned}
\hat{\pi}_{(1,1)} &= \frac{277}{277+152} = 0.646, \\
\hat{\pi}_{(0,0)} &= \frac{109}{109+322} = 0.253, \\
\hat{\pi}_{(1,0)} &= 1 - 0.646 - 0.253 = 0.101.
\end{aligned}$$

The sample means of the outcome for surviving patients are

$$\begin{aligned}
\hat{E}(Y \mid Z=1, M=1) &= \frac{54}{322} = 0.168, \\
\hat{E}(Y \mid Z=0, M=1) &= \frac{59}{277} = 0.213.
\end{aligned}$$

The estimates for the bounds on $E\{Y(1) \mid M(1)=1, M(0)=1\}$ are

$$\left[\frac{(0.646+0.101)\times 0.168 - 0.101}{0.101}, \frac{(0.646+0.101)\times 0.168}{0.101}\right] = [0.037, 0.194],$$

so the bounds on $\tau(1,1)$ are

$$[0.037 - 0.213, 0.194 - 0.213] = [-0.176, -0.019].$$

Incorporating the sampling uncertainty based on the bootstrap, the confidence interval based on Imbens and Manski (2004) is $[-0.265, 0.040]$, which covers 0. The `R` code is shown below.

```
> ## truncation by death example
> ## data from Yang and Small (2016)
> Z = c(rep(1, 322+109), rep(0, 277+152))
> M = c(rep(1, 322), rep(0, 109),
+       rep(1, 277), rep(0, 152))
> Y = c(rep(1, 54), rep(0, 268), rep(NA, 109),
+       rep(1, 59), rep(0, 218), rep(NA, 152))
> yangsmall = SACE01(Z, M, Y)
> round(yangsmall, 3)
         est    se confidence limit
lower -0.176 0.054           -0.265
upper -0.019 0.036            0.040
```

26.4.4 Extensions

Zhang and Rubin (2003) started the literature of large-sample bounds. Imai (2008a) and Lee (2009) were two follow-up papers. Cheng and Small (2006) derived the bounds with multiple treatment arms. Yang and Small (2016) used a secondary outcome and Yang and Ding (2018a) used detailed survival information to sharpen the bounds on the survivor average causal effect.

26.5 Principal score method

Without additional assumptions, we can only derive bounds on the causal effects within principal strata, but cannot identify them in general. We must impose additional assumptions to achieve nonparametric identification of the $\tau(m_1, m_0)$'s. There is no consensus on the choice of the assumptions. Those additional assumptions are not testable, and their plausibility depends on the application.

A line of research based on the *principal score* under the *principal ignorability* assumption parallels causal inference with observational studies based on the propensity score under the ignorability assumption. For simplicity, I focus on the case with strong monotonicity.

26.5.1 Principal score method under strong monotonicity

Assumption 26.3 (strong monotonicity) $M(0) = 0$.

Similar to the ignorability assumption, we now assume the *principal ignorability* assumption.

Assumption 26.4 (principal ignorability) *We have*

$$E\{Y(0) \mid M(1) = 1, X\} = E\{Y(0) \mid M(1) = 0, X\}.$$

Assumption 26.4 implies that $E\{Y(0) \mid M(1), X\} = E\{Y(0) \mid X\}$ or, equivalently,

$$E\{Y(0)M(1) \mid X\} = E\{Y(0) \mid X\}E\{M(1) \mid X\}, \tag{26.6}$$

that is, $Y(0)$ and $M(1)$ are mean independent (or uncorrelated) given X. Assumption 26.4 is not testable because $Y(0)$ and $M(1)$ are not jointly observable for all units.

Assumptions 26.1, 26.3, and 26.4 ensure nonparametric identification of the causal effects within principal strata, as summarized by Theorem 26.2 below.

Theorem 26.2 *Under Assumptions 26.1, 26.3, and 26.4,*

1. *the conditional and marginal probabilities of* $M(1)$, $\pi(X) = \text{pr}\{M(1) = 1 \mid X\}$ *and* $\pi = \text{pr}\{M(1) = 1\}$, *can be identified by*

$$\pi(X) = \text{pr}(M = 1 \mid Z = 1, X)$$

and

$$\pi = \text{pr}(M = 1 \mid Z = 1),$$

respectively;

2. the principal stratification average causal effects can be identified by

$$\tau(1,0) = E(Y \mid Z=1, M=1) - E\{\pi(X)Y \mid Z=0\}/\pi$$

and

$$\tau(0,0) = E(Y \mid Z=1, M=0) - E\{(1-\pi(X))Y \mid Z=0\}/(1-\pi).$$

The conditional probability $\pi(X) = \mathrm{pr}\{M(1)=1 \mid X\}$ is called the *principal score.* Theorem 26.2 states that $\tau(1,0)$ and $\tau(0,0)$ can be identified by the difference in means with appropriate weights depending on the principal score.

Proof of Theorem 26.2: I will only prove the result for $\tau(1,0)$. We have

$$E(Y \mid Z=1, M=1) = E\{Y(1) \mid Z=1, M(1)=1\} = E\{Y(1) \mid M(1)=1\}.$$

Moreover,

$$\begin{aligned}
& E\{\pi(X)Y \mid Z=0\}/\pi \\
= \; & E\{\pi(X)Y(0) \mid Z=0\}/\pi \\
= \; & E\{\pi(X)Y(0)\}/\pi \qquad \text{(Assumption 26.1)} \\
= \; & E\left[E\{\pi(X)Y(0) \mid X\}\right]/\pi \qquad \text{(tower property)} \\
= \; & E\left[\pi(X)E\{Y(0) \mid X\}\right]/\pi \\
= \; & E\left[E\{M(1) \mid X\}E\{Y(0) \mid X\}\right]/\pi \\
= \; & E\left[E\{M(1)Y(0) \mid X\}\right]/\pi \qquad \text{(by (26.6))} \\
= \; & E\{M(1)Y(0)\}/\mathrm{pr}\{M(1)=1\} \\
= \; & E\{Y(0) \mid M(1)=1\}.
\end{aligned}$$

I relegate the proof of the result for $\tau(0,0)$ as Problem 26.1. □

Theorem 26.2 motivates the following simple estimators for $\tau(1,0)$ and $\tau(0,0)$, respectively:

1. fit a logistic regression of M on X using only data from the treated group to obtain $\hat{\pi}(X_i)$;
2. estimate π by $\hat{\pi} = \sum_{i=1}^n Z_i M_i / \sum_{i=1}^n Z_i$;
3. obtain moment estimators:

$$\hat{\tau}(1,0) = \frac{\sum_{i=1}^n Z_i M_i Y_i}{\sum_{i=1}^n Z_i M_i} - \frac{\sum_{i=1}^n (1-Z_i)\hat{\pi}(X_i)Y_i}{\hat{\pi}\sum_{i=1}^n (1-Z_i)}$$

and

$$\hat{\tau}(0,0) = \frac{\sum_{i=1}^n Z_i(1-M_i)Y_i}{\sum_{i=1}^n Z_i(1-M_i)} - \frac{\sum_{i=1}^n (1-Z_i)(1-\hat{\pi}(X_i)\}Y_i}{(1-\hat{\pi})\sum_{i=1}^n (1-Z_i)};$$

4. use the bootstrap to approximate the variances of $\hat{\tau}(1,0)$ and $\hat{\tau}(0,0)$.

26.5.2 An example

The following function can compute the point estimates of $\hat{\tau}(1,0)$ and $\hat{\tau}(0,0)$.

```
psw = function(Z, M, Y, X) {
  ## probabilities of 10 and 00
  pi.10 = mean(M[Z==1])
  pi.00 = 1 - pi.10

  ## conditional probabilities of 10 and 00
  ps.10 = glm(M ~ X, family = binomial,
              weights = Z)$fitted.values
  ps.00 = 1 - ps.10

  ## PCEs 10 and 00
  tau.10 = mean(Y[Z==1&M==1]) - mean(Y[Z==0]*ps.10[Z==0])/pi.10
  tau.00 = mean(Y[Z==1&M==0]) - mean(Y[Z==0]*ps.00[Z==0])/pi.00
  c(tau.10, tau.00)
}
```

The following function can compute the point estimators as well as the bootstrap standard errors.

```
psw.boot = function(Z, M, Y, X, n.boot = 500){
  ## point estimates
  point.est = psw(Z, M, Y, X)
  ## bootstrap standard errors
  n = length(Z)
  boot.est = replicate(n.boot, {
    id.boot = sample(1:n, n, replace = TRUE)
    psw(Z[id.boot], M[id.boot], Y[id.boot], X[id.boot, ])
  })
  boot.se     = apply(boot.est, 1, sd)
  ## results
  res         = rbind(point.est, boot.se)
  rownames(res) = c("est", "se")
  colnames(res) = c("tau10", "tau00")
  return(res)
}
```

Revisit the data used in Section 21.5. Previously, we assumed exclusion restriction for the IV analysis. Now, we drop the exclusion restriction but assume principal ignorability. The results are below.

```
> jobsdata = read.csv("jobsdata.csv")
> getX     = lm(treat ~ sex + age + marital
+                + nonwhite + educ + income,
+                data = jobsdata)
> X = model.matrix(getX)[, -1]
> Z = jobsdata$treat
> M = jobsdata$comply
> Y = jobsdata$job_seek
> table(Z, M)
   M
Z     0   1
  0 299   0
  1 228 372
> psw.boot(Z, M, Y, X)
        tau10       tau00
est 0.1694979 -0.09904909
se  0.1042405  0.15612983
```

The point estimator $\hat{\tau}(1,0)$ does not differ much from the one based on the IV analysis. The point estimator $\hat{\tau}(0,0)$ is close to zero with a large standard error. In this example, exclusion restriction seems a reasonable assumption.

26.5.3 Extensions

Follmann (2000), Hill et al. (2002), Jo and Stuart (2009), Jo et al. (2011), and Stuart and Jo (2015) started the literature of using the principal score to identify causal effects within principal strata. Ding and Lu (2017) provided a theoretical foundation for this strategy. They proved Theorem 26.2 as well as a more general version under monotonicity; see Problem 26.2.

26.6 Other methods for principal stratification

To estimate principal stratification average causal effects without the exclusion restriction assumption, Zhang et al. (2009) proposed to use the Normal mixture models. However, the inference based on the Normal mixture models can be quite fragile. A strategy is to use additional information to improve the inference under some restrictions (Ding et al., 2011; Mealli and Pacini, 2013; Mattei et al., 2013; Jiang et al., 2016).

Conceptually, the principal stratification framework works for general M. A multi-valued M generates many latent principal strata, and a continuous M generates infinitely many latent principal strata. In those cases, identifying the probability of the principal strata is non-trivial in the first place let alone identifying the principal stratification average causal effects. Jiang and Ding (2021) reviewed some useful strategies, and Lu et al. (2023) reported some recent progress in dealing with continuous M.

26.7 Homework problems

26.1 Complete the proof of Theorem 26.2

Prove the result for $\tau(0,0)$ in Theorem 26.2.

26.2 Principal score method under monotonicity

This problem extends Theorem 26.2, with Assumption 26.3 replaced by Assumption 26.2 and Assumption 26.4 replaced by Assumption 26.5 below.

Assumption 26.5 (principal ignorability) *We have*

$$E\{Y(1) \mid M(1)=1, M(0)=0, X\} = E\{Y(1) \mid M(1)=1, M(0)=1, X\}$$

and

$$E\{Y(0) \mid M(1)=1, M(0)=0, X\} = E\{Y(0) \mid M(1)=0, M(0)=0, X\}.$$

Theorem 26.3 *Under Assumptions 26.1, 26.2, and 26.5,*

1. the conditional and marginal principal scores can be identified by

$$\begin{aligned}\pi_{(0,0)}(X) &= \text{pr}(M=0 \mid Z=1, X),\\ \pi_{(1,1)}(X) &= \text{pr}(M=1 \mid Z=0, X),\\ \pi_{(1,0)}(X) &= \text{pr}(M=1 \mid Z=1, X) - \text{pr}(M=1 \mid Z=0, X)\end{aligned}$$

and

$$\begin{aligned}\pi_{(0,0)} &= \text{pr}(M=0 \mid Z=1),\\ \pi_{(1,1)} &= \text{pr}(M=1 \mid Z=0),\\ \pi_{(1,0)} &= \text{pr}(M=1 \mid Z=1) - \text{pr}(M=1 \mid Z=0),\end{aligned}$$

respectively;

2. the principal stratification average causal effects can be identified by

$$\begin{aligned}\tau(1,0) &= E\left\{w_{1,(1,0)}(X)Y \mid Z=1, M=1\right\}\\ &\quad -E\left\{w_{0,(1,0)}(X)Y \mid Z=0, M=0\right\},\\ \tau(0,0) &= E(Y \mid Z=1, M=0) - E\left\{w_{0,(0,0)}(X)Y \mid Z=0, M=0\right\},\\ \tau(1,1) &= E\left\{w_{1,(1,1)}(X)Y \mid Z=1, M=1\right\} - E(Y \mid Z=0, M=1)\end{aligned}$$

with

$$\begin{aligned}w_{1,(1,0)}(X) &= \frac{\pi_{(1,0)}(X)}{\pi_{(1,0)}(X)+\pi_{(1,1)}(X)} \Big/ \frac{\pi_{(1,0)}}{\pi_{(1,0)}+\pi_{(1,1)}},\\ w_{0,(1,0)}(X) &= \frac{\pi_{(1,0)}(X)}{\pi_{(1,0)}(X)+\pi_{(0,0)}(X)} \Big/ \frac{\pi_{(1,0)}}{\pi_{(1,0)}+\pi_{(0,0)}},\\ w_{0,(0,0)}(X) &= \frac{\pi_{(0,0)}(X)}{\pi_{(1,0)}(X)+\pi_{(0,0)}(X)} \Big/ \frac{\pi_{(0,0)}}{\pi_{(1,0)}+\pi_{(0,0)}},\\ w_{1,(1,1)}(X) &= \frac{\pi_{(1,1)}(X)}{\pi_{(1,0)}(X)+\pi_{(1,1)}(X)} \Big/ \frac{\pi_{(1,1)}}{\pi_{(1,0)}+\pi_{(1,1)}}.\end{aligned}$$

Prove Theorem 26.3.

Remark: Based on Theorem 26.3, we can construct weighting estimators. Theorem 26.3 is Proposition 2 in Ding and Lu (2017), which also provided more details for the estimation.

26.3 Principal score method in observational studies

This problem extends Theorem 26.2, with Assumption 26.1 replaced by the ignorability assumption below.

Assumption 26.6 $Z \perp\!\!\!\perp \{M(1), M(0), Y(1), Y(0)\} \mid X$.

Recall the definition of the propensity score $e(X) = \text{pr}(Z=1 \mid X)$. We have the following identification result.

Theorem 26.4 *Under Assumptions 26.6, 26.3, and 26.4,*

1. the conditional and marginal probabilities of $M(1)$, $\pi(X) = \text{pr}\{M(1)=1 \mid X\}$ *and* $\pi = \text{pr}\{M(1)=1\}$, *can be identified by*

$$\pi(X) = \text{pr}(M=1 \mid Z=1, X)$$

and

$$\pi = E\{\text{pr}(M=1 \mid Z=1, X)\},$$

respectively;

2. *the principal stratification average causal effects can be identified by*

$$\tau(1,0) = E\left\{\frac{M}{\pi}\frac{Z}{e(X)}Y\right\} - E\left\{\frac{\pi(X)}{\pi}\frac{1-Z}{1-e(X)}Y\right\}$$

and

$$\tau(0,0) = E\left\{\frac{1-M}{1-\pi}\frac{Z}{e(X)}Y\right\} - E\left\{\frac{1-\pi(X)}{1-\pi}\frac{1-Z}{1-e(X)}Y\right\}.$$

Prove Theorem 26.4.

26.4 General principal score method

Extend Theorems 26.3 and 26.4 under Assumptions 26.6, 26.2, and 26.5.

Remark: See Jiang et al. (2022).

26.5 Recommended reading

Frangakis and Rubin (2002) proposed the principal stratification framework. Zhang and Rubin (2003) derived large-sample bounds on the survivor average causal effect. Jiang and Ding (2021) reviewed various strategies to identify the causal effects within principal strata. Jiang et al. (2022) gave a unified discussion of this strategy for observational studies and proposed multiply robust estimators for causal effects within principal strata.

27

Mediation Analysis: Natural Direct and Indirect Effects

With an intermediate variable M between the treatment Z and outcome Y, the causal effects within principal strata defined by $U = \{M(1), M(0)\}$ can assess the treatment effect heterogeneity across latent groups U. When M is indeed on the causal pathway from Z to Y, causal effects within some principal strata, $\tau(1,1)$ and $\tau(0,0)$, can give information about the direct effect of Z on Y. However, these direct effects are only for two latent groups. The causal effects within the other two principal strata, $\tau(1,0)$ and $\tau(0,1)$, contain both direct and indirect effects. Fundamentally, principal stratification does not provide any information about the indirect effect of Z on Y through M because it does not even assume that M can be intervened.

In the above discussion, I use the notions of "direct effect" and "indirect effect" in a casual way. When M lies on the pathway from Z to Y, researchers often want to assess the extent to which the effect of Z on Y is through M and the extent to which the effect is through other pathways. This is called *mediation analysis*. It is the topic of this chapter.

27.1 Motivating examples

In mediation analysis, we have a treatment Z, an outcome Y, a mediator M, and some background covariates X. Figure 27.1 illustrates their relationship. Below I give some concrete examples.

Example 27.1 *VanderWeele et al. (2012) conducted a mediation analysis to assess the extent to which the effect of variants on chromosome 15q25.1 on lung cancer is mediated through smoking and the extent to which it operates through other causal pathways.*[1] *The exposure levels correspond to changes from 0 to 2 C alleles, smoking intensity is measured by the square root of the number of cigarettes smoked per day, and the outcome is the lung cancer indicator. The study by VanderWeele et al. (2012) contained many sociodemographic covariates.*

Example 27.2 *Rudolph et al. (2018) studied the causal mechanism from neighborhood poverty to adolescent substance use, mediated by the school and peer environment. They used data from the National Comorbidity Survey Replication Adolescent Supplement, a nationally representative survey of U.S. adolescents conducted during 2001–2004. The treatment is the binary indicator of neighborhood disadvantage, defined as living in the lowest tertile of neighborhood socioeconomic status based on data from the 2000 U.S. Census. Four binary mediators are measures of school and peer environments, and six binary outcomes*

[1] Recall Chapter 17 for Fisher's hypothesis of a common genetic factor that affects the smoking behavior and lung cancer simultaneously.

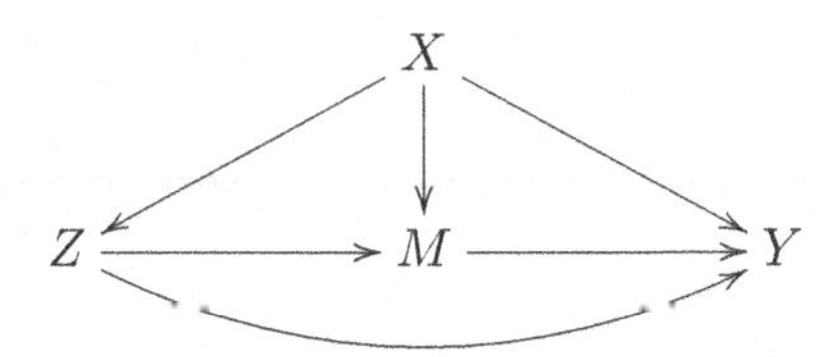

FIGURE 27.1: Causal diagram for mediation analysis

are measures of substance use. Baseline covariates included the adolescent's sex, age, race, immigration generation, family income, etc.

Example 27.3 *The* `mediation` *package in* `R` *contains a dataset called* `jobs`*, which is from JOBS II, a randomized field experiment that investigates the efficacy of a job training intervention on unemployed workers. We used this dataset in Section 21.5 to illustrate the idea of IV. The program is designed to not only increase reemployment among the unemployed but also enhance the mental health of the job seekers. It is therefore of interest to assess the indirect effect of the intervention on mental health through job search efficacy and its direct effect acting through other pathways. We will revisit this example in Section 27.4.2 later.*

27.2 Nested potential outcomes

27.2.1 Natural direct and indirect effects

Below we drop the index i for unit i and assume all random variables are IID draws from a superpopulation. For simplicity, we focus on a binary treatment Z.

We first consider the hypothetical intervention on z and define potential mediators and outcomes corresponding to the intervention on z:

$$\{M(z), Y(z) : z = 0, 1\}.$$

We then consider hypothetical intervention on both z and m and define potential outcomes corresponding to the interventions on z and m:

$$\{Y(z, m) : z = 0, 1; m \in \mathcal{M}\},$$

where $\mathcal{M}$ contains all possible values of m. Robins and Greenland (1992) and Pearl (2001) further consider the nested potential outcomes corresponding to intervention on z and $m = M(z')$:

$$\{Y(z, M_{z'}) : z = 0, 1; z' = 0, 1\}$$

where we write $M(z')$ as $M_{z'}$ to avoid excessive parentheses. The notation $Y(z, M_{z'})$ is the hypothetical outcome if the treatment were set at level z and the mediator were set at its potential level $M(z')$ under treatment z'. Importantly, z and z' can be different. With a binary treatment, we have four nested potential outcomes in total:

$$\{Y(1, M_1), Y(1, M_0), Y(0, M_1), Y(0, M_0)\}.$$

The nested potential outcome $Y(1, M_1)$ is the hypothetical outcome if the treatment were set at $z = 1$ and the mediator were set at what would happen under $z = 1$. Similarly, $Y(0, M_0)$

is the outcome if the treatment were set at $z = 0$ and the mediator were set at what would happen under $z = 0$. It would be surprising if $Y(1, M_1) \neq Y(1)$ or $Y(0, M_0) \neq Y(0)$. Therefore, we make the following composition assumption throughout this chapter.

Assumption 27.1 (composition) $Y(z, M_z) = Y(z)$ *for* $z = 0, 1$.

The composition assumption cannot be proved. It is indeed an assumption. Without causing philosophical debates, we can even define $Y(1)$ as $Y(1, M_1)$, and define $Y(0)$ as $Y(0, M_0)$. Then by definition, Assumption 27.1 holds.

The nested potential outcome $Y(1, M_0)$ is the hypothetical outcome if the unit received treatment 1 but its mediator were set at its natural value M_0 without the treatment. Similarly, $Y(0, M_1)$ is the hypothetical outcome if the unit received control 0 but its mediator were set at its natural value M_1 under the treatment. They are two cross-world counterfactual terms and are useful for defining the direct and indirect effects.

Definition 27.1 (total, direct, and indirect effects) *Define the total effect of the treatment on the outcome as*

$$\tau = E\{Y(1) - Y(0)\}.$$

Define the natural direct effect as

$$\text{NDE} = E\{Y(1, M_0) - Y(0, M_0)\}.$$

Define the natural indirect effect as

$$\text{NIE} = E\{Y(1, M_1) - Y(1, M_0)\}.$$

The total effect is the standard average causal effect of Z on Y. The natural direct effect measures the effect of the treatment on the outcome if the mediator were set at the natural value M_0 without the intervention. The natural indirect effect measures the effect of the treatment through changing the mediator if the treatment itself were set at $z = 1$. Under the composition assumption, the natural direct and indirect effects reduce to

$$\text{NDE} = E\{Y(1, M_0) - Y(0)\}, \quad \text{NIE} = E\{Y(1) - Y(1, M_0)\},$$

and therefore, we can decompose the total effect as the sum of the natural direct and indirect effects.

Proposition 27.1 *By Definition 27.1 and Assumption 27.1, we have*

$$\tau = \text{NDE} + \text{NIE}.$$

We can also define the natural indirect effect as

$$\text{NIE}' = E\{Y(0, M_1) - Y(0, M_0)\},$$

where the treatment is fixed at 0. Then we need to modify the natural direct effect as

$$\text{NDE}' = E\{Y(1, M_1) - Y(0, M_1)\}$$

to ensure that the decomposition

$$\tau = \text{NDE}' + \text{NIE}' \tag{27.1}$$

holds. This chapter focus on the decomposition in Proposition 27.1. Analogous discussion holds for the decomposition 27.1.

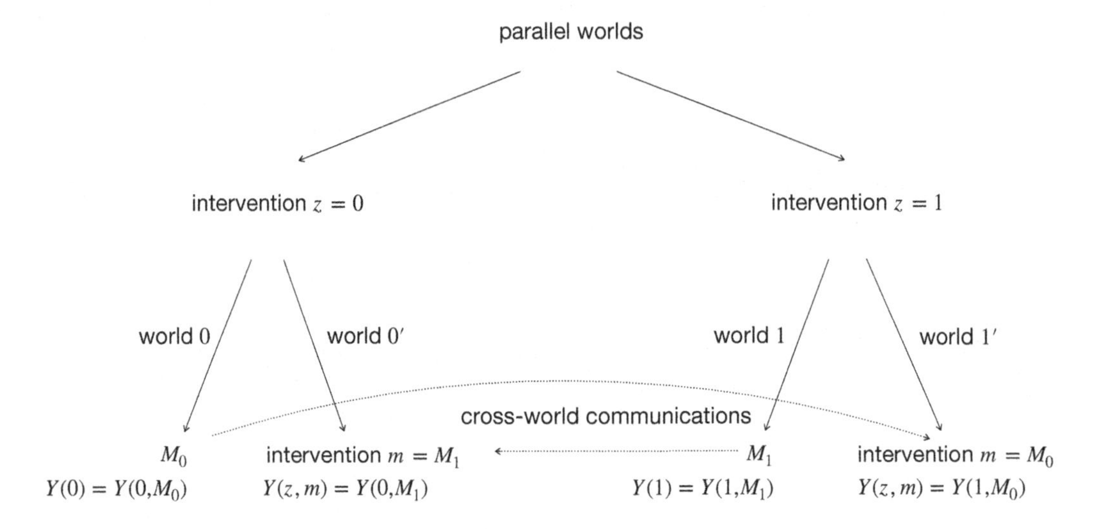

FIGURE 27.2: Cross-world potential outcomes $Y(1, M_0)$ and $Y(0, M_1)$

Unfortunately, the nest potential outcome $Y(1, M_0)$ is not an easy quantity to understand due to the cross-world nature of the interventions: the treatment is set at $z = 1$ but the mediator is set at its natural value M_0 under treatment $z = 0$. Clearly, these two interventions on the treatment cannot simultaneously happen in any realized experiment. To understand the cross-world potential outcome $Y(1, M_0)$, we need to imagine the existence of parallel worlds as shown in Figure 27.2. Let's focus on $Y(1, M_0)$. When the treatment is set at $z = 1$, the mediator must take value M_1. If at the same time, we want to set the mediator at $m = M_0$, we must know the value of M_0 for the same unit from another experiment in the parallel world. This can be an unrealistic physical experiment because it requires that the same unit is intervened at two different levels of the treatment. Under some strong assumptions about the homogeneity of units, we may use another unit's mediator value under the control as a proxy for M_0.

27.2.2 Metaphysics or science

Causal inference is hard, and there is no agreement even on its mathematical notation. Robins and Greenland (1992) and Pearl (2001) used the nested potential outcomes to define the natural direct and indirect effects. However, Frangakis and Rubin (2002) called $Y(1, M_0)$ and $Y(0, M_1)$ *a priori counterfactuals* because we can not observe them in any physical experiments. In this sense, they do not exist a priori. According to Popper (1963), a way to distinguish science and metaphysics is the falsifiability of the statements. That is, if a statement is not falsifiable based on any physical experiments or observations, then it is not a scientific but rather a metaphysical statement. Because we can not observe $Y(1, M_0)$ and $Y(0, M_1)$ in any experiments, we can not falsify any statements involving them except for the trivial ones (e.g., some outcomes are binary, or continuous, or bounded). Therefore, a strict Popperian statistician would view mediation analysis as metaphysics.

More strikingly, Dawid (2000) criticized the potential outcomes framework to be metaphysical, and he called Rubin's "Science Table" defined in Section 2.2 a "metaphysical array."[2] This is a critique on not only the a priori counterfactuals $Y(1, M_0)$ and $Y(0, M_1)$

[2] In a recent interview (Vovk and Shafer, 2023), Dawid recalled his first encounter with Rubin's work

but also the simpler potential outcomes $Y(1)$ and $Y(0)$. Dawid (2000) argued that because we can never observe $Y(1)$ and $Y(0)$ jointly, then introducing the notation $\{Y(1), Y(0)\}$ is a metaphysical activity. He is correct about the metaphysical nature of the joint distribution of $\text{pr}\{Y(1), Y(0)\}$, but he is incorrect about the marginal distributions. Based on the observed data, we indeed can falsify some statements about the marginal distributions, although we cannot falsify any statements about the joint distribution.[3] Therefore, even according to Popper (1963), Rubin's Science Table is not metaphysical because it has some nontrivial falsifiable implications although not all implications are falsifiable. This is the fundamental difference between $\{Y(1), Y(0)\}$ and $\{Y(1, M_0), Y(0, M_1)\}$.

27.3 The mediation formula

The mediation formula in Pearl (2001) relies on the following four assumptions. The first three essentially assume that the treatment and the mediator are both randomized conditional on observed covariates.

Assumption 27.2 *There is no treatment-outcome confounding:*

$$Z \perp\!\!\!\perp Y(z, m) \mid X$$

for all z and m.

Assumption 27.3 *There is no mediator-outcome confounding:*

$$M \perp\!\!\!\perp Y(z, m) \mid (X, Z)$$

for all z and m.

Assumptions 27.2 and 27.3 together are often called *sequential ignorability*. They are equivalent to the assumption that (Z, M) are jointly randomized conditioning on X:

$$(Z, M) \perp\!\!\!\perp Y(z, m) \mid X \tag{27.2}$$

for all z and m. I leave the proof of (27.2) as Problem 27.2.

Assumption 27.4 *There is no treatment-mediator confounding:*

$$Z \perp\!\!\!\perp M(z) \mid X$$

for all z.

---

based on potential outcomes in the 1970s and thought "this is absolutely the wrong way to go about things." Ironically, Dawid himself used potential outcomes in his own papers (e.g., Dawid et al., 2017). Once I asked him at a conference why he used potential outcomes given his strong opinion against them, he said that "I will get rid of them one day." His view is very thought-provoking, but so far, it is not productive for statistical practice.

[3] Given the marginal distributions of $\text{pr}(Y(1) \leq y_1)$ and $\text{pr}(Y(0) \leq y_0)$, we can bound the joint distribution of $\text{pr}(Y(1) \leq y_1, Y(0) \leq y_0)$ by the Frechet–Hoeffding inequality:

$$\begin{aligned} & \max\{0, \text{pr}(Y(1) \leq y_1) + \text{pr}(Y(0) \leq y_0) - 1\} \\ \leq \; & \text{pr}(Y(1) \leq y_1, Y(0) \leq y_0) \\ \leq \; & \min\{\text{pr}(Y(1) \leq y_1), \text{pr}(Y(0) \leq y_0)\}. \end{aligned}$$

It is a well known result in probability theory. However, it is often a loose inequality. Unfortunately, we do not have any information beyond this inequality without imposing additional assumptions.

The last assumption is the cross-world independence.

Assumption 27.5 *There is no cross-world dependence between the potential outcomes and potential mediators:*

$$Y(z, m) \perp\!\!\!\perp M(z') \mid X$$

for all z, z', and m.

Assumptions 27.2–27.4 are very strong, but at least they hold under experiments with randomized treatment and mediator. Assumption 27.5 is stronger because no physical experiment can ensure it. Because we can never observe $Y(z, m)$ and $M(z')$ in any experiment if $z \neq z'$, Assumption 27.5 can never be validated so it is fundamentally metaphysical.

I give an example below in which Assumptions 27.2–27.5 all hold.

Example 27.4 *Given X, we generate*

$$\begin{aligned} Z &= I\{g_Z(X, \varepsilon_Z) \geq 0\}, \\ M(z) &= I\{g_M(X, z, \varepsilon_M) \geq 0\}, \\ Y(z, m) &= g_Y(X, z, m, \varepsilon_Y), \end{aligned}$$

for $z, m = 0, 1$, where the g's are general functions and $\varepsilon_Z, \varepsilon_M, \varepsilon_Y$ are independent random errors. Consequently, we generate the observed values of M and Y from

$$\begin{aligned} M &= M(Z) = I\{g_M(X, Z, \varepsilon_M) \geq 0\}, \\ Y &= Y(Z, M) = g_Y(X, Z, M, \varepsilon_Y). \end{aligned}$$

We can verify that Assumptions 27.2–27.5 hold under this data-generating process. On the contrary, if we allow ε_M and ε_Y to be $\varepsilon_M(z)$ and $\varepsilon_Y(z, m)$, then Assumptions 27.2–27.5 can fail. See Problem 27.3 for more details.

Pearl (2001) proved the following key result for mediation analysis.

Theorem 27.1 *Under Assumptions 27.2–27.5, we have*

$$\begin{aligned} & E\{Y(z, M_{z'}) \mid X = x\} \\ = & \sum_m E(Y \mid Z = z, M = m, X = x)\text{pr}(M = m \mid Z = z', X = x) \end{aligned}$$

and therefore,

$$\begin{aligned} & E\{Y(z, M_{z'})\} \\ = & \sum_x E\{Y(z, M_{z'}) \mid X = x\}\text{pr}(X = x) \\ = & \sum_x \sum_m E(Y \mid Z = z, M = m, X = x)\text{pr}(M = m \mid Z = z', X = x)\text{pr}(X = x). \end{aligned}$$

Theorem 27.1 assumes that both M and X are discrete. With general M and X, the mediation formulas become

$$\begin{aligned} & E\{Y(z, M_{z'}) \mid X = x\} \\ = & \int E(Y \mid Z = z, M = m, X = x) f(m \mid Z = z', X = x) \text{d}m \end{aligned}$$

and

$$\begin{aligned} & E\{Y(z, M_{z'})\} \\ = & \int E\{Y(z, M_{z'}) \mid X = x\} f(x) \mathrm{d}x \\ = & \int\int E(Y \mid Z = z, M = m, X = x) f(m \mid Z = z', X = x) f(x) \mathrm{d}m \mathrm{d}x. \end{aligned}$$

From Theorem 27.1, the identification formulas for the means of the nested potential outcomes depend on the conditional mean of the outcome given the treatment, mediator, and covariates, as well as the conditional mean of the mediator given the treatment and covariates. We need to evaluate these two conditional means at different treatment levels if the nested potential outcome involves cross-world interventions.

If we drop the cross-world independence assumption, we can modify the definition of the natural direct and indirect effects and the same formulas hold. See Problem 27.11 for more details.

I give the proof below.

Proof of Theorem 27.1: By the tower property, $E\{Y(z, M_{z'})\} = E[E\{Y(z, M_{z'}) \mid X\}]$, so we need only to prove the formula for $E\{Y(z, M_{z'}) \mid X = x\}$. Starting with the law of total probability, we have

$$\begin{aligned} & E\{Y(z, M_{z'}) \mid X = x\} \\ = & \sum_m E\{Y(z, M_{z'}) \mid M_{z'} = m, X = x\} \mathrm{pr}(M_{z'} = m \mid X = x) \\ = & \sum_m E\{Y(z, m) \mid M_{z'} = m, X = x\} \mathrm{pr}(M_{z'} = m \mid X = x) \\ = & \sum_m \underbrace{E\{Y(z, m) \mid X = x\}}_{\text{Assumption 27.5}} \underbrace{\mathrm{pr}(M = m \mid Z = z', X = x)}_{\text{Assumption 27.4}} \\ = & \sum_m \underbrace{E(Y \mid Z = z, M = m, X = x)}_{\text{Assumptions 27.2 and 27.3}} \mathrm{pr}(M = m \mid Z = z', X = x). \end{aligned}$$

□

The above proof is trivial from a mathematical perspective. It illustrates the necessity of Assumptions 27.2–27.5.

Conditional on $X = x$, the mediation formulas for $Y(1, M_1)$ and $Y(0, M_0)$ simplify to

$$\begin{aligned} & E\{Y(1, M_1) \mid X = x\} \\ = & \sum_m E(Y \mid Z = 1, M = m, X = x) \mathrm{pr}(M = m \mid Z = 1, X = x) \\ = & E(Y \mid Z = 1, X = x) \end{aligned}$$

and

$$\begin{aligned} & E\{Y(0, M_0) \mid X = x\} \\ = & \sum_m E(Y \mid Z = 0, M = m, X = x) \mathrm{pr}(M = m \mid Z = 0, X = x) \\ = & E(Y \mid Z = 0, X = x) \end{aligned}$$

based on the law of total probability; the mediation formula for $Y(1, M_0)$ simplifies to

$$\begin{aligned} & E\{Y(1, M_0) \mid X = x\} \\ = & \sum_m E(Y \mid Z = 1, M = m, X = x) \mathrm{pr}(M = m \mid Z = 0, X = x), \end{aligned}$$

where the conditional expectation of the outcome is given $Z = 1$ but the conditional distribution of the mediator is given $Z = 0$. This leads to the identification formulas of the natural direct and indirect effects.

Corollary 27.1 *Under Assumptions 27.2–27.5, the conditional natural direct and indirect effects are identified by*

$$\begin{aligned}\text{NDE}(x) &= E\{Y(1, M_0) - Y(0, M_0) \mid X = x\} \\ &= \sum_m \{E(Y \mid Z = 1, M = m, X = x) - E(Y \mid Z = 0, M = m, X = x)\} \\ &\qquad \times \text{pr}(M = m \mid Z = 0, X = x)\end{aligned}$$

and

$$\begin{aligned}\text{NIE}(x) &= E\{Y(1, M_1) - Y(1, M_0) \mid X = x\} \\ &= \sum_m E(Y \mid Z = 1, M = m, X = x) \\ &\qquad \times \{\text{pr}(M = m \mid Z = 1, X = x) - \text{pr}(M = m \mid Z = 0, X = x)\};\end{aligned}$$

the unconditional ones can be identified by $\text{NDE} = \sum_x \text{NDE}(x)\text{pr}(X = x)$ *and* $\text{NIE} = \sum_x \text{NIE}(x)\text{pr}(X = x)$.

As a special case, with a binary M, the formula of the NIE reduces to a product form below.

Corollary 27.2 *Under Assumptions 27.2–27.5, for a binary mediator M, we have*

$$\text{NIE}(x) = \tau_{Z \to M}(x)\tau_{M \to Y}(1, x)$$

and $\text{NIE} = E\{\text{NIE}(X)\}$, *where*

$$\tau_{Z \to M}(x) = \text{pr}(M = 1 \mid Z = 1, X = x) - \text{pr}(M = 1 \mid Z = 0, X = x)$$

and

$$\tau_{M \to Y}(z, x) = E(Y \mid Z = z, M = 1, X = x) - E(Y \mid Z = z, M = 0, X = x).$$

I leave the proof of Corollary 27.2 as Problem 27.6. Corollary 27.2 gives a simple formula in the case of a binary M. With randomized Z conditional on X, we can view $\tau_{Z \to M}(x)$ as the conditional average causal effect of Z on M. With randomized M conditional on (X, Z), we can view $\tau_{M \to Y}(z, x)$ as the conditional average causal effect of M on Y. The conditional natural indirect effect equals their product. This is coherent with our intuition that the indirect effect acts from Z to M and then from M to Y.

27.4 The mediation formula under linear models

Theorem 27.1 gives the nonparametric identification formula for mediation analysis. It allows us to derive various formulas for mediation analysis under different models. I will introduce the famous Baron–Kenny method under linear models below. VanderWeele (2015) gives explicit formulas for the natural direct and indirect effects for many commonly used models. I relegate the details of other models to Section 27.6.

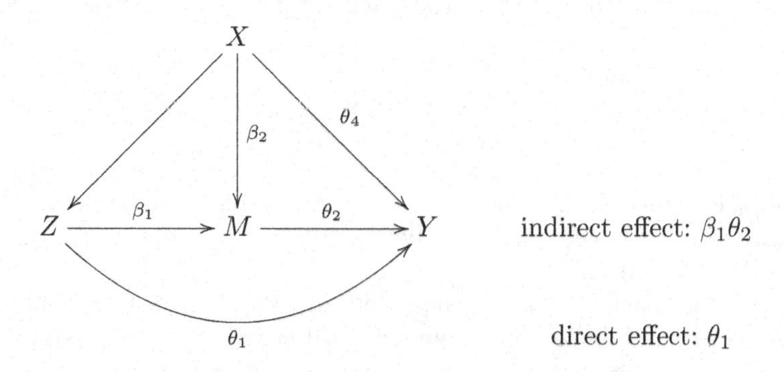

FIGURE 27.3: The Baron–Kenny method for mediation under linear models

27.4.1 The Baron–Kenny method

The Baron–Kenny method assumes the following linear models for the mediator and outcome given the treatment and covariates.

Assumption 27.6 (linear models for the Baron–Kenny method) *Both the mediator and outcome follow linear models:*

$$\begin{cases} E(M \mid Z, X) &= \beta_0 + \beta_1 Z + \beta_2^\intercal X, \\ E(Y \mid Z, M, X) &= \theta_0 + \theta_1 Z + \theta_2 M + \theta_4^\intercal X. \end{cases}$$

Under these linear models, the formulas for the natural direct and indirect effects simplify to functions of the coefficients.

Corollary 27.3 (Baron–Kenny formulas for mediation) *Under Assumptions 27.2–27.5 and 27.6, we have*

$$\begin{aligned} \text{NDE} &= \theta_1, \\ \text{NIE} &= \theta_2\beta_1. \end{aligned}$$

The formulas in Corollary 27.3 are intuitive based on Figure 27.3. The direct effect equals the coefficient on the path $Z \to Y$, and the indirect effect equals the product of the coefficients on the path $Z \to M \to Y$. I give the proof below based on Theorem 27.1.
Proof of Corollary 27.3: The conditional NDE equals

$$\text{NDE}(x) = \sum_m \theta_1 \text{pr}(M = m \mid Z = 0, X = x) = \theta_1$$

and the conditional NIE equals

$$\begin{aligned} \text{NIE}(x) &= \sum_m (\theta_0 + \theta_1 + \theta_2 m + \theta_4^\intercal x) \\ &\qquad \times \{\text{pr}(M = m \mid Z = 1, X = x) - \text{pr}(M = m \mid Z = 0, X = x)\} \\ &= \theta_2 \{E(M = m \mid Z = 1, X = x) - E(M = m \mid Z = 0, X = x)\} \\ &= \theta_2\beta_1, \end{aligned}$$

which do not depend on x. Therefore, they are also the formulas for the unconditional natural direct and indirect effects. □

If we obtain OLS estimators of these coefficients, we can estimate the direct and indirect effects by

$$\begin{aligned}\hat{\text{NDE}} &= \hat{\theta}_1,\\ \hat{\text{NIE}} &= \hat{\theta}_2\hat{\beta}_1,\end{aligned}$$

which is called the Baron–Kenny method (Judd and Kenny, 1981; Baron and Kenny, 1986) although it had several antecedents (e.g., Hyman, 1955; Alwin and Hauser, 1975; Judd and Kenny, 1981; Sobel, 1982).

Standard software packages report the standard error of $\hat{\text{NDE}}$ from OLS. Sobel (1982, 1986) used the delta method to obtain the standard error of $\hat{\text{NIE}}$. Based on the formula in Example A.2, the asymptotic variance of $\hat{\theta}_2\hat{\beta}_1$ equals $\text{var}(\hat{\theta}_2)\beta_1^2+\theta_2^2\text{var}(\hat{\beta}_1)$. So the estimated variance is

$$\hat{\text{var}}(\hat{\theta}_2)\hat{\beta}_1^2 + \hat{\theta}_2^2\hat{\text{var}}(\hat{\beta}_1).$$

Testing the null hypothesis of NIE based on $\hat{\theta}_2\hat{\beta}_1$ and the estimated variance above is called *Sobel's test* in the literature of mediation analysis.

27.4.2 An example

We can easily implement the Baron–Kenny method via the following code.

```
library("car")
BKmediation = function(Z, M, Y, X)
{
  ## two regressions and coefficients
  mediator.reg    = lm(M ~ Z + X)
  mediator.Zcoef  = mediator.reg$coef[2]
  mediator.Zse    = sqrt(hccm(mediator.reg)[2, 2])

  outcome.reg     = lm(Y ~ Z + M + X)
  outcome.Zcoef   = outcome.reg$coef[2]
  outcome.Zse     = sqrt(hccm(outcome.reg)[2, 2])
  outcome.Mcoef   = outcome.reg$coef[3]
  outcome.Mse     = sqrt(hccm(outcome.reg)[3, 3])

  ## Baron-Kenny point estimates
  NDE = outcome.Zcoef
  NIE = outcome.Mcoef*mediator.Zcoef

  ## Sobel's variance estimate based the delta method
  NDE.se = outcome.Zse
  NIE.se = sqrt(outcome.Mse^2*mediator.Zcoef^2 +
                outcome.Mcoef^2*mediator.Zse^2)

  res = matrix(c(NDE, NIE,
                 NDE.se, NIE.se,
                 NDE/NDE.se, NIE/NIE.se),
               2, 3)
  rownames(res) = c("NDE", "NIE")
  colnames(res) = c("est", "se", "t")

  res
}
```

Revisiting Example 27.3, we obtain the following estimates for the direct and indirect effects:

```
> jobsdata = read.csv("jobsdata.csv")
> Z = jobsdata$treat
> M = jobsdata$job_seek
> Y = jobsdata$depress2
> getX      = lm(treat ~ econ_hard + depress1 +
+                sex + age + occp + marital +
+                nonwhite + educ + income,
+                data = jobsdata)
> X = model.matrix(getX)[, -1]
> res = BKmediation(Z, M, Y, X)
> round(res, 3)
       est    se      t
NDE -0.037 0.042 -0.885
NIE -0.014 0.009 -1.528
```

Both the estimates for the direct and indirect effects are negative although they are insignificant.

27.5 Sensitivity analysis

Mediation analysis relies on strong and untestable assumptions. One crucial assumption is that there is no unmeasured confounding among the treatment, mediator, and outcome. However, even if the treatment in randomized, the mediator is often not randomized, leaving the mediator-outcome confounding not full controlled. Moreover, mediation analysis often relies on observational studies with non-randomized treatment and mediator. Therefore, it is possible that an unmeasured confounder U can affect all of Z, M and Y, even conditional on the observed covariates X. Figure 27.4 illustrates the causal mechanism.

With unmeasured confounding, the estimators discussed in this chapter are all biased in general. Various sensitivity analysis methods appeared in the literature. In particular, Ding and VanderWeele (2016b) proposed Cornfield-type sensitivity bounds, and Zhang and Ding (2022) proposed a sensitivity analysis method tailored to the Baron–Kenny method based on linear structural equation models. These are beyond the scope of this book.

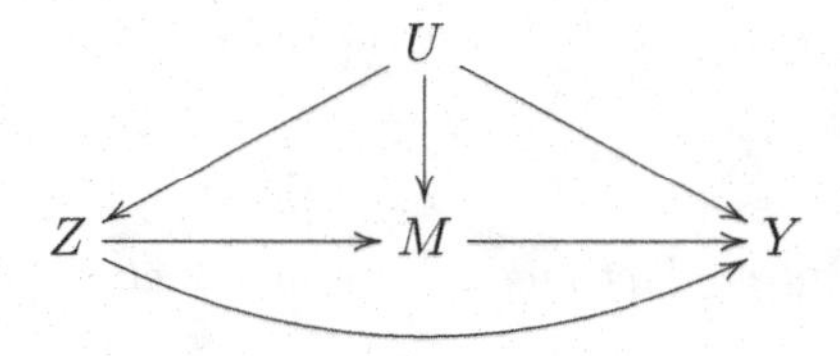

FIGURE 27.4: Causal diagram for mediation analysis with unmeasured confounding: the unmeasured confounder U can affect all of Z, M and Y, even conditional on the observed covariates X

27.6 Homework problems

27.1 Clarification of the units based on nested potential outcomes

Assume (Z, M, Y) are all binary. Based on the joint value of

$$M(1), M(0), Y(1, M_1), Y(1, M_0), Y(0, M_1), Y(0, M_0),$$

how many types of units do we have?

If we assume monotonicity of Z on M, that is,

$$M(1) \geq M(0),$$

then how many types of units do we have?

If we further assume monotonicity of Z and M on Y, that is,

$$Y(1, m) \geq Y(0, m), \quad Y(z, 1) \geq Y(z, 0)$$

for all z and m, then how many types of units do we have?

Remark: With binary (Z, M, Y), Robins and Greenland (1992) classified the units into latent types to discuss direct and indirect effects.

27.2 Sequential randomization and joint randomization

Show (27.2) is equivalent to Assumptions 27.2 and 27.3.

27.3 Verifying the assumptions for mediation analysis

Show that Assumptions 27.2–27.5 hold under the data generating process in Example 27.4. Also show that if we allow ε_M and ε_Y to be $\varepsilon_M(z)$ and $\varepsilon_Y(z, m)$, then Assumptions 27.2–27.5 can fail.

27.4 Another set of assumptions for the mediation formula

Imai et al. (2010) invoked the following set of assumptions to derive the mediation formula.

Assumption 27.7 *We have*

$$\{Y(z, m), M(z')\} \perp\!\!\!\perp Z \mid X$$

and

$$Y(z, m) \perp\!\!\!\perp M(z') \mid (Z = z', X)$$

for all z, z', m.

Theorem 27.2 *Under Assumption 27.7, the mediation formula in Theorem 27.1 holds.*

Prove Theorem 27.2.

27.5 Difference method and product method are identical

In the main text, we first obtain the OLS fits

$$\begin{cases} \hat{M}_i &= \hat{\beta}_0 + \hat{\beta}_1 Z_i + \hat{\beta}_2^{\top} X_i, \\ \hat{Y}_i &= \hat{\theta}_0 + \hat{\theta}_1 Z_i + \hat{\theta}_2 M_i + \hat{\theta}_4^{\top} X_i. \end{cases}$$

The estimate for the NIE is the product $\hat{\theta}_2\hat{\beta}_1$, which is called the *product method.*

Alternatively, we can first obtain OLS fits:

$$\begin{cases} \hat{Y}_i &= \hat{\alpha}_0 + \hat{\alpha}_1 Z_i + \hat{\alpha}_2^\top X_i, \\ \hat{Y}_i &= \hat{\theta}_0 + \hat{\theta}_1 Z_i + \hat{\theta}_2 M_i + \hat{\theta}_4^\top X_i. \end{cases}$$

Another estimate for the NIE is the difference $\hat{\alpha}_1 - \hat{\theta}_1$, which is called the *difference method.*

Show that $\hat{\alpha}_1 - \hat{\theta}_1 = \hat{\theta}_2\hat{\beta}_1$.

Remark: Recall Cochran's formula in Problem 16.2.

27.6 Natural indirect effect with a binary mediator

Prove Corollary 27.2.

27.7 With treatment-outcome interaction on the outcome

VanderWeele (2015) suggested using the following linear models:

$$\begin{cases} E(M \mid Z, X) &= \beta_0 + \beta_1 Z + \beta_2^\top X, \\ E(Y \mid Z, M, X) &= \theta_0 + \theta_1 Z + \theta_2 M + \theta_3 ZM + \theta_4^\top X, \end{cases}$$

where the outcome model has the interaction term between the treatment and the mediator.

Under the above linear models, show that

$$\text{NDE} = \theta_1 + \theta_3\{\beta_0 + \beta_2^\top E(X)\}, \qquad \text{NIE} = (\theta_2 + \theta_3)\beta_1.$$

How do we estimate NDE and NIE with IID data?

Remark: Consider the simple case with a binary Z and binary M. Under the linear models, the average causal effect of Z of M equals β_1, and the average causal effect of M on Y equals $\theta_2 + \theta_3 E(Z)$. Therefore, it is possible that both of these effects are positive, but the natural indirect effect is negative. For instance:

$$\beta_1 = 1, \quad \theta_2 = 1, \quad \theta_3 = -1.5, \quad E(Z) = 0.5.$$

This is somewhat paradoxical and can be called the *mediator paradox.* Chen et al. (2007) reported a related *surrogate endpoint paradox* or *intermediate variable paradox.*

27.8 Mediation analysis with continuous mediator and binary outcome

Consider the following Normal linear model for the mediator and logistic model for the binary outcome:

$$\begin{cases} M \mid Z, X &\sim \mathrm{N}(\beta_0 + \beta_1 Z + \beta_2^\top X, \sigma_M^2), \\ \mathrm{logit}\{\mathrm{pr}(Y = 1 \mid Z, M, X)\} &= \theta_0 + \theta_1 Z + \theta_2 M + \theta_4^\top X, \end{cases}$$

where $\mathrm{logit}(w) = \log\{w/(1-w)\}$ with inverse $\mathrm{expit}(w) = (1 + e^{-w})^{-1}$. Express NDE and NIE in terms of the model parameters and the distribution of X. How do we estimate NDE and NIE with IID data?

27.9 Mediation analysis with binary mediator and continuous outcome

Consider the following logistic model for the binary mediator and linear model for the outcome:

$$\begin{cases} \mathrm{logit}\{\mathrm{pr}(M = 1 \mid Z, X)\} &= \beta_0 + \beta_1 Z + \beta_2^\top X, \\ E(Y \mid Z, M, X) &= \theta_0 + \theta_1 Z + \theta_2 M + \theta_4^\top X. \end{cases}$$

Under these models, show that

$$\text{NDE} = \theta_1, \qquad \text{NIE} = \theta_2 E\left\{\text{expit}(\beta_0 + \beta_1 + \beta_2^{\mathrm{T}} X) - \text{expit}(\beta_0 + \beta_2^{\mathrm{T}} X)\right\}.$$

How do we estimate NDE and NIE with IID data?

27.10 Mediation analysis with binary mediator and outcome

Consider the following logistic models for the binary mediator and outcome:

$$\begin{cases} \text{logit}\{\text{pr}(M = 1 \mid Z, X)\} & = & \beta_0 + \beta_1 Z + \beta_2^{\mathrm{T}} X, \\ \text{logit}\{\text{pr}(Y = 1 \mid Z, M, X)\} & = & \theta_0 + \theta_1 Z + \theta_2 M + \theta_4^{\mathrm{T}} X. \end{cases}$$

Express NDE and NIE in terms of the model parameters and the distribution of X. How do we estimate NDE and NIE with IID data?

27.11 Modify the definitions to drop the cross-world independence

Define

$$Y(z, F_{M_{z'}|X}) = \int Y(z, m) f(M_{z'} = m \mid X) \mathrm{d}m$$

as the potential outcome under treatment z and a random draw from the distribution of $M_{z'} \mid X$. With a discrete M, the definition simplifies to

$$Y(z, F_{M_{z'}|X}) = \sum_m Y(z, m) \text{pr}(M_{z'} = m \mid X).$$

The key difference between $Y(z, M_{z'})$ and $Y(z, F_{M_{z'}|X})$ is that $M_{z'}$ is the potential mediator for the same unit whereas $F_{M_{z'}|X}$ is a random draw from the conditional distribution of the potential mediator in the whole population. Define the natural direct and indirect effects as

$$\text{NDE} = E\{Y(1, F_{M_0|X}) - Y(0, F_{M_0|X})\}, \quad \text{NIE} = E\{Y(1, F_{M_1|X}) - Y(1, F_{M_0|X})\}.$$

Show that under Assumptions 27.2–27.4, the identification formulas for NDE and NIE remain the same as in the main text.

Remark: Modifying the definitions of the nested potential outcomes allows us to relax the strong cross-world independence assumption but weakens the interpretation of the natural direct and indirect effects. See VanderWeele (2015) for more discussion and VanderWeele and Tchetgen Tchetgen (2017) for an application to a more complex setting with time-varying treatment and mediator.

27.12 Connections between principal stratification and mediation analysis

VanderWeele (2008) and Forastiere et al. (2018) reviewed and compared principal stratification and mediation analysis.

28

Controlled Direct Effect

The formulation of mediation analysis in Chapter 27 relies on the nested potential outcomes, and fundamentally, some nested potential outcomes are not observable in any physical experiments. If we stick to the Popperian philosophy of science reviewed in Section 27.2.2, we should only define causal parameters in terms of quantities that are observable under some experiments. This chapter discusses an alternative view of causal inference with an intermediate variable. In this view, we can only define the direct effect but can not define the indirect effect.

28.1 Definition of the controlled direct effect

We view Z and M as two treatment factors that can be manipulated, and define potential outcomes $Y(z, m)$ for $z = 0, 1$ and $m \in \mathcal{M}$. Based on these potential outcomes, we can define the *controlled direct effect* (CDE) below.

Definition 28.1 (CDE) *Define*

$$\text{CDE}(m) = E\{Y(1, m) - Y(0, m)\}.$$

By definition, $\text{CDE}(m)$ is the average causal effect of the treatment if the intermediate variable is fixed at m. The parameter $\text{CDE}(m)$ can capture the direct effect of the treatment holding the mediator at m. However, this formulation cannot capture the indirect effect. For example, the parameter $E\{Y(z, 1) - Y(z, 0)\}$ only measures the effect of the mediator on the outcome holding the treatment at z. This is not a meaningful definition of the indirect effect.

28.2 Identification and estimation of the controlled direct effect

To identify $\text{CDE}(m)$, we need the following assumption, which requires that Z and M are jointly randomized given X.

Assumption 28.1 *Sequential ignorability requires*

$$Z \perp\!\!\!\perp Y(z, m) \mid X, \quad M \perp\!\!\!\perp Y(z, m) \mid (Z, X)$$

or, equivalently (see Problem 27.2),

$$(Z, M) \perp\!\!\!\perp Y(z, m) \mid X.$$

I will focus on the case with a binary Z and M. Mathematically, we can just view this problem as an observational study with four treatment levels

$$(z, m) \in \{(0,0),(0,1),(1,0),(1,1)\}.$$

The following theorem extends the results for observational studies with a binary treatment, identifying

$$\mu_{zm} = E\{Y(z,m)\}$$

based on outcome regression, IPW, and doubly robust estimation.

Define

$$\mu_{zm}(x) = E(Y \mid Z = z, M = m, X = x)$$

as the outcome mean conditional on the treatment, mediator, and covariates. Define

$$\begin{aligned} e_{zm}(x) &= \text{pr}(Z = z, M = m \mid X = x) \\ &= \text{pr}(Z = z \mid X = x)\text{pr}(M = m \mid Z = z, X = x) \end{aligned}$$

as the probability of the joint value of Z and M conditional on the covariates.

Theorem 28.1 *Under Assumption 28.1, we have*

$$\mu_{zm} = E\{\mu_{zm}(X)\}$$

based on the outcome model, and

$$\mu_{zm} = E\left\{\frac{I(Z = z, M = m)Y}{e_{zm}(X)}\right\}$$

based on the propensity score model. Moreover, based on the working models $e_{zm}(X, \alpha)$ and $\mu_{zm}(X, \beta)$ for $e_{zm}(X)$ and $\mu_{zm}(X)$, respectively, we have the doubly robust formula for μ_{zm}:

$$\mu_{zm}^{\text{dr}} = E\{\mu_{zm}(X,\beta)\} + E\left[\frac{I(Z = z, M = m)\{Y - \mu_{zm}(X,\beta)\}}{e_{zm}(X,\alpha)}\right],$$

which equals μ_{zm} if either $e_{zm}(X,\alpha) = e_{zm}(X)$ or $\mu_{zm}(X,\beta) = \mu_{zm}(X)$.

The proof of Theorem 28.1 is similar to those for the standard unconfounded observational studies. Problem 28.2 gives a general result. Based on the outcome mean model, we can obtain $\hat{\mu}_{zm}(x)$ for $\mu_{zm}(x)$. Based on the treatment model, we can obtain $\hat{e}_z(x)$ for $\text{pr}(Z = z \mid X = x)$; based on the intermediate variable model, we can obtain $\hat{e}_m(z, x)$ for $\text{pr}(M = m \mid Z = z, X = x)$. We can then estimate μ_{zm} by outcome regression

$$\hat{\mu}_{zm}^{\text{reg}} = n^{-1}\sum_{i=1}^{n} \hat{\mu}_{zm}(X_i),$$

by IPW

$$\begin{aligned} \hat{\mu}_{zm}^{\text{ht}} &= n^{-1}\sum_{i=1}^{n} \frac{I(Z_i = z, M_i = m)Y_i}{\hat{e}_z(X_i)\hat{e}_m(z,X_i)}, \\ \hat{\mu}_{zm}^{\text{haj}} &= \sum_{i=1}^{n} \frac{I(Z_i = z, M_i = m)Y_i}{\hat{e}_z(X_i)\hat{e}_m(z,X_i)} \Big/ \sum_{i=1}^{n} \frac{I(Z_i = z, M_i = m)}{\hat{e}_z(X_i)\hat{e}_m(z,X_i)}, \end{aligned}$$

or by augmented IPW

$$\hat{\mu}^{\text{dr}}_{zm} = \hat{\mu}^{\text{reg}}_{zm} + n^{-1} \sum_{i=1}^{n} \frac{I(Z_i = z, M_i = m)\{Y_i - \hat{\mu}_{zm}(X_i)\}}{\hat{e}_z(X_i)\hat{e}_m(z, X_i)}.$$

We can then estimate $\text{CDE}(m)$ by $\hat{\mu}^*_{1m} - \hat{\mu}^*_{0m}$ ($* = \text{reg}, \text{ht}, \text{haj}, \text{dr}$) and use the bootstrap to approximate the standard error.

If we are willing to assume a linear outcome model, the controlled direct effect simplifies to the coefficient of the treatment. Example 28.1 below gives the details.

Example 28.1 *Under Assumption 28.1 and a linear outcome model,*

$$E(Y \mid Z, M, X) = \theta_0 + \theta_1 Z + \theta_2 M + \theta_4^{\text{T}} X,$$

we can use Theorem 28.1 to show that

$$\begin{aligned} \text{CDE}(m) &= E\{\mu_{1m}(X)\} - E\{\mu_{0m}(X)\} \\ &= E(\theta_0 + \theta_1 + \theta_2 m + \theta_4^{\text{T}} X) - E(\theta_0 + \theta_2 m + \theta_4^{\text{T}} X) \\ &= \theta_1, \end{aligned}$$

which coincides with the natural direct effect in the Baron–Kenny method.

28.3 Discussion

The formulation of the controlled direct effect does not involve the nested or a priori counterfactual potential outcomes, and its identification does not require the cross-world counterfactual independence assumption. The parameter $\text{CDE}(m)$ can capture the direct effect of the treatment holding the mediator at m. However, this formulation cannot capture the indirect effect.

The mediation analysis framework can decompose the total effect into natural direct and indirect effects, but it requires nested potential outcomes and cross-world independence. The principal stratification and controlled direct effect frameworks cannot define indirect effects but they do not involve nested potential outcomes and cross-world independence. Moreover, the principal stratification framework does not necessarily require that M lies on the causal pathway from the treatment to the outcome. However, its identification and estimation involves disentangling mixture distributions, which is a nontrivial task in statistics.

I summarize the causal frameworks for intermediate variables in Table 28.1.

TABLE 28.1: Causal frameworks for intermediate variables

chapter	framework	direct effect	indirect effect
26	principal stratification	$\tau(1,1)$, $\tau(0,0)$	?
27	mediation analysis	NDE	NIE
28	controlled direct effect	$\text{CDE}(m)$	?

28.4 Homework problems

28.1 CDE *and* NDE

Show that under cross world independence $Y(z,m) \perp\!\!\!\perp M(z') \mid X$ for all z, z' and m, the conditional controlled direct effect $\text{CDE}(m \mid x) = E\{Y(1,m) - Y(0,m) \mid X = x\}$ and the conditional natural direct effect $\text{NDE}(x) = E\{Y(1, M_0) - Y(0, M_0) \mid X = x\}$ have the following relationship:

$$\text{NDE}(x) = \sum_m \text{CDE}(m \mid x)\text{pr}(M_0 = m \mid X = x)$$

for a discrete M. Without the cross-world independence, does this relationship still hold in general?

28.2 Observational studies with a multi-valued treatment

Theorem 28.1 is a special case of the following theorem for unconfounded observational studies with multiple treatment levels (Imai and Van Dyk, 2004; Cattaneo, 2010). Below, I state the general problem and theorem.

Consider an observational study with a multi-valued treatment $Z \in \{1, \ldots, K\}$, covariates X, and outcome Y. Unit i has K potential outcomes $Y_i(1), \ldots, Y_i(K)$ corresponding to the K treatment levels. In general, we can define causal effect in terms of contrasts of the potential outcomes:

$$\tau_C = \sum_{k=1}^{K} C_k E\{Y(k)\}$$

where $\sum_{k=1}^{K} C_k = 0$. The canonical choice of the pairwise comparison

$$\tau_{k,k'} = E\{Y(k) - Y(k')\}.$$

Therefore, the key is to identify and estimate the means of the potential outcomes $\mu_k = E\{Y(k)\}$ under the ignorability and overlap assumptions below based on IID data of $(Z_i, X_i, Y_i)_{i=1}^n$.

Assumption 28.2 *$Z \perp\!\!\!\perp \{Y(1), \ldots, Y(K)\} \mid X$ and $\text{pr}(Z = k \mid X) > 0$ for $k = 1, \ldots, K$.*

Define the generalized propensity score as

$$e_k(X) = \text{pr}(Z = k \mid X),$$

and define the conditional outcome mean as

$$\mu_k(X) = E(Y \mid Z = k, X)$$

for $k = 1, \ldots, K$. We have the following theorem.

Theorem 28.2 *Under Assumption 28.2, we have*

$$\mu_k = E\{\mu_k(X)\} = E\left\{\frac{I(Z = k)Y}{e_k(X)}\right\}.$$

Moreover, based on the working models $e_k(X, \alpha)$ and $\mu_k(X, \beta)$ for $e_k(X)$ and $\mu_k(X)$, respectively, we have the doubly robust formula for μ_k:

$$\mu_k^{\text{dr}} = E\{\mu_k(X, \beta)\} + E\left[\frac{I(Z = k)\{Y - \mu_k(X, \beta)\}}{e_k(X, \alpha)}\right],$$

which equals μ_k if either $e_k(X, \alpha) = e_k(X)$ or $\mu_k(X, \beta) = \mu_k(X)$.

Prove Theorem 28.2.

Remark: Theorem 28.1 is a special case of Theorem 28.2 if we view the (Z, M) in Theorem 28.1 as a treatment with four levels. The $\text{CDE}(m)$ is a special case of τ_C.

28.3 CDE *in the linear outcome model*

This problem extends Example 28.1.

Show that under Assumption 28.1, if $E(Y \mid Z, M, X) = \theta_0 + \theta_1 Z + \theta_2 M + \theta_3 ZM + \theta_4^{\top} X$, then

$$\text{CDE}(m) = \theta_1 + \theta_3 m$$

for all m.

28.4 CDE *in the logistic outcome model*

Show that for a binary outcome, under Assumption 28.1, if

$$\text{logit}\{\text{pr}(Y = 1 \mid Z, M, X)\} = \theta_0 + \theta_1 Z + \theta_2 M + \theta_4^{\top} X,$$

then

$$\text{CDE}(m) = E\{\text{expit}(\theta_0 + \theta_1 + \theta_2 m + \theta_4^{\top} X) - \text{expit}(\theta_0 + \theta_2 m + \theta_4^{\top} X)\};$$

if

$$\text{logit}\{\text{pr}(Y = 1 \mid Z, M, X)\} = \theta_0 + \theta_1 Z + \theta_2 M + \theta_3 ZM + \theta_4^{\top} X,$$

then

$$\text{CDE}(m) = E\{\text{expit}(\theta_0 + \theta_1 + \theta_2 m + \theta_3 m + \theta_4^{\top} X) - \text{expit}(\theta_0 + \theta_2 m + \theta_4^{\top} X)\}.$$

28.5 Recommended reading

Nguyen et al. (2021) provided a friendly review of of the topics in Chapters 27 and 28.

29

Time-Varying Treatment and Confounding

Studies with time-varying treatments are common in biomedical and social sciences. James Robins championed the research in biostatistics. A classic example is that HIV patients may take azidothymidine, an antiretroviral medication, on and off over time (Robins et al., 2000; Hernán et al., 2000). Similar problems also exist in other fields. In education, a classic example is that students may receive different types of instructions over time (Hong and Raudenbush, 2008). In political science, a classic example is that candidates continuously recalibrate their campaign strategy based on time-varying polls and opponent actions (Blackwell, 2013).

Causal inference with time-varying treatments is not a simple extension of causal inference with a treatment at a single time point. The main challenge is time-varying confounding. Even if we assume all time-varying confounders are observed, we still face statistical challenges in adjusting for those confounders. On the one hand, we should stratify on these confounders to adjust for confounding; on the other hand, stratifying on post-treatment variables will cause bias. Due to these two conflicting goals, causal inference with time-varying treatments and confounding requires more sophisticated statistical methods. It is the main topic of this chapter.

To minimize the notational burden, I will use the setting with treatments at two time points to convey the most important ideas. Extensions to treatments at multiple time points can be conceptually straightforward although technical complexities will arise in finite samples. I will discuss the complexities and relegate general results to Problems 29.6–29.9.

29.1 Basic setup and sequential ignorability

Start with treatments at two time points. The temporal order (not the causal diagram) of the variables with two time points is below:

$$X_0 \to Z_1 \to X_1 \to Z_2 \to Y$$

where

- X_0 denotes the baseline pretreatment covariates;
- Z_1 denotes the treatment at time point 1;
- X_1 denotes the time-varying covariates between the treatments at time points 1 and 2;
- Z_2 denotes the treatment at time point 2;
- Y denotes the outcome.

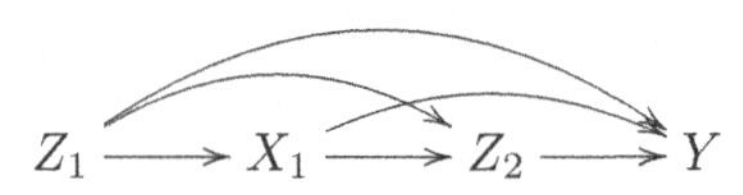

FIGURE 29.1: Assumption 29.1 holds without unmeasured confounding U between X_1 and Y. The causal diagram conditions on the pretreatment covariates X_0.

With binary treatments (Z_1, Z_2), each unit has four potential outcomes $Y(z_1, z_2)$ for $z_1, z_2 = 0, 1$. The observed outcome equals

$$Y = Y(Z_1, Z_2) = \sum_{z_1=0,1} \sum_{z_2=0,1} I(Z_1 = z_1) I(Z_2 = z_2) Y(z_1, z_2).$$

I will focus on the canonical setting with sequential ignorability, that is, the treatments are sequentially randomized given the observed history.

Assumption 29.1 (sequential ignorability) *(1) Z_1 is randomized given X_0:*

$$Z_1 \perp\!\!\!\perp Y(z_1, z_2) \mid X_0 \text{ for } z_1, z_2 = 0, 1.$$

(2) Z_2 is randomized given (Z_1, X_1, X_0):

$$Z_2 \perp\!\!\!\perp Y(z_1, z_2) \mid (Z_1, X_1, X_0) \text{ for } z_1, z_2 = 0, 1.$$

Figure 29.1 is a simple causal diagram corresponding to Assumption 29.1, which does not contain any unmeasured confounding.

Figure 29.2 is a more complex causal diagram corresponding to Assumption 29.1. Sequential ignorability rules out only the confounding between the treatment (Z_1, Z_2) and the outcome Y, but allows for unmeasured confounding between the time-varying covariate X_1 and the outcome Y. The possible existence of U causes many subtle issues even under sequential ignorability.

29.2 g-formula and outcome modeling

Recall the identification formula based on outcome modeling with a treatment at a single time point:

$$E\{Y(z)\} = E\{E(Y \mid Z = z, X)\}.$$

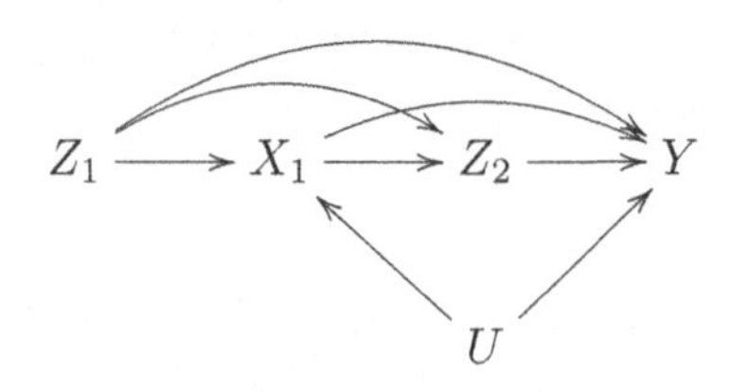

FIGURE 29.2: Assumption 29.1 holds with unmeasured confounding between X_1 and Y. The causal diagram conditions on the pretreatment covariates X_0.

With discrete X, it reduces to

$$E\{Y(z)\} = \sum_x E(Y \mid Z = z, X = x)\text{pr}(X = x);$$

with continuous X, it reduces to

$$E\{Y(z)\} = \int E(Y \mid Z = z, X = x)f(x)\text{d}x.$$

The following result extends it to the setting with treatments at two time points.

Theorem 29.1 *Under Assumption 29.1, we have*

$$E\{Y(z_1, z_2)\} = E\Big[E\{E(Y \mid z_2, z_1, X_1, X_0) \mid z_1, X_0\}\Big]. \tag{29.1}$$

In Theorem 29.1, I simplify the notation "$Z_2 = z_2$" to "z_2". To avoid complex formulas in this chapter, I will use the lowercase letter to represent the event that the random variable takes the corresponding value. With discrete X_0 and X_1, the identification formula (29.1) reduces to

$$E\{Y(z_1, z_2)\} = \sum_{x_0}\sum_{x_1} E(Y \mid z_2, z_1, x_1, x_0)\text{pr}(x_1 \mid z_1, x_0)\text{pr}(x_0); \tag{29.2}$$

with continuous X_0 and X_1, the identification formula (29.1) reduces to

$$E\{Y(z_1, z_2)\} = \int\int E(Y \mid z_2, z_1, x_1, x_0)f(x_1 \mid z_1, x_0)f(x_0)\text{d}x_1\text{d}x_0. \tag{29.3}$$

Compare (29.2) with the formula based on the law of total probability to gain more insights:

$$\begin{aligned} E(Y) &= \sum_{x_0}\sum_{z_1}\sum_{x_1}\sum_{z_2} E(Y \mid z_2, z_1, x_1, x_0) \\ &\quad \text{pr}(z_2 \mid z_1, x_1, x_0)\text{pr}(x_1 \mid z_1, x_0)\text{pr}(z_1 \mid x_0)\text{pr}(x_0). \end{aligned} \tag{29.4}$$

Erasing the probabilities of Z_2 and Z_1 in (29.4), we can obtain the formula (29.3). This is intuitive because the potential outcome $Y(z_1, z_2)$ has the meaning of fixing Z_1 and Z_2 at z_1 and z_2, respectively.

Robins called (29.2) and (29.3) the g-formulas. Now I will prove Theorem 29.1.

Proof of Theorem 29.1: By the tower property,

$$E\{Y(z_1, z_2)\} = E\Big[E\{Y(z_1, z_2) \mid X_0\}\Big],$$

so I will focus on $E\{Y(z_1, z_2) \mid X_0\}$. By Assumption 29.1(1) and the tower property,

$$\begin{aligned} E\{Y(z_1, z_2) \mid X_0\} &= E\{Y(z_1, z_2) \mid z_1, X_0\} \\ &= E\Big[E\{Y(z_1, z_2) \mid z_1, X_1, X_0\} \mid z_1, X_0\Big]. \end{aligned}$$

By Assumption 29.1(2),

$$\begin{aligned} E\{Y(z_1, z_2) \mid X_0\} &= E\Big[E\{Y(z_1, z_2) \mid z_2, z_1, X_1, X_0\} \mid z_1, X_0\Big] \\ &= E\Big[E\{Y \mid z_2, z_1, X_1, X_0\} \mid z_1, X_0\Big]. \end{aligned}$$

The formula (29.1) follows. □

29.2.1 Plug-in estimation based on outcome modeling

The g-formulas (29.2) and (29.3) suggest that to estimate the means of the potential outcomes, we need to model $E(Y \mid z_2, z_1, x_1, x_0)$ and $\text{pr}(x_1 \mid z_1, x_0)$. With these fitted models, we can plug them into the g-formulas to obtain the estimator

$$\hat{E}^{\text{reg}}\{Y(z_1, z_2)\} = \frac{1}{n}\sum_{i=1}^{n}\sum_{x_1}\hat{E}(Y_i \mid z_2, z_1, x_1, X_{i0})\hat{\text{pr}}(x_1 \mid z_1, X_{i0}).$$

With some special functional forms, this task can be simplified. Example 29.1 below gives the results under a linear model for the outcome.

Example 29.1 *Assume a linear outcome model*

$$E(Y \mid z_2, z_1, x_1, x_0) = \beta_0 + \beta_1 z_2 + \beta_2 z_1 + \beta_3 x_1 + \beta_4 x_0.$$

We can verify that

$$\begin{aligned} E\{Y(z_1, z_2)\} &= \sum_{x_0}\sum_{x_1}(\beta_0 + \beta_1 z_2 + \beta_2 z_1 + \beta_3 x_1 + \beta_4 x_0)\text{pr}(x_1 \mid z_1, x_0)\text{pr}(x_0) \\ &= \beta_0 + \beta_1 z_2 + \beta_2 z_1 + \beta_3 \sum_{x_0} E(X_1 \mid z_1, x_0)\text{pr}(x_0) + \beta_4 E(X_0). \end{aligned}$$

Define

$$E\{X_1(z_1)\} = \sum_{x_0} E(X_1 \mid z_1, x_0)\text{pr}(x_0) \tag{29.5}$$

to simplify the formula as

$$E\{Y(z_1, z_2)\} = \beta_0 + \beta_1 z_2 + \beta_2 z_1 + \beta_3 E\{X_1(z_1)\} + \beta_4 E(X_0).$$

In (29.5), *I introduce the potential outcome of X_1 under the treatment $Z_1 = z_1$ at time point 1. It is reasonable because the right-hand side of* (29.5) *is the identification formula of $E\{X_1(z_1)\}$ under ignorability $X_1(z_1) \perp\!\!\!\perp Z_1 \mid X_0$ for $z_1 = 0, 1$. We do not really need the potential outcome $X_1(z_1)$ and the ignorability, but it is a convenient notation and matches our previous discussion.*

Define $\tau_{Z_1 \to X_1} = E\{X_1(1) - X_1(0)\}$. We can verify that

$$\begin{aligned} E\{Y(1,0) - Y(0,0)\} &= \beta_2 + \beta_3 \tau_{Z_1 \to X_1}, \\ E\{Y(0,1) - Y(0,0)\} &= \beta_1, \\ E\{Y(1,1) - Y(0,0)\} &= \beta_1 + \beta_2 + \beta_3 \tau_{Z_1 \to X_1}. \end{aligned}$$

Therefore, we can estimate the effect of (Z_1, Z_2) on Y based on the above formulas by first estimating the regression coefficients βs and the average causal effect of Z_1 on X_1 using standard methods.

If we further assume a linear model for X_1:

$$E(X_1 \mid z_1, x_0) = \gamma_0 + \gamma_1 z_1 + \gamma_2 x_0,$$

then $\tau_{Z_1 \to X_1} = \gamma_1$ and

$$E\{Y(1,0) - Y(0,0)\} = \beta_2 + \beta_3 \gamma_1, \tag{29.6}$$
$$E\{Y(0,1) - Y(0,0)\} = \beta_1, \tag{29.7}$$
$$E\{Y(1,1) - Y(0,0)\} = \beta_1 + \beta_2 + \beta_3 \gamma_1. \tag{29.8}$$

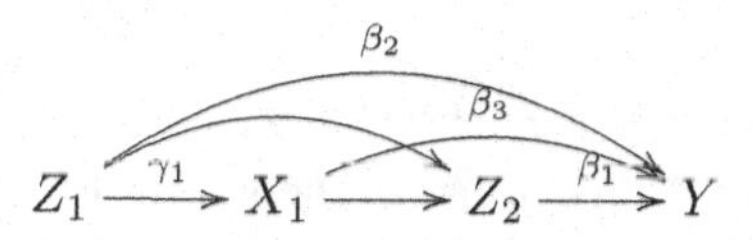

FIGURE 29.3: A linear causal diagram with coefficients on the arrows, conditional on the pretreatment covariates X_0.

The formulas in (29.6)–(29.8) *are intuitive based on Figure 29.3 with regression coefficients on the arrows. In* (29.6), *the effect of* Z_1 *equals the sum of the coefficients on the paths* $Z_1 \to Y$ *and* $Z_1 \to X_1 \to Y$; *in* (29.7), *the effect of* Z_2 *equals the coefficient on the path* $Z_2 \to Y$; *in* (29.8), *the total effect of* (Z_1, Z_2) *equals the sum of the coefficients on the paths* $Z_1 \to Y$, $Z_1 \to X_1 \to Y$ *and* $Z_2 \to Y$.

However, Robins and Wasserman (1997) pointed out a surprising drawback of the plug-in estimation based on outcome modeling. They showed that under the causal diagram in Figure 29.4, with model misspecification in the plug-in estimation, data analyzers may falsely reject the null hypothesis of zero causal effect of (Z_1, Z_2) on Y even when the true effect is zero in the data-generating process. They called it the *g-null paradox*. McGrath et al. (2021) revisited this paradox. Murray et al. (2017) and Campbell and Gustafson (2018) offered different views based on numerical examples.

29.2.2 Recursive estimation based on outcome modeling

The plug-in estimation in Section 29.2.1 involves modeling the time-varying confounder X_1 and causes the unpleasant g-null paradox. It is not a desirable method.

Recall the outcome regression estimator with a treatment at a single time based on $E\{Y(z)\} = E\{E(Y \mid Z = z, X)\}$. We first fit a model of Y on X using the subset of the data with $Z = z$, and obtain the fitted values $\hat{Y}_i(z)$ for all units. We then obtain the estimator

$$\hat{E}\{Y(z)\} = n^{-1} \sum_{i=1}^{n} \hat{Y}_i(z).$$

Similarly, the recursive expectation formula in (29.1) motivates a simpler method for estimation. Start from the inner conditional expectation, denoted by

$$\tilde{Y}_2(z_1, z_2) = E(Y \mid Z_2 = z_2, Z_1 = z_1, X_1, X_0).$$

We can fit a model of Y on (X_1, X_0) using the subset of the data with $(Z_2 = z_2, Z_1 = z_1)$, and

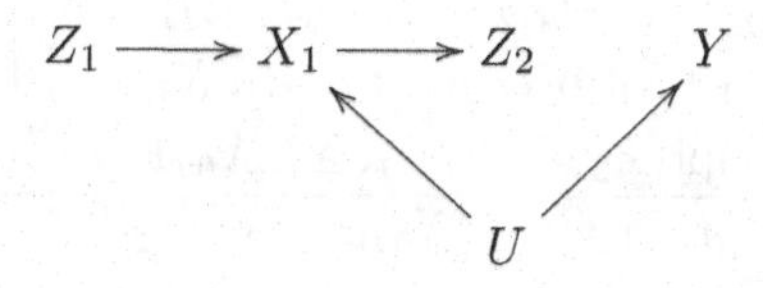

FIGURE 29.4: With unmeasured confounding between X_1 and Y. The causal diagram ignores the pretreatment covariates X_0.

obtain the fitted values $\hat{Y}_{i2}(z_1, z_2)$ for all units. Move on to outer conditional expectation, denoted by

$$\tilde{Y}_1(z_1, z_2) = E\{\tilde{Y}_2(z_1, z_2) \mid Z_1 = z_1, X_0\}.$$

We can fit a model of $\hat{Y}_2(z_1, z_2)$ on X_0 using the subset of data with $Z_1 = z_1$, and obtain the fitted values $\hat{Y}_{i1}(z_1, z_2)$ for all units. The final estimator for $E\{Y(z_1, z_2)\}$ is then

$$\hat{E}^{\text{rec}}\{Y(z_1, z_2)\} = n^{-1} \sum_{i=1}^{n} \hat{Y}_{i1}(z_1, z_2).$$

The above recursive estimation does not involve fitting a model for X_1 and avoids the g-null paradox. However, the estimator based on recursive regression is not easy to implement because it involves modeling variables that do not correspond to the natural structure of the causal diagram, e.g., $\tilde{Y}_2(z_1, z_2)$.

29.3 Inverse propensity score weighting

Recall the IPW identification formula with a treatment at a single time point:

$$E\{Y(z)\} = E\left\{ \frac{I(Z = z)Y}{\text{pr}(Z = z \mid X)} \right\}.$$

The following result extends it to the setting with a treatment at two time points. Define

$$e(z_1, X_0) = \text{pr}(Z_1 = z_1 \mid X_0)$$

and

$$e(z_2, Z_1, X_1, X_0) = \text{pr}(Z_2 = z_2 \mid Z_1, X_1, X_0)$$

as the propensity scores at time points 1 and 2, respectively.

Theorem 29.2 *Under Assumption 29.1, we have*

$$E\{Y(z_1, z_2)\} = E\left\{ \frac{I(Z_1 = z_1)I(Z_2 = z_2)Y}{e(z_1, X_0)e(z_2, Z_1, X_1, X_0)} \right\}. \tag{29.9}$$

Theorem 29.2 reveals the omitted overlap assumption:

$$0 < e(z_1, X_0) < 1, \quad 0 < e(z_2, Z_1, X_1, X_0) < 1$$

for all z_1 and z_2. If some propensity scores are 0 or 1, then the identification formula (29.9) blows up to infinity.

Proof of Theorem 29.2: Conditioning on (Z_1, X_1, X_0) and using Assumption 29.1(2), we can simplify the right-hand side of (29.9) as

$$\begin{aligned}
& E\left\{ \frac{I(Z_1 = z_1)I(Z_2 = z_2)Y(z_1, z_2)}{\text{pr}(Z_1 = z_1 \mid X_0)\text{pr}(Z_2 = z_2 \mid Z_1, X_1, X_0)} \right\} \\
= \; & E\left\{ \frac{I(Z_1 = z_1)\text{pr}(Z_2 = z_2 \mid Z_1, X_1, X_0)E(Y(z_1, z_2) \mid Z_1, X_1, X_0)}{\text{pr}(Z_1 = z_1 \mid X_0)\text{pr}(Z_2 = z_2 \mid Z_1, X_1, X_0)} \right\} \\
= \; & E\left\{ \frac{I(Z_1 = z_1)}{\text{pr}(Z_1 = z_1 \mid X_0)} E(Y(z_1, z_2) \mid Z_1, X_1, X_0) \right\} \\
= \; & E\left\{ \frac{I(Z_1 = z_1)}{\text{pr}(Z_1 = z_1 \mid X_0)} Y(z_1, z_2) \right\},
\end{aligned} \tag{29.10}$$

where (29.10) follows from the tower property.

Conditioning on X_0 and using Assumption 29.1(1), we can simplify the right-hand side of (29.10) as

$$\begin{aligned} & E\left\{\frac{\text{pr}(Z_1 = z_1 \mid X_0)}{\text{pr}(Z_1 = z_1 \mid X_0)} E(Y(z_1, z_2) \mid X_0)\right\} \\ = & E\{E(Y(z_1, z_2) \mid X_0)\} \\ = & E\{Y(z_1, z_2)\}, \end{aligned}$$

where, again, the last line follows from the tower property. □

29.3.1 IPW estimators based on estimated propensity scores

The estimator based on IPW is much simpler which only involves modeling two binary treatment indicators. First, we can fit a model of Z_1 on X_0 to obtain the fitted values $\hat{e}(z_1, X_{i0})$ and fit a model of Z_2 on (Z_1, X_1, X_0) to obtain the fitted values $\hat{e}(z_2, Z_{i1}, X_{i1}, X_{i0})$ for all units. Then, we obtain the following IPW estimator:

$$\hat{E}^{\text{ht}}\{Y(z_1, z_2)\} = n^{-1} \sum_{i=1}^{n} \frac{I(Z_{i1} = z_1) I(Z_{i2} = z_2) Y_i}{\hat{e}(z_1, X_{i0}) \hat{e}(z_2, Z_{i1}, X_{i1}, X_{i0})}.$$

Similar to the discussion in Chapter 11, the HT estimator is not invariant to the location shift of the outcome and suffers from instability in finite samples. A modified Hajek-type estimator is

$$\hat{E}^{\text{haj}}\{Y(z_1, z_2)\} = \frac{\hat{E}^{\text{ht}}\{Y(z_1, z_2)\}}{\hat{1}^{\text{ht}}(z_1, z_2)},$$

where

$$\hat{1}^{\text{ht}}(z_1, z_2) = n^{-1} \sum_{i=1}^{n} \frac{I(Z_{i1} = z_1) I(Z_{i2} = z_2)}{\hat{e}(z_1, X_{i0}) \hat{e}(z_2, Z_{i1}, X_{i1}, X_{i0})}.$$

29.3.2 Simple numeric examples

The following R code generates data from a linear outcome model and logistic propensity score models. The outcome regression estimator based on Example 29.1 and the Hajek estimator are close on average over simulation. The Hajek estimator has a much larger variance than the outcome regression estimator.

```
> EST = replicate(10^4, {
+     n = 2000
+     x0 = rnorm(n)
+     e1 = 1/(1 + exp(-x0))
+     z1 = rbinom(n, 1, e1)
+     x1 = z1 + x0 + rnorm(n)
+     e2 = 1/(1 + exp(-(-0.5 + z1 + x1 + 0.5*x0)))
+     z2 = rbinom(n, 1, e2)
+     y = z2 + z1 + x1 + x0 + rnorm(n)
+     ## OLS estimation of (1,1) v.s. (0,0)
+     ## based on the formula in Example 29.1
+     ols.y = lm(y ~ z2 + z1 + x1 + x0)$coef
+     ols.x1 = lm(x1 ~ z1 + x0)$coef
+     est1 = ols.y[2] + ols.y[3] + ols.y[4]*ols.x1[2]
+     ## IPW - Hajek form
```

```
+     e.z1 = glm(z1 ~ x0, family = binomial)$fitted.values
+     e.z2 = glm(z2 ~ z1 + x1 + x0, family = binomial)$fitted.values
+     mean11 = mean(z1*z2*y/e.z1/e.z2)/
+               mean(z1*z2/e.z1/e.z2)
+     mean00 = mean((1-z1)*(1-z2)*y/(1-e.z1)/(1-e.z2))/
+               mean((1-z1)*(1-z2)/(1-e.z1)/(1-e.z2))
+     est2 = mean11 - mean00
+
+     c(est1, est2)
+  })
>
> apply(EST, 1, mean)
      z2
3.000374 3.060660
> apply(EST, 1, sd)
       z2
0.0805125 0.4422689
```

The following R code modifies the above data-generating process to allow for the presence of unmeasured confounder U between X_1 and Y; see Figure 29.2. The outcome regression estimator based on Example 29.1 and the Hajek estimator are close on average over simulation although the outcome model is misspecified. Again, the Hajek estimator has a much larger variance than the outcome regression estimator.

```
> EST = replicate(10^4, {
+   n = 2000
+   um = rnorm(n)
+   x0 = rnorm(n)
+   e1 = 1/(1 + exp(-x0))
+   z1 = rbinom(n, 1, e1)
+   x1 = z1 + x0 + um + rnorm(n)
+   e2 = 1/(1 + exp(-(-0.5 + z1 + x1 + 0.5*x0)))
+   z2 = rbinom(n, 1, e2)
+   y = z2 + z1 + x1 + x0 + um +rnorm(n)
+   ## OLS estimation of (1,1) v.s. (0,0)
+   ## based on the formula in Example 29.1
+   ols.y = lm(y ~ z2 + z1 + x1 + x0)$coef
+   ols.x1 = lm(x1 ~ z1 + x0)$coef
+   est1 = ols.y[2] + ols.y[3] + ols.y[4]*ols.x1[2]
+   ## IPW - Hajek form
+   e.z1 = glm(z1 ~ x0, family = binomial)$fitted.values
+   e.z2 = glm(z2 ~ z1 + x1 + x0, family = binomial)$fitted.values
+   mean11 = mean(z1*z2*y/e.z1/e.z2)/
+     mean(z1*z2/e.z1/e.z2)
+   mean00 = mean((1-z1)*(1-z2)*y/(1-e.z1)/(1-e.z2))/
+     mean((1-z1)*(1-z2)/(1-e.z1)/(1-e.z2))
+   est2 = mean11 - mean00
+
+   c(est1, est2)
+ })
>
> apply(EST, 1, mean)
      z2
2.997013 3.092963
> apply(EST, 1, sd)
       z2
0.1300347 0.6205916
```

29.4 Multiple time points

Extending the estimation strategies in Sections 29.2 and 29.3 is not immediate with multiple time points. Even with a binary treatment and K time points, the number of treatment combinations grows exponentially with K (for example, $2^5 = 32$ and $2^{10} = 1024$). Consequently, the outcome regression and IPW estimators in Sections 29.2 and 29.3 are not feasible in finite samples because they require enough data for every combination of the treatment levels.

29.4.1 Marginal structural model

A powerful approach is based on the marginal structural model (MSM) (Robins et al., 2000; Hernán et al., 2000). For simplicity of notation, I will only present the MSM with $K = 2$ although its main use is in the general case.

Definition 29.1 (MSM) *The marginal mean of* $Y(z_1, z_2)$ *equals*

$$E\{Y(z_1, z_2)\} = f(z_1, z_2; \beta).$$

A leading example of Definition 29.1 is $E\{Y(z_1, z_2)\} = \beta_0 + \beta_1 z_1 + \beta_2 z_2$. It is also straightforward to include the baseline covariates in the model. Definition 29.2 below extends Definition 29.1.

Definition 29.2 (MSM with baseline covariates) *The mean of* $Y(z_1, z_2)$ *conditional on* X_0 *equals*

$$E\{Y(z_1, z_2) \mid X_0\} = f(z_1, z_2, X_0; \beta).$$

A leading example of Definition 29.2 is

$$E\{Y(z_1, z_2) \mid X_0\} = \beta_0 + \beta_1 z_1 + \beta_2 z_2 + \beta_3^{\mathrm{T}} X_0. \tag{29.11}$$

If we observe all the potential outcomes, we can solve β from the following minimization problem:

$$\beta = \arg\min_b \sum_{z_2} \sum_{z_1} E\{Y(z_1, z_2) - f(z_1, z_2, X_0; b)\}^2. \tag{29.12}$$

For simplicity, I focus on the least squares formulation. We can also extend the discussion to general models; see Problem 29.3 for an example of the logistic model.

Under sequential ignorability, we can solve β from the following minimization problem that only involves observables.

Theorem 29.3 (IPW under MSM) *Under Assumption 29.1 and Definition 29.2, the* β *in* (29.12) *equals*

$$\beta = \arg\min_b \sum_{z_2} \sum_{z_1} E\left[\frac{I(Z_1 = z_1) I(Z_2 = z_2)}{e(z_1, X_0) e(z_2, Z_1, X_1, X_0)} \{Y - f(z_1, z_2, X_0; b)\}^2 \right].$$

The proof of Theorem 29.3 is similar to that of Theorem 29.2. I relegate it to Problem 29.2.

Theorem 29.3 implies a simple estimation strategy based on weighted regressions. For instance, under (29.11), we can use the following two-step procedure to estimate β:

1. obtain the fitted values of the propensity scores $\hat{e}(z_1, X_{i0})$ and $\hat{e}(z_2, Z_{i1}, X_{i1}, X_{i0})$, for example, by running logistic regressions of Z_{i1} on $(1, X_{i0})$ and Z_{i2} on $(1, Z_{i1}, X_{i1}, X_{i0})$, respectively;
2. obtain $\hat{\beta}$ as the coefficient of the WLS fit of Y_i on $(1, Z_{i2}, Z_{i1}, X_{i0})$ with weights $\hat{e}^{-1}(Z_{i1}, X_{i0})\hat{e}^{-1}(Z_{i2}, Z_{i1}, X_{i1}, X_{i0})$.

29.4.2 Structural nested model

A key problem of IPW is that it is not applicable if the overlap assumption is violated. To address this challenge, Robins proposed the structural nested model. Again, to simplify the presentation, I only review the version with two time points.

Definition 29.3 (structural nested model) *The conditional effect at time point 1 is*

$$E\{Y(z_1, 0) - Y(0, 0) \mid Z_1 = z_1, X_0\} = g_1(z_1, X_0; \beta)$$

for all z_1, and the conditional effect at time point 2 is

$$E\{Y(z_1, z_2) - Y(z_1, 0) \mid Z_2 = z_2, Z_2 = z_1, X_1, X_0\} = g_2(z_2, z_1, X_1, X_0; \beta)$$

for all z_1, z_2.

In Definition 29.3, two logical restrictions are

$$g_1(0, X_0; \beta) = 0$$

and

$$g_2(0, z_1, X_1, X_0; \beta) = 0 \text{ for all } z_1.$$

Two leading choices of Definition 29.3 are below.

Example 29.2 *Assume*

$$\begin{cases} g_1(z_1, X_0; \beta) = \beta_1 z_1, \\ g_2(z_2, z_1, X_1, X_0; \beta) = (\beta_2 + \beta_3 z_1) z_2. \end{cases}$$

Example 29.3 *Assume*

$$\begin{cases} g_1(z_1, X_0; \beta) = (\beta_1 + \beta_2^{\top} X_0) z_1, \\ g_2(z_2, z_1, X_1, X_0; \beta) = (\beta_3 + \beta_4 z_1 + \beta_5^{\top} X_1) z_2. \end{cases}$$

Compare Definitions 29.2 and 29.3. The structural nested model allows for adjusting for the baseline covariates as well as the time-varying covariates whereas the marginal structural model only allows for adjusting for the baseline covariates.

The estimation under Definition 29.3 is more involved. A strategy is to estimate the parameter based on estimating equations. I first introduce two important building blocks for discussing the estimation. Define

$$U_2(\beta) = Y - g_2(Z_2, Z_1, X_1, X_0; \beta)$$

and

$$U_1(\beta) = Y - g_2(Z_2, Z_1, X_1, X_0; \beta) - g_1(Z_1, X_0; \beta).$$

They are not directly computable from the data because they depend on the true value of the parameter β. At the true value, they have the following properties.

Lemma 29.1 *Under Assumption 29.1 and Definition 29.3, we have*

$$\begin{aligned} E\{U_2(\beta) \mid Z_2, Z_1, X_1, X_0\} &= E\{U_2(\beta) \mid Z_1, X_1, X_0\} \\ &= E\{Y(Z_1, 0) \mid Z_1, X_1, X_0\} \end{aligned}$$

and

$$\begin{aligned} E\{U_1(\beta) \mid Z_1, X_0\} &= E\{U_1(\beta) \mid X_0\} \\ &= E\{Y(0,0) \mid X_0\}. \end{aligned}$$

Lemma 29.1 involves a subtle notation $Y(Z_1, 0)$ because Z_1 is random. It should be read as $Y(Z_1, 0) = Z_1 Y(1,0) + (1 - Z_1) Y(0,0)$. Based on the definitions and Lemma 29.1, $U_1(\beta)$ acts as the control potential outcome before receiving any treatment and $U_2(\beta)$ acts as the control potential outcome after receiving the treatment at time point 1.
Proof of Lemma 29.1: Part 1. We have

$$\begin{aligned} & E\{U_2(\beta) \mid Z_2 = 1, Z_1, X_1, X_0\} \\ = \; & E\{Y(Z_1, 1) - g_2(1, Z_1, X_1, X_0; \beta) \mid Z_2 = 1, Z_1, X_1, X_0\} \\ = \; & E\{Y(Z_1, 0) \mid Z_2 = 1, Z_1, X_1, X_0\} \end{aligned}$$

and

$$\begin{aligned} & E\{U_2(\beta) \mid Z_2 = 0, Z_1, X_1, X_0\} \\ = \; & E\{Y(Z_1, 0) - g_2(0, Z_1, X_1, X_0; \beta) \mid Z_2 = 0, Z_1, X_1, X_0\} \\ = \; & E\{Y(Z_1, 0) \mid Z_2 = 0, Z_1, X_1, X_0\} \end{aligned}$$

so

$$\begin{aligned} E\{U_2(\beta) \mid Z_2, Z_1, X_1, X_0\} &= E\{Y(Z_1, 0) \mid Z_2, Z_1, X_1, X_0\} \\ &= E\{Y(Z_1, 0) \mid Z_1, X_1, X_0\} \end{aligned}$$

where the last identity follows from $Z_2 \perp\!\!\!\perp Y(Z_1, 0) \mid (Z_1, X_1, X_0)$ by sequential ignorability. Since the last term does not depend on Z_2, we also have

$$E\{U_2(\beta) \mid Z_2, Z_1, X_1, X_0\} = E\{U_2(\beta) \mid Z_1, X_1, X_0\}.$$

Part 2. Using the above results, we have

$$\begin{aligned} & E\{U_1(\beta) \mid Z_1, X_0\} \\ = \; & E\{U_2(\beta) - g_1(Z_1, X_0; \beta) \mid Z_1, X_0\} \qquad \text{(Definition 29.3)} \\ = \; & E\left[E\{U_2(\beta) - g_1(Z_1, X_0; \beta) \mid X_1, Z_1, X_0\} \mid Z_1, X_0\right] \qquad \text{(tower property)} \\ = \; & E\left[E\{Y(Z_1, 0) - g_1(Z_1, X_0; \beta) \mid X_1, Z_1, X_0\} \mid Z_1, X_0\right] \qquad \text{(part 1)} \\ = \; & E\{Y(Z_1, 0) - g_1(Z_1, X_0; \beta) \mid Z_1, X_0\} \qquad \text{(tower property)} \\ = \; & E\{Y(0,0) \mid Z_1, X_0\} \qquad \text{(Definition 29.3)} \\ = \; & E\{Y(0,0) \mid X_0\} \qquad \text{(sequential ignorability).} \end{aligned}$$

Since the last term does not depend on Z_1, we also have

$$E\{U_1(\beta) \mid Z_1, X_0\} = E\{U_1(\beta) \mid X_0\}.$$

□

With Lemma 29.1, we can prove Theorem 29.4 below.

Theorem 29.4 *Under Assumption 29.1 and Definition 29.3, we have*

$$\begin{aligned} E\Big[h_2(Z_1, X_1, X_0)\{Z_2 - e(1, Z_1, X_1, X_0)\}U_2(\beta)\Big] &= 0, \\ E\Big[h_1(X_0)\{Z_1 - e(1, X_0)\}U_1(\beta)\Big] &= 0 \end{aligned}$$

for any functions h_1 and h_2, provided that the moments exist.

Proof of Theorem 29.2: Use the tower property by conditioning on (Z_2, Z_1, X_1, X_0) and Lemma 29.1 to obtain

$$\begin{aligned} & E\left[h_2(Z_1, X_1, X_0)\{Z_2 - e(1, Z_1, X_1, X_0)\}E\{U_2(\beta) \mid Z_2, Z_1, X_1, X_0\}\right] \\ = \; & E\left[h_2(Z_1, X_1, X_0)\{Z_2 - e(1, Z_1, X_1, X_0)\}E\{U_2(\beta) \mid Z_1, X_1, X_0\}\right]. \end{aligned}$$

Use the tower property by conditioning on (Z_1, X_1, X_0) to show that the last identity equals 0 because $E\{Z_2 - e(1, Z_1, X_1, X_0) \mid Z_1, X_1, X_0\} = 0$.

Similarly, use the tower property by conditioning on (Z_1, X_0) and Lemma 29.1 to obtain

$$E\left[h_1(X_0)\{Z_1 - e(1, X_0)\}E\{U_1(\beta) \mid Z_1, X_0\}\right] = E\left[h_1(X_0)\{Z_1 - e(1, X_0)\}E\{U_1(\beta) \mid X_0\}\right].$$

Use the tower property by conditioning on X_0 to show that the last identity equals 0 because $E\{Z_1 - e(1, X_0) \mid X_0\} = 0$. □

To use Theorem 29.4, we must specify h_1 and h_2 to ensure that there are enough equations for solving β. Example 29.4 below revisits Example 29.2.

Example 29.4 *Under Example 29.2, we can choose $h_1 = 1$ and $h_2 = (1, Z_1)$ to obtain*

$$\begin{aligned} E\left[\{Z_2 - e(1, Z_1, X_1, X_0)\}\{Y - (\beta_2 + \beta_3 Z_1)Z_2\}\right] &= 0, \\ E\left[Z_1\{Z_2 - e(1, Z_1, X_1, X_0)\}\{Y - (\beta_2 + \beta_3 Z_1)Z_2\}\right] &= 0, \\ E\left[\{Z_1 - e(1, X_0)\}\{Y - (\beta_2 + \beta_3 Z_1)Z_2 - \beta_1 Z_1\}\right] &= 0. \end{aligned}$$

We can then solve for the β's from the above linear equations; see Problem 29.5. A natural question is whether alternative choices of (h_1, h_2) can lead to more efficient estimators. The answer is yes. For example, we can choose many (h_1, h_2) and use the generalized method of moment (Hansen, 1982). The technical details are beyond this book.

Naimi et al. (2017) and Vansteelandt and Joffe (2014) provided tutorials on the structural nested models.

29.5 Discussion

Causal inference with time-varying treatment and confounding is a very non-trivial problem. Due to the space limit, this chapter only introduces some basic ideas in this area. To fully understand the problem, we should develop the theory and methods in parallel with the simple causal inference problem discussed from Chapter 10 to Chapter 20.

Section 29.3.2 presents two simple simulation studies. I encourage the readers to conduct more simulation studies to evaluate the finite-sample properties of other methods discussed in this chapter. Moreover, it will be fruitful to apply the ideas in this chapter to solve problems from various areas.

29.6 Homework problems

29.1 Extending Example 29.1

Assume linear models for Y and X_1:

$$E(Y \mid z_2, z_1, x_1, x_0) = \beta_0 + \beta_1 z_2 + \beta_2 z_1 + \beta_3 z_2 z_1 + \beta_4 x_1 + \beta_5 x_0$$

and

$$E(X_1 \mid z_1, x_0) = \gamma_0 + \gamma_1 z_1 + \gamma_2 x_0.$$

Compared with Example 29.1, the outcome model allows for the interaction between z_1 and z_2. Prove that

$$\begin{aligned} E\{Y(1,0) - Y(0,0)\} &= \beta_2 + \beta_4 \gamma_1, \\ E\{Y(0,1) - Y(0,0)\} &= \beta_1, \\ E\{Y(1,1) - Y(0,0)\} &= \beta_1 + \beta_2 + \beta_3 + \beta_4 \gamma_1. \end{aligned}$$

29.2 IPW under MSM

Prove Theorem 29.3.

29.3 A nonlinear example of Definition 29.2

Another leading example of Definition 29.2 is

$$\text{logit}\left[\text{pr}\{Y(z_1, z_2) = 1 \mid X_0\}\right] = \beta_0 + \beta_1 z_1 + \beta_2 z_2 + \beta_3^{\mathsf{T}} X_0. \tag{29.13}$$

If we observe all potential outcomes, we can solve β by minimizing the expectation of the negative log-likelihood function (see Section B.6.2 for a simpler version):

$$\beta = \arg\min_b \sum_{z_2} \sum_{z_1} E\left\{\log(1 + e^{\ell}) - Y(z_1, z_2)\ell\right\} \tag{29.14}$$

where $\ell = \beta_0 + \beta_1 z_1 + \beta_2 z_2 + \beta_3^{\mathsf{T}} X_0$. Under sequential ignorability, we can solve β from the following minimization problem that only involves observables.

Theorem 29.5 (IPW under MSM) *Under Assumption 29.1 and Definition 29.2, the β in (29.14) equals*

$$\beta = \arg\min_b \sum_{z_2} \sum_{z_1} E\left[\frac{I(Z_1 = z_1) I(Z_2 = z_2)}{e(z_1, X_0) e(z_2, Z_1, X_1, X_0)} \left\{\log(1 + e^{\ell}) - Y(z_1, z_2)\ell\right\}\right].$$

Prove Theorem 29.5.

Remark: Theorem 29.5 implies a simple estimation strategy based on weighted regressions. For instance, under (29.13), we can use the following two-step procedure to estimate β:

1. obtain the fitted values of the propensity scores $\hat{e}(z_1, X_{i0})$ and $\hat{e}(z_2, Z_{i1}, X_{i1}, X_{i0})$, for example, by running logistic regressions of Z_{i1} on $(1, X_{i0})$ and Z_{i2} on $(1, Z_{i1}, X_{i1}, X_{i0})$, respectively;
2. obtain $\hat{\beta}$ as the coefficient of the weighted logistic regression of Y_i on $(1, Z_{i2}, Z_{i1}, X_{i0})$ with weights $\hat{e}^{-1}(Z_{i1}, X_{i0}) \hat{e}^{-1}(Z_{i2}, Z_{i1}, X_{i1}, X_{i0})$.

29.4 Structural nested model with a single time point

Recall the standard setting of observational studies with IID data drawn from $\{X, Z, Y(1), Y(0)\}$. Define the propensity score as $e(X) = \text{pr}(Z = 1 \mid X)$. Assume

$$Z \perp\!\!\!\perp Y(0) \mid X$$

and the following structural nested model.

Definition 29.4 (structural nested model with a single time point) *The conditional mean of the individual effect is*

$$E\{Y(z) - Y(0) \mid Z = z, X\} = g(z, X; \beta).$$

In Definition 29.4, a logical restriction is $g(0, X; \beta) = 0$. Prove the following results.

1. We have

$$E\{Y - g(Z, X; \beta) \mid X, Z\} = E\{Y - g(Z, X; \beta) \mid X\} = E\{Y(0) \mid X\}.$$

2. We have

$$E\Big[h(X)\{Z - e(X)\}\{Y - g(Z, X; \beta)\}\Big] = 0 \tag{29.15}$$

for any function h, provided that the moment exists.

Remark: Equation (29.15) is the basis for parameter estimation. I give two examples below.

Example 29.5 *Consider a special case of Definition 29.4 with $g(z, X; \beta) = \beta z$. Choose $h(X) = 1$ to obtain*

$$E\{(Z - e(X))(Y - \beta Z)\} = 0.$$

Solve for β to obtain

$$\beta = \frac{E\{(Z - e(X))Y\}}{E\{(Z - e(X))Z\}}.$$

Therefore, β equals the coefficient of Z in the TSLS of Y on Z with $Z - e(X)$ being the IV for Z. With some basic calculations, we can also show that

$$\beta = \frac{\text{cov}\{Z - e(X), Y\}}{\text{var}\{Z - e(X)\}}.$$

Therefore, β equals the coefficient of $Z - e(X)$ in the OLS of Y on $Z - e(X)$, which appeared in Section 14.1 before.

Example 29.6 *Consider another special case of Definition 29.4 with $g(z, X; \beta) = (\beta_0 + \beta_1^{\top} X)z$. Choose $h(X) = (1, X)$ to obtain*

$$E\left\{\begin{pmatrix} Z - e(X) \\ (Z - e(X))X \end{pmatrix} (Y - \beta_0 Z - \beta_1^{\top} XZ)\right\} = 0.$$

That is, (β_0, β_1) equal the coefficients in the TSLS of Y on (Z, XZ) with $(Z - e(X), (Z - e(X))X)$ being the IV for (Z, XZ).

29.5 Estimation under Example 29.4

We can estimate the β's by solving the empirical version of the estimating equations in Example 29.4. We first estimate the two propensity scores and obtain the centered treatment

$$\check{Z}_{i1} = Z_{i1} - \hat{e}(1, X_{i0})$$

at time point 1 and

$$\check{Z}_{i2} = Z_{i2} - \hat{e}(1, Z_{i1}, X_{i1}, X_{i0})$$

at time point 2.

Show that we can estimate β_2 and β_3 by running TSLS of Y_i on $(Z_{i2}, Z_{i1}Z_{i2})$ with $(\check{Z}_{i2}, Z_{i1}\check{Z}_{i2})$ as the IV for $(Z_{i2}, Z_{i1}Z_{i2})$, and then we can estimate β_1 by running TSLS of $Y_i - (\hat{\beta}_2 + \hat{\beta}_3 Z_{i1})Z_{i2}$ on Z_{i1} with $\check{Z}_{i1}$ as the IV for Z_{i1}.

29.6 g-formula with a treatment at multiple time points

Extend the discussion to the setting with K time points. The temporal ordering (but not the causal diagram) of the variables is

$$X_0 \to Z_1 \to X_1 \to Z_2 \to \cdots \to X_{K-1} \to Z_K.$$

Introduce the notation $\overline{Z}_k = (Z_1, \ldots, Z_k)$ and $\overline{X}_k = (X_0, X_1, \ldots, X_k)$ with lower case $\overline{z}_k$ and $\overline{x}_k$ denoting the corresponding realized values. With $k = 0$, we have $\overline{X}_0 = X_0$ and $\overline{Z}_0$ is empty. Each unit has 2^K potential outcomes:

$$Y(\overline{z}_K) \text{ for all } z_1, \ldots, z_K = 0, 1.$$

Assume sequential ignorability below.

Assumption 29.2 (sequential ignorability at multiple time points) *We have*

$$Z_k \perp\!\!\!\perp Y(\overline{z}_K) \mid (\overline{Z}_{k-1}, \overline{X}_{k-1})$$

for all $k = 1, \ldots, K$ and all $z_1, \ldots, z_K = 0, 1$.

Prove Theorem 29.6 below.

Theorem 29.6 (g-formula with multiple time points) *Under Assumption 29.2, we have*

$$E\{Y(\overline{z}_K)\} = E\left[\cdots E\{E(Y \mid \overline{z}_K, \overline{X}_{K-1}) \mid \overline{z}_{K-1}, \overline{X}_{K-2}\} \cdots \mid z_1, X_0\right].$$

Remark: In Theorem 29.6, I use the simplified notation "$\overline{z}_k$" for "$\overline{Z}_k = \overline{z}_k$." With discrete X, Theorem 29.6 reduces to

$$E\{Y(\overline{z}_K)\} = \sum_{x_0} \sum_{x_1} \cdots \sum_{x_{K-1}} E(Y \mid \overline{z}_K, \overline{x}_{K-1}) \cdot \text{pr}(x_{K-1} \mid \overline{z}_{K-1}, \overline{x}_{K-2}) \cdots \text{pr}(x_1 \mid z_1, x_0)\text{pr}(x_0);$$

with continuous X, Theorem 29.6 reduces to

$$E\{Y(\overline{z}_K)\} = \int E(Y \mid \overline{z}_K, \overline{x}_{K-1}) \cdot f(x_{K-1} \mid \overline{z}_{K-1}, \overline{x}_{K-2}) \cdots f(x_1 \mid z_1, x_0) f(x_0) \mathrm{d}\overline{x}_{K-1}.$$

To use the g-formulas to construct estimators for $E\{Y(\overline{z}_K)\}$, we need model Y and the X_k's.

29.7 IPW with treatments at multiple time points

Inherit the setting of Problem 29.6. Define the propensity score at K time points as

$$\begin{aligned} e(z_1, X_0) &= \text{pr}(Z_1 = z_1 \mid X_0), \\ &\vdots \\ e(z_k, \overline{Z}_{k-1}, \overline{X}_{k-1}) &= \text{pr}(Z_k = z_k \mid \overline{Z}_{k-1}, \overline{X}_{k-1}), \\ &\vdots \\ e(z_K, \overline{Z}_{K-1}, \overline{X}_{K-1}) &= \text{pr}(Z_K = z_K \mid \overline{Z}_{K-1}, \overline{X}_{K-1}). \end{aligned}$$

Prove Theorem 29.8 below assuming overlap implicitly.

Theorem 29.7 (IPW with multiple time points) *Under Assumption 29.2,*

$$E\{Y(\overline{z}_K)\} = E\left\{ \frac{I(Z_1 = z_1) \cdots I(Z_K = z_K) Y}{e(z_1, X_0) \cdots e(z_K, \overline{Z}_{K-1}, \overline{X}_{K-1})} \right\}.$$

Remark: Based on Theorem 29.8, we can construct the Horvitz–Thompson and Hajek estimators with estimated propensity scores.

29.8 MSM with treatments at multiple time points

The number of potential outcomes grows exponentially with K. The formulas in Problems 29.6 and 29.7 are not directly applicable in finite samples. We can impose the following structural assumptions on the potential outcomes.

Definition 29.5 (MSM with multiple time points) *Assume*

$$E\{Y(\overline{z}_K) \mid X_0\} = f(\overline{z}_K, X_0; \beta).$$

Two leading examples of Definition 29.5 are

$$E\{Y(\overline{z}_K) \mid X_0\} = \beta_0 + \beta_1 \sum_{k=1}^{K} z_k + \beta_2^{\mathsf{T}} X_0$$

and

$$E\{Y(\overline{z}_K) \mid X_0\} = \beta_0 + \sum_{k=1}^{K} \beta_k z_k + \beta_{K+1}^{\mathsf{T}} X_0.$$

If we know all the potential outcomes, we can solve β from the following minimization problem:

$$\beta = \arg\min_b \sum_{\overline{z}_K} E\{Y(\overline{z}_K) - f(\overline{z}_K, X_0; \beta)\}^2.$$

Theorem 29.8 below shows that under Assumption 29.2, we can solve β from a minimization problem that only involves observables.

Theorem 29.8 (IPW for MSM with multiple time points) *Under Assumption 29.2, we have*

$$\beta = \arg\min_b \sum_{\overline{z}_K} E\left[\frac{I(Z_1 = z_1) \cdots I(Z_K = z_K)}{e(z_1, X_0) \cdots e(z_K, \overline{Z}_{K-1}, \overline{X}_{K-1})} \{Y - f(\overline{z}_K, X_0; \beta)\}^2 \right].$$

Prove Theorem 29.8.

Remark: Based on Theorem 29.8, we can estimate the parameters in the MSM by WLS with weights proportional to the product of the inverse of the estimated propensity scores.

29.9 Structural nested model with treatments at multiple time points

Inherit the setting from Problem 29.6 and the notation from Problem 29.7. This problem presents a general structural nested model.

Definition 29.6 (structural nested model with multiple time points) *The conditional effect at time k is*

$$E\{Y(\overline{z}_k, 0) - Y(\overline{z}_{k-1}, 0) \mid \overline{z}_k, \overline{X}_{k-1}\} = g_k(\overline{z}_k, \overline{X}_{k-1}; \beta)$$

for all $\overline{z}_k$ and all $k = 1, \ldots, K$.

In Definition 29.6, a logical restriction is

$$g_k(0, \overline{z}_{k-1}, \overline{X}_{k-1}; \beta) = 0$$

for all $\overline{z}_{k-1}$ and all $k = 1, \ldots, K$.

Define

$$U_k(\beta) = Y - \sum_{s=1}^{k} g_s(\overline{Z}_s, \overline{X}_{s-1}; \beta)$$

for all $k = 1, \ldots, K$. Theorem 29.9 below extends Theorem 29.4.

Theorem 29.9 *Under Assumption 29.2 and Definition 29.6, we have*

$$E\left[h_k(\overline{Z}_{k-1}, \overline{X}_{k-1})\{Z_k - e(1, \overline{Z}_{k-1}, \overline{X}_{k-1})\}U_k(\beta)\right] = 0$$

for any functions h_k ($k = 1, \ldots, K$), provided that the moment exists.

Prove Theorem 29.9.

Remark: Choosing appropriate h_k's, we can estimate β by solving the empirical version of Theorem 29.9.

29.10 Recommended reading

Robins et al. (2000) reviewed the MSM. Naimi et al. (2017) reviewed the g-methods.

Part VII

Appendices

A

Probability and Statistics

This book assumes that the readers have basic knowledge of probability theory and statistical inference. Therefore, this chapter is not a comprehensive review of probability and statistics. For easy reference, I review the key concepts that are crucial for the main text.

A.1 Probability

A.1.1 Pearson correlation coefficient and squared multiple correlation coefficient

For two random variables Y and X, define the Pearson correlation coefficient as

$$\rho_{YX} = \frac{\text{cov}(Y, X)}{\sqrt{\text{var}(Y)\text{var}(X)}}$$

which measures the linear dependence of Y on X. The definition is symmetric in Y and X in that

$$\rho_{YX} = \rho_{XY}.$$

With a random variable Y and a random vector X, define the squared multiple correlation coefficient as

$$\begin{aligned} R^2_{YX} &= \text{corr}^2(Y, X) \\ &= \frac{\text{cov}(Y, X)\text{cov}(X)^{-1}\text{cov}(X, Y)}{\text{var}(Y)} \end{aligned}$$

where $\text{cov}(Y, X)$ is a row vector and $\text{cov}(X, Y)$ is a column vector. It also measures the linear dependence of Y on X. But this definition is not symmetric in Y and X.

A.1.2 Multivariate Normal random vector

A multivariate Normal random vector $\text{N}(\mu, \Sigma)$ is determined by its mean vector μ and covariance matrix Σ. Partition it into two parts:

$$\begin{pmatrix} Y_1 \\ Y_2 \end{pmatrix} \sim \text{N}\left(\begin{pmatrix} \mu_1 \\ \mu_2 \end{pmatrix}, \begin{pmatrix} \Sigma_{11} & \Sigma_{12} \\ \Sigma_{21} & \Sigma_{22} \end{pmatrix} \right).$$

First, the marginal distributions are Normal:

$$\begin{aligned} Y_1 &\sim \text{N}(\mu_1, \Sigma_{11}), \\ Y_2 &\sim \text{N}(\mu_2, \Sigma_{22}). \end{aligned}$$

Second, if Σ_{22} is positive definite, then the conditional distribution is also Normal:

$$Y_1 \mid Y_2 = y_2 \sim \text{N}\left(\mu_1 + \Sigma_{12}\Sigma_{22}^{-1}(y_2 - \mu_2), \Sigma_{11} - \Sigma_{12}\Sigma_{22}^{-1}\Sigma_{21}\right).$$

A.1.3 χ^2 and t distributions

Assume $X_1, \ldots, X_n$ are IID $\mathrm{N}(0,1)$. Then

$$\sum_{i-1}^{n} X_i^2 \sim \chi_n^2$$

follows a χ_n^2 distribution with degrees of freedom n. The χ_n^2 distribution has mean n and variance $2n$.

Assume $X \sim \mathrm{N}(0,1), Q_n \sim \chi_n^2$ and $X \perp\!\!\!\perp Q_n$. Then

$$\frac{X}{\sqrt{Q_n/n}} \sim t_n$$

follows a t_n distribution with degrees of freedom n. The t_n distribution has mean 0 if $n > 1$. When $n = 1$, the t_1 distribution is also called the Cauchy distribution, which does not have a finite mean.

With IID $X_1, \ldots, X_n \sim \mathrm{N}(\mu, \sigma^2)$, the t distribution arises naturally from

$$\frac{\bar{X} - \mu}{\sqrt{\hat{\sigma}^2/n}} \sim t_{n-1}, \qquad \text{(A.1)}$$

where the sample mean $\bar{X} = n^{-1}\sum_{i=1}^{n} X_i$ and the sample variance $\hat{\sigma}^2 = (n-1)^{-1}\sum_{i=1}^{n}(X_i - \bar{X})^2$ are independent.

A.1.4 Cauchy–Schwarz inequality

The Cauchy–Schwarz inequality has many forms. With two random variables A and B, we have

$$|E(AB)| \leq \sqrt{E(A^2)E(B^2)}$$

with equality holding when $B = \beta A$ for some β. Centering A and B to have mean 0, we have

$$|\mathrm{cov}(A, B)| \leq \sqrt{\mathrm{var}(A)\mathrm{var}(B)}$$

with equality holding when $B = \alpha + \beta A$ for some α and β.

When A and B are uniform random variables over finite sets $\{a_1, \ldots, a_n\}$ and $\{b_1, \ldots, b_n\}$, respectively, we have

$$\left|\sum_{i=1}^{n} a_i b_i\right| \leq \sqrt{\sum_{i=1}^{n} a_i^2 \sum_{i=1}^{n} b_i^2}$$

with equality holding if there exists β such that $b_i = \beta a_i$ for all i's.

A.1.5 Tower property and variance decomposition

I first focus on discrete random variables. For random variable A, we can define its expectation as $E(A) = \sum_a a \cdot \mathrm{pr}(A = a)$. For random variables A and B, we can define the conditional expectation of A given $B = b$ as $E(A \mid B = b) = \sum_a a \cdot \mathrm{pr}(A = a \mid B = b)$, which is a function of b, denoted by $g_1(b)$. Furthermore, we can define $E(A \mid B) = g_1(B)$ as a function of B, which is a random variable because B is a random variable. With these definitions, we can verify the *tower property* that

$$E(A) = E\{E(A \mid B)\}, \qquad \text{(A.2)}$$

which essentially says that

$$E(A) = \sum_b E(A \mid B = b)\text{pr}(B = b) = \sum_b \sum_a a \cdot \text{pr}(A = a \mid B = b)\text{pr}(B = b).$$

It follows from the law of total probability.

Moreover, for a random variable A, we can define its variance as $\text{var}(A) = \sum_a \{a - E(A)\}^2 \cdot \text{pr}(A = a)$. For random variables A and B, we can define the conditional variance of A given $B = b$ as $\text{var}(A \mid B = b) = \sum_a \{a - E(A \mid B = b)\}^2 \cdot \text{pr}(A = a \mid B = b)$, which is a function of b, denoted by $g_2(b)$. Furthermore, we can define $\text{var}(A \mid B) = g_2(B)$ as a function of B, which is a random variable because B is a random variable. With these definitions, we can verify that

$$\text{var}(A) = E\{\text{var}(A \mid B)\} + \text{var}\{E(A \mid B)\}, \tag{A.3}$$

which essentially says that

$$\text{var}(A) = \sum_b \text{var}(A \mid B = b) \cdot \text{pr}(B = b) + \sum_b \{E(A \mid B = b) - E(A)\}^2 \cdot \text{pr}(B = b).$$

It follows from decomposing the variance of A into the "within-group variances" measured by the weighted average of the $\text{var}(A \mid B = b)$'s and the "between-group variance" measured by the variance of $E(A \mid B = b)$'s across groups of B.

The formulas (A.2) and (A.3) are fundamental for many proofs in this book. When I was in graduate school, my professors Carl Morris and Joe Blitzstein called (A.3) the *Eve's Law* due to the "EVVE" form of the formula. They then went back to call (A.2) the *Adam's Law*.

The above discussion is clear for discrete random variables A and B. More interestingly, the same formulas (A.2) and (A.3) also hold for general, continuous random variables. Subtle issues do arise for general random variables because the event $B = b$ can be zero probability. In those cases, we can interpret the conditional expectation $E(A \mid B = b)$ and the conditional variance $\text{var}(A \mid B = b)$ as the limiting values of $E(A \mid b - \varepsilon \leq B \leq b + \varepsilon)$ and $\text{var}(A \mid b - \varepsilon \leq B \leq b + \varepsilon)$, respectively, as ε goes to 0. More rigorous definitions of conditional expectation and variance require the knowledge of measure theory. Nevertheless, we can simply treat all variables as discrete in the proofs of this book.

Formulas (A.2) and (A.3) have several important extensions. First, with random variables or vectors A, B, C, we have

$$E(A \mid C) = E\{E(A \mid B, C) \mid C\}$$

and

$$\text{var}(A \mid C) = E\{\text{var}(A \mid B, C) \mid C\} + \text{var}\{E(A \mid B, C) \mid C\}$$

which extend (A.2) and (A.3) with $E(\cdot)$ and $\text{var}(\cdot)$ replaced by $E(\cdot \mid C)$ and $\text{var}(\cdot \mid C)$, respectively. Second, we also have analogous formulas for the covariance:

$$\text{cov}(A_1, A_2) = E\{\text{cov}(A_1, A_2 \mid B)\} + \text{cov}\{E(A_1 \mid B), E(A_2 \mid B)\}$$

and

$$\text{cov}(A_1, A_2 \mid C) = E\{\text{cov}(A_1, A_2 \mid B, C) \mid C\} + \text{cov}\{E(A_1 \mid B, C), E(A_2 \mid B, C) \mid C\}.$$

A.1.6 Limiting theorems

Definition A.1 (convergence in probability) *A sequence of random variables $(X_n)_{n\geq 1}$ converges to X in probability, if for every $\varepsilon > 0$, we have*

$$\mathrm{pr}(|X_n - X| > \varepsilon) \to 0$$

as $n \to \infty$.

Definition A.2 (convergence in distribution) *A sequence of random variables $(X_n)_{n\geq 1}$ converges to X in distribution, if*

$$\mathrm{pr}(X_n \leq x) \to \mathrm{pr}(X \leq x)$$

for all continuity point x of $\mathrm{pr}(X \leq x)$, as $n \to \infty$.

Convergence in probability is stronger than convergence in distribution. Definitions A.1 and A.2 are useful for stating the following two fundamental theorems on the sample average of independent and identically distributed (IID) random variables.

Theorem A.1 (law of large numbers) *If $X_1, \ldots, X_n \overset{\mathrm{IID}}{\sim} X$ with $E|X| < \infty$, then $\bar{X} = n^{-1}\sum_{i=1}^n X_i \to E(X)$ in probability.*

The law of large numbers in Theorem A.1 states that the sample average is close to the population mean in the limit.

Theorem A.2 (central limit theorem (CLT)) *If $X_1, \ldots, X_n \overset{\mathrm{IID}}{\sim} X$ with $\mathrm{var}(X) < \infty$, then*

$$\frac{\bar{X} - E(X)}{\sqrt{\mathrm{var}(X)/n}} \to \mathrm{N}(0, 1)$$

in distribution.

The CLT in Theorem A.2 states that the standardized sample average is close to a standard Normal random variable in the limit.

Theorems A.1 and A.2 assume IID random variables for convenience. There are also many laws of large numbers and CLTs for the sample mean of independent or weakly dependent random variables (e.g., Durrett, 2019).

A.1.7 Delta method

The delta method is a power tool to derive the asymptotic Normality of nonlinear functions of an asymptotically Normal random vector. I review a special case of the delta method below.

Theorem A.3 (delta method) *Assume $\sqrt{n}(X_n - \mu) \to \mathrm{N}(0, \Sigma)$ in distribution and the function $g(x)$ has non-zero derivative $\nabla g(\mu)$ at μ. Then*

$$\sqrt{n}\{g(X_n) - g(\mu)\} \to \mathrm{N}(0, (\nabla g(\mu))^{\mathsf{T}} \Sigma \nabla g(\mu))$$

in distribution.

I will omit the proof of Theorem A.3. It is intuitive based on the first-order Taylor expansion:

$$g(X_n) - g(\mu) \approx (\nabla g(\mu))^{\mathsf{T}}(X_n - \mu).$$

So $\sqrt{n}\{g(X_n) - g(\mu)\}$ is close to the linear transformation of $\mathrm{N}(0, \Sigma)$, which is $\mathrm{N}(0, (\nabla g(\mu))^{\mathsf{T}}\Sigma\nabla g(\mu))$.

As illustrations, we can use the delta method to obtain the asymptotic Normality of the ratio and product.

Example A.1 (asymptotic Normality of the ratio) *Assume*

$$\sqrt{n}\begin{pmatrix} Y_n - \mu_Y \\ X_n - \mu_X \end{pmatrix} \to \mathrm{N}\left(\begin{pmatrix} 0 \\ 0 \end{pmatrix}, \begin{pmatrix} \sigma_Y^2 & \sigma_{YX} \\ \sigma_{YX} & \sigma_X^2 \end{pmatrix}\right) \tag{A.4}$$

in distribution with $\mu_X \neq 0$. Apply Theorem A.3 to obtain that

$$\sqrt{n}\left(\frac{Y_n}{X_n} - \frac{\mu_Y}{\mu_X}\right) \to \mathrm{N}\left(0, \frac{\sigma_Y^2}{\mu_X^2} + \frac{\mu_Y^2\sigma_X^2}{\mu_X^4} - \frac{2\mu_Y\sigma_{YX}}{\mu_X^3}\right) \tag{A.5}$$

in distribution. In the special case that X_n and Y_n are asymptotically independent with $\sigma_{YX} = 0$, the asymptotic variance of Y_n/X_n simplifies to $\sigma_Y^2/\mu_X^2 + \mu_Y^2\sigma_X^2/\mu_X^4$. I leave the details to Problem A.2.

The asymptotic variance in Example A.1 is a little cumbersome. An easier way to memorize it is based on the following approximation:

$$\begin{aligned} \frac{Y_n}{X_n} - \frac{\mu_Y}{\mu_X} &= \frac{Y_n - \mu_Y/\mu_X \cdot X_n}{X_n} \\ &\approx \frac{Y_n - \mu_Y/\mu_X \cdot X_n}{\mu_X}, \end{aligned} \tag{A.6}$$

so the asymptotic variance of the ratio equals the asymptotic variance of

$$\frac{Y_n - \mu_Y/\mu_X \cdot X_n}{\mu_X},$$

which is a linear combination of Y_n and X_n. Slutsky's theorem can make the approximation in (A.6) rigorous but it is beyond this book.

Example A.2 (asymptotic Normality of the product) *Assume* (A.4). *Apply Theorem A.3 to obtain that*

$$\sqrt{n}\,(X_nY_n - \mu_X\mu_Y) \to \mathrm{N}\left(0, \mu_Y^2\sigma_X^2 + \mu_X^2\sigma_Y^2 + 2\mu_X\mu_Y\sigma_{XY}\right) \tag{A.7}$$

in distribution. In the special case that X_n and Y_n are asymptotically independent with $\sigma_{YX} = 0$, the asymptotic variance of X_nY_n simplifies to $\mu_Y^2\sigma_X^2 + \mu_X^2\sigma_Y^2$. I leave the details to Problem A.3.

A.2 Statistical inference

A.2.1 Point estimation

Assume that θ is the parameter of interest. Oftentimes, the problem also contains other parameters not of interest, denoted by η. Statisticians call η the *nuisance parameter*. Based on the data, we can compute an estimator $\hat{\theta}$. Throughout this book, we take the frequentists' perspective by assuming that θ is a fixed number and $\hat{\theta}$ is random due to the randomness of data. Two basic requirements for an estimator are below.

Definition A.3 (unbiasedness) *The estimator $\hat{\theta}$ is unbiased for θ if*

$$E(\hat{\theta}) = \theta$$

for all possible values of θ and η.

Definition A.4 (consistency) *The estimator $\hat{\theta}$ is consistent for θ if*

$$\hat{\theta} \to \theta$$

in probability as the sample size approaches infinity, for all possible values of θ and η.

Unbiasedness requires that the mean of the estimator is identical to the parameter of interest. Consistency requires that the estimator is close to the true parameter in the limit. Unbiasedness does not imply consistency, and consistency does not imply unbiasedness either. Unbiasedness can be restrictive because it is impossible even in some simple statistics problems. Consistency is often the basic requirement in most statistics problems.

A.2.2 Confidence interval

A point estimator $\hat{\theta}$ is a random variable that differs from the true parameter θ. Statisticians are often interested in finding an interval that covers the true parameter with a certain given probability. This interval is computed based on the data, and it is random.

Definition A.5 (confidence interval) *A data-dependent interval $[\hat{\theta}_{\text{L}}, \hat{\theta}_{\text{U}}]$ is a confidence interval for θ with coverage probability at least $1-\alpha$ if*

$$\text{pr}(\hat{\theta}_{\text{L}} \leq \theta \leq \hat{\theta}_{\text{U}}) \geq 1-\alpha$$

for all possible values of θ and η.

Definition A.6 (asymptotic confidence interval) *A data-dependent interval $[\hat{\theta}_{\text{L}}, \hat{\theta}_{\text{U}}]$ is an asymptotic confidence interval for θ with coverage probability at least $1-\alpha$ if*

$$\text{pr}(\hat{\theta}_{\text{L}} \leq \theta \leq \hat{\theta}_{\text{U}}) \to 1-\alpha', \qquad \text{as } n \to \infty$$

with $\alpha' \leq \alpha$, for all possible values of θ and η.

A standard choice is $\alpha = 0.05$. In Definitions A.5 and A.6, the coverage probabilities can be larger than the nominal level $1-\alpha$. That is, the definitions allow for over-coverage but do not allow for under-coverage. With over-coverage, we say that the confidence interval is conservative. Of course, we hope the confidence interval to be as narrow as possible. Otherwise, the definition of the confidence interval can be arbitrary.

A.2.3 Hypothesis testing

Many applied problems can be formulated as testing a hypothesis:

$$H_0 : \theta = 0.$$

The decision rule ϕ is a binary function of the data: $\phi = 1$ if we reject H_0; $\phi = 0$ if we fail to reject H_0. The type one error rate of the test is the probability of rejection if the null hypothesis holds. I review the definition below.

Definition A.7 (type one error rate) *When H_0 holds, define the type one error rate of the test ϕ as the maximum value of the probability*

$$\text{pr}(\phi = 1)$$

over all possible values of θ and η.

A standard choice is to make sure that the type one error rate is below $\alpha = 0.05$. The type two error rate of the test is the probability of no rejection if the null hypothesis does not hold. I review the definition below.

Definition A.8 (type two error rate) *When H_0 does not hold, define the type two error rate of the test ϕ as the maximum value of the probability*

$$\text{pr}(\phi = 0)$$

over all possible values of θ and η.

Given the control of the type one error rate under H_0, we hope the type two error rate is as low as possible when H_0 does not hold.

A.2.4 Wald-type confidence interval and test

Many statistics problems have the following structure. The parameter of interest is θ. We first find a consistent estimator $\hat{\theta}$ that converges in probability to θ, and show that it is asymptotically Normal with mean θ and variance v which may depend on θ as well as the nuisance parameter η. We then find a consistent estimator $\hat{v}$ for v, based on analytic formulas or the bootstrap reviewed in Section A.6 later. The square root of $\hat{v}$ is called the *standard error*. We finally construct the Wald-type confidence interval for θ as

$$\hat{\theta} \pm z_{1-\alpha/2}\sqrt{\hat{v}}$$

where $z_{1-\alpha/2}$ is the $1 - \alpha/2$ upper quantile of the standard Normal random variable. It covers θ with probability approximately $1 - \alpha$. When this interval excludes a particular c, for example, $c = 0$, we reject the null hypothesis $H_0(c) : \theta = c$, which is called the Wald test.

A.2.5 Duality between constructing confidence sets and testing null hypotheses

Consider the statistical inference problem for a scalar parameter θ. A fundamental result in statistics is that constructing confidence sets for θ is equivalent to testing null hypotheses about θ. This is often called the duality between constructing confidence sets and testing null hypotheses.

Section A.2.4 has reviewed the duality based on the Wald-type confidence interval and test. The duality also holds in general. Assume that $\hat{\Theta}$ is a $(1 - \alpha)$-level confidence set for θ:

$$\text{pr}(\theta \in \hat{\Theta}) \geq 1 - \alpha.$$

Then we can reject the null hypothesis $H_0(c) : \theta = c$ if c is not in the set $\hat{\Theta}$. This is a valid test because when θ indeed equals c, we have the correct type one error rate $\text{pr}(\theta \notin \hat{\Theta}) \leq \alpha$. Conversely, if we test a sequence of null hypotheses $H_0(c) : \theta = c$, we can obtain the corresponding p-values, $p(c)$, as a function of c. Then the values of c that we fail to reject at level α form a confidence set for θ:

$$\hat{\Theta} = \{c : p(c) \geq \alpha\} = \{c : \text{ fail to reject } H_0(c) \text{ at level } \alpha\}.$$

It is a valid confidence set because

$$\text{pr}(\theta \in \hat{\Theta}) = \text{pr}\{\text{fail to reject } H_0(\theta) \text{ at level } \alpha\} \geq 1 - \alpha.$$

Here I use "confidence set" instead of "confidence interval" because $\hat{\Theta}$ based on inverting tests may not be an interval. See the use of the duality in Section A.4.2 and Section 3.6.1.

A.3 Inference with two-by-two tables

A.3.1 Fisher's exact test

Fisher proposed an exact test for $H_0 : p_1 = p_0$ under the statistical model:

$$\begin{aligned} n_{11} &\sim \text{Binomial}(n_1, p_1), \\ n_{01} &\sim \text{Binomial}(n_0, p_0), \\ n_{11} &\perp\!\!\!\perp n_{01}. \end{aligned}$$

The table below summarizes the data.

	response 1	response 0	row sum
sample 1	n_{11}	n_{10}	n_1
sample 0	n_{01}	n_{00}	n_0
column sum	$n_{\cdot 1}$	$n_{\cdot 0}$	n

He argued that the sums $n_{11} + n_{01} = n_{\cdot 1}$ and $n_{10} + n_{00} = n_{\cdot 0}$ contain little information about the difference between p_1 and p_0, and conditional on them, n_{11} follows a Hypergeometric distribution that does not depend on the unknown parameter $p_1 = p_0$ under H_0:

$$\text{pr}(n_{11} = k \mid n_{\cdot 1}, n_{\cdot 0}) = \frac{\binom{n_{\cdot 1}}{k}\binom{n - n_{\cdot 1}}{n_1 - k}}{\binom{n}{n_1}}.$$

In R, the function `fisher.test` implements this test.

A.3.2 Estimation with two-by-two tables

Based on the model in Section A.3.1, we can estimate the parameters p_1 and p_0 by sample frequencies:

$$\begin{aligned} \hat{p}_1 &= \frac{n_{11}}{n_1}, \\ \hat{p}_0 &= \frac{n_{01}}{n_0}. \end{aligned}$$

Therefore, we can estimate the risk difference, log risk ratio, and log odds ratio

$$\begin{aligned} \text{RD} &= p_1 - p_0, \\ \log \text{RR} &= \log \frac{p_1}{p_0}, \\ \log \text{OR} &= \log \frac{p_1/(1-p_1)}{p_0/(1-p_0)} \end{aligned}$$

by the sample analogs

$$\begin{aligned} \hat{\text{RD}} &= \hat{p}_1 - \hat{p}_0, \\ \log \hat{\text{RR}} &= \log \frac{\hat{p}_1}{\hat{p}_0}, \\ \log \hat{\text{OR}} &= \log \frac{\hat{p}_1/(1-\hat{p}_1)}{\hat{p}_0/(1-\hat{p}_0)} = \log \frac{n_{11}n_{00}}{n_{10}n_{01}}. \end{aligned}$$

Based on the asymptotic approximation (see Problem A.5), the estimated variance for the above parameters are

$$\begin{aligned} &\frac{\hat{p}_1(1-\hat{p}_1)}{n_1} + \frac{\hat{p}_0(1-\hat{p}_0)}{n_0}, \\ &\frac{1-\hat{p}_1}{n_1\hat{p}_1} + \frac{1-\hat{p}_0}{n_0\hat{p}_0}, \\ &\frac{1}{n_1\hat{p}_1(1-\hat{p}_1)} + \frac{1}{n_0\hat{p}_0(1-\hat{p}_0)}, \end{aligned}$$

respectively. The log transformation above yields better Normal approximations because the risk ratio and odds ratio are always positive.

A.4 Two famous problems in statistics

A.4.1 Behrens–Fisher problem

Consider the two-sample problem with n_1 units under the treatment and n_0 units under the control, respectively. Assume the outcomes under the treatment $\{Y_i : Z_i = 1\}$ are IID from $\text{N}(\mu_1, \sigma_1^2)$ and the outcomes under the control $\{Y_i : Z_i = 0\}$ are IID from $\text{N}(\mu_0, \sigma_0^2)$, respectively. The goal is to test $H_0 : \mu_1 = \mu_0$.

Start with the easier case with $\sigma_1^2 = \sigma_0^2$. Coherent with Chapter 3, let

$$\begin{aligned} \hat{\bar{Y}}(1) &= n_1^{-1} \sum_{Z_i=1} Y_i, \\ \hat{\bar{Y}}(0) &= n_0^{-1} \sum_{Z_i=0} Y_i \end{aligned}$$

denote the sample means of the outcomes under the treatment and control, respectively. A standard result is that

$$t_{\text{equal}} = \frac{\hat{\bar{Y}}(1) - \hat{\bar{Y}}(0)}{\sqrt{\frac{n}{n_1 n_0 (n-2)} \left[\sum_{Z_i=1} \{Y_i - \hat{\bar{Y}}(1)\}^2 + \sum_{Z_i=0} \{Y_i - \hat{\bar{Y}}(0)\}^2 \right]}} \sim t_{n-2}.$$

Based on t_{equal}, we can construct a test for H_0.

Now consider the more difficult case with possibly different σ_1^2 and σ_0^2. The distribution of t_{equal} is no longer t_{n-2}. Estimating the variances separately, we can also define

$$t_{\text{unequal}} = \frac{\hat{\bar{Y}}(1) - \hat{\bar{Y}}(0)}{\sqrt{\frac{\hat{S}^2(1)}{n_1} + \frac{\hat{S}^2(0)}{n_0}}},$$

where

$$\begin{aligned}\hat{S}^2(1) &= (n_1 - 1)^{-1} \sum_{Z_i=1} \{Y_i - \hat{\bar{Y}}(1)\}^2, \\ \hat{S}^2(0) &= (n_0 - 1)^{-1} \sum_{Z_i=0} \{Y_i - \hat{\bar{Y}}(0)\}^2\end{aligned}$$

are the sample variances of the outcomes under the treatment and control, respectively. Unfortunately, the exact distribution of t_{unequal} depends on the known variances. Testing H_0 without assuming equal variances is the famous Behrens–Fisher problem. With large sample sizes n_1 and n_0, the CLT ensures that t_{unequal} is approximately $\text{N}(0,1)$. So we can construct an approximate test for H_0. By duality, a large-sample Wald-type confidence interval for $\mu_1 - \mu_0$ is

$$\hat{\bar{Y}}(1) - \hat{\bar{Y}}(0) \pm z_{1-\alpha/2}\sqrt{\frac{\hat{S}^2(1)}{n_1} + \frac{\hat{S}^2(0)}{n_0}}$$

where $z_{1-\alpha/2}$ is the $1 - \alpha/2$ upper quantile of the standard Normal random variable.

A.4.2 Fieller–Creasy problem

Consider the two-sample problem with n_1 units under the treatment and n_0 units under the control, respectively. Assume the outcomes under the treatment $\{Y_i : Z_i = 1\}$ are IID from $\text{N}(\mu_1, 1)$ and the outcomes under the control $\{Y_i : Z_i = 0\}$ are IID from $\text{N}(\mu_0, 1)$, respectively. The goal is to estimate $\gamma = \mu_1/\mu_0$. We can use $\hat{\gamma} = \hat{\bar{Y}}(1)/\hat{\bar{Y}}(0)$ to estimate γ. However, the point estimator has a complicated distribution, which does not yield a simple procedure to construct the confidence interval for γ.

Fieller's confidence interval can be formulated as inverting tests for a sequence of null hypotheses: $H_0(c) : \gamma = c$. Under $H_0(c)$, we have

$$\frac{\hat{\bar{Y}}(1) - c\hat{\bar{Y}}(0)}{\sqrt{1/n_1 + c^2/n_0}} \sim \text{N}(0,1)$$

which motivates the confidence interval

$$\left\{c : \left(\frac{\hat{\bar{Y}}(1) - c\hat{\bar{Y}}(0)}{\sqrt{1/n_1 + c^2/n_0}}\right)^2 \le z_{1-\alpha/2}^2\right\}$$

where $z_{1-\alpha/2}$ is the $1 - \alpha/2$ upper quantile of the standard Normal random variable.

Fieller's confidence interval has been an intriguing topic in the early history of mathematical statistics. With different realized values of the Y_i's, Fieller's confidence interval can be a finite bounded interval, can be the union of two open intervals, and can even be an empty set. The theory states that Fieller's confidence interval covers γ with probability

$1-\alpha$, whereas it is possible that the interval itself is an empty set. Different statisticians have different opinions about Fieller's confidence interval. Some of them feel that an empty confidence interval is illogical and even go further to challenge the foundation of frequentists' statistics. Some of them feel that the frequentists' theory allows for not covering the parameter with probability α so it is totally fine to have an empty confidence interval and moreover, an empty confidence interval may suggest something unusual in the data. Sections 21.4 and 23.6.3 discuss related problems in the setting with instrumental variables.

A.5 Monte Carlo method in statistics

The Monte Carlo method is a powerful tool in statistics. I will review its basic use in approximating expectations or averages, which is fundamental in understanding the idea of FRT introduced in Chapter 3.

If our goal is to calculate

$$\theta = E\{g(Y)\}$$

with Y being a random variable, we can simply draw IID samples $Y_1, \ldots, Y_n$ from the distribution of Y and obtain the moment estimator for θ:

$$\hat{\theta} = n^{-1} \sum_{i=1}^{n} g(Y_i).$$

The moment estimator $\hat{\theta}$ is unbiased for θ, and by the law of large numbers, it is consistent for θ.

As a special case, Y is a uniform distribution over $\{y_1, \ldots, y_N\}$ and $\theta = N^{-1} \sum_{i=1}^{N} g(y_i)$. We can draw n IID samples $\{Y_1, \ldots, Y_n\}$ from $\{y_1, \ldots, y_N\}$ to obtain the moment estimator $\hat{\theta}$ defined above. This is called *sampling with replacement*, which is different from *sampling without replacement* reviewed in Appendix C.

A.6 Bootstrap

It is often very tedious to derive the variance formulas for complex estimators. Efron (1979) proposed the bootstrap as a general tool for variance estimation. There are many versions of the bootstrap (Davison and Hinkley, 1997). In this book, we only need the most basic one: the nonparametric bootstrap, which will be simply called the bootstrap.

Consider the generic setting with

$$Y_1, \ldots, Y_n \stackrel{\text{IID}}{\sim} Y,$$

where Y_i can be a general random element denoting the observed data for unit i. An estimator $\hat{\theta}$ is a function of the observed data: $\hat{\theta} = T(Y_1, \ldots, Y_n)$. When T is a complex function, it may not be easy to obtain the variance or asymptotic variance of $\hat{\theta}$.

The uncertainty of $\hat{\theta}$ is driven by the IID sampling of $Y_1, \ldots, Y_n$ from the true distribution. Although the true distribution is unknown, it can be well approximated by its empirical version

$$\hat{F}_n(y) = n^{-1} \sum_{i=1}^{n} I(Y_i \leq y),$$

when the sample size n is large. If we believe this approximation, we can simulate $\hat{\theta}$ by sampling

$$(Y_1^*, \ldots, Y_n^*) \stackrel{\text{IID}}{\sim} \hat{F}_n(y).$$

Because $\hat{F}_n(y)$ is a discrete distribution with mass $1/n$ on each observed data point, the simulation of $\hat{\theta}$ reduces to the following procedure:

1. sample $(Y_1^*, \ldots, Y_n^*)$ from $\{Y_1, \ldots, Y_n\}$ with replacement;
2. compute $\hat{\theta}^* = T(Y_1^*, \ldots, Y_n^*)$;
3. repeat the above two steps B times to obtain the bootstrap replicates $\{\hat{\theta}_1^*, \ldots, \hat{\theta}_B^*\}$.

We can then approximate the (asymptotic) variance of $\hat{\theta}$ by the sample variance of the bootstrap replicates:

$$\hat{V}_{\text{boot}} = (B-1)^{-1} \sum_{b=1}^{B} (\hat{\theta}_b^* - \bar{\theta}^*)^2,$$

where $\bar{\theta}^* = B^{-1} \sum_{b=1}^{B} \hat{\theta}_b^*$. The bootstrap confidence interval based on the Normal approximation is then

$$\hat{\theta} \pm z_{1-\alpha/2} \sqrt{\hat{V}_{\text{boot}}},$$

where $z_{1-\alpha/2}$ is the $1-\alpha/2$ upper quantile of the standard Normal random variable.

I use the following simple example to illustrate the idea of the bootstrap. With $n = 100$ IID observations Y_i's from $\text{N}(1,1)$, the sample mean should be close to 1 with variance $1/100 = 0.01$. Over 500 simulations, the classic variance estimator and the bootstrap variance estimator with $B = 200$ both have average values close to 0.01.

```
> ## sample size
> n   = 100
> ## number of Monte Carlo simulations
> MC  = 500
> ## number of bootstrap replicates B
> n.boot  = 200
> simulation  = replicate(MC, {
+    Y = rnorm(n, 1, 1)
+    boot.mu.hat  = replicate(n.boot, {
+      index  = sample(1:n, n, replace = TRUE)
+      mean(Y[index])
+    })
+    c(mean(Y), var(Y)/n, var(boot.mu.hat))
+ })
> ## summarize the results
> apply(simulation, 1, mean)
[1] 0.997602961 0.010006303 0.009921895
```

A.7 Homework problems

A.1 Independent but not IID data

Assume that the Y_i's are independent with mean μ_i and variances σ_i^2 for $i = 1, \ldots, n$. The parameter of interest is $\mu = n^{-1} \sum_{i=1}^{n} \mu_i$. Show that $\hat{\mu} = n^{-1} \sum_{i=1}^{n} Y_i$ is unbiased for μ and find its variance. Show that the usual variance estimator for IID data

$$\hat{v} = \{n(n-1)\}^{-1} \sum_{i=1}^{n} (Y_i - \hat{\mu})^2$$

is a conservative estimator for the variance of $\hat{\mu}$ in the sense that

$$E(\hat{v}) - \text{var}(\hat{\mu}) = \{n(n-1)\}^{-1} \sum_{i=1}^{n} (\mu_i - \mu)^2 \geq 0.$$

Remark: Consider a simpler case with $\mu_i = \mu$ and $\sigma_i^2 = \sigma^2$ for all $i = 1, \ldots, n$. The sample mean is unbiased for μ with variance σ^2/n. Moreover, an unbiased estimator for the variance σ^2/n is $\hat{\sigma}^2/n = \hat{v}$, where $\hat{\sigma}^2 = (n-1)^{-1} \sum_{i=1}^{n} (Y_i - \hat{\mu})^2$. This problem states a more general result for the case with independent but not identically distributed observations. The result has an important implication for Theorem 7.1 for the matched-pairs experiment discussed in Chapter 7.

A.2 Asymptotic Normality of ratio

Prove (A.5).

A.3 Asymptotic Normality of product

Prove (A.7).

A.4 Product of two independent Normals

Assume $X \sim \text{N}(\mu_X, \sigma_X^2), Y \sim \text{N}(\mu_Y, \sigma_Y^2)$ and $X \perp\!\!\!\perp Y$. Show that

$$\text{var}(XY) = \sigma_X^2 \sigma_Y^2 + \mu_X^2 \sigma_Y^2 + \mu_Y^2 \sigma_X^2.$$

Remark: This problem gives a non-asymptotic form of Example A.2.

A.5 Variance estimators in two-by-two tables

Use the delta method to show the variance estimators in Section A.3.2.

A.6 Meta-analysis and Fisher weighting

Assume that we have p independent unbiased estimators $\hat{\theta}_1, \ldots, \hat{\theta}_p$ for a common parameter θ:

$$E(\hat{\theta}_j) = \theta, \quad (j = 1, \ldots, p).$$

The estimator $\hat{\theta}_j$ has variance v_j $(j = 1, \ldots, p)$.

Construct a new estimator

$$\hat{\theta} = \sum_{j=1}^{p} w_j \hat{\theta}_j$$

with nonrandom constants w_j's. Show that $\hat{\theta}$ is unbiased for estimating θ if $\sum_{j=1}^{p} w_j = 1$. Show that the optimal w_j's such that $\hat{\theta}$ has the smallest variance under the constraint $\sum_{j=1}^{p} w_j = 1$ are

$$w_j^* = \frac{1/v_j}{\sum_{j'=1}^{p} 1/v_{j'}}, \quad (j = 1, \ldots, p).$$

Show that the resulting estimator $\hat{\theta}^* = \sum_{j=1}^p w_j^* \hat{\theta}_j$ has variance

$$\text{var}(\hat{\theta}^*) = \frac{1}{\sum_{j=1}^p 1/v_j}.$$

Remark: The optimal weights are proportional to the inverses of the variances. This is called Fisher weighting which is commonly used in meta-analysis that summaries multiple independent estimates for the same parameter of interest. The result is useful for the discussion in Chapter 25.

B

Linear and Logistic Regressions

Appendix B here only reviews the basics of linear and logistic regressions, which are the minimal requirements for understanding the main text of the book. I have a companion book (Ding, 2024), *Linear Model and Extensions,* to cover more detailed properties of linear and logistic regressions.

B.1 Population ordinary least squares

Assume that $(x_i, y_i)_{i=1}^n \stackrel{\text{IID}}{\sim} (x, y)$, where x is a p-dimensional random scalar or vector and y is a random scalar. Below I will use (x, y) to denote a general observation, dropping the subscript i for simplicity. Define the population ordinary least squares (OLS) coefficient as

$$\beta = \arg\min_b E\left\{(y - x^\top b)^2\right\}.$$

The objective function is quadratic in b, so we can show that the minimizer is

$$\beta = \left\{E\left(xx^\top\right)\right\}^{-1} E\left(xy\right)$$

if the moments exist and $E(xx^\top)$ is invertible.

With β, we can define $x^\top\beta$ as the *linear projection* of y on x, and define

$$\varepsilon = y - x^\top\beta \tag{B.1}$$

as the *population residual.* By the definition of β, we can verify that

$$\begin{aligned} E(x\varepsilon) &= E\left\{x(y - x^\top\beta)\right\} \\ &= E(xy) - E(xx^\top)\beta \\ &= 0. \end{aligned}$$

Example B.1 (population OLS with the intercept) *If we include* 1 *as a component of x, then*

$$E(\varepsilon) = E(y - x^\top\beta) = 0$$

which further implies that $\text{cov}(x, \varepsilon) = 0$. *So with an intercept in β, the mean of the population residual must be zero, and it is uncorrelated with other covariates by construction.*

Example B.2 (univariate population OLS with the intercept) *An important special case is that for scalars x and y, we can define*

$$(\alpha, \beta) = \arg\min_{a,b} E\{(y - a - bx)^2\}$$

which have explicit formulas

$$\beta = \frac{\text{cov}(x, y)}{\text{var}(x)},$$
$$\alpha = E(y) - \beta E(x).$$

Example B.3 (univariate population OLS without an intercept) *Without intercept, we can define*

$$\gamma = \arg\min_c E\{(y - cx)^2\}$$

which equals

$$\gamma = \frac{E(xy)}{E(x^2)}.$$

When x has mean zero, γ equals the β in Example B.2.

We can also rewrite (B.1) as

$$y = x^{\mathsf{T}}\beta + \varepsilon, \tag{B.2}$$

which holds by the definition of the population OLS coefficient and residual without any modeling assumption. We call (B.2) the population OLS decomposition.

B.2 Sample ordinary least squares

Based on data $(x_i, y_i)_{i=1}^n \overset{\text{IID}}{\sim} (x, y)$, we can easily obtain the moment estimator for the population OLS coefficient

$$\hat{\beta} = \left(n^{-1}\sum_{i=1}^n x_i x_i^{\mathsf{T}}\right)^{-1}\left(n^{-1}\sum_{i=1}^n x_i y_i\right),$$

and the residuals $\hat{\varepsilon}_i = y_i - x_i^{\mathsf{T}}\hat{\beta}$. This is called the sample OLS or simply the OLS. The OLS coefficient $\hat{\beta}$ minimizes the residual sum of squares

$$\hat{\beta} = \arg\min_b n^{-1}\sum_{i=1}^n (y_i - x_i^{\mathsf{T}} b)^2,$$

so it must satisfy

$$\sum_{i=1}^n x_i(y_i - x_i^{\mathsf{T}}\hat{\beta}) = 0,$$

which is sometimes called the *Normal equation*. The fitted values, also called the *linear projections* of y_i on x_i, equal

$$\hat{y}_i = x_i^{\mathsf{T}}\hat{\beta} \quad (i = 1, \ldots, n).$$

Using the matrix notation

$$X = \begin{pmatrix} x_1^{\mathsf{T}} \\ \vdots \\ x_n^{\mathsf{T}} \end{pmatrix}, \quad Y = \begin{pmatrix} y_1 \\ \vdots \\ y_n \end{pmatrix},$$

we can write the OLS coefficient as

$$\hat{\beta} = (X^{\mathsf{T}}X)^{-1}X^{\mathsf{T}}Y$$

and the fitted vector as

$$\hat{Y} = X\hat{\beta} = X(X^{\mathsf{T}}X)^{-1}X^{\mathsf{T}}Y.$$

Define the hat matrix as

$$H = X(X^{\mathsf{T}}X)^{-1}X^{\mathsf{T}}.$$

Then we also have $\hat{Y} = HY$, justifying the name "hat matrix." The diagonal elements of H, h_{ii}'s, are often called the *leverage scores.*

Assuming finite fourth moments of (x, y), we can use the law of large numbers and the CLT to show that

$$\sqrt{n}(\hat{\beta} - \beta) \to \mathrm{N}(0, V)$$

in distribution with $V = B^{-1}MB^{-1}$, where $B = E(xx^{\mathsf{T}})$ and $M = E(\varepsilon^2 xx^{\mathsf{T}})$. So a moment estimator for the asymptotic variance of $\hat{\beta}$ is

$$\hat{V}_{\mathrm{EHW}} = n^{-1}\left(n^{-1}\sum_{i=1}^{n} x_i x_i^{\mathsf{T}}\right)^{-1}\left(n^{-1}\sum_{i=1}^{n} \hat{\varepsilon}_i^2 x_i x_i^{\mathsf{T}}\right)\left(n^{-1}\sum_{i=1}^{n} x_i x_i^{\mathsf{T}}\right)^{-1}, \tag{B.3}$$

which is called the Eicker–Huber–White (EHW) robust covariance estimator (Eicker, 1967; Huber, 1967; White, 1980). We can show that $n\hat{V}_{\mathrm{EHW}} \to V$ in probability. Based on $\hat{\beta}$ and $\hat{V}_{\mathrm{EHW}}$, we can make inference about the population OLS coefficient β.

There are many variants of the EHW robust covariance estimator based on the leverage scores (Long and Ervin, 2000). In particular, the HC1 variant modifies $\hat{\varepsilon}_i^2$ to $\hat{\varepsilon}_i^2/(n-p)$, the HC2 variant modifies $\hat{\varepsilon}_i^2$ to $\hat{\varepsilon}_i^2/(1-h_{ii})$, and the HC3 variant modifies $\hat{\varepsilon}_i^2$ to $\hat{\varepsilon}_i^2/(1-h_{ii})^2$, in the definition of $\hat{V}_{\mathrm{EHW}}$.

B.3 Frisch–Waugh–Lovell Theorem

The Frisch–Waugh–Lovell (FWL) theorem has two versions: one at the population level and the other at the sample level. It reduces multivariate OLS to univariate OLS and therefore facilitates the understanding and calculation of the OLS coefficients. Below I will present special cases of the FWL theorem which are enough for this book.

Theorem B.1 (population FWL) *The coefficient of x_1 in the OLS fit of y on $(x_1, x_2, \ldots, x_p)$ equals the coefficient of $\tilde{x}_1$ in the OLS fit of y or $\tilde{y}$ on $\tilde{x}_1$, where $\tilde{y}$ is the residual from the OLS fit of y on $(x_2, \ldots, x_p)$ and $\tilde{x}_1$ is the residual from the OLS fit of x_1 on $(x_2, \ldots, x_p)$.*

In Theorem B.1, residualizing x_1 is crucial but residualizing y is not.

Theorem B.2 (sample FWL) *With data $(Y, X_1, X_2, \ldots, X_p)$ containing column vectors, the coefficient of X_1 in the OLS fit of Y on $(X_1, X_2, \ldots, X_p)$ equals the coefficient of $\tilde{X}_1$ in the OLS fit of Y or $\tilde{Y}$ on $\tilde{X}_1$, where $\tilde{Y}$ is the residual vector from the OLS fit of Y on $(X_2, \ldots, X_p)$ and $\tilde{X}_1$ is the residual vector from the OLS fit of X_1 on $(X_2, \ldots, X_p)$.*

Again, in Theorem B.2, residualizing X_1 is crucial but residualizing Y is not. Ding (2021) gives more numerical properties related to the sample FWL theorem.

B.4 Linear model

Sometimes, we impose a stronger model assumption which requires the conditional mean of y given x is linear:

$$E(y \mid x) = x^{\mathsf{T}} \beta$$

or, equivalently,

$$y = x^{\mathsf{T}}\beta + \varepsilon \qquad \text{with} \qquad E(\varepsilon \mid x) = 0,$$

which is called the restricted mean model. Under this model, the population OLS coefficient is the true parameter of interest:

$$\begin{aligned} \left\{E(xx^{\mathsf{T}})\right\}^{-1} E(xy) &= \left\{E(xx^{\mathsf{T}})\right\}^{-1} E\left\{xE(y \mid x)\right\} \\ &= \left\{E(xx^{\mathsf{T}})\right\}^{-1} E(xx^{\mathsf{T}}\beta) \\ &= \beta. \end{aligned}$$

Moreover, the population OLS coefficient does not depend on the distribution of x. The asymptotic inference in Section B.1 applies to this model too.

In the special case with $\text{var}(\varepsilon \mid x) = \sigma^2$, the asymptotic variance of the OLS coefficient reduces to

$$V = \sigma^2\{E(xx^{\mathsf{T}})\}^{-1}$$

so a simpler moment estimator for the asymptotic variance of $\hat{\beta}$ is

$$\hat{V}_{\text{OLS}} = \hat{\sigma}^2 \left(\sum_{i=1}^{n} x_i x_i^{\mathsf{T}} \right)^{-1} \tag{B.4}$$

where $\hat{\sigma}^2 = (n-p)^{-1}\sum_{i=1}^{n} \hat{\varepsilon}_i^2$ is an unbiased estimator for σ^2, recalling that n denotes the sample size and p denotes the dimension of x. This is the standard covariance estimator from the `lm` function.

Based on the `BostonHousing` data, we first obtain the standard output from the `lm` function.

```
> library("mlbench")
> data(BostonHousing)
> ols.fit = lm(medv ~ ., data = BostonHousing)
> summary(ols.fit)

Call:
lm(formula = medv ~ ., data = BostonHousing)

Residuals:
     Min       1Q   Median       3Q      Max
-15.595   -2.730   -0.518    1.777   26.199

Coefficients:
               Estimate Std. Error t value Pr(>|t|)
(Intercept)  3.646e+01  5.103e+00   7.144 3.28e-12 ***
crim        -1.080e-01  3.286e-02  -3.287 0.001087 **
zn           4.642e-02  1.373e-02   3.382 0.000778 ***
indus        2.056e-02  6.150e-02   0.334 0.738288
chas1        2.687e+00  8.616e-01   3.118 0.001925 **
```

```
nox          -1.777e+01  3.820e+00   -4.651 4.25e-06 ***
rm            3.810e+00  4.179e-01    9.116  < 2e-16 ***
age           6.922e-04  1.321e-02    0.052 0.958229
dis          -1.476e+00  1.995e-01   -7.398 6.01e-13 ***
rad           3.060e-01  6.635e-02    4.613 5.07e-06 ***
tax          -1.233e-02  3.760e-03   -3.280 0.001112 **
ptratio      -9.527e-01  1.308e-01   -7.283 1.31e-12 ***
b             9.312e-03  2.686e-03    3.467 0.000573 ***
lstat        -5.248e-01  5.072e-02  -10.347  < 2e-16 ***
```

In R, the lm function can compute $\hat{\beta}$, and the hccm function in the package car can compute $\hat{V}_{\text{EHW}}$ as well as its variants. Below we compare the t-statistics based on different choices of the standard errors. In this example, the EHW standard errors differ a lot for some regression coefficients.

```
> library("car")
> ols.fit.hc0 = sqrt(diag(hccm(ols.fit, type = "hc0")))
> ols.fit.hc1 = sqrt(diag(hccm(ols.fit, type = "hc1")))
> ols.fit.hc2 = sqrt(diag(hccm(ols.fit, type = "hc2")))
> ols.fit.hc3 = sqrt(diag(hccm(ols.fit, type = "hc3")))
> tvalues = summary(ols.fit)$coef[,1]/
+    cbind(summary(ols.fit)$coef[,2],
+          ols.fit.hc0,
+          ols.fit.hc1,
+          ols.fit.hc2,
+          ols.fit.hc3)
> colnames(tvalues) = c("ols", "hc0", "hc1", "hc2", "hc3")
> round(tvalues, 2)
               ols   hc0   hc1   hc2   hc3
(Intercept)   7.14  4.62  4.56  4.48  4.33
crim         -3.29 -3.78 -3.73 -3.48 -3.17
zn            3.38  3.42  3.37  3.35  3.27
indus         0.33  0.41  0.41  0.41  0.40
chas1         3.12  2.11  2.08  2.05  2.00
nox          -4.65 -4.76 -4.69 -4.64 -4.53
rm            9.12  4.57  4.51  4.43  4.28
age           0.05  0.04  0.04  0.04  0.04
dis          -7.40 -6.97 -6.87 -6.81 -6.66
rad           4.61  5.05  4.98  4.91  4.76
tax          -3.28 -4.65 -4.58 -4.54 -4.43
ptratio      -7.28 -8.23 -8.11 -8.06 -7.89
b             3.47  3.53  3.48  3.44  3.34
lstat       -10.35 -5.34 -5.27 -5.18 -5.01
```

B.5 Weighted least squares

Assume that $(w_i, x_i, y_i) \overset{\text{IID}}{\sim} (w, x, y)$ with $w \neq 0$. At the population level, we can define the weighted least squares (WLS) coefficient as

$$\beta_w = \arg\min_b E\{w(y - x^{\top}b)^2\},$$

which satisfies

$$E\{wx(y - x^{\top}\beta_w)\} = 0$$

and thus equals

$$\beta_w = \{E(wxx^{\top})\}^{-1}E(wxy)$$

if $E(wxx^{\top})$ is invertible.

At the sample level, we can define the WLS coefficient as

$$\hat{\beta}_w = \arg\min_b \sum_{i=1}^n w_i(y_i - x_i^{\top}b)^2,$$

which satisfies

$$\sum_{i=1}^n w_i x_i (y_i - x_i^{\top}\hat{\beta}_w) = 0$$

and thus equals

$$\hat{\beta}_w = \left(n^{-1}\sum_{i=1}^n w_i x_i x_i^{\top}\right)^{-1}\left(n^{-1}\sum_{i=1}^n w_i x_i y_i\right)$$

if $\sum_{i=1}^n w_i x_i x_i^{\top}$ is invertible.

In `R`, we can specify `weights` in the `lm` function to implement WLS.

B.6 Logistic regression

B.6.1 Model

Technically, we can apply the OLS procedure even if the outcome y is binary. However, it is a little awkward to have predicted probabilities outside the range of $[0, 1]$. This motivates us to consider the following model:

$$\text{pr}(y_i = 1 \mid x_i) = g(x_i^{\top}\beta),$$

where $g(\cdot) : \mathbb{R} \to [0, 1]$ is a monotone function, and its inverse is often called the *link function*. The $g(\cdot)$ function can be any distribution function of a random variable, but we will focus on the logistic form:

$$g(z) = \frac{e^z}{1 + e^z} = (1 + e^{-z})^{-1}.$$

We can also write the logistic model as

$$\text{pr}(y_i = 1 \mid x_i) = \frac{e^{x_i^{\top}\beta}}{1 + e^{x_i^{\top}\beta}},$$

or, equivalently, with the definition $\text{logit}(z) = \log \frac{z}{1-z}$, we have

$$\text{logit}\{\text{pr}(y_i = 1 \mid x_i)\} = x_i^{\top}\beta.$$

Assume that x_{i1} is binary. Under the logistic model, we have

$$\begin{aligned}\beta_1 &= \text{logit}\{\text{pr}(y_i = 1 \mid x_{i1} = 1, \ldots)\} - \text{logit}\{\text{pr}(y_i = 1 \mid x_{i1} = 0, \ldots)\} \\ &= \log \frac{\text{pr}(y_i = 1 \mid x_{i1} = 1, \ldots)/\text{pr}(y_i = 0 \mid x_{i1} = 1, \ldots)}{\text{pr}(y_i = 1 \mid x_{i1} = 0, \ldots)/\text{pr}(y_i = 0 \mid x_{i1} = 0, \ldots)},\end{aligned}$$

where $\ldots$ contains all other regressor $x_{i2}, \ldots, x_{ip}$. Therefore, the coefficient β_1 equals the log of the odds ratio of x_{i1} on y_i conditional on other regressors.

B.6.2 Maximum likelihood estimate

Let $\text{pr}(y_i = 1 \mid x_i) = \pi(x_i, \beta)$. To estimate the parameter β, we can maximize the following likelihood function:

$$\begin{aligned} L(b) &= \prod_{i=1}^{n} \{\pi(x_i, b)\}^{y_i} \{1 - \pi(x_i, b)\}^{1-y_i} \\ &= \prod_{i=1}^{n} \left\{ \frac{\pi(x_i, b)}{1 - \pi(x_i, b)} \right\}^{y_i} \{1 - \pi(x_i, b)\} \\ &= \prod_{i=1}^{n} \left(e^{x_i^\top b} \right)^{y_i} \frac{1}{1 + e^{x_i^\top b}} \\ &= \prod_{i=1}^{n} \frac{e^{y_i x_i^\top b}}{1 + e^{x_i^\top b}}. \end{aligned}$$

Let $\hat{\beta}$ denote the maximizer, which is called the maximum likelihood estimate (MLE). Taking the log of $L(b)$ and differentiating it with respect to b, we can show that the MLE must satisfy the first-order condition:

$$\sum_{i=1}^{n} x_i \{y_i - \pi(x_i, \hat{\beta})\} = 0.$$

If x_i contains the intercept, the MLE must satisfy

$$\sum_{i=1}^{n} \{y_i - \pi(x_i, \hat{\beta})\} = 0,$$

that is, the average of the observed y_i's must be identical to the average of the fitted probabilities $\pi(x_i, \hat{\beta})$'s.

Using the general theory for the MLE, we can show that it is consistent for the true parameter β and is asymptotically Normal:

$$\sqrt{n}(\hat{\beta} - \beta) \to \text{N}(0, V)$$

in distribution, where $V = E[\pi(x_i, \beta)\{1 - \pi(x_i, \beta)\}xx^\top]$. So we can approximate the covariance matrix of $\hat{\beta}$ by

$$n^{-1} \sum_{i=1}^{n} \pi(x_i, \hat{\beta})\{1 - \pi(x_i, \hat{\beta})\} x_i x_i^\top.$$

In `R`, the `glm` function can find the MLE and report the estimated covariance matrix. We use the `lalonde` data to illustrate the logistic regression with the binary outcome indicating positive real earnings in 1978.

```
> library(Matching)
> data(lalonde)
> logit.re78 = glm(I(re78>0) ~ ., family = binomial,
+                  data = lalonde)
> summary(logit.re78)

Call:
glm(formula = I(re78 > 0) ~ ., family = binomial, data = lalonde)
```

```
Deviance Residuals:
    Min        1Q    Median        3Q       Max
-2.1789   -1.3170    0.7568    0.9413    1.0882

Coefficients:
              Estimate Std. Error z value Pr(>|z|)
(Intercept)  1.910e+00  1.241e+00   1.539   0.1238
age         -2.812e-03  1.533e-02  -0.183   0.8545
educ        -2.179e-02  7.831e-02  -0.278   0.7808
black       -1.060e+00  5.041e-01  -2.103   0.0354 *
hisp         2.741e-01  6.967e-01   0.393   0.6940
married      7.577e-02  3.057e-01   0.248   0.8042
nodegr      -1.984e-01  3.460e-01  -0.573   0.5664
re74         7.857e-06  3.173e-05   0.248   0.8044
re75         4.016e-05  6.058e-05   0.663   0.5074
u74         -6.177e-02  4.095e-01  -0.151   0.8801
u75          1.505e-02  3.518e-01   0.043   0.9659
treat        5.412e-01  2.222e-01   2.435   0.0149 *
```

B.6.3 Extension to the case-control study

In case-control studies, sampling is conditional on the binary outcome, that is, units with outcomes $y_i = 1$ and $y_i = 0$ are sampled with different probabilities. Let s_i be the sampling indicator. In case-control studies, we have

$$\text{pr}(s_i = 1 \mid x_i, y_i) = \text{pr}(s_i = 1 \mid y_i)$$

as a function of y_i, and we only observe units with $s_i = 1$. Case-control studies are very common in epidemiology. Most disease outcomes of interest in epidemiology have small probabilities so it is more reasonable to over sample the units with diseases.

Based on the model and the sampling mechanism of the case-control study, it does not seem obvious whether or not we can still use logistic regression to estimate the coefficients. Nevertheless, Prentice and Pyke (1979) proved a positive result. In case-control studies, logistic regression can still consistently estimate all the coefficients except for the intercept. See Problem B.7 for a more formal justification of using the logistic regression in case-control studies.

B.6.4 Logistic regression with weights

Sometimes, unit i has weight w_i. Then we can fit a weighted logistic regression by solving

$$\sum_{i=1}^{n} w_i x_i \{y_i - \pi(x_i, \hat{\beta})\} = 0.$$

In R, we can specify **weights** in the **glm** function to implement weighted logistic regression.

B.7 Homework problems

B.1 Sample WLS with the intercept

Assume the regressor x_i contains the intercept. Show that

$$\bar{y}_w = \bar{x}_w^{\top} \hat{\beta}_w \tag{B.5}$$

where $\bar{x}_w = \sum_{i=1}^n w_i x_i / \sum_{i=1}^n w_i$ and $\bar{y}_w = \sum_{i=1}^n w_i y_i / \sum_{i=1}^n w_i$ are the weighted averages of the x_i's and y_i's.

B.2 Population OLS with a binary regressor

Assume x is binary. Define the population OLS:

$$(\alpha, \beta) = \arg\min_{(a,b)} E\{(y - a - bx)^2\}.$$

Show that

$$\beta = E(y \mid x = 1) - E(y \mid x = 0),$$

and

$$\alpha = E(y \mid x = 0).$$

B.3 Univariate WLS

As a special case of WLS, define

$$(\hat{\alpha}_w, \hat{\beta}_w) = \arg\min_{(a,b)} \sum_{i=1}^n w_i (y_i - a - bx_i)^2$$

where $w_i \geq 0$. Show that

$$\hat{\beta}_w = \frac{\sum_{i=1}^n w_i (x_i - \bar{x}_w)(y_i - \bar{y}_w)}{\sum_{i=1}^n w_i (x_i - \bar{x}_w)^2} \tag{B.6}$$

and

$$\hat{\alpha}_w = \bar{y}_w - \hat{\beta}_w \bar{x}_w, \tag{B.7}$$

where $\bar{x}_w = \sum_{i=1}^n w_i x_i / \sum_{i=1}^n w_i$ and $\bar{y}_w = \sum_{i=1}^n w_i y_i / \sum_{i=1}^n w_i$ are the weighted averages of the x_i's and y_i's.

Further assume that the x_i's are binary. Show that

$$\hat{\beta}_w = \frac{\sum_{i=1}^n w_i x_i y_i}{\sum_{i=1}^n w_i x_i} - \frac{\sum_{i=1}^n w_i (1 - x_i) y_i}{\sum_{i=1}^n w_i (1 - x_i)}. \tag{B.8}$$

That is, if the regressor is binary in the univariate WLS, the coefficient of the regressor equals the difference in the weighted means.

Remark: To prove (B.8), use an appropriate reparametrization of the WLS problem. Otherwise, the derivation can be tedious.

B.4 OLS with orthogonal regressors

Consider sample OLS fit of an n-vector Y on an $n \times p$ matrix X, with coefficient $\hat\beta$. Partition X into $X = (X_1, X_2)$, where X_1 is an $n \times k$ matrix and X_2 is an $n \times l$ matrix, with $p = k + l$. Correspondingly, partition $\hat\beta$ into

$$\hat\beta = \begin{pmatrix} \hat\beta_1 \\ \hat\beta_2 \end{pmatrix}.$$

Assume X_1 and X_2 are orthogonal, that is, $X_1^\top X_2 = 0$. Show that $\hat\beta_1$ equals the coefficient from OLS of Y on X_1 and $\hat\beta_2$ equals the coefficient from OLS of Y on X_2, respectively.

B.5 OLS with a non-degenerate transformation of the regressors

Define $\hat\beta$ as the coefficient from the sample OLS fit of an n-vector Y on an $n \times p$ matrix X. Let Γ be a $p \times p$ non-degenerate matrix, and define $X' = X\Gamma$. Define $\hat\beta'$ as the coefficient from the sample OLS fit of Y on X'.

Show that

$$\hat\beta = \Gamma \hat\beta'.$$

B.6 Variances of the OLS estimator

Assume $(x_i, y_i)_{i=1}^n \overset{\text{IID}}{\sim} (x, y)$ and

$$\begin{aligned} E(y \mid x) &= x^\top \beta, \\ \text{var}(y \mid x) &= \sigma^2(x). \end{aligned}$$

Show that the OLS estimator $\hat\beta$ has conditional mean

$$E(\hat\beta \mid x_1, \ldots, x_n) = \beta$$

and conditional variance

$$\text{var}(\hat\beta \mid x_1, \ldots, x_n) = \left(\sum_{i=1}^n x_i x_i^\top\right)^{-1} \left(\sum_{i=1}^n \sigma^2(x_i) x_i x_i^\top\right) \left(\sum_{i=1}^n x_i x_i^\top\right)^{-1}.$$

Remark: This problem provides some intuition for the variance estimators (B.3) and (B.4).

B.7 Logistic regression in case-control studies

Prove Theorem B.3 below.

Theorem B.3 *Assume IID (x_i, y_i, s_i) with*

$$\text{pr}(y_i = 1 \mid x_i) = \frac{e^{\beta_0 + x_i^\top \beta}}{1 + e^{\beta_0 + x_i^\top \beta}}$$

and

$$\begin{aligned} \text{pr}(s_i = 1 \mid x_i, y_i) &= \text{pr}(s_i = 1 \mid y_i) \\ &= \begin{cases} p_1, & \text{if } y_i = 1, \\ p_0, & \text{if } y_i = 0. \end{cases} \end{aligned}$$

Then

$$\text{pr}(y_i = 1 \mid x_i, s_i = 1) = \frac{e^{\log(p_1/p_0) + \beta_0 + x_i^\top \beta}}{1 + e^{\log(p_1/p_0) + \beta_0 + x_i^\top \beta}}.$$

Remark: In Theorem B.3, p_1 and p_0 are often unknown. Theorem B.3 states that if $y_i \mid x_i$ follows a logistic model, then $y_i \mid x_i, s_i = 1$ also follows a logistic model with the same slopes but a different intercept. Therefore, even if the sampling depends on the outcomes as in case-control studies, logistic regression can still estimate the slopes consistently. This justifies the use of the logistic regression in case-control studies.

C

Some Useful Lemmas for Simple Random Sampling From a Finite Population

C.1 Lemmas

Simple random sampling is a basic topic in standard survey sampling textbooks (e.g., Cochran, 1953). Below I review some results for simple random sampling that are useful for design-based inference in the CRE in Chapters 3 and 4. I define simple random sampling based on the distribution of the inclusion indicators.

Definition C.1 (simple random sampling) *A simple random sample of size n_1 consists of a subset from a finite population of n units indexed by $i = 1, \ldots, n$. Let $\boldsymbol{Z} = (Z_1, \ldots, Z_n)$ be the inclusion indicators of the n units with $Z_i = 1$ if unit i is sampled and $Z_i = 0$ otherwise. The vector $\boldsymbol{Z}$ can take $\binom{n}{n_1}$ possible permutations of a vector of n_1 1's and n_0 0's, and each has equal probability.*

By Definition C.1, simple random sampling is a special form of *sampling without replacement* because it does not allow for repeatedly sampling the same unit. Lemma C.1 below summarizes the first two moments of the inclusion indicators.

Lemma C.1 *Under simple random sampling, we have*

$$\begin{aligned} E(Z_i) &= \frac{n_1}{n}, \\ \mathrm{var}(Z_i) &= \frac{n_1 n_0}{n^2}, \\ \mathrm{cov}(Z_i, Z_j) &= -\frac{n_1 n_0}{n^2(n-1)}. \end{aligned}$$

In more compact forms, we have

$$\begin{aligned} E(\boldsymbol{Z}) &= \frac{n_1}{n}\mathbf{1}_n, \\ \mathrm{cov}(\boldsymbol{Z}) &= \frac{n_1 n_0}{n(n-1)}\boldsymbol{P}_n, \end{aligned}$$

where $\mathbf{1}_n$ is an n-dimensional vector of 1's, and $\boldsymbol{P}_n = \boldsymbol{I}_n - n^{-1}\mathbf{1}_n\mathbf{1}_n^{\mathsf{T}}$ is the $n \times n$ projection matrix orthogonal to $\mathbf{1}_n$.

Let $\{c_1, \ldots, c_n\}$ be a finite population with mean $\bar{c} = \sum_{i=1}^n c_i/n$ and variance

$$S_c^2 = (n-1)^{-1}\sum_{i=1}^n (c_i - \bar{c})^2;$$

let $\{d_1, \ldots, d_n\}$ be another finite population with mean $\bar{d} = \sum_{i=1}^n d_i/n$ and variance

$$S_d^2 = (n-1)^{-1} \sum_{i=1}^n (d_i - \bar{d})^2;$$

their covariance is

$$S_{cd} = (n-1)^{-1} \sum_{i=1}^n (c_i - \bar{c})(d_i - \bar{d}).$$

Under simple random sampling, the sample means are

$$\begin{aligned} \hat{\bar{c}} &= n_1^{-1} \sum_{i=1}^n Z_i c_i, \\ \hat{\bar{d}} &= n_1^{-1} \sum_{i=1}^n Z_i d_i; \end{aligned}$$

sample variances are

$$\begin{aligned} \hat{S}_c^2 &= (n_1 - 1)^{-1} \sum_{i=1}^n Z_i (c_i - \hat{\bar{c}})^2, \\ \hat{S}_d^2 &= (n_1 - 1)^{-1} \sum_{i=1}^n Z_i (d_i - \hat{\bar{d}})^2; \end{aligned}$$

the sample covariance is

$$\hat{S}_{cd} = (n_1 - 1)^{-1} \sum_{i=1}^n Z_i (c_i - \hat{\bar{c}})(d_i - \hat{\bar{d}}).$$

Lemma C.2 below gives the moments of the sample means $\hat{\bar{c}}$ and $\hat{\bar{d}}$.

Lemma C.2 *Under simple random sampling, the sample means are unbiased for the population means:*

$$\begin{aligned} E(\hat{\bar{c}}) &= \bar{c}, \\ E(\hat{\bar{d}}) &= \bar{d}. \end{aligned}$$

Their variances and covariance are

$$\begin{aligned} \mathrm{var}\left(\hat{\bar{c}}\right) &= \frac{n_0}{n n_1} S_c^2, \\ \mathrm{var}\left(\hat{\bar{d}}\right) &= \frac{n_0}{n n_1} S_d^2, \\ \mathrm{cov}\left(\hat{\bar{c}}, \hat{\bar{d}}\right) &= \frac{n_0}{n n_1} S_{cd}. \end{aligned}$$

In the variance formulas in Lemma C.2, the coefficient

$$\frac{n_0}{n n_1} = \frac{1}{n_1} \times \left(1 - \frac{n_1}{n}\right)$$

is different from $1/n_1$ under IID sampling. The additional term $1 - n_1/n = n_0/n$ is called the *finite population correction* factor. Because the finite population correction factor is smaller than or equal to 1, the variances under simple random sampling are smaller than the corresponding variances under IID sampling.[1]

Lemma C.3 below gives the unbiasedness of the sample variances and covariance for estimating the population analogs.

Lemma C.3 *Under simple random sampling, the sample variances and covariance are unbiased for their population versions:*

$$\begin{aligned} E(\hat{S}_c^2) &= S_c^2, \\ E(\hat{S}_d^2) &= S_d^2, \\ E(\hat{S}_{cd}) &= S_{cd}. \end{aligned}$$

An important practical question is to make inference about $\bar{c}$ based on the simple random sample. This requires a more precise characterization of the distribution of its unbiased estimator $\hat{\bar{c}}$. The finite-sample exact distribution of $\hat{\bar{c}}$ depends on the whole finite population $\{c_1, \ldots, c_n\}$, which is unknown. The following finite population CLT characterizes the asymptotic distribution of $\hat{\bar{c}}$ based on its first two moments.

Lemma C.4 (finite population CLT) *Under simple random sampling, as $n \to \infty$, if*

$$\frac{\max_{1\leq i\leq n}(c_i - \bar{c})^2}{\min(n_1, n_0)S_c^2} \to 0,$$

then

$$\frac{\hat{\bar{c}} - \bar{c}}{\sqrt{\frac{n_0}{nn_1} S_c^2}} \to \mathrm{N}(0,1)$$

in distribution, and $\hat{S}_c^2 / S_c^2 \to 1$ in probability.

Lemma C.4 justifies the Wald-type $1-\alpha$ confidence interval for $\bar{c}$:

$$\hat{\bar{c}} \pm z_{1-\alpha/2} \sqrt{\frac{n_0}{nn_1} \hat{S}_c^2}$$

where $z_{1-\alpha/2}$ is the $1-\alpha/2$ upper quantile of the standard Normal random variable.

C.2 Proofs

Proof of Lemma C.1: By symmetry, the Z_i's have the same mean, so

$$n_1 = \sum_{i=1}^n Z_i = E\left(\sum_{i=1}^n Z_i\right) = nE(Z_i),$$

which implies that

$$E(Z_i) = n_1/n$$

[1]Hoeffding (1963, Theorem 4) proved a general inequality to compare simple random sampling and IID sampling. Lei and Ding (2021, supplementary material) reviewed the results on simple random sampling that are particularly useful for causal inference.

for all i's. Because Z_i is a Bernoulli random variable, its variance is

$$\text{var}(Z_i) = \frac{n_1}{n}\left(1 - \frac{n_1}{n}\right) = \frac{n_1 n_0}{n^2}.$$

By symmetry again, the Z_i's have the same variance and the pairs (Z_i, Z_j)'s have the same covariance, so

$$0 = \text{var}\left(\sum_{i=1}^n Z_i\right) = n\text{var}(Z_i) + n(n-1)\text{cov}(Z_i, Z_j)$$

which implies that

$$\text{cov}(Z_i, Z_j) = -\frac{n_1 n_0}{n^2(n-1)}$$

for all $i \neq j$. □

Proof of Lemma C.2: The unbiasedness of the sample mean follows from linearity. For example,

$$\begin{aligned} E(\hat{\bar{c}}) &= E\left(\frac{1}{n_1}\sum_{i=1}^n Z_i c_i\right) \\ &= \frac{1}{n_1}\sum_{i=1}^n E(Z_i) c_i \\ &= \bar{c}. \end{aligned}$$

The covariance of the sample means is

$$\begin{aligned} &\text{cov}(\hat{\bar{c}}, \hat{\bar{d}}) \\ = \;& \text{cov}\left\{\frac{1}{n_1}\sum_{i=1}^n Z_i(c_i - \bar{c}), \frac{1}{n_1}\sum_{i=1}^n Z_i(d_i - \bar{d})\right\} \\ = \;& \frac{1}{n_1^2}\left[\sum_{i=1}^n \text{var}(Z_i)(c_i - \bar{c})(d_i - \bar{d}) + \sum_{i\neq j}\text{cov}(Z_i, Z_j)(c_i - \bar{c})(d_j - \bar{d})\right] \\ = \;& \frac{1}{n_1^2}\left[\frac{n_1 n_0}{n^2}\sum_{i=1}^n (c_i - \bar{c})(d_i - \bar{d}) - \frac{n_1 n_0}{n^2(n-1)}\sum_{i\neq j}(c_i - \bar{c})(d_j - \bar{d})\right]. \end{aligned}$$

Because

$$0 = \sum_{i=1}^n (c_i - \bar{c})\sum_{i=1}^n (d_i - \bar{d}) = \sum_{i=1}^n (c_i - \bar{c})(d_i - \bar{d}) + \sum_{i\neq j}(c_i - \bar{c})(d_j - \bar{d}),$$

the covariance of the sample means reduces to

$$\begin{aligned} &\text{cov}(\hat{\bar{c}}, \hat{\bar{d}}) \\ = \;& \frac{1}{n_1^2}\left[\frac{n_1 n_0}{n^2}\sum_{i=1}^n (c_i - \bar{c})(d_i - \bar{d}) + \frac{n_1 n_0}{n^2(n-1)}\sum_{i=1}^n (c_i - \bar{c})(d_i - \bar{d})\right] \\ = \;& \frac{n_0}{n n_1} S_{cd}. \end{aligned}$$

The variance formulas are just special cases with $\hat{\bar{c}} = \hat{\bar{d}}$. □

Proof of Lemma C.3: We prove only the sample covariance term because the formulas for sample variances are special cases. We have the following decomposition:

$$
\begin{aligned}
(n_1 - 1)\hat{S}_{cd} &= \sum_{i=1}^{n} Z_i(c_i - \hat{\bar{c}})(d_i - \hat{\bar{d}}) \\
&= \sum_{i=1}^{n} Z_i\{(c_i - \bar{c}) - (\hat{\bar{c}} - \bar{c})\}\{(d_i - \bar{d}) - (\hat{\bar{d}} - \bar{d})\} \\
&= \sum_{i=1}^{n} Z_i(c_i - \bar{c})(d_i - \bar{d}) - n_1(\hat{\bar{c}} - \bar{c})(\hat{\bar{d}} - \bar{d}).
\end{aligned}
$$

Taking expectations on both sides, we have

$$
\begin{aligned}
E\{(n_1 - 1)\hat{S}_{cd}\} &= \sum_{i=1}^{n} E(Z_i)(c_i - \bar{c})(d_i - \bar{d}) - n_1 E\{(\hat{\bar{c}} - \bar{c})(\hat{\bar{d}} - \bar{d})\} \\
&= \frac{n_1}{n}\sum_{i=1}^{n}(c_i - \bar{c})(d_i - \bar{d}) - n_1\frac{n_0}{nn_1}S_{cd} \\
&= S_{cd}\left\{\frac{n_1(n-1)}{n} - \frac{n_0}{n}\right\} \\
&= (n_1 - 1)S_{cd},
\end{aligned}
$$

and the conclusion follows by dividing both sides by $n_1 - 1$. □

Proof of Lemma C.4: Hájek (1960) gave a proof of the CLT for simple random sampling, and Lehmann (1975) gave a more accessible version of the proof. Li and Ding (2017) modified the CLT as presented in Lemma C.4, and proved the consistency of the sample variance. Due to the technical complexities, I omit the proof. □

C.3 Homework problems

C.1 Sampling without replacement and the Hypergeometric distribution

Consider a special case of Lemma C.2 with c_i's being binary. Assume the total number of 1's equals T so the total number of 0's equals $n - T$. Let

$$t = \sum_{i=1}^{n} Z_i c_i$$

denote the total number of 1's in the sample of size n_1.

Find the distribution, mean, and variance of t.

Remark: t follows a Hypergeometric distribution.

C.2 Vector form of Lemma C.2

Assume the c_i's are vectors and define

$$
\begin{aligned}
S_c^2 &= (n-1)^{-1}\sum_{i=1}^{n}(c_i - \bar{c})(c_i - \bar{c})^{\mathsf{T}}, \\
\hat{S}_c^2 &= (n_1-1)^{-1}\sum_{i=1}^{n} Z_i(c_i - \hat{\bar{c}})(c_i - \hat{\bar{c}})^{\mathsf{T}}.
\end{aligned}
$$

Show that

$$
\begin{aligned}
E(\hat{\bar{c}}) &= \bar{c}, \\
\mathrm{cov}(\hat{\bar{c}}) &= \frac{n_0}{nn_1} S_c^2, \\
E(\hat{S}_c^2) &= S_c^2.
\end{aligned}
$$

C.3 Recommended reading

Survey sampling and experimental design have been deeply connected ever since the seminal work by Neyman (1934, 1935). Li and Ding (2017) and Mukerjee et al. (2018) made many theoretical ties between these two areas.

Bibliography

Abadie, A., Athey, S., Imbens, G. W., and Wooldridge, J. M. (2020). Sampling-based versus design-based uncertainty in regression analysis. *Econometrica*, 88:265–296.

Abadie, A. and Imbens, G. W. (2006). Large sample properties of matching estimators for average treatment effects. *Econometrica*, 74:235–267.

Abadie, A. and Imbens, G. W. (2008). On the failure of the bootstrap for matching estimators. *Econometrica*, 76:1537–1557.

Abadie, A. and Imbens, G. W. (2011). Bias-corrected matching estimators for average treatment effects. *Journal of Business and Economic Statistics*, 29:1–11.

Abadie, A. and Imbens, G. W. (2016). Matching on the estimated propensity score. *Econometrica*, 84:781–807.

Alwin, D. F. and Hauser, R. M. (1975). The decomposition of effects in path analysis. *American Sociological Review*, 40:37–47.

Amarante, V., Manacorda, M., Miguel, E., and Vigorito, A. (2016). Do cash transfers improve birth outcomes? Evidence from matched vital statistics, program, and social security data. *American Economic Journal: Economic Policy*, 8:1–43.

Anderson, T. W. and Rubin, H. (1950). The asymptotic properties of estimates of the parameters of a single equation in a complete system of stochastic equations. *Annals of Mathematical Statistics*, 21:570–582.

Angrist, J., Lang, D., and Oreopoulos, P. (2009). Incentives and services for college achievement: Evidence from a randomized trial. *American Economic Journal: Applied Economics*, 1:136–163.

Angrist, J. and Lavy, V. (2009). The effects of high stakes high school achievement awards: Evidence from a randomized trial. *American Economic Review*, 99:1384–1414.

Angrist, J. D. (1990). Lifetime earnings and the Vietnam era draft lottery: Evidence from social security administrative records. *American Economic Review*, 80:313–336.

Angrist, J. D. (1998). Estimating the labor market impact of voluntary military service using social security data on military applicants. *Econometrica*, 66:249–288.

Angrist, J. D. (2022). Empirical strategies in economics: Illuminating the path from cause to effect. *Econometrica*, 90:2509–2539.

Angrist, J. D. and Evans, W. N. (1998). Children and their parents' labor supply: Evidence from exogenous variation in family size. *American Economic Review*, 88:450–477.

Angrist, J. D. and Imbens, G. W. (1995). Two-stage least squares estimation of average causal effects in models with variable treatment intensity. *Journal of the American Statistical Association*, 90:431–442.

Angrist, J. D., Imbens, G. W., and Rubin, D. B. (1996). Identification of causal effects using instrumental variables (with discussion). *Journal of the American Statistical Association*, 91:444–455.

Angrist, J. D. and Krueger, A. B. (1991). Does compulsory school attendance affect schooling and earnings? *Quarterly Journal of Economics*, 106:979–1014.

Angrist, J. D. and Pischke, J.-S. (2008). *Mostly Harmless Econometrics: An Empiricist's Companion*. Princeton: Princeton University Press.

Angrist, J. D. and Pischke, J.-S. (2014). *Mastering 'Metrics: The Path from Cause to Effect*. Princeton: Princeton University Press.

Aronow, P. M., Green, D. P., and Lee, D. K. K. (2014). Sharp bounds on the variance in randomized experiments. *Annals of Statistics*, 42:850–871.

Asher, S. and Novosad, P. (2020). Rural roads and local economic development. *American Economic Review*, 110:797–823.

Baker, S. G. and Lindeman, K. S. (1994). The paired availability design: A proposal for evaluating epidural analgesia during labor. *Statistics in Medicine*, 13:2269–2278.

Balke, A. and Pearl, J. (1997). Bounds on treatment effects from studies with imperfect compliance. *Journal of the American Statistical Association*, 92:1171–1176.

Ball, S., Bogatz, G., Rubin, D., and Beaton, A. (1973). Reading with television: An evaluation of the Electric Company. A report to the Children's Television Workshop. Volumes 1 and 2.

Bang, H. and Robins, J. M. (2005). Doubly robust estimation in missing data and causal inference models. *Biometrics*, 61:962–973.

Barnard, G. A. (1947). Significance tests for 2×2 tables. *Biometrika*, 34:123–138.

Baron, R. M. and Kenny, D. A. (1986). The moderator-mediator variable distinction in social psychological research: Conceptual, strategic, and statistical considerations. *Journal of Personality and Social Psychology*, 51:1173–1182.

Basmann, R. L. (1957). A generalized classical method of linear estimation of coefficients in a structural equation. *Econometrica*, 25:77–83.

Bazzano, L. A., He, J., Muntner, P., Vupputuri, S., and Whelton, P. K. (2003). Relationship between cigarette smoking and novel risk factors for cardiovascular disease in the United States. *Annals of Internal Medicine*, 138:891–897.

Berk, R., Pitkin, E., Brown, L., Buja, A., George, E., and Zhao, L. (2013). Covariance adjustments for the analysis of randomized field experiments. *Evaluation Review*, 37:170–196.

Bertrand, M. and Mullainathan, S. (2004). Are Emily and Greg more employable than Lakisha and Jamal? A field experiment on labor market discrimination. *American Economic Review*, 94:991–1013.

Bickel, P. J., Hammel, E. A., and O'Connell, J. W. (1975). Sex bias in graduate admissions: Data from Berkeley. *Science*, 187:398–404.

Bickel, P. J., Klaassen, C. A. J., Ritov, Y., and Wellner, J. A. (1993). *Efficient and Adaptive Estimation for Semiparametric Models.* Baltimore: Johns Hopkins University Press.

Bind, M.-A. C. and Rubin, D. B. (2020). When possible, report a Fisher-exact P value and display its underlying null randomization distribution. *Proceedings of the National Academy of Sciences of the United States of America*, 117:19151–19158.

Blackwell, M. (2013). A framework for dynamic causal inference in political science. *American Journal of Political Science*, 57:504–520.

Bloniarz, A., Liu, H., Zhang, C. H., Sekhon, J., and Yu, B. (2016). Lasso adjustments of treatment effect estimates in randomized experiments. *Proceedings of the National Academy of Sciences of the United States of America*, 113:7383–7390.

Bloom, H. S. (1984). Accounting for no-shows in experimental evaluation designs. *Evaluation Review*, 8:225–246.

Bor, J., Moscoe, E., Mutevedzi, P., Newell, M.-L., and Bärnighausen, T. (2014). Regression discontinuity designs in epidemiology: Causal inference without randomized trials. *Epidemiology*, 25:729.

Bowden, J., Davey Smith, G., and Burgess, S. (2015). Mendelian randomization with invalid instruments: Effect estimation and bias detection through Egger regression. *International Journal of Epidemiology*, 44:512–525.

Bowden, J., Spiller, W., Fabiola Del Greco M., Sheehan, N., Thompson, J., Minelli, C., and Davey Smith, G. (2018). Improving the visualization, interpretation and analysis of two-sample summary data mendelian randomization via the radial plot and radial regression. *International Journal of Epidemiology*, 47:1264–1278.

Box, G. E. P. (1979). Robustness in the strategy of scientific model building. In Launer, R. L. and Wilkinson, G. N., editors, *Robustness in Statistics*, pages 201–236. New York: Academic Press, Inc.

Box, G. E. P., Hunter, W. H., and Hunter, S. (1978). *Statistics for Experimenters: An Introduction to Design, Data Analysis, and Model Building.* New York: John Wiley and Sons.

Bradford Hill, A. (1965). The environment and disease: Association or causation? *Proceedings of the Royal Society of Medicine*, 58:295–300.

Bradford Hill, A. (2020). The environment and disease: Association or causation? (with discussion). *Observational Studies*, 6:1–65.

Bruhn, M. and McKenzie, D. (2009). In pursuit of balance: Randomization in practice in development field experiments. *American Economic Journal: Applied Economics*, 1:200–232.

Brumback, B. A. (2022). *Fundamentals of Causal Inference: With R.* New York: Chapman & Hall.

Butler, C. C. (1969). A test for symmetry using the sample distribution function. *Annals of Mathematical Statistics*, 40:2209–2210.

Campbell, H. and Gustafson, P. (2018). The validity and efficiency of hypothesis testing in observational studies with time-varying exposures. *Observational Studies*, 4:260–291.

Cao, W., Tsiatis, A. A., and Davidian, M. (2009). Improving efficiency and robustness of the doubly robust estimator for a population mean with incomplete data. *Biometrika*, 96:723–734.

Card, D. (1993). Using geographic variation in college proximity to estimate the return to schooling. Technical report, National Bureau of Economic Research.

Carpenter, C. and Dobkin, C. (2009). The effect of alcohol consumption on mortality: Regression discontinuity evidence from the minimum drinking age. *American Economic Journal: Applied Economics*, 1:164–182.

Cattaneo, M. D. (2010). Efficient semiparametric estimation of multi-valued treatment effects under ignorability. *Journal of Econometrics*, 155:138–154.

Cattaneo, M. D., Frandsen, B. R., and Titiunik, R. (2015). Randomization inference in the regression discontinuity design: An application to party advantages in the US Senate. *Journal of Causal Inference*, 3:1–24.

Chan, K. C. G., Yam, S. C. P., and Zhang, Z. (2016). Globally efficient non-parametric inference of average treatment effects by empirical balancing calibration weighting. *Journal of the Royal Statistical Society: Series B (Statistical Methodology)*, 78:673–700.

Chapin, F. S. (1947). *Experimental Designs in Sociological Research.* New York: Harper and Brothers.

Charig, C. R., Webb, D. R., Payne, S. R., and Wickham, J. E. (1986). Comparison of treatment of renal calculi by open surgery, percutaneous nephrolithotomy, and extracorporeal shockwave lithotripsy. *British Medical Journal*, 292:879–882.

Chen, H., Geng, Z., and Jia, J. (2007). Criteria for surrogate end points. *Journal of the Royal Statistical Society: Series B (Statistical Methodology)*, 69:919–932.

Cheng, J. and Small, D. S. (2006). Bounds on causal effects in three-arm trials with non-compliance. *Journal of the Royal Statistical Society: Series B (Statistical Methodology)*, 68:815–836.

Chernozhukov, V., Chetverikov, D., Demirer, M., Duflo, E., Hansen, C., Newey, W., and Robins, J. (2018). Double/debiased machine learning for treatment and structural parameters. *Econometrics Journal*, 21:C1–C68.

Chong, A., Cohen, I., Field, E., Nakasone, E., and Torero, M. (2016). Iron deficiency and schooling attainment in Peru. *American Economic Journal: Applied Economics*, 8:222–55.

Chung, E. and Romano, J. (2013). Exact and asymptotically robust permutation tests. *Annals of Statistics*, 41:484–507.

Cochran, W. G. (1938). The omission or addition of an independent variate in multiple linear regression. *Supplement to the Journal of the Royal Statistical Society*, 5:171–176.

Cochran, W. G. (1953). *Sampling Techniques.* New York: Wiley.

Cochran, W. G. (1957). Analysis of covariance: Its nature and uses. *Biometrics*, 13:261–281.

Cochran, W. G. (1965). The planning of observational studies of human populations (with discussion). *Journal of the Royal Statistical Society: Series A (General)*, 128:234–266.

Cochran, W. G. (1968). The effectiveness of adjustment by subclassification in removing bias in observational studies. *Biometrics*, 24:295–313.

Cochran, W. G. (2015). Observational studies. *Observational Studies*, 1:126–136.

Cochran, W. G. and Rubin, D. B. (1973). Controlling bias in observational studies: A review. *Sankhyā*, 35:417–446.

Cornfield, J., Haenszel, W., Hammond, E. C., Lilienfeld, A. M., Shimkin, M. B., and Wynder, E. L. (1959). Smoking and lung cancer: recent evidence and a discussion of some questions. *Journal of the National Cancer Institute*, 22:173–203.

Cox, D. R. (1982). Randomization and concomitant variables in the design of experiments. In G. Kallianpur, P. R. Krishnaiah and Ghosh, J. K., editors, *Statistics and Probability: Essays in Honor of C. R. Rao*, pages 197–202. North-Holland, Amsterdam.

Cox, D. R. (2007). On a generalization of a result of W. G. Cochran. *Biometrika*, 94:755–759.

Crump, R. K., Hotz, V. J., Imbens, G. W., and Mitnik, O. A. (2009). Dealing with limited overlap in estimation of average treatment effects. *Biometrika*, 96:187–199.

Cunningham, S. (2021). *Causal Inference: The Mixtape.* New Haven: Yale University Press.

Cuzick, J., Edwards, R., and Segnan, N. (1997). Adjusting for non-compliance and contamination in randomized clinical trials. *Statistics in Medicine*, 16:1017–1029.

D'Amour, A., Ding, P., Feller, A., Lei, L., and Sekhon, J. (2021). Overlap in observational studies with high-dimensional covariates. *Journal of Econometrics*, 221:644–654.

Dasgupta, T., Pillai, N. S., and Rubin, D. B. (2015). Causal inference from 2^K factorial designs by using potential outcomes. *Journal of the Royal Statistical Society: Series B (Statistical Methodology)*, 77:727–753.

Davey Smith, G. and Ebrahim, S. (2003). "Mendelian randomization": Can genetic epidemiology contribute to understanding environmental determinants of disease? *International Journal of Epidemiology*, 32:1–22.

Davison, A. C. and Hinkley, D. V. (1997). *Bootstrap Methods and Their Application.* Cambridge: Cambridge University Press.

Dawid, A. P. (1979). Conditional independence in statistical theory. *Journal of the Royal Statistical Society: Series B (Methodological)*, 41:1–15.

Dawid, A. P. (2000). Causal inference without counterfactuals (with discussion). *Journal of the American Statistical Association*, 95:407–424.

Dawid, A. P., Musio, M., and Murtas, R. (2017). The probability of causation. *Law, Probability and Risk*, 16:163–179.

Dehejia, R. H. and Wahba, S. (1999). Causal effects in nonexperimental studies: Reevaluating the evaluation of training programs. *Journal of the American statistical Association*, 94:1053–1062.

Ding, P. (2016). A paradox from randomization-based causal inference (with discussion). *Statistical Science*, 32:331–345.

Ding, P. (2021). The Frisch–Waugh–Lovell theorem for standard errors. *Statistics and Probability Letters*, 168:108945.

Ding, P. (2024). *Linear Model and Extensions*. Chapman & Hall/CRC.

Ding, P. and Dasgupta, T. (2016). A potential tale of two by two tables from completely randomized experiments. *Journal of American Statistical Association*, 111:157–168.

Ding, P. and Dasgupta, T. (2017). A randomization-based perspective on analysis of variance: A test statistic robust to treatment effect heterogeneity. *Biometrika*, 105:45–56.

Ding, P., Feller, A., and Miratrix, L. (2016). Randomization inference for treatment effect variation. *Journal of the Royal Statistical Society: Series B (Statistical Methodology)*, 78:655–671.

Ding, P., Feller, A., and Miratrix, L. (2019). Decomposing treatment effect variation. *Journal of the American Statistical Association*, 114:304–317.

Ding, P., Geng, Z., Yan, W., and Zhou, X.-H. (2011). Identifiability and estimation of causal effects by principal stratification with outcomes truncated by death. *Journal of the American Statistical Association*, 106:1578–1591.

Ding, P. and Li, F. (2018). Causal inference: A missing data perspective. *Statistical Science*, 33:214–237.

Ding, P., Li, X., and Miratrix, L. W. (2017a). Bridging finite and super population causal inference. *Journal of Causal Inference*, 5:20160027.

Ding, P. and Lu, J. (2017). Principal stratification analysis using principal scores. *Journal of the Royal Statistical Society: Series B (Statistical Methodology)*, 79:757–777.

Ding, P. and Miratrix, L. W. (2015). To adjust or not to adjust? Sensitivity analysis of M-bias and butterfly-bias. *Journal of Causal Inference*, 3:41–57.

Ding, P. and Miratrix, L. W. (2019). Model-free causal inference of binary experimental data. *Scandinavian Journal of Statistics*, 46:200–214.

Ding, P. and VanderWeele, T. J. (2014). Generalized Cornfield conditions for the risk difference. *Biometrika*, 101:971–977.

Ding, P. and VanderWeele, T. J. (2016a). Sensitivity analysis without assumptions. *Epidemiology*, 27:368–377.

Ding, P. and VanderWeele, T. J. (2016b). Sharp sensitivity bounds for mediation under unmeasured mediator-outcome confounding. *Biometrika*, 103:483–490.

Ding, P., VanderWeele, T. J., and Robins, J. M. (2017b). Instrumental variables as bias amplifiers with general outcome and confounding. *Biometrika*, 104:291–302.

Doll, R. and Hill, A. B. (1950). Smoking and carcinoma of the lung. *British Medical Journal*, 2:739.

Dorn, H. F. (1953). Philosophy of inferences from retrospective studies. *American Journal of Public Health and the Nations Health*, 43:677–683.

Durrett, R. (2019). *Probability: Theory and Examples*. Cambridge: Cambridge University Press.

Efron, B. (1979). Bootstrap methods: Another look at the jackknife. *The Annals of Statistics*, 7:1–26.

Efron, B. and Feldman, D. (1991). Compliance as an explanatory variable in clinical trials (with discussion). *Journal of the American Statistical Association*, 86:9–17.

Eicker, F. (1967). Limit theorems for regressions with unequal and dependent errors. In *Proceedings of the Fifth Berkeley Symposium on Mathematical Statistics and Probability*, volume 1, pages 59–82. Berkeley, CA: University of California Press.

Fan, J. and Gijbels, I. (1996). *Local Polynomial Modelling and Its Applications*. New York: Chapman and Hall/CRC.

Fieller, E. C. (1954). Some problems in interval estimation. *Journal of the Royal Statistical Society: Series B (Methodological)*, 16:175–185.

Firth, D. and Bennett, K. E. (1998). Robust models in probability sampling (with discussion). *Journal of the Royal Statistical Society: Series B (Statistical Methodology)*, 60:3–21.

Fisher, R. A. (1925). *Statistical Methods for Research Workers*. Edinburgh by Oliver and Boyd, 1st edition.

Fisher, R. A. (1935). *The Design of Experiments*. Edinburgh, London: Oliver and Boyd, 1st edition.

Fisher, R. A. (1957). Dangers of cigarette smoking [letter]. *British Medical Journal*, 2:297–298.

Fogarty, C. B. (2018a). On mitigating the analytical limitations of finely stratified experiments. *Journal of the Royal Statistical Society. Series B (Statistical Methodology)*, 80:1035–1056.

Fogarty, C. B. (2018b). Regression assisted inference for the average treatment effect in paired experiments. *Biometrika*, 105:994–1000.

Follmann, D. A. (2000). On the effect of treatment among would-be treatment compliers: An analysis of the multiple risk factor intervention trial. *Journal of the American Statistical Association*, 95:1101–1109.

Forastiere, L., Mattei, A., and Ding, P. (2018). Principal ignorability in mediation analysis: Through and beyond sequential ignorability. *Biometrika*, 105:979–986.

Frangakis, C. E. and Rubin, D. B. (2002). Principal stratification in causal inference. *Biometrics*, 58:21–29.

Freedman, D. A. (2008a). On regression adjustments in experiments with several treatments. *Annals of Applied Statistics*, 2:176–196.

Freedman, D. A. (2008b). On regression adjustments to experimental data. *Advances in Applied Mathematics*, 40:180–193.

Freedman, D. A. (2008c). Randomization does not justify logistic regression. *Statistical Science*, 23:237–249.

Freedman, D. A. (2009). *Statistical Models: Theory and Practice*. Cambridge: Cambridge University Press.

Freedman, D. A. and Berk, R. A. (2008). Weighting regressions by propensity scores. *Evaluation Review*, 32:392–409.

Frumento, P., Mealli, F., Pacini, B., and Rubin, D. B. (2012). Evaluating the effect of training on wages in the presence of noncompliance, nonemployment, and missing outcome data. *Journal of the American Statistical Association*, 107:450–466.

Funk, M. J., Westreich, D., Wiesen, C., Stürmer, T., Brookhart, M. A., and Davidian, M. (2011). Doubly robust estimation of causal effects. *American Journal of Epidemiology*, 173:761–767.

Gastwirth, J. L., Krieger, A. M., and Rosenbaum, P. R. (1998). Cornfield's inequality. In Armitage, P. and Colton, T., editors, *Encyclopedia of Biostatistics*. New York: Wiley.

Gerber, A. S. and Green, D. P. (2012). *Field Experiments: Design, Analysis, and Interpretation*. WW Norton.

Gerber, A. S., Green, D. P., and Larimer, C. W. (2008). Social pressure and voter turnout: Evidence from a large-scale field experiment. *American Political Science Review*, 102:33–48.

Gilbert, P. B. and Hudgens, M. G. (2008). Evaluating candidate principal surrogate endpoints. *Biometrics*, 64:1146–1154.

Gould, A. L. (1998). Multi-centre trial analysis revisited. *Statistics in Medicine*, 17:1779–1797.

Greenwood, E. (1945). *Experimental Sociology*. New York: Columbia University Press.

Greevy, R., Lu, B., Silber, J. H., and Rosenbaum, P. (2004). Optimal multivariate matching before randomization. *Biostatistics*, 5:263–275.

Guo, K. and Basse, G. (2023). The generalized Oaxaca–Blinder estimator. *Journal of American Statistical Association*, 118:524–536.

Guo, K. and Rothenhäusler, D. (2023). On the statistical role of inexact matching in observational studies. *Biometrika*, 110:631–644.

Hahn, J. (1998). On the role of the propensity score in efficient semiparametric estimation of average treatment effects. *Econometrica*, 66:315–331.

Hahn, J., Todd, P., and Van der Klaauw, W. (2001). Identification and estimation of treatment effects with a regression-discontinuity design. *Econometrica*, 69:201–209.

Hahn, P. R., Murray, J. S., and Carvalho, C. M. (2020). Bayesian regression tree models for causal inference: Regularization, confounding, and heterogeneous effects. *Bayesian Analysis*, 15:965–1056.

Hainmueller, J. (2012). Entropy balancing for causal effects: A multivariate reweighting method to produce balanced samples in observational studies. *Political Analysis*, 20:25–46.

Hájek, J. (1960). Limiting distributions in simple random sampling from a finite population. *Publications of the Mathematics Institute of the Hungarian Academy of Science*, 5:361–74.

Hájek, J. (1971). Comment on "an essay on the logical foundations of survey sampling, part one". *The foundations of survey sampling*, 236.

Hammond, E. C. and Horn, D. (1958). Smoking and death rates: Report on forty four months of follow-up of 187,783 men. *Journal of the American Medicial Association*, 166:1159–1172, 1294–1308.

Hansen, B. B. and Bowers, J. (2009). Attributing effects to a cluster-randomized get-out-the-vote campaign. *Journal of the American Statistical Association*, 104:873–885.

Hansen, L. P. (1982). Large sample properties of generalized method of moments estimators. *Econometrica*, 50:1029–1054.

Hartley, H. O., Rao, J. N. K., and Kiefer, G. (1969). Variance estimation with one unit per stratum. *Journal of the American Statistical Association*, 64:841–851.

Hausman, J. A. (1978). Specification tests in econometrics. *Econometrica*, 46:1251–1271.

Hearst, N., Newman, T. B., and Hulley, S. B. (1986). Delayed effects of the military draft on mortality. *New England Journal of Medicine*, 314:620–624.

Heckman, J. and Navarro-Lozano, S. (2004). Using matching, instrumental variables, and control functions to estimate economic choice models. *Review of Economics and Statistics*, 86:30–57.

Heckman, J. J. (1979). Sample selection bias as a specification error. *Econometrica*, 47:153–161.

Heckman, J. J., Ichimura, H., and Todd, P. E. (1997). Matching as an econometric evaluation estimator: Evidence from evaluating a job training programme. *Review of Economic Studies*, 64:605–654.

Hennessy, J., Dasgupta, T., Miratrix, L., Pattanayak, C., and Sarkar, P. (2016). A conditional randomization test to account for covariate imbalance in randomized experiments. *Journal of Causal Inference*, 4:61–80.

Hernán, M. Á., Brumback, B., and Robins, J. M. (2000). Marginal structural models to estimate the causal effect of zidovudine on the survival of hiv-positive men. *Epidemiology*, 11:561–570.

Hernán, M. A. and Robins, J. M. (2020). *Causal Inference: What If.* Boca Raton: Chapman & Hall/CRC.

Hill, J., Waldfogel, J., and Brooks-Gunn, J. (2002). Differential effects of high-quality child care. *Journal of Policy Analysis and Management*, 21:601–627.

Hill, J. L. (2011). Bayesian nonparametric modeling for causal inference. *Journal of Computational and Graphical Statistics*, 20:217–240.

Hirano, K. and Imbens, G. W. (2001). Estimation of causal effects using propensity score weighting: An application to data on right heart catheterization. *Health Services and Outcomes Research Methodology*, 2:259–278.

Hirano, K., Imbens, G. W., Rubin, D. B., and Zhou, X. H. (2000). Assessing the effect of an influenza vaccine in an encouragement design. *Biostatistics*, 1:69–88.

Ho, D. E., Imai, K., King, G., and Stuart, E. A. (2007). Matching as nonparametric preprocessing for reducing model dependence in parametric causal inference. *Political Analysis*, 15:199–236.

Ho, D. E., Imai, K., King, G., and Stuart, E. A. (2011). MatchIt: Nonparametric preprocessing for parametric causal inference. *Journal of Statistical Software*, 42:1–28.

Hodges, J. L. and Lehmann, E. L. (1962). Rank methods for combination of independent experiments in analysis of variance. *Annals of Mathematical Statistics*, 33:482–497.

Hoeffding, W. (1963). Probability inequalities for sums of bounded random variables. *Journal of the American Statistical Association*, 58:13–30.

Holland, P. W. (1986). Statistics and causal inference (with discussion). *Journal of the American Statistical Association*, 81:945–960.

Hong, G. and Raudenbush, S. W. (2008). Causal inference for time-varying instructional treatments. *Journal of Educational and Behavioral Statistics*, 33:333–362.

Horvitz, D. G. and Thompson, D. J. (1952). A generalization of sampling without replacement from a finite universe. *Journal of the American statistical Association*, 47:663–685.

Huber, M. (2023). *Causal Analysis: Impact Evaluation and Causal Machine Learning with Applications in R.* Cambridge: MIT Press.

Huber, P. J. (1967). The behavior of maximum likelihood estimates under nonstandard conditions. In Cam, L. M. L. and Neyman, J., editors, *Proceedings of the Fifth Berkeley Symposium on Mathematical Statistics and Probability*, volume 1, pages 221–233. Berkeley, California: University of California Press.

Hudgens, M. G. and Halloran, M. E. (2008). Toward causal inference with interference. *Journal of the American Statistical Association*, 103:832–842.

Huntington-Klein, N. (2022). *The effect: An introduction to research design and causality.* New York: Chapman & Hall.

Hyman, H. H. (1955). *Survey Design and Analysis: Principles, Cases, and Procedures.* Glencoe, IL: Free Press.

Imai, K. (2008a). Sharp bounds on the causal effects in randomized experiments with "truncation-by-death". *Statistics and Probability Letters*, 78:144–149.

Imai, K. (2008b). Variance identification and efficiency analysis in randomized experiments under the matched-pair design. *Statistics in Medicine*, 27:4857–4873.

Imai, K., Keele, L., and Yamamoto, T. (2010). Identification, inference and sensitivity analysis for causal mediation effects. *Statistical Science*, 25:51–71.

Imai, K. and Van Dyk, D. A. (2004). Causal inference with general treatment regimes: Generalizing the propensity score. *Journal of the American Statistical Association*, 99:854–866.

Imbens, G. (2020). Potential outcome and directed acyclic graph approaches to causality: Relevance for empirical practice in economics. *Journal of Economic Literature*, 58:1129–1179.

Imbens, G. W. (2003). Sensitivity to exogeneity assumptions in program evaluation. *American Economic Review*, 93:126–132.

Imbens, G. W. (2004). Nonparametric estimation of average treatment effects under exogeneity: A review. *Review of Economics and Statistics*, 86:4–29.

Imbens, G. W. (2014). Instrumental variables: An econometrician's perspective. *Statistical Science*, 29:323–358.

Imbens, G. W. (2015). Matching methods in practice: Three examples. *Journal of Human Resources*, 50:373–419.

Imbens, G. W. and Angrist, J. D. (1994). Identification and estimation of local average treatment effects. *Econometrica*, 62:467–475.

Imbens, G. W. and Lemieux, T. (2008). Regression discontinuity designs: A guide to practice. *Journal of Econometrics*, 142:615–635.

Imbens, G. W. and Manski, C. F. (2004). Confidence intervals for partially identified parameters. *Econometrica*, 72:1845–1857.

Imbens, G. W. and Rubin, D. B. (1997). Estimating outcome distributions for compliers in instrumental variables models. *Review of Economic Studies*, 64:555–574.

Imbens, G. W. and Rubin, D. B. (2015). *Causal Inference for Statistics, Social, and Biomedical Sciences: An Introduction.* Cambridge: Cambridge University Press.

Imbens, G. W. and Wooldridge, J. M. (2009). Recent developments in the econometrics of program evaluation. *Journal of Economic Literature*, 47:5–86.

Investigators, I. T. et al. (2014). Endovascular or open repair strategy for ruptured abdominal aortic aneurysm: 30 day outcomes from improve randomised trial. *British Medical Journal*, 348:f7661.

Ioannidis, J. P. A., Tan, Y. J., and Blum, M. R. (2019). Limitations and misinterpretations of E-values for sensitivity analyses of observational studies. *Annals of Internal Medicine*, 170:108–111.

Jackson, L. A., Jackson, M. L., Nelson, J. C., Neuzil, K. M., and Weiss, N. S. (2006). Evidence of bias in estimates of influenza vaccine effectiveness in seniors. *International Journal of Epidemiology*, 35:337–344.

Janssen, A. (1997). Studentized permutation tests for non-iid hypotheses and the generalized Behrens-Fisher problem. *Statistics and Probability Letters*, 36:9–21.

Jiang, Z. and Ding, P. (2020). Measurement errors in the binary instrumental variable model. *Biometrika*, 107:238–245.

Jiang, Z. and Ding, P. (2021). Identification of causal effects within principal strata using auxiliary variables. *Statistical Science*, 36:493–508.

Jiang, Z., Ding, P., and Geng, Z. (2016). Principal causal effect identification and surrogate end point evaluation by multiple trials. *Journal of the Royal Statistical Society: Series B (Statistical Methodology)*, 78:829–848.

Jiang, Z., Yang, S., and Ding, P. (2022). Multiply robust estimation of causal effects under principal ignorability. *Journal of the Royal Statistical Society - Series B (Statistical Methodology)*, 84:1423–1445.

Jin, H. and Rubin, D. B. (2008). Principal stratification for causal inference with extended partial compliance. *Journal of the American Statistical Association*, 103:101–111.

Jo, B. and Stuart, E. A. (2009). On the use of propensity scores in principal causal effect estimation. *Statistics in Medicine*, 28:2857–2875.

Jo, B., Stuart, E. A., MacKinnon, D. P., and Vinokur, A. D. (2011). The use of propensity scores in mediation analysis. *Multivariate Behavioral Research*, 46:425–452.

Judd, C. M. and Kenny, D. A. (1981). Process analysis estimating mediation in treatment evaluations. *Evaluation Review*, 5:602–619.

Kang, J. D. Y. and Schafer, J. L. (2007). Demystifying double robustness: A comparison of alternative strategies for estimating a population mean from incomplete data. *Statistical Science*, 22:523–539.

Katan, M. B. (1986). Apoupoprotein E isoforms, serum cholesterol, and cancer. *Lancet*, 327:507–508.

King, G. and Zeng, L. (2006). The dangers of extreme counterfactuals. *Political Analysis*, 14:131–159.

Kitagawa, T. (2015). A test for instrument validity. *Econometrica*, 83:2043–2063.

Koenker, R. and Xiao, Z. (2002). Inference on the quantile regression process. *Econometrica*, 70:1583–1612.

Künzel, S. R., Sekhon, J. S., Bickel, P. J., and Yu, B. (2019). Metalearners for estimating heterogeneous treatment effects using machine learning. *Proceedings of the National Academy of Sciences of the United States of America*, 116:4156–4165.

Kurth, T., Walker, A. M., Glynn, R. J., Chan, K. A., Gaziano, J. M., Berger, K., and Robins, J. M. (2005). Results of multivariable logistic regression, propensity matching, propensity adjustment, and propensity-based weighting under conditions of nonuniform effect. *American Journal of Epidemiology*, 163:262–270.

LaLonde, R. J. (1986). Evaluating the econometric evaluations of training programs with experimental data. *American Economic Review*, 76:604–620.

Leamer, E. (1978). *Specification Searches: Ad Hoc Inference with Nonexperimental Data.* John Wiley and Sons.

Lee, D. S. (2008). Randomized experiments from non-random selection in US House elections. *Journal of Econometrics*, 142:675–697.

Lee, D. S. (2009). Training, wages, and sample selection: Estimating sharp bounds on treatment effects. *Review of Economic Studies*, 76:1071–1102.

Lee, D. S. and Lemieux, T. (2010). Regression discontinuity designs in economics. *Journal of Economic Literature*, 48:281–355.

Lee, M.-J. (2018). Simple least squares estimator for treatment effects using propensity score residuals. *Biometrika*, 105:149–164.

Lee, W.-C. (2011). Bounding the bias of unmeasured factors with confounding and effect-modifying potentials. *Statistics in Medicine*, 30:1007–1017.

Lehmann, E. L. (1975). *Nonparametrics: Statistical Methods Based on Ranks.* California: Holden-Day, Inc.

Lei, L. and Ding, P. (2021). Regression adjustment in completely randomized experiments with a diverging number of covariates. *Biometrika*, 108:815–828.

Lewis, D. (1973). Causation. *Journal of Philosophy*, 70:556–567.

Li, F., Ding, P., and Mealli, F. (2023). Bayesian causal inference: a critical review. *Philosophical Transactions of the Royal Society A*, 381:20220153.

Li, F., Mattei, A., and Mealli, F. (2015). Evaluating the causal effect of university grants on student dropout: evidence from a regression discontinuity design using principal stratification. *Annals of Applied Statistics*, 9:1906–1931.

Li, F., Morgan, K. L., and Zaslavsky, A. M. (2018a). Balancing covariates via propensity score weighting. *Journal of the American Statistical Association*, 113:390–400.

Li, F., Thomas, L. E., and Li, F. (2019). Addressing extreme propensity scores via the overlap weights. *American Journal of Epidemiology*, 188:250–257.

Li, X. and Ding, P. (2016). Exact confidence intervals for the average causal effect on a binary outcome. *Statistics in Medicine*, 35:957–960.

Li, X. and Ding, P. (2017). General forms of finite population central limit theorems with applications to causal inference. *Journal of the American Statistical Association*, 112:1759–1769.

Li, X. and Ding, P. (2020). Rerandomization and regression adjustment. *Journal of the Royal Statistical Society, Series B (Statistical Methodology)*, 82:241–268.

Li, X., Ding, P., and Rubin, D. B. (2018b). Asymptotic theory of rerandomization in treatment-control experiments. *Proceedings of the National Academy of Sciences of the United States of America*, 115:9157–9162.

Lin, W. (2013). Agnostic notes on regression adjustments to experimental data: Reexamining Freedman's critique. *Annals of Applied Statistics*, 7:295–318.

Lin, Z., Ding, P., and Han, F. (2023). Estimation based on nearest neighbor matching: From density ratio to average treatment effect. *Econometrica*, 91:2187–2217.

Lind, J. (1753). A treatise of the scurvy. *Three Parts. Containing an Inquiry into the Nature, Causes and Cure, of that Disease. Together with a Critical and Chronological View of what has been Published on the Subject.*

Lipsitch, M., Tchetgen Tchetgen, E., and Cohen, T. (2010). Negative controls: A tool for detecting confounding and bias in observational studies. *Epidemiology*, 21:383–388.

Little, R. and An, H. (2004). Robust likelihood-based analysis of multivariate data with missing values. *Statistica Sinica*, 14:949–968.

Liu, H. and Yang, Y. (2020). Regression-adjusted average treatment effect estimates in stratified randomized experiments. *Biometrika*, 107:935–948.

Long, J. S. and Ervin, L. H. (2000). Using heteroscedasticity consistent standard errors in the linear regression model. *American Statistician*, 54:217–224.

Lu, S. and Ding, P. (2023). Flexible sensitivity analysis for causal inference in observational studies subject to unmeasured confounding. *Biometrics*.

Lu, S., Jiang, Z., and Ding, P. (2023). Principal stratification with continuous post-treatment variables: Nonparametric identification and semiparametric estimation. *arXiv preprint arXiv:2309.12425.*

Lumley, T., Shaw, P. A., and Dai, J. Y. (2011). Connections between survey calibration estimators and semiparametric models for incomplete data. *International Statistical Review*, 79:200–220.

Lunceford, J. K. and Davidian, M. (2004). Stratification and weighting via the propensity score in estimation of causal treatment effects: A comparative study. *Statistics in Medicine*, 23:2937–2960.

Luo, X., Dasgupta, T., Xie, M., and Liu, R. Y. (2021). Leveraging the Fisher randomization test using confidence distributions: Inference, combination and fusion learning. *Journal of the Royal Statistical Society: Series B (Statistical Methodology)*, 83:777–797.

Manski, C. F. (1990). Nonparametric bounds on treatment effects. *American Economic Review*, 2:319–323.

Manski, C. F. (2003). *Partial Identification of Probability Distributions.* New York: Springer.

Mattei, A., Li, F., and Mealli, F. (2013). Exploiting multiple outcomes in bayesian principal stratification analysis with application to the evaluation of a job training program. *Annals of Applied Statistics*, 7:2336–2360.

McCrary, J. (2008). Manipulation of the running variable in the regression discontinuity design: A density test. *Journal of Econometrics*, 142:698–714.

McDonald, C. J., Hui, S. L., and Tierney, W. M. (1992). Effects of computer reminders for influenza vaccination on morbidity during influenza epidemics. *MD Computing: Computers in Medical Practice*, 9:304–312.

McGrath, S., Young, J. G., and Hernán, M. A. (2021). Revisiting the g-null paradox. *Epidemiology*, 33:114–120.

Mealli, F. and Pacini, B. (2013). Using secondary outcomes to sharpen inference in randomized experiments with noncompliance. *Journal of the American Statistical Association*, 108:1120–1131.

Meinert, C. L., Knatterud, G. L., Prout, T. E., and Klimt, C. R. (1970). A study of the effects of hypoglycemic agents on vascular complications in patients with adult-onset diabetes. II. Mortality results. *Diabetes*, 19:789–830.

Mercatanti, A. and Li, F. (2014). Do debit cards increase household spending? Evidence from a semiparametric causal analysis of a survey. *Annals of Applied Statistics*, 8:2485–2508.

Ming, K. and Rosenbaum, P. R. (2000). Substantial gains in bias reduction from matching with a variable number of controls. *Biometrics*, 56:118–124.

Ming, K. and Rosenbaum, P. R. (2001). A note on optimal matching with variable controls using the assignment algorithm. *Journal of Computational and Graphical Statistics*, 10:455–463.

Miratrix, L. W., Sekhon, J. S., Theodoridis, A. G., and Campos, L. F. (2018). Worth weighting? How to think about and use weights in survey experiments. *Political Analysis*, 26:275–291.

Miratrix, L. W., Sekhon, J. S., and Yu, B. (2013). Adjusting treatment effect estimates by post-stratification in randomized experiments. *Journal of the Royal Statistical Society: Series B (Statistical Methodology)*, 75:369–396.

Morgan, K. L. and Rubin, D. B. (2012). Rerandomization to improve covariate balance in experiments. *Annals of Statistics*, 40:1263–1282.

Morgan, S. L. and Winship, C. (2015). *Counterfactuals and Causal Inference.* Cambridge: Cambridge University Press.

Mukerjee, R., Dasgupta, T., and Rubin, D. B. (2018). Using standard tools from finite population sampling to improve causal inference for complex experiments. *Journal of the American Statistical Association*, 113:868–881.

Murray, E. J., Robins, J. M., Seage, G. R., Freedberg, K. A., and Hernán, M. A. (2017). A comparison of agent-based models and the parametric g-formula for causal inference. *American Journal of Epidemiology*, 186:131–142.

Naimi, A. I., Cole, S. R., and Kennedy, E. H. (2017). An introduction to g methods. *International Journal of Epidemiology*, 46:756–762.

Negi, A. and Wooldridge, J. M. (2021). Revisiting regression adjustment in experiments with heterogeneous treatment effects. *Econometric Reviews*, 40:504–534.

Newey, W. K. and McFadden, D. (1994). Large sample estimation and hypothesis testing. *Handbook of Econometrics*, 4:2111–2245.

Neyman, J. (1923). On the application of probability theory to agricultural experiments. Essay on principles (with discussion). Section 9 (translated). reprinted ed. *Statistical Science*, 5:465–472.

Neyman, J. (1934). On the two different aspects of the representative method: The method of stratified sampling and the method of purposive selection (with discussion). *Journal of the Royal Statistical Society*, 97:558–625.

Neyman, J. (1935). Statistical problems in agricultural experimentation (with discussion). *Supplement to the Journal of the Royal Statistical Society*, 2:107–180.

Nguyen, T. Q., Schmid, I., Ogburn, E. L., and Stuart, E. A. (2021). Clarifying causal mediation analysis for the applied researcher: Effect identification via three assumptions and five potential outcomes. *Psychological Methods*, 26:255–271.

Otsu, T. and Rai, Y. (2017). Bootstrap inference of matching estimators for average treatment effects. *Journal of the American Statistical Association*, 112:1720–1732.

Pashley, N. E. and Miratrix, L. W. (2021). Insights on variance estimation for blocked and matched pairs designs. *Journal of Educational and Behavioral Statistics*, 46:271–296.

Pearl, J. (1995). Causal diagrams for empirical research (with discussion). *Biometrika*, 82:669–688.

Pearl, J. (2000). *Causality: Models, Reasoning and Inference.* Cambridge: Cambridge University Press.

Pearl, J. (2001). Direct and indirect effects. In Breese, J. S. and Koller, D., editors, *Proceedings of the 17th Conference on Uncertainty in Artificial Intelligence*, pages 411–420. pp. 411–420. San Francisco: Morgan Kaufmann Publishers Inc.

Pearl, J. (2010a). On a class of bias-amplifying variables that endanger effect estimates. In Grunwald, P. and Spirtes, P., editors, *Proceedings of the Twenty-Sixth Conference on Uncertainty in Artificial Intelligence (UAI 2010)*, Corvallis, OR: 425–432. Association for Uncetainty in Artificial Intelligence.

Pearl, J. (2010b). On the consistency rule in causal inference: Axiom, definition, assumption, or theorem? *Epidemiology*, 21:872–875.

Pearl, J. (2011). Invited commentary: Understanding bias amplification. *American Journal of Epidemiology*, 174:1223–1227.

Pearl, J. (2018). Does obesity shorten life? Or is it the soda? On non-manipulable causes. *Journal of Causal Inference*, 6:20182001.

Pearl, J. and Bareinboim, E. (2014). External validity: From do-calculus to transportability across populations. *Statistical Science*, 29:579–595.

Pearl, J. and Mackenzie, D. (2018). *The Book of Why: The New Science of Cause and Effect.* Basic Books.

Permutt, T. and Hebel, J. R. (1989). Simultaneous-equation estimation in a clinical trial of the effect of smoking on birth weight. *Biometrics*, 45:619–622.

Phipson, B. and Smyth, G. K. (2010). Permutation P-values should never be zero: Calculating exact P-values when permutations are randomly drawn. *Statistical Applications in Genetics and Molecular Biology*, 9:Article 39.

Pimentel, S. D., Yoon, F., and Keele, L. (2015). Variable-ratio matching with fine balance in a study of the Peer Health Exchange. *Statistics in Medicine*, 34:4070–4082.

Poole, C. (2010). On the origin of risk relativism. *Epidemiology*, 21:3–9.

Popper, K. (1963). *Conjectures and Refutations: The Growth of Scientific Knowledge.* Routledge.

Powers, D. E. and Swinton, S. S. (1984). Effects of self-study for coachable test item types. *Journal of Educational Psychology*, 76:266–278.

Prentice, R. L. and Pyke, R. (1979). Logistic disease incidence models and case-control studies. *Biometrika*, 66:403–411.

Rao, C. R. (1970). Estimation of heteroscedastic variances in linear models. *Journal of the American Statistical Association*, 65:161–172.

Reichenbach, H. (1957). *The Direction of Time.* University of California Press.

Rigdon, J. and Hudgens, M. G. (2015). Randomization inference for treatment effects on a binary outcome. *Statistics in Medicine*, 34:924–935.

Robins, J., Sued, M., Lei-Gomez, Q., and Rotnitzky, A. (2007). Comment: Performance of double-robust estimators when inverse probability weights are highly variable. *Statistical Science*, 22:544–559.

Robins, J. M. (1999). Association, causation, and marginal structural models. *Synthese*, 121:151–179.

Robins, J. M. and Greenland, S. (1992). Identifiability and exchangeability for direct and indirect effects. *Epidemiology*, 3:143–155.

Robins, J. M., Hernan, M. A., and Brumback, B. (2000). Marginal structural models and causal inference in epidemiology. *Epidemiology*, 11:550–560.

Robins, J. M., Mark, S. D., and Newey, W. K. (1992). Estimating exposure effects by modelling the expectation of exposure conditional on confounders. *Biometrics*, 48:479–495.

Robins, J. M. and Wasserman, L. A. (1997). Estimation of effects of sequential treatments by reparameterizing directed acyclic graphs. In *Proceedings of the Thirteenth conference on Uncertainty in artificial intelligence*, volume 409–420.

Rosenbaum, P. R. (1984). The consequences of adjustment for a concomitant variable that has been affected by the treatment. *Journal of the Royal Statistical Society. Series A*, 147:656–666.

Rosenbaum, P. R. (1987a). Model-based direct adjustment. *Journal of the American Statistical Association*, 82:387–394.

Rosenbaum, P. R. (1987b). Sensitivity analysis for certain permutation inferences in matched observational studies. *Biometrika*, 74:13–26.

Rosenbaum, P. R. (1989). The role of known effects in observational studies. *Biometrics*, 45:557–569.

Rosenbaum, P. R. (2002a). Covariance adjustment in randomized experiments and observational studies (with discussion). *Statistical Science*, 17:286–327.

Rosenbaum, P. R. (2002b). *Observational Studies.* New York: Springer, 2nd edition.

Rosenbaum, P. R. (2010). *Design of Observational Studies.* New York: Springer.

Rosenbaum, P. R. (2015). Two R packages for sensitivity analysis in observational studies. *Observational Studies*, 1:1–17.

Rosenbaum, P. R. (2018). Sensitivity analysis for stratified comparisons in an observational study of the effect of smoking on homocysteine levels. *Annals of Applied Statistics*, 12:2312–2334.

Rosenbaum, P. R. (2020). Modern algorithms for matching in observational studies. *Annual Review of Statistics and Its Application*, 7:143–176.

Rosenbaum, P. R. and Rubin, D. B. (1983a). Assessing sensitivity to an unobserved binary covariate in an observational study with binary outcome. *Journal of the Royal Statistical Society - Series B (Statistical Methodology)*, 45:212–218.

Rosenbaum, P. R. and Rubin, D. B. (1983b). The central role of the propensity score in observational studies for causal effects. *Biometrika*, 70:41–55.

Rosenbaum, P. R. and Rubin, D. B. (1984). Reducing bias in observational studies using subclassification on the propensity score. *Journal of the American statistical Association*, 79:516–524.

Rosenbaum, P. R. and Rubin, D. B. (2023). Propensity scores in the design of observational studies for causal effects. *Biometrika*, 110:1–13.

Rothman, E. D. and Woodroofe, M. (1972). A Cramér von-Mises type statistic for testing symmetry. *Annals of Mathematical Statistics*, 43:2035–2038.

Rothman, K. J., Greenland, S., Lash, T. L., et al. (2008). *Modern Epidemiology.* Wolters Kluwer Health/Lippincott Williams & Wilkins Philadelphia, 3rd edition.

Rubin, D. B. (1974). Estimating causal effects of treatments in randomized and nonrandomized studies. *Journal of Educational Psychology*, 66:688–701.

Rubin, D. B. (1975). Bayesian inference for causality: The importance of randomization. In *The Proceedings of the social statistics section of the American Statistical Association*, volume 233, page 239. American Statistical Association Alexandria, VA.

Rubin, D. B. (1978). Bayesian inference for causal effects: The role of randomization. *Annals of Statistics*, 6:34–58.

Rubin, D. B. (1980). Comment on "Randomization analysis of experimental data: The Fisher randomization test" by D. Basu. *Journal of American Statistical Association*, 75:591–593.

Rubin, D. B. (2001). Comment: Self-experimentation for causal effects. *CHANCE*, 14:16–17.

Rubin, D. B. (2005). Causal inference using potential outcomes: Degisn, modeling, decisions. *Journal of American Statistical Association*, 100:322–331.

Rubin, D. B. (2006a). Causal inference through potential outcomes and principal stratification: Application to studies with "censoring" due to death (with discussion). *Statistical Science*, 21:299–309.

Rubin, D. B. (2006b). *Matched Sampling for Causal Effects.* Cambridge: Cambridge University Press.

Rubin, D. B. (2007). The design versus the analysis of observational studies for causal effects: Parallels with the design of randomized trials. *Statistics in Medicine*, 26:20–36.

Rubin, D. B. (2008). For objective causal inference, design trumps analysis. *Annals of Applied Statistics*, 2:808–840.

Rudolph, K. E., Goin, D. E., Paksarian, D., Crowder, R., Merikangas, K. R., and Stuart, E. A. (2018). Causal mediation analysis with observational data: Considerations and illustration examining mechanisms linking neighborhood poverty to adolescent substance use. *American Journal of Epidemiology*, 188:598–608.

Sabbaghi, A. and Rubin, D. B. (2014). Comments on the Neyman–Fisher controversy and its consequences. *Statistical Science*, 29:267–284.

Salsburg, D. (2001). *The Lady Tasting Tea: How Statistics Revolutionized Science in the Twentieth Century.* Henry Holt and Company.

Sanders, E. Gustafson, P. and Karim, M. E. (2021). Incorporating partial adherence into the principal stratification analysis framework. *Statistics in Medicine*, 40:3625–3644.

Sanderson, E., Macdonald-Wallis, C., and Davey Smith, G. (2017). Negative control exposure studies in the presence of measurement error: Implications for attempted effect estimate calibration. *International Journal of Epidemiology*, 47:587–596.

Scharfstein, D. O., Rotnitzky, A., and Robins, J. M. (1999). Adjusting for nonignorable drop-out using semiparametric nonresponse models. *Journal of the American Statistical Association*, 94:1096–1120.

Schlesselman, J. J. (1978). Assessing effects of confounding variables. *American Journal of Epidemiology*, 108:3–8.

Schochet, P. Z., Burghardt, J., and McConnell, S. (2008). Does job corps work? Impact findings from the national job corps study. *American Economic Review*, 98:1864–1886.

Sekhon, J. S. (2009). Opiates for the matches: Matching methods for causal inference. *Annual Review of Political Science*, 12:487–508.

Sekhon, J. S. (2011). Multivariate and propensity score matching software with automated balance optimization: The matching package for R. *Journal of Statistical Software*, 47:1–52.

Sekhon, J. S. and Titiunik, R. (2017). On interpreting the regression discontinuity design as a local experiment. In *Regression Discontinuity Designs*, volume 38. Emerald Publishing Limited.

Shinozaki, T. and Matsuyama, Y. (2015). Doubly robust estimation of standardized risk difference and ratio in the exposed population. *Epidemiology*, 26:873–877.

Sjölander, A. and Greenland, S. (2022). Are E-values too optimistic or too pessimistic? Both and neither! *International Journal of Epidemiology*, 51:355–363.

Small, D. S. (2015). Introduction to Observational Studies and the Reprint of Cochran's paper "Observational Studies" and Comments. *Observational Studies*, 1:124–125.

Sobel, M. E. (1982). Asymptotic confidence intervals for indirect effects in structural equation models. *Sociological Methodology*, 13:290–312.

Sobel, M. E. (1986). Some new results on indirect effects and their standard errors in covariance structure models. *Sociological Methodology*, 16:159–186.

Sommer, A. and Zeger, S. L. (1991). On estimating efficacy from clinical trials. *Statistics in Medicine*, 10:45–52.

Stefanski, L. A. and Boos, D. D. (2002). The calculus of m-estimation. *American Statistician*, 56:29–38.

Stock, J. H. and Trebbi, F. (2003). Retrospectives: Who invented instrumental variable regression? *Journal of Economic Perspectives*, 17:177–194.

Stuart, E. A. (2010). Matching methods for causal inference: A review and a look forward. *Statistical Science*, 25:1–21.

Stuart, E. A. and Jo, B. (2015). Assessing the sensitivity of methods for estimating principal causal effects. *Statistical Methods in Medical Research*, 24:657–674.

Su, F., Mou, W., Ding, P., and Wainwright, M. (2023). When is the estimated propensity score better? High-dimensional analysis and bias correction. *arXiv preprint arXiv:2303.17102*.

Tao, Y. and Fu, H. (2019). Doubly robust estimation of the weighted average treatment effect for a target population. *Statistics in Medicine*, 38:315–325.

Theil, H. (1953). Estimation and simultaneous correlation in complete equation systems. Technical report, Central Planning Bureau, The Hague.

Thistlethwaite, D. L. and Campbell, D. T. (1960). Regression-discontinuity analysis: An alternative to the ex post facto experiment. *Journal of Educational Psychology*, 51:309.

Thistlewaite, D. L. and Campbell, D. T. (2016). Regression-discontinuity analysis: An alternative to the ex-post facto experiment (with discussion). *Observational Studies*, 2:119–209.

Tibshirani, R. (1996). Regression shrinkage and selection via the lasso. *Journal of the Royal Statistical Society: Series B (Methodological)*, 58:267–288.

Titterington, D. (2013). *Biometrika* highlights from volume 28 onwards. *Biometrika*, 100:17–73.

Valeri, L. and Vanderweele, T. J. (2014). The estimation of direct and indirect causal effects in the presence of misclassified binary mediator. *Biostatistics*, 15:498–512.

Van der Laan, M. J. and Rose, S. (2011). *Targeted Learning: Causal Inference for Observational and Experimental Data.* New York: Springer.

Van der Vaart, A. W. (2000). *Asymptotic Statistics.* Cambridge: Cambridge University Press.

Van Elteren, P. (1960). On the combination of independent two-sample tests of wilcoxon. *Bulletin of the Institute of International Statistics*, 37:351–361.

VanderWeele, T. J. (2008). Simple relations between principal stratification and direct and indirect effects. *Statistics and Probability Letters*, 78:2957–2962.

VanderWeele, T. J. (2015). *Explanation in Causal Inference: Methods for Mediation and Interaction.* Oxford: Oxford University Press.

VanderWeele, T. J., Asomaning, K., and Tchetgen Tchetgen, E. J. (2012). Genetic variants on 15q25.1, smoking, and lung cancer: An assessment of mediation and interaction. *American Journal of Epidemiology*, 175:1013–1020.

VanderWeele, T. J. and Ding, P. (2017). Sensitivity analysis in observational research: Introducing the E-value. *Annals of Internal Medicine*, 167:268–274.

VanderWeele, T. J. and Shpitser, I. (2011). A new criterion for confounder selection. *Biometrics*, 67:1406–1413.

VanderWeele, T. J. and Tchetgen Tchetgen, E. J. (2017). Mediation analysis with time varying exposures and mediators. *Journal of the Royal Statistical Society: Series B (Statistical Methodology)*, 79:917–938.

VanderWeele, T. J., Tchetgen Tchetgen, E. J., Cornelis, M., and Kraft, P. (2014). Methodological challenges in Mendelian randomization. *Epidemiology*, 25:427.

Vansteelandt, S. and Daniel, R. M. (2014). On regression adjustment for the propensity score. *Statistics in Medicine*, 33:4053–4072.

Vansteelandt, S. and Dukes, O. (2022). Assumption-lean inference for generalised linear model parameters (with discussion). *Journal of the Royal Statistical Society, Series B (Statistical Methodology)*, 84:657–685.

Vansteelandt, S. and Joffe, M. (2014). Structural nested models and G-estimation: The partially realized promise. *Statistical Science*, 29:707–731.

Vermeulen, K. and Vansteelandt, S. (2015). Bias-reduced doubly robust estimation. *Journal of the American Statistical Association*, 110:1024–1036.

Voight, B. F., Peloso, G. M., Orho-Melander, M., Frikke-Schmidt, R., Barbalic, M., Jensen, M. K., Hindy, G., Hólm, H., Ding, E. L., and Johnson, T. (2012). Plasma HDL cholesterol and risk of myocardial infarction: A Mendelian randomisation study. *The Lancet*, 380:572–580.

Vovk, V. and Shafer, G. (2023). A Conversation with A. Philip Dawid. *Statistical Science*, page in press.

Wager, S. and Athey, S. (2018). Estimation and inference of heterogeneous treatment effects using random forests. *Journal of the American Statistical Association*, 113:1228–1242.

Wager, S., Du, W., Taylor, J., and Tibshirani, R. J. (2016). High-dimensional regression adjustments in randomized experiments. *Proceedings of the National Academy of Sciences of the United States of America*, 113:12673–12678.

Wald, A. (1940). The fitting of straight lines if both variables are subject to error. *Annals of Mathematical Statistics*, 11:284–300.

Wang, L., Zhang, Y., Richardson, T. S., and Zhou, X.-H. (2020). Robust estimation of propensity score weights via subclassification. *arXiv preprint arXiv:1602.06366*.

Wang, Y. and Li, X. (2022). Rerandomization with diminishing covariate imbalance and diverging number of covariates. *Annals of Statistics*, 50:3439–3465.

White, H. (1980). A heteroskedasticity-consistent covariance matrix estimator and a direct test for heteroskedasticity. *Econometrica*, 48:817–838.

Wooldridge, J. (2016). Should instrumental variables be used as matching variables? *Research in Economics*, 70:232–237.

Wooldridge, J. M. (2015). Control function methods in applied econometrics. *Journal of Human Resources*, 50:420–445.

Wu, J. and Ding, P. (2021). Randomization tests for weak null hypotheses in randomized experiments. *Journal of the American Statistical Association*, 116:1898–1913.

Yang, F. and Ding, P. (2018a). Using survival information in truncation by death problems without the monotonicity assumption. *Biometrics*, 74:1232–1239.

Yang, F. and Small, D. S. (2016). Using post-outcome measurement information in censoring-by-death problems. *Journal of the Royal Statistical Society: Series B (Statistical Methodology)*, 78:299–318.

Yang, S. and Ding, P. (2018b). Asymptotic causal inference with observational studies trimmed by the estimated propensity scores. *Biometrika*, 105:487–493.

Zelen, M. (1979). A new design for randomized clinical trials. *New England Journal of Medicine*, 300:1242–1245.

Zhang, J. L. and Rubin, D. B. (2003). Estimation of causal effects via principal stratification when some outcomes are truncated by "death". *Journal of Educational and Behavioral Statistics*, 28:353–368.

Zhang, J. L., Rubin, D. B., and Mealli, F. (2009). Likelihood-based analysis of causal effects of job-training programs using principal stratification. *Journal of the American Statistical Association*, 104:166–176.

Zhang, M. and Ding, P. (2022). Interpretable sensitivity analysis for the Baron–Kenny approach to mediation with unmeasured confounding. *arXiv preprint arXiv:2205.08030.*

Zhao, A. and Ding, P. (2021). Covariate-adjusted Fisher randomization tests for the average treatment effect. *Journal of Econometrics*, 225:278–294.

Zhao, A. and Ding, P. (2024). No star is good news: A unified look at rerandomization based on p-values from covariate balance tests. *Journal of Econometrics*, 241:105724.

Zhao, Q., Wang, J., Hemani, G., Bowden, J., and Small, D. (2020). Statistical inference in two-sample summary-data Mendelian randomization using robust adjusted profile score. *Annals of Statistics*, 48:1742–1769.

Index

Printed in the United States
by Baker & Taylor Publisher Services